Non-Conventional Energy Resources

Non-Conventional Energy Resources

Shobh Nath Singh

Professor Emeritus
Department of Electrical Engineering
Indian Institute of Technology (Banaras Hindu University)
Varanasi, India

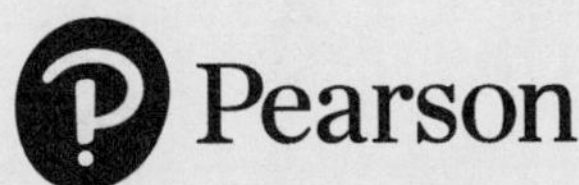

ISBN 9789332543577
First Impression, 2015
Twelfth Impression, 2024
Thirteenth Impression, 2025

Published by Pearson India Education Services Pvt. Ltd, CIN: U72200TN2005PTC057128.

Head Office: 1st Floor, Berger Tower, Plot No. C-001A/2, Sector 16B, Noida - 201 301, Uttar Pradesh, India.
Registered Office: Featherlite, 'The Address' 5th Floor, Survey No 203/10B, 200 Ft MMRD Road, Zamin Pallavaram, Chennai – 600 044.
Website: in.pearson.com; Email: companysecretary.india@pearson.com

Printed in India at: Manipal Technologies Limited, Manipal

In the loving memory of my parents

Late (Shri) Brahm Deo Singh

Late (Smt) Sumitra

Dedicated to my grandsons

Sahil

Priyanshu

Shubham

Isham

Ishant

and

Future Generation Engineers

Brief Contents

Contents

Chapter 2: Energy from the Sun

Chapter 3: Solar Thermal Energy Collectors

Chapter 9: Biomass Energy

Chapter 10: Biogas Energy

Preface

Energy is a key input for techno-economic growth, and the prosperity of a nation and its people is indexed by the per capita energy consumptions. Nevertheless, as a result of rapidly increasing population, financial scarcity, improper coordination, unsystematic approach, etc. for accepting the challenges of energy scarcity, narrowing on the energy demand–supply gap has become the necessity of time.

World energy crisis has started creeping in the year 1973 with oil crisis. Since then, energy has continued to remain in the news of current interests. The need of developing and conserving alternative energy option (renewable energy resources) has led to considerable R&D work in this field. Non-conventional energy resources (renewable energy) are mostly concentrated on the following four aspects:

1. Minimization of energy cost
2. Improved performances including increased efficiency of conversion devices
3. Easy applicability and environmental and social acceptability
4. Energy conservation and management

Energy resources, which are defined as any material object containing energy in abundance and transferrable to usable energy form and which is being given significant importance, are non-conventional energy resources.

Aspects related with principle, working, utilization, availability of energy resources, and related topics have been included in the book. Solved numerical problems are provided at appropriate places throughout the book. Multiple choice questions with answers and several energy-related terms and definitions are also provided for the benefit of the reader. Despite the fact that this book is prepared on the basis of UPTU syllabus, it also covers the syllabus and curriculum topics of IITs, NITs, and various Central and State Universities.

Sincere thanks are also due to Prof. N. K. Samaria, Prof. D. N. Vishwakarma, Prof. Arun K. Kapoor, Mr Avinash Kumar Singh, and others for their cooperation, moral support, and technical assistance in the preparation of the text of the book.

The author also acknowledges with thanks to all the authors and experts in the field of renewable energy and also the publishers whose works have been cited in the book with clear motives to provide the readers complete information at one place.

My sincere thanks to editorial staff of Pearson publications particularly Ms Anita Yadav and Mr M. Balakrishnan for their continued support in preparation of the book.

Although utmost care has been taken to eliminate typographical and technical errors, esteemed readers are requested to communicate the mistakes and errors, if noticed, through email at shobhnpv@gmail.com. Those timely and relevant information will be duly acknowledged in the future editions of the book.

Shobh Nath Singh

About the Author

S. N. Singh was born in Varanasi in the year 1943. He obtained B.Sc. Engg. (Electrical Engg.) and M.Sc. Engg. (Electrical Machines) from Banaras Hindu University in the year 1967 and 1969, respectively. He obtained his Ph.D. in Electrical Drives and Power Electronics in 1983 from Roorkee University.

As a specialist in Power Electronics and Drives, Electrical Machines, Energy Technology and Management, he has served for about four decades in teaching, research and administration in various capacities like Lecturer, Reader, Professor and Head at IT-BHU, Roorkee University, and Director at VBS Purvanchal University, Jaunpur, GL Bajaj ITM, IECCET, etc.

He has 80 plus technical papers published and presented in both International and National Journals and Conferences. He has also authored books titled *A text Book on Power Electronics* and *Electrical Engineering*.

He is a recipient of Dr A. N. Khosla Research Silver Medal, which includes both cash prize and certificates (twice) and Commendation Certificate of Merit from the Institution of Engineers (India). He is a Chartered Engineer (India), Fellow of IE (India), Life Member of ISTE, IAQR, and a consultant and expert member to several organizations like Indian Railways, BHEL, small scale industries, and several government committees like AICTE, UGC, UPPSC, etc.

After superannuation from IT-BHU in the year 2005, he has served as Director for G. L. Bajaj Institute of Technology and Management, Executive Director Apostel Institute, Director (Planning and Development) for IIMT College of Engineering at Greater Noida, and Director for Aryan Institute of Technology, Ghaziabad.

He is presently associated with Department of Electrical Engineering, IIT (BHU) Varanasi, as Professor Emeritus.

Chapter 1

NCER—An Overview

Energy is the capacity of a physical system to perform work. It exists in several forms such as heat, mechanical (potential and kinetic), light, electrical, or other forms of energy. According to the law of conservation of energy, the total energy of a system remains constant. However, energy may transform from one form into another form. The SI unit of energy is the joule (J) or Newton-meter (N × m). Joule is also the SI unit of work.

Energy is now well recognized as an essential parameter of socio-techno-economic development of any system and it is presently used as the measure of the standard of living, quality of life, civilization, and culture of a country. Both the Industrial Revolution in the 19th century, specifically after World War II, and the awareness of energy importance have revolutionized the fuller utilization of all the known and unknown energy reservoirs. As a result of this process, energy conversion, management, resource recovery, and storage have attracted the attention of the scientists, engineers, and technologists, which made them accept the challenges of rising demand of energy requirements.

Energy consumption dictates social inequalities and disparities between one country and the other. Conventional energy sources such as fossil fuels (coal, natural gas, and oil) and nuclear fuels (uranium and thorium) have been a resource of world energy needs since the Industrial Revolution. However, the fast depleting rate of conventional energy resources, climate change and environmental problems, and 1973 oil crisis forced the countries to redefine their energy development objective to use clean, affordable, and secure sources of energy to continue eco-friendly technological progress. Non-conventional energy resources

KEY CONCEPTS

- World population and daily per capita energy consumption through human history
- Energy system model and system acceptability index
- Causes of energy scarcity (shortages) and its solution
- factors affecting energy resource development
- Definition and classification of energy resources
- Energy transfer frame and energy conversion
- Direct and indirect methods of energy conversions
- Brief introduction to conven-tional and non-conventional energy resources
- Brief introduction to different methods of energy storage

from the sun (such as solar light, heat, and wind), the earth (such as geothermal, micro-hydroelectric, and bio-mass and bio gas), and the sea (such as tides, waves, and ocean thermal) are just some examples of energy sources that meet such objectives, and these energy resources have risen to the challenge of meeting the increasing energy demands of the present and future generations.

Oil shales are perhaps the oldest energy resource used in the ancient days for producing shale oil very much like petroleum and diesel. This important energy resource is available in abundance in many countries but high cost of shale oil production and associated environmental problem have restricted their development.

This chapter outlines the various aspects related with all the possible conventional and non-conventional resources, principles of conversion, pros and cons, applications, and storage of energy resources.

1.1 HISTORY OF HUMAN CIVILIZATION

Human civilization has taken a long way to reach the present technology made society. This movement is based on consuming or providing energy for social obligations. The following five indicators of social development (such as quality of life, lifestyle, and standard of living) have always been considered as bare minimum. Techno-economic development maintained pace with social development:

1. Food (Roti)
2. Cloth (Kapada)
3. Shelter (Makan)
4. Health (Swasthya)
5. Education (Siksha)

It may be difficult to find out exactly how people's senses of values have changed through history. However, to some extent, it can calculate how their material wealth has changed, and how much energy has been used for that purpose (see Figs. 1.1 and 1.2). For not only can it be worked out how much energy has been consumed as food, but also know that as material wealth increases, increasing amounts of energy will be needed to create that wealth.

1.1.1 Primitive Man

About two million years ago, there appeared a primitive man—an ape man who could walk on two legs and uses his hands and had the intelligence to make simple tools. His energy consumption was no more than was needed for basic biological survival and is believed to have been around 3,000 kilocalories a day (see Figs. 1.1 and 1.2). Therefore, his energy needs were no more than those of any other animal, and nothing could have been further from modern materialistic society than this totally natural existence of our early forerunner.

Human civilization is considered with no fire and no hunting. Only requirement was food. The earliest humans probably lived primarily on *scavenging*, not actual hunting. Early humans in the *Lower Paleolithic* lived in mixed habitats that allowed them to collect seafood, eggs, nuts, and fruits besides scavenging. Rather than killing large animals for meat, they used carcasses of large animals killed by other predators or carcasses from animals that died by natural causes.

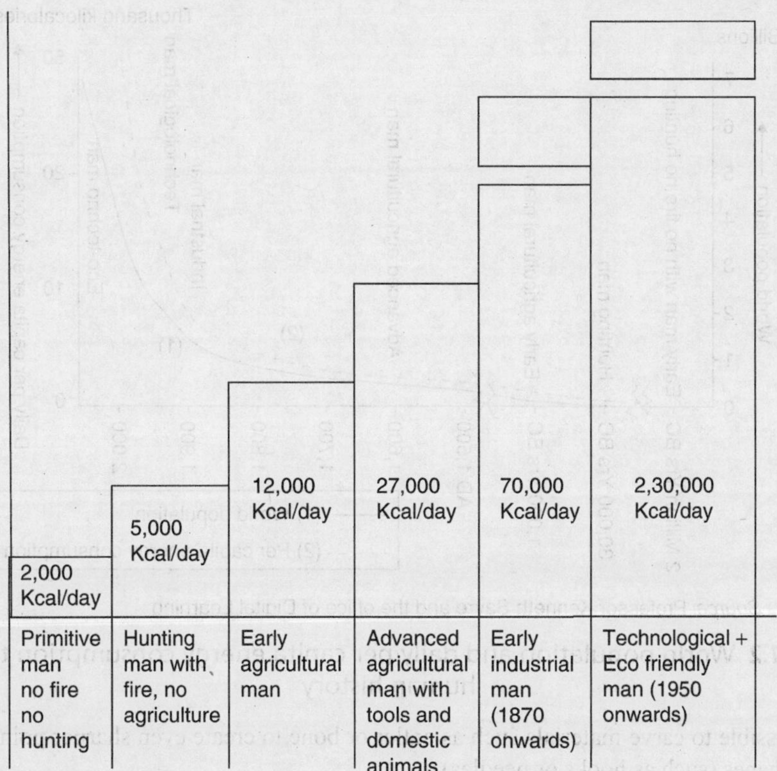

Figure 1.1 Bar chart for daily energy used by a man through human history

1.1.2 Hunting Man

A hunter–gatherer or forager society is one in which most or all food is obtained from wild plants and animals in contrast to *agricultural* societies that rely mainly on *domesticated* species. Hunting and gathering was the ancestral *subsistence mode* of *homo*, and all *modern humans* were hunter–gatherers until around 10,000 years ago. Hunter–gatherer societies tend to be relatively mobile, given their reliance upon the ability of a given natural environment to provide sufficient resources in order to sustain their population and the variable availability of these resources owing to local climatic and seasonal conditions. Individual *band societies* tend to be small in number (10–30 individuals), but these may gather together seasonally to temporarily from a large group (100 or more) when resources are abundant. In a few places where the environment is especially productive, such as that of the Pacific Northwest coast or Jomon-era Japan, hunter–gatherers are able to settle permanently.

In the early periods, a single tool is usually made from the core of the flint, resulting in an instrument that can be used in a fairly rough manner for either cutting or scraping. Hundreds of thousands of years later, craftsmen have become skilled at forming the flakes themselves into implements of various kinds, producing specialist tools for cutting, scraping, gouging, or boring, as well as sharp points for arrow and spear heads. These sophisticated stone tools, in their turn,

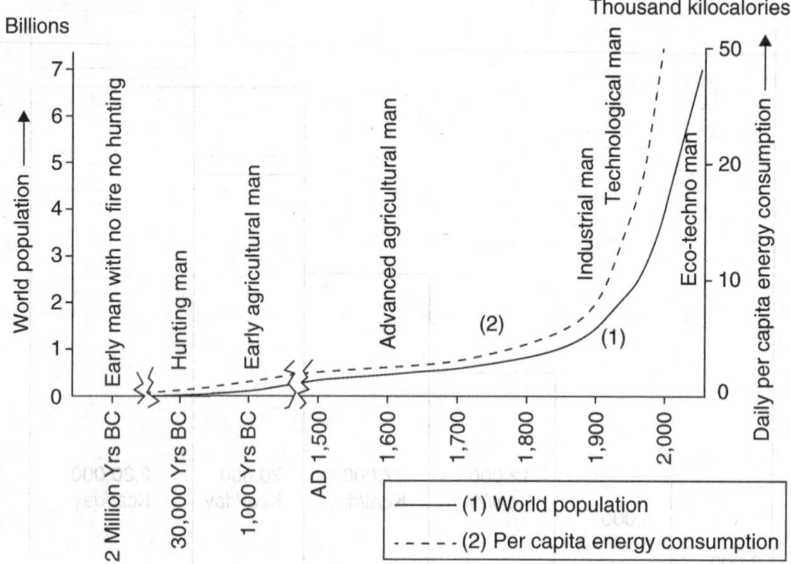

Source: Professor Kenneth Sayre and the office of Digital Learning

Figure 1.2 World population and daily per capita energy consumption through human history

make it possible to carve materials such as antler or bone to create even sharper points or more complex shapes (such as hooks or needles).

The predominant use of stone as the material for tools has caused this period to be known as the Stone Age. The specialization of work also involved creating specialized tools such as fishing nets and hooks and bone harpoons.

1.1.3 Early Agricultural Man

Following the development of agriculture, hunters–gatherers have been displaced by farming or pastoralist groups in most parts of the world. Only a few contemporary societies are classified as hunters–gatherers, and many supplements, sometimes extensively, their foraging activity with farming and/or keeping animals. The transition into the subsequent Neolithic period is chiefly defined by the unprecedented development of nascent agricultural practices. Agriculture originated and spread in several different areas including the Middle East, Asia, Mesoamerica, and the Andes beginning as early as 10,000 years ago. Forest gardening was also being used as a food production system in various parts of the world over this period. Forest gardens originated in prehistoric times along jungle-clad river banks and in the wet foothills of monsoon regions.

1.1.4 Advanced Agricultural Man

Following these revolutionary changes in life and culture in places such as Mesopotamia and parts of Africa, world population rose again to 30 million; daily per capita energy consumption reached 12,000 kilocalories. Therefore, the transition from hunting man to early agricultural man, as we might call this next stage, was relatively rapid when compared with the previous pace of change. Even so, it still took around 20,000 years. They are considered as using advanced agricultural activities together with tools and domestic animals. In the gradual process of families improving their immediate environment, useful trees, and vine species were identified, protected, and improved, whilst undesirable species were eliminated. Eventually, superior foreign species were selected and incorporated into the gardens. Many groups continued their hunter–gatherer ways of life, although their numbers have perpetually declined partly as a result of pressure from growing agricultural and pastoral communities. Many of them reside in arid regions and tropical forests in the developing world. Areas that were formerly available to hunters–gatherers were—and continue to be—encroached upon by the settlements of agriculturalists. In the resulting competition for land use, hunter–gatherer societies either adopted these practices or moved to other areas. In addition, Jared Diamond has blamed a decline in the availability of wild foods, particularly animal resources. In North and South America, for example, most large mammal species had gone extinct by the end of the Pleistocene, according to Diamond, because of overexploitation by humans although the overkill hypothesis he advocates is strongly contested. As the number and size of agricultural societies increased, they expanded into lands traditionally used by hunters–gatherers. This process of agriculture-driven expansion led to the development of the first forms of government in agricultural centres such as the Fertile Crescent, Ancient India, Ancient China, Olmec, Sub-Saharan Africa, and Norte Chico.

As a result of the now near-universal human reliance upon agriculture, the few contemporary hunter–gatherer cultures usually live in areas unsuitable for agricultural use. As agricultural production increased, technology and culture progressed; villages became towns, and towns became cities. As organizational structures to unite these groups of settlers were developed, city-states, or small states based on the city as a unit, came into being. The Babylonian dynasty, which had started out from one of these city-states, is believed to have united all of the city-states of Mesopotamia into one nation under its control in 2169 BC. This was yet another step toward political sophistication by the time of the birth of Jesus Christ; constant improvements in agriculture and other skills had contributed to an increase in the world's human population to 250 million. This doubled to 500 million by about 1650, when daily per capita energy consumption had raised again to 26,000 kilocalories. It had taken 5,000 years for early agricultural man to change into 'advanced agricultural man', the state of human civilization on the threshold of the industrial revolution.

1.1.5 Industrial Man

In the late 18th century in England, coal was first used to provide the energy to make steam, which in its turn was used to drive engines. These early engines drove spinning machines for cotton, which meant that what had always been manual work was suddenly transformed into work that could be done by a machine. The industrial revolution really did revolutionize industry. Large factories, capable of producing large quantities, sprang up. Society, which for so long had been based on agriculture, invested vast amounts in German Professor J. Liebig's discovery of how to enhance agricultural yield by using artificial fertilizer that can be produced artificially in a factory. Once this came into use in agriculture, the food production increased with an increased population. Further, even greater numbers could work in the factories. The industrial production of fertilizers and the subsequent rapid rise in agricultural productivity made the first stage of the industrial revolution was complete.

1.1.6 Technological Man

As industrial activity expanded, society became more and more complex and more and more materialistic. However, while mankind had previously always existed in a kind of harmony with the natural environment, the danger arose that those natural bounds might be overstepped. World population soared: one billion (i.e., a thousand million, or a one followed by nine zeroes) in 1850; one and a half billion in 1900. As a result, energy consumption also soared to 77,000 kilocalories per person per day. 'industrial man' had taken only a few thousand years to replace agricultural man. Expanding populations and growing materialism have led to clashes among nations, including the two World Wars of the 20th century. These wars themselves have spawned yet more advances in science and technology based on yet more inventions and discoveries. Greater and greater amounts of energy from coal, oil, and nuclear power have been consumed. Thus, mankind armed with so much technical knowledge, numbered 2 billion in 1930, or half a billion more than just 30 years before that. In 1981, there were 4.6 billion of us, or more than double the number 50 years ago. At 2,30,000 kilocalories, a modern person's daily consumption of energy had reached 100 times that of our primitive forebears. Industrial man had been replaced by 'technological man'. By calculating the difference between the energy required for biological survival and the total consumed by the technological man and then by comparing that amount with that consumed by primitive man and the difference is almost infinite. Further, remember that it took only hundred years or more for the technological man to change from industrial man.

1.1.7 Eco-friendly Technological Man

There are now about 6 billion of us, each requiring incomparably more energy than any other animal. When we multiply the daily per capita consumption by the total population, we have an incredibly huge amount, which is bound to rise yet further as we become more materialistic, and as the population continues to increase. The world is no longer big enough for us. The demands of individuals and nations are straining spaceship. Earth needs yet another new kind of human being to support this all-consuming society, to find and guarantee new resources, and to use them effectively. This new type of person must be not only a technocrat, but must also have a determination to restore harmony with the natural environment. Such human population may be considered as eco-friendly technological man.

1.2 WORLD POPULATION AND ENERGY CONSUMPTION PATTERN PROJECTION

From the study of human history the worldwide population and per capita energy consumption projection is shown in Figures 1.1–1.3. From the general trend to a more detailed analysis of the actual study of yearly changes, it can be seen that the growth was, in fact, irregular. There have been periods (World War I and II) when the general pattern has been reversed. There has been notable decreasing energy consumption. The annual percentage rates of change show slight fluctuation from year to year, but if these short-term perturbations are overlooked, one can pick up three main tendencies in energy consumption.

1. Fast almost exponential growth from 1850: beginning of World War I with average annual average rate of growth = 3.6% and annual average rate of per capital growth = 4.3%.

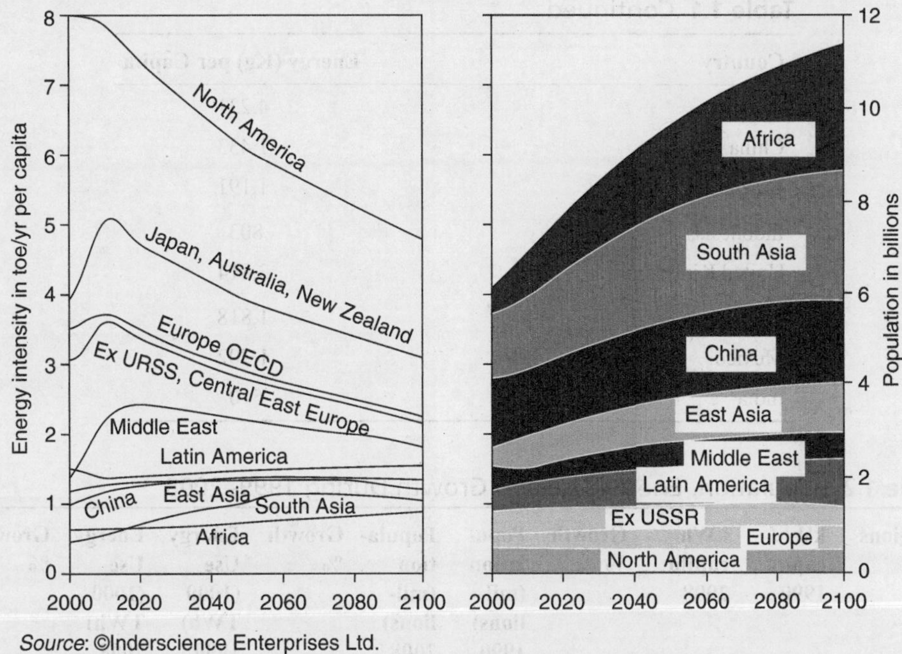

Figure 1.3 World population and per capita energy consumption projection through history

2. Slow growth during World War I up to World War II (1910–1950) with average annual percentage rate of aggregated consumption = 1.5% and annual per capita consumption growth = 1%.
3. Greatly accelerated, almost exponential growth from World War II till date. With average annual average rate of growth = 4.9% and annual average rate of per capital growth = 2.9%.

Although an increase in energy consumption was noted throughout the world, there was a big difference in the level of energy requirements and rate of growth in different countries. Table 1.1 shows the per capita energy consumption of different countries that indicates the level of techno-socioeconomics status of one nation as compared to others. In order to give an idea about how India compares with other economies in energy consumption, we can say that India used 510 kg of energy when compared to U.S.A, which consumes 7,778 kg of energy per capita. The world average of energy consumption is close to 1,818 kg, as shown in Tables 1.1 and 1.2.

Table 1.1 Energy Consumption of a Few Countries (2006)

Country	Energy (Kg) per Capita
Australia	5,917
Singapore	6,968
United States	7,778
Russia	4,745

(Continued)

Table 1.1 Continued

Country	Energy (Kg) per Capita
Germany	4,231
China	1,433
Brazil	1,191
Indonesia	803
United Kingdom	3,814
World Average	1,818
Mexico	1,702
India	510

Table 1.2 Population, Energy Use and Growth During 1999–2008

Regions	kWh/ capita 1990	kWh/ capita 2008	Growth %	Population (millions) 1990	Population (millions) 2008	Growth %	Energy Use (1000 TWh) 1990	Energy Use (1000 TWh) 2008	Growth %
USA	89,021	87,216	−2.0	250	305	22	22.3	26.6	20
India	4,419	6,280	42	850	1140	34	3.8	7.2	91
Latin America	11,281	14,421	28.0	355	462	30.0	4.0	6.7	66
Middle East	19,422	34,774	79.0	132	199	51	2.6	6.9	170.0
Africa	7,094	7,792	10.0	634	984	55.0	4.5	7.7	70.0
China	8,839	18,608	111.0	1141	1333	17.0	10.1	24.8	146
EU-27	40,240	40,821	1.0	473	499	5.0	19.0	20.4	7.0
Others	25,217	23,871	−5.33	1430	1766	23	36.1	42.2	17
The World	19,422	21,283	10	5265	6688	27	102.3	142.3	39

Others include countries in Asia and Australia. Energy use varies between country to country in others category.

1.3 ENERGY SYSTEMS MODEL

A system is a big black box having input and output pairs related by parameters that permit to relate an input, an output, and a state. When the black-box view is applied to energy systems' modelling and planning, an energy activity sector (such aes industrial, transportation, domestic, and agriculture), or a city region or country can be represented as a system shown in Figure 1.4. It is clear from the figure that system is characterized by the following five parameters:

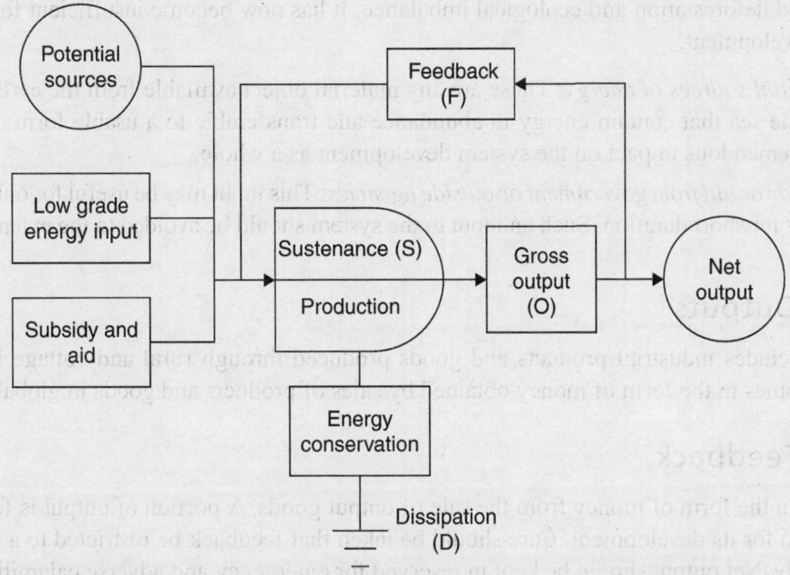

Figure 1.4 Energy system's model

1. Production and sustenance
2. Inputs
3. Outputs
4. Feedback
5. Dissipation

1.3.1 Production and Sustenance Activities

Every society or system is comprised of two important activities: sustenance and production. Sustenance activities relate with the social development that includes standard of living and lifestyle and to fulfil the indicator of quality of life, such as food, cloth, shelter, education, and health. However, these activities are not self-sustaining. They have to be sustained through continuous inputs either from within the system itself or from the outside or as subsidy from government and other agencies. Production activities relate with techno-economic development of the system. It is an important parameter of a system since it is self-sustaining and creates employment. Competitiveness in global market is a constraint for production activities.

1.3.2 Inputs

Inputs to the system are as follows:

1. *Low grade energy*: These are available at almost zero cost from agricultural land and forests. Early men could sustain their survival with such a meagre amount of energy. This has also

caused deforestation and ecological imbalance. It has now become insufficient for any kind of development.

2. *Potential sources of energy*: These are any material object available from the earth, the sun, and the sea that contain energy in abundance and transferable to a usable form. Further, it has tremendous impact on the system development as a whole.
3. *Subsidy or aid from government or outside agencies*: This input may be useful for only in emergency for short duration. Such an input to the system should be avoided to the extent possible.

1.3.3 Outputs

Output includes industrial products and goods produced through rural and cottage industries. Energy comes in the form of money obtained by sales of products and goods in global markets.

1.3.4 Feedback

It comes in the form of money from the sale of output goods. A portion of output is fed back to the system for its development. Care should be taken that feedback be restricted to a portion of output only. Net output should be kept in reserved for emergency and adverse calamities.

1.3.5 Dissipation

Dissipation will always be present in any energy activity system. This is also referred to as degraded energy output of the system. It leaves the system as garbage, low-grade heat, smoke, etc., that damage the environment in terms of pollution. Waste energy recovery system converts waste into wealth. However, dissipation cannot be completely avoided by energy conversion system; further, it can be minimized to a large extent.

Expressing all the material objects of the energy system in same energy unit, the system can be expressed as follows:

$$\text{Inputs} + \text{Feedback} = \text{Dissipation} + \text{Sustenance} + \text{Outputs}$$

or

$$I + F = D + S + O \tag{1.1}$$

From Equation (1.1), it can be analysed that the objective of improving standard of living conditions, lifestyles, and quality of life in any system, energy for sustenance must be raised.

Equation (1.1) also depicts that for techno-socio-economical development of system, input energy (I) to the system should be increased and dissipation (D) should be minimized to the extent possible. Thus, all the enhanced potential sources of energy inputs and waste energy recycling provide energy for sustenance (S) and outputs (O) resulting in overall development of system considered.

1.4 SYSTEM ACCEPTABILITY INDEX (δ)

System acceptability index can also be taken as the measure of gross efficiency of the system and can be expressed as

$$\delta = \frac{\text{Energy inputs to the system} - \text{Degraded energy ouytputs of the system(dissipation)}}{\text{Energy inputs to the system}}$$

Degraded energy outputs leave the system as garbage, low grade heat, smoke, etc., which damage the environment in terms of pollution. Therefore, it must be kept as low as possible. The numerical value of index, δ, is a measure of the system's acceptability. A value of $\delta = 1$ can be approached by properly conducting any energy management program to avoid dissipation in the system.

1.5 CAUSES OF ENERGY SCARCITY

While the whole world is in the grip of energy scarcity, several countries, including India also, are facing various associated difficulties for its techno-socio-economic development because of energy shortages and many more things. However, they have been further complicated by the energy dependence on the other countries. Energy use scenario, as shown in Table 1.3, indicates that how equality (social and economical) can be achieved, when 30% population is utilizing 70% of energy and 70% population is forced to live with the 30% of the remaining energy.

Following points may be considered as the principal causes of energy scarcity.

1.5.1 Increasing Population

Undoubtedly, only 40–45% population constitutes child producing groups, worldwide population is increasing at an alarming rate. It is extrapolated that by the turn of 21st century, population will increase manifold (Malthusian population model). These populations are unevenly distributed worldwide. Africa shares the largest population growth rate, followed by South Asia and then by Europe.

1.5.2 Increasing Energy Usage or Consumption

The movement of civilization from early man to the present technological man was totally based on energy usage. Energy is constantly used at home, at work, and for leisure period of enjoyment. Energy maintains techno-socio-economic development. Energy provides the society with heat and electricity daily and motive power to industry, transportation, and modern way of life.

1. In homes, for lighting and cooking, domestic appliances, televisions, computers, etc.
2. In industry to power the manufacture of the products.
3. In transport system to power cars, trucks, ships, and aeroplanes for transporting peoples and goods.

An increase in the world population and consequent increase in energy consumption increases energy demands manifolds. World Energy Council has provided the most reliable prediction as

Table 1.3 Energy Use Scenario

% of population	70%	30%
% of energy usage	30%	70%

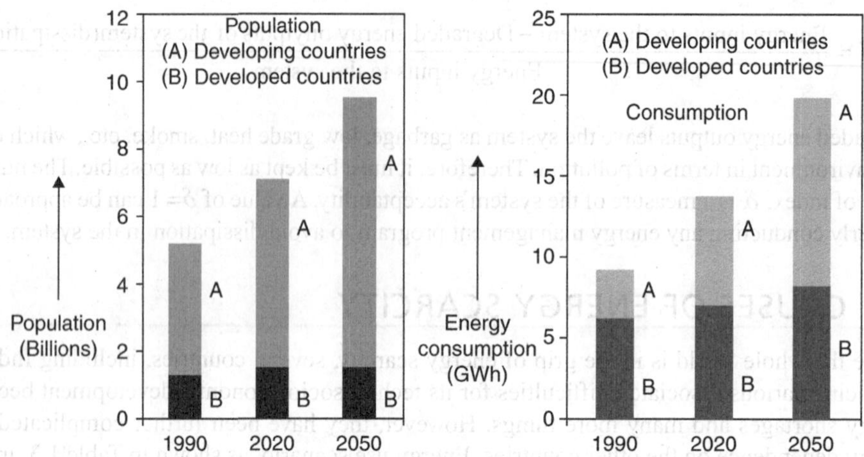

Figure 1.5 Population and energy consumption (Energy council)

shown in Figure 1.5. This indicates that by 2050, the world population will nearly be doubled from the present level and will rise to about 10 billion. Likewise, energy demand is projected to be at least double than the present level (Energy council).

1.5.3 Uneven Distribution of Energy Resources

It is well understood that very few wealthy countries have access to and actually use the largest part of the world's energy and material resources. The generation of environmental and social instability in several area of globe can be discussed in relation to the existence of disparity. Uneven distribution of energy and resource trade among countries is of paramount importance to environmental and political stability. For example, Middle East countries are full of crude oil reserves, but they are forced to involve in conflicts and wars and their energy reserves are forcefully used by wealthy countries. Geographical distribution is the main consideration for an unevenly distribution of fossil fuels (coal, oil, gas, and nuclear). Renewable energy flows are also spread out unevenly. Cloudiness in equatorial regions reduces solar radiation. Whole stretches of the continent have insufficient wind. There are very few sites with the best potential for geothermal, tides, or ocean thermal. In fact, a few densely populated region or area have no significant locally available energy sources at all.

1.5.4 Lacks of Technical Knowhow

Despite the fact that several countries or regions are having energy in abundance, they are not able to fully utilize them due to the lack of knowledge of conversion, transmission, distribution, and utilization. Because of the lack of technical knowledge, resources are mined and processed in resource enriched countries and then refined and used in developed countries. The price of exported resources is normally inadequate to compensate for the depletion of energy reserves and the environmental burden that is generated by resource extraction and primary processing in energy enriched countries. However, resources drive significant economic and environmental benefits in techno-economically developed countries.

1.6 SOLUTION TO ENERGY CRISIS OR SCARCITY

Owing to the growing importance of energy awareness, efforts should be systematically diverted in the following directions to tackle the gigantic energy crunch problems:

1. Minimizing population growth exploitation and harnessing the large utilization of known and unknown energy reservoirs.
2. Development of energy conversion techniques to convert basic energy available from energy reservoirs (primary energy resources) to usable form of energy (secondary energy resources). Usable energy form should be such that it is easy to generate, control, transport, and utilize. Electrical energy being the one and only usable form of energy to meet all these at present. Hydrogen energy and heat energy are other usable energy forms that are also being projected.
3. Keep the new energy system pollution free as far as possible, thereby environmentally acceptable to human beings.
4. The development of cheap and reliable energy storage systems. Maintaining new energy development program that is independent of foreign impact to the extent is possible.
5. Energy management.

1.7 FACTORS AFFECTING ENERGY RESOURCE DEVELOPMENT

An impartial examination of certain basic principles of energy availability studies reveals the following five factors that make energy resource development more difficult than normally realized.

1.7.1 Energy or Fuel Substitution or Scale of Shift

Today, there is no readily available energy resources that is large enough to substitute for fossil fuels (coal, oil, gas, and nuclear) at requisite scale. Undoubtedly, solar energy is several orders of magnitude larger than any conceivable global energy demand (about 10^{17} w). Practical conversion to electricity using photovoltaic or large scale industrial heat are quite negligible.

1.7.2 Energy Density

The amount of energy contained in a unit of material object (energy resource) is termed as energy density. Air-dry crop residue (mostly straw and agricultural waste) contain only 12–15 MJ/kg. For example, the energy density of good quality coal is twice as high (i.e., 25–30 MJ/kg) as that of crude oil (i.e., 42–45 MJ/kg). In order to obtain an equivalent output, replacement of a unit of fossil fuels with approximately 2 kg of phytomass will be needed to substitute solid biofuel. The ratio would be about 1.5 times when substituting plant-derived ethanol for petrol. These realities would be reflected in the reserve capacity, cost, and operation of the required infrastructure.

1.7.3 Power Density

Power density refers to the rate of energy production per unit of earth's area and usually expressed in watts per square meters (w/m^2). Owing to lengthy period of formation (from biomass to coal and then from coal to hydrocarbons), fossil fuel deposits are an extraordinarily concentrated source of high quality energy. They are commonly produced with power densities of 10^2 or 10^3 w/m^2 of coal or hydrocarbon field, and hence, only small land areas are required to supply enormous energy flows. In contrast, biomass energy production has densities below 1 w/m^2, while density of electricity produced by water and wind is below 10 w/m^2. Only photovoltaic electricity generation can deliver larger than 20 w/m^2, although the cost and performance are the constraints of mass utilization.

1.7.4 Intermittency

Growing demand for fuels, energy, and electricity fluctuates daily and seasonally in modern civilization. Further, the base load, which is defined as the minimum energy required meeting the demand of the day, has been increasing. Easily storable high-energy density fossil fuels and thermal electricity generating stations that are capable of operating with high load factors (775% for the coal-fired stations, 790% for nuclear plants) meet these needs.

On the other hand, wind and direct solar radiation are intermittent and far from practicable. They can never deliver such high load factors. Photovoltaic electric generation is still so negligible to offer any meaningful averages. The annual load factors of wind generation in countries with relatively large capacities are 20–25%. Unfortunately, we still lack the means for storing wind or solar-generated electricity on a large scale.

1.7.5 Geographical Energy Distribution

As already mentioned, there are uneven distributions of fossil fuels and the non-fossil fuels (solar, wind, etc.). Cloudiness in the equatorial zone reduces direct solar radiation. Whole stretches of continent has insufficient wind. There are very few sites with the best potential for geothermal, tidal, or ocean energy conversions. Based on the abovementioned five basic considerations, energy sources can be considered possible, probable, and practicable as given in Table 1.4.

Table 1.4 Energy Options

Sources	Possible	Probable	Practicable
Solar light	Yes	Yes	Yes
Solar heat	Yes	Yes	Yes
Wind	Yes	Yes	Yes
Water power	Yes	Yes	Yes
Fusion	Yes	Yes	Yes
Fission	Yes	Yes	No
MHD	Yes	Yes	No

(Continued)

Table 1.4 Continued

Sources	Possible	Probable	Practicable
Geothermal	Yes	Yes	Yes
Biomass	Yes	Yes	Yes
Sea waves	Yes	Yes	Yes
Sea tides	Yes	Yes	Yes
OTEC	Yes	Yes	No
Super conducting Gen. and Transformer	Yes	Yes	No
Thermionic	Yes	Yes	No
Thermoelectric	Yes	Yes	Yes

1.8 QUALITY OF ENERGY FORM

Nowadays, there seems to be a lot of confusion between energy and technology. Society seems to depend so much on technology with an assumption that it can solve anything. However, energy is part of the fundamental fabric of the universe. As such, it is not dependant on technology.

According to the second law of thermodynamics, energy flows from an organized form to a disorganized form (the concept of entropy) or from a concentrated form to a more dispersed form. Obviously, more work can be done with energy in concentrated form (as it is easier to convert to other usable form, transport, distribute, and utilize). This gives rise to the concept of energy quality.

To illustrate this, let us assume two large containers with each containing 10 units of energy (20 units in total) in the form of compressed air. One container is at a high pressure, and the other at a low pressure. If a valve is installed between the two containers and is opened, energy (in the form of pressured gas) will flow from the container with the high pressure to the one with the low pressure, as the pressures in the two containers will try to equilibrate. Let us consider transfer of one unit of energy from the high-pressure container to the other. If an energy balance is made on the containers, 20 units of energy are still there. However, if transfer of one unit of energy back to the high-pressure container is required, the same will need a compressor to do it. Therefore, even though it is still there, nothing can be done with that unit of energy, as it has lost its quality.

Similarly, heat flows from a hotter to a cooler medium. In fact, all energy flows from a higher quality to a lower one, as the universe is a cold quasi-void and energy tries to fill it and equilibrate itself (have the same quality everywhere). This is why no process is perfectly reversible; there will always be losses to go back to the highest form of energy, like the energy consumption of the compressor in the preceding example. The branch of thermodynamics concerned with the usefulness of a given energy form is called energy. Electricity is the usable energy form with the highest quality (the one that requires the biggest compressor to produce). Heat has a much low quality, but its quality increases with temperature for applications requiring heat. Heat can be produced from electricity at 100% efficiency by passing a current in a resistance; however, if conversion of heat into the electricity is attempted, there will be large losses (>60% losses). Further, low losses occur if the temperature is high. Using higher quality energy for lower quality applications makes no sense in the context of thermodynamics. For example, when designing a heating system, it makes no sense to produce heat at 100°C, if the system requires heat at 40°C. Similarly, producing electricity (and suffering the high losses required to do so) and converting it to heat for space heating is a very inefficient use of energy. For example, converting solar radiation to heat with solar collectors will give about 6 times more energy than converting it to electricity using photovoltaic. The electricity route may be convenient for transport, but it is definitely inefficient. Oil is a very high quality form of energy. It is very concentrated and easy to transport. It took about 85 tons of vegetation to produce one barrel of oil. This means that in the last 100 years, we have consumed oil that is equivalent to 13,000 years of plant growth (or photosynthesis) on the planet. Society has become used at depending on high quality forms of energy to accomplish its everyday tasks, as the planet is enriched with a good supply of them. As these are fast depleting and becoming less abundant, there will be need to replace them with more dispersed ones, like concentrating solar energy to a usable form. Converting dispersed energy into a more concentrated form will always be inefficient, that is, a basic law and technology will not allow anyone to bypass it.

1.9 ENERGY RESOURCES AND CLASSIFICATION

The following sections deals with the classification of promising energy resources of immediate interests.

1.9.1 Primary and Secondary Energy Resources

1. Primary energy resources are derived directly from natural reserve. Examples are chemical fuels, solar, wind, geothermal, nuclear hydropower, etc. They are used either in basic raw energy form or by converting them to usable form (secondary energy).
2. Secondary energy resources are usable forms of energy generated by means of suitable plants to convert the primary energy. Examples are electrical energy, steam power, hot water power, hydrogen energy, etc.

Usable form of energy is cost effective, highly efficient with improved performance, environmentally acceptable and system acceptability index approaching to unity is achievable during conversion, transportation, distribution, and end use. From the abovementioned viewpoints, electrical energy will continue to be dominant and will also be a usable form of energy till the turn of the century.

Primary energy resources may be further sub-classified as follows:

1. *Conventional and non-conventional energy resources*: (a) Conventional energy resources and their technical knowledge are known to mankind to a great extent. They are the energy stored within the earth and the sea. They include both fossil fuels (coal, oil, and gas) and nuclear energy (uranium and thorium) and required human intervention to release the energy from them. These sources have formed over hundreds of millions of years ago and when they are used, there will be no more for future generations. They are also known as finite energy resources. (b) Non-conventional energy resources are also known as infinite energy resources. Their technical knowledge is little known and they need full exploitation and improved technical understanding. However, it may be mentioned that owing to the cost factor and overall performance, one may think of utilizing all these energy resources only

when all the conventional energy resources have been fully exploited and utilized. They are obtained from the energy flowing through the natural environment. It is necessary to note that the energy is passing through the environment as a current or as a flow and whether there is an artificial device there to intercept and harness the power or not. Further, it is important to know the rate at which useful energy can be obtained from these sources.

2. *Renewable and non-renewable energy resources*: (a) Renewable energy resources are continuously restored by nature. Examples are solar, water, wind, etc. (b) Non-renewable energy resources are the reserve that is once accumulated in nature has practically ceased to form under new geological conditions. They are also known as expendable energy. Examples are coal, oil, gas, nuclear, etc. Therefore, energy resources may be represented as shown in Table 1.5.

1.9.2 Oil

Oil companies estimate that the world's proven oil reserves are about 1,050 thousand million barrels (BP 2002). This is equivalent to about 6.4 × 1,021 J or 6,400 exajoules. Estimates of reserves are always subject to uncertainty and change. There is a very uneven distribution of oil reserves across the world, with some 71% of proven oil reserves being in the Middle East. The ultimately recoverable and unconventional reserves are very much more difficult to specify. The estimates of additional reserves that will be found and the growth to existing fields will vary widely. However, there is general agreement that crude oil is a finite resource that will run short and may sometime become very expensive in the first half of this century.

Table 1.5 Classification of Energy Resources

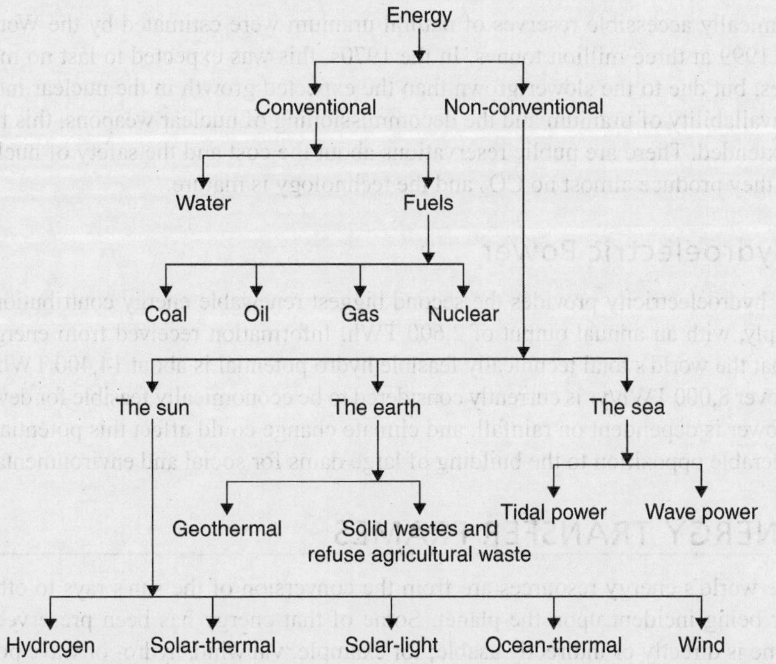

1.9.3 Natural Gas

The proven reserves of natural gas are presently some 152 trillion cubic meters (about 5.9 × 1,021 Joules or 5,900 exajoules). This is about the same as the reserves of oil. However, because gas is more difficult to transport and trade, there has not been as much effort is put into finding gas when compared with that of finding oil. There are some prospective regions of the world that have not been fully explored. Technologies for extracting gas constantly improve, thus making it difficult to estimate the sizes of the gas fields. The 2001 world gas consumption rate of 2.5 trillion cubic meters per annum (The World Fact book) has doubled over the last 30 years, while oil consumption has only increased some 30%.

1.9.4 Coal

In 1999, the proved recoverable reserves of coal is around one million tonnes [The World Energy Council estimates]. There is much more coal than any other fossil fuel. This is equivalent to about 3×10^{22} J or 30,000 exajoules; this is enough to sustain present production for more than 200 years. The world's consumption of coal is still rising (at less than 1% a year), but most industrial countries over recent decades have decreased their dependence on coal. The use of coal is limited more by environmental considerations than by the size of the resource. Modern techniques for burning coal using liquefaction and gasification processes can greatly reduce some of the pollutants from coal. However, coal always produces a great deal of carbon dioxide (greenhouse gas). There had been no cost-effective way developed for capturing and sequestering this carbon dioxide, but extensive research programs are underway.

1.9.5 Uranium

The economically accessible reserves of natural uranium were estimated by the World Energy Council in 1999 at three million tonnes. In the 1970s, this was expected to last no more than a few decades, but due to the slower grown than the expected growth in the nuclear industry and increased availability of uranium and the decommissioning of nuclear weapons, this time frame has been extended. There are public reservations about the cost and the safety of nuclear power plants, but they produce almost no CO_2 and the technology is mature.

1.9.6 Hydroelectric Power

At present, hydroelectricity provides the second biggest renewable energy contribution to world energy supply, with an annual output of 2,600 TWh. Information received from energy sources indicates that the world's total technically feasible hydro potential is about 14,400 TWh/yr, out of which just over 8,000 TWh/yr is currently considered to be economically feasible for development.

Hydropower is dependent on rainfall, and climate change could affect this potential. There is also considerable opposition to the building of large dams for social and environmental reasons.

1.10 ENERGY TRANSFER FRAMES

Most of the world's energy resources are from the conversion of the sun's rays to other energy forms after being incident upon the planet. Some of that energy has been preserved as fossil energy; some is directly or indirectly usable; for example, via wind, hydro- or wave power.

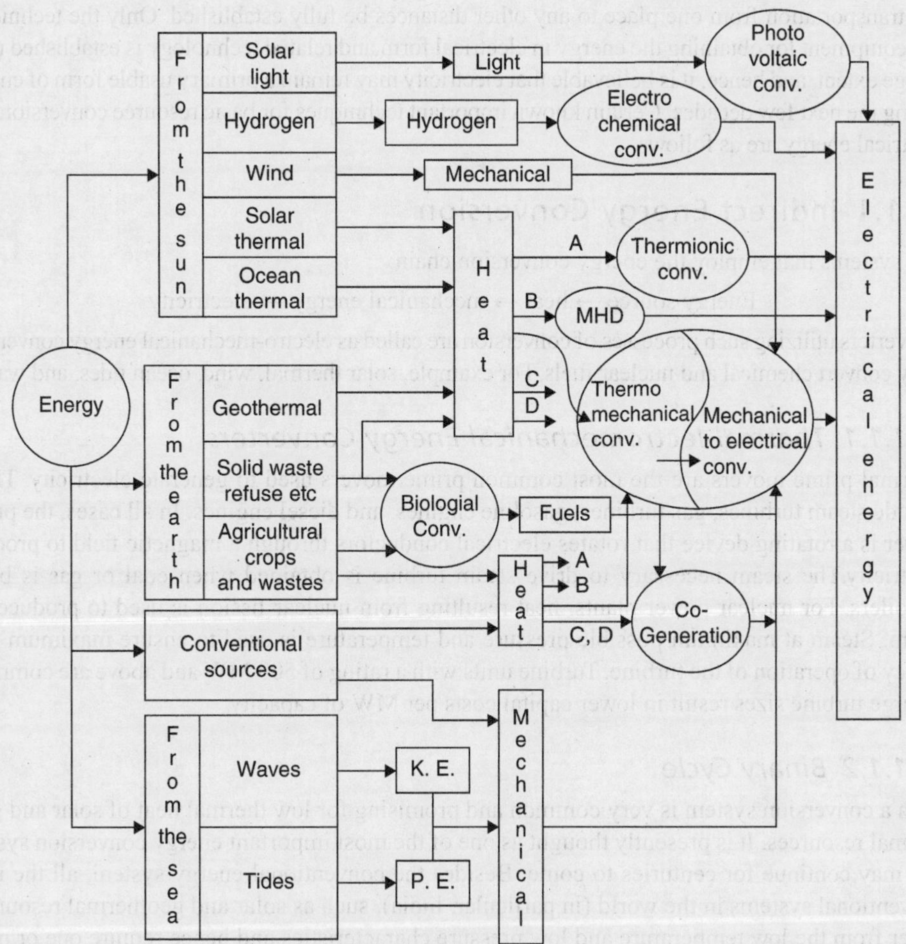

Figure 1.6 Energy transfer frame

In order to have better understanding of the energy conversion storage and related aspects of immediate importance, it is essential to know the promising energy resources and their nature and characteristics in detail, which will lead to the development of economical and feasible energy storage and conversion techniques. Discussions of all these energy resources are beyond the scope of this chapter. However, these resources and the techniques for conversion to usable electrical energy form and intermediary product forms, which are capable of being stored, can be well understood from Figure 1.6. It represents the transformation of basic energy resource to usable form and intermediary products. Following Figure 1.6, a suitable energy conversion and storage schemes can be selected and designed to meet the requirements.

1.11 ENERGY CONVERSION

In whatever form the basic energy may be available from various conventional and non-conventional energy reservoirs, the development of techniques and systems for their conversion to usable form, and transportation from one place to any other distances be fully established. Only the techniques and equipment for obtaining the energy in electrical form and related technology is established up to a large extent, and hence, it is believable that electricity may remain a primary usable form of energy during the next few decades. Certain known important techniques for basic resource conversion into electrical energy are as follows.

1.11.1 Indirect Energy Conversion

The systems that employ the energy conversion chain:

Energy source → heat → mechanical energy → electricity

Converters utilizing such processes of conversion are called as electro-mechanical energy converters. They convert chemical and nuclear fuels. For example, solar thermal, wind, ocean tides, and waves.

1.11.1.1 Thermal-lectromechanical Energy Converters

Thermal prime movers are the most common prime movers used to generate electricity. These include steam turbines, gas turbines, gasoline engines, and diesel engines. In all cases, the prime mover is a rotating device that rotates electrical conductors through a magnetic field to produce electricity.The steam necessary to drive steam turbine is obtained when coal or gas is burnt in boilers. For nuclear power plants, heat resulting from nuclear fission is used to produce the steam. Steam at maximum possible pressure and temperature is used to ensure maximum efficiency of operation of the turbine. Turbine units with a rating of 500 MW and above are common, as large turbine sizes result in lower capital costs per MW of capacity.

1.11.1.2 Binary Cycle

Such a conversion system is very common and promising for low thermal heat of solar and geothermal resources. It is presently thought as one of the most important energy conversion system that may continue for centuries to come. Besides the conventional energy system, all the non-conventional systems in the world (in particular, India), such as solar and geothermal resources, suffer from the low temperature and low pressure characteristics and hence require one or more types of binary cycle capable of assisting in the conversion of these energy resources into usable electrical energy form. A schematic arrangement of a binary cycle used for the purpose may be understood from Figure 1.7.

It consists of the following parts:

1. *Hot water tank or well*: It either stores solar heat with the help of concentrating collectors or geothermal fluid or similar basic energy resources.
2. *Heat exchanger*: This is normally required to transfer the collected heat in the tank to the working fluid.
3. *Power unit*: The heated working fluid is then used to drive turbine to produce mechanical energy through coupled generators' air engine; the heat exchanger and power unit are combined in one, but a closed vapour cycle system needs a condenser and a feed pump in addition to all the abovementioned components. Further, in binary cycle, hot water is transferred to another. Each of these components may consist of one or more parts. For example, in hot liquid, isobutene or pentene or Freon gas is used. This secondary fluid vaporizes at

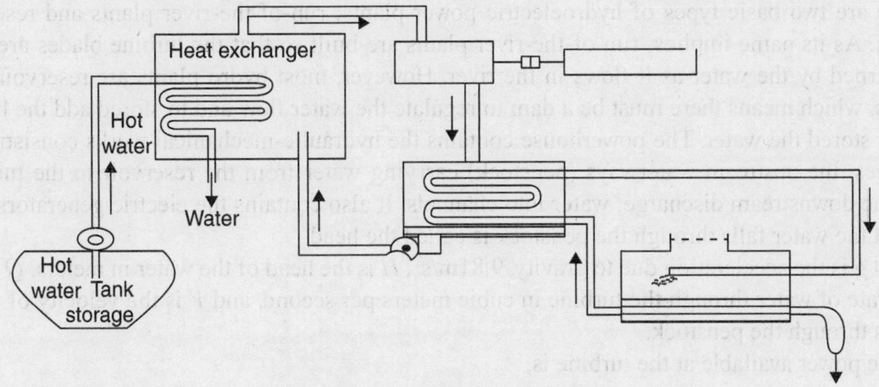

Figure 1.7 Binary cycle

an adequate pressure for turbine drive and could be condensed by surface water, which a tower could then cool. Heat transfers from hot water to secondary fluid occur without phase change, thus eliminate problems from evaporation-caused deposits.

1.11.1.3 Cogeneration

Cogeneration is a term used for representing the coincident generation of steam and electricity by an industrial complex with or without the involvement of a utility or by the utility itself. Hence, a cogeneration plant may

1. Be owned and operated by the individual industry or the central supply systems.
2. Serve one or more individual users or can have isolated operation.
3. Be an integral part of the local supply grid.

However, each cogeneration schemes and their use should be carefully analysed in the context of individual industrial process, social, economical, and environmental interferences. Cogneration schemes, if implemented, may result in as high a thermal efficiency of 80% as compared to fossil-fuel fired integrated stations operating at the maximum thermal efficiency of 30–40%. The two attractive thermal cycle used in cogeneration schemes is the topping cycle and bottoming cycle that refer to the points in the thermal cycles at which electrical or mechanical energy are produced.

In topping cycle, fuel is burnt to produce electrical or mechanical power; the waste heat from the power production plant serve as process heat, which may be reutilized for electrical or mechanical power. In bottoming cycle, the excess process heat is then converted into electrical or mechanical power. With such an efficient conversion of thermal heat into kilowatt hour, cogeneration appears to be a solution of conserving energy.

1.11.2 Hydroelectromechanical Energy Converters

Hydroelectric power plants and tidal power plants both use hydraulic turbines as prime movers, which convert the potential energy of an elevated body of water to rotating kinetic energy.

There are two basic types of hydroelectric power plants: run-of-the-river plants and reservoir plants. As its name implies, run-of-the-river plants are built so that the turbine blades are simply turned by the water as it flows in the river. However, most hydro plants are reservoir-type plants, which means there must be a dam to regulate the water flow and to stored add the height to the stored the water. The powerhouse contains the hydraulic-mechanical works consisting of turbines, the upstream waterways (penstock) carrying water from the reservoir to the turbine, and the downstream discharge, water into channels. It also contains the electric generators. The height the water falls through the penstock is called the head.

Let g is the acceleration due to gravity, 9.81m/s^2, H is the head of the water in meters, Q is the flow rate of water through the turbine in cubic meters per second, and V is the velocity of water in m/s through the penstock.

The power available at the turbine is,

$$P = gHQ \text{ kW} \tag{1.2}$$

Since a dam converts potential energy into kinetic energy, we get

$$mgH = \tfrac{1}{2}mV^2 \tag{1.3}$$

From which,

$$V = \sqrt{(2gH)} \tag{1.4}$$

The velocity of the water through the penstock can be converted to flow rate in cubic meters per second through

$$Q = (\pi D^2/4)V \tag{1.5}$$

The power produced is proportional to the head and the flow rate. Dams are roughly classified into low head (6 to 30 m), medium head (30 to 200 m), and high head (above 220 m). To transfer water downstream, low-head plants utilize dams and high-head plants use penstocks.

There are three types of hydraulic turbines available

1. Kaplan turbines are used for heads up to 60 m.
2. Francis turbines are used for heads from 30 to 300 m and of ratings exceeding 500 MW have been built.
3. Pelton wheels, for heads larger than 90 m and of ratings of 40 MW are in use.

Maximum efficiencies of hydraulic turbines are between 85 and 95%. Hydraulic turbines can be started almost instantaneously from rest, and they have the obvious advantage that no losses are incurred when at a standstill. Thus, working in parallel with thermal power stations, hydroelectric plants can meet peak loads at minimum operating cost.

1.11.3 Direct Energy Conversion

Several conversion processes in which the heat-to-mechanical energy transformation link is not essential. In this, systems employ the energy conversion chain:

Energy source→Electricity

In such processes, the source energy is converted directly to electricity. Hence, the definition of direct energy conversion is a process of conversion of one form of energy (such as sunlight) to another (such as electricity) without going through an intermediate stage (such as steam to spin generator turbines).

The direct energy conversion devices are known as direct energy converters. They convert solar, thermal, chemical, and nuclear energies into electricity without involving a rotating or reciprocating mechanical prime mover. The following are the various direct energy converters:

1. Fuel cells and batteries
2. Photovoltaic
3. Photoelectric
4. Electrostatic generators
5. Thermionic
6. Thermoelectric
7. Ferroelectric generators
8. Magnetohydrodynamic generators
9. Piezoelectric generators

It is projected that, of all the direct energy converters, only solar photovoltaic and fuel cells will contribute to the production of any significant amount of electrical power in the near future.

1.11.3.1 Photovoltaic Conversions

The photoelectric effect was first noted by a French physicist, Edmund Becquerel, in 1839, who found that certain materials would produce small amounts of electric current when exposed to light. In 1905, Albert Einstein described the nature of light and the photoelectric effect on which photovoltaic technology is based, for which he later won a Nobel Prize in physics. The first photovoltaic module was built by Bell Laboratories in 1954. It was billed as a solar battery and was mostly just a curiosity as it was too expensive to gain widespread use. In the 1960s, the space industry began to make the first serious use of the technology to provide power aboard spacecraft. Through the space programs, the technology advanced; its reliability was established, and the cost began to decline. During the energy crisis in the 1970s, photovoltaic technology gained recognition as a source of power for non-space applications.

Figure 1.8 illustrates the operation of a basic photovoltaic cell. It is also called a solar cell. They are made of some kinds of semiconductor materials, like silicon, used in the microelectronics industry. For solar cells, a thin semiconductor wafer is specially treated to form an electric field, positive on one side and negative on the other. When light energy strikes the solar cell, electrons are knocked loose from the atoms in the semiconductor material. If electrical conductors are attached to the positive and negative sides, forming an electrical circuit, the electrons can be captured in the form of an electric current, that is, electricity. This electricity can then be used to power a load, such as a light or a tool.

Photovoltaic conversion is thus considered a method of generating electrical power by converting solar radiation into direct current electricity using semiconductors that exhibit the photovoltaic effect.

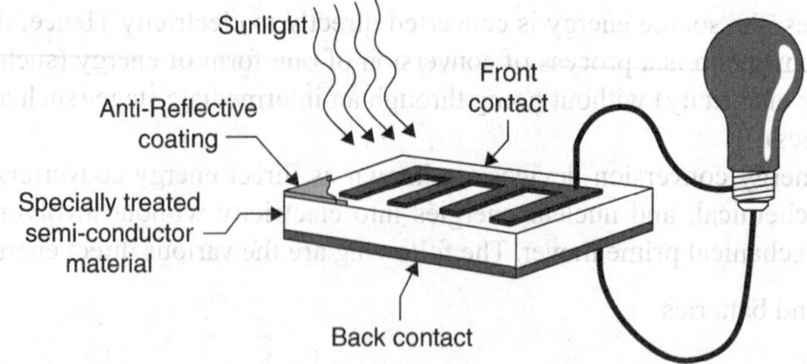

Figure 1.8 Operation of a basic photovoltaic cell

1.11.3.2 Working of Solar Cells

As already mentioned, solar cell is the other name for photovoltaic. These cells are responsible for producing energy out of sunlight it receives. They are made of special materials that are semiconductors. These semiconductors produce electricity when sunlight falls onto its surface. Solar electric cells are simple cells to use, they do not require anything but sunlight to operate; they are long lasting, reliable, and easy to maintain. Normally, solar panels' life time is 25 years.

A solar cell (see Fig. 1.9) is essentially a PN junction with a large surface area .The P-type material is thick (about 200–300 μm). The N-type material is kept thin (of the order of 0.2–0.3 μm) to allow light to pass through to the PN junction. Light travels in packets of energy called photons. The energy (E) of photon is given by

$$E = hc/\lambda \tag{1.6}$$

where

h = Planck's constant, 6.62×10^{-27} erg/s
c = velocity of light, 3×10^{8} m/s
λ = wavelength of light, μm

From which, we get

$$E = 1.24/\lambda \tag{1.7}$$

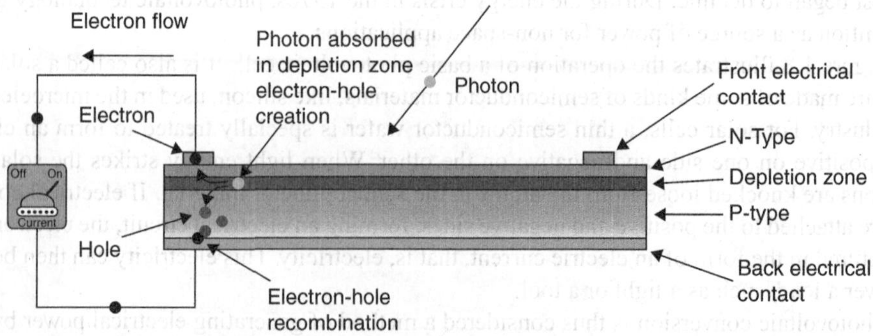

Figure 1.9 Physical configuration of a typical solar cell

When a photon of light is absorbed by one of these atoms in the N-type silicon, it will dislodge an electron, creating a free electron and a hole. The free electron and hole have sufficient energy to jump out of the depletion zone. If a wire is connected from the cathode (N-type silicon) to the anode (P-type silicon), electrons will flow through the wire. The electron is attracted to the positive charge of the P-type material and travels through the external load (meter) creating a flow of electric current.

The hole created by the dislodged electron is attracted to the negative charge of N-type material and is migrated to the back electrical contact layer. As the electron enters the P-type silicon from the back electrical contact, it combines with the hole restoring the electrical neutrality. A number of solar cells electrically connected to each other and mounted in a support structure or frame is called a photovoltaic module. Modules are designed to supply electricity at a certain voltage, such as a common 12-V system. The current produced is directly dependent on how much light strikes the module.

1.11.3.3 Thermionic Conversion

Thermionic Converter or thermionic generator is a device for the direct conversion of thermal energy into electrical energy on the basis of the phenomenon of thermionic emission. It consists of a hot electrode that thermionically emits electrons over a potential energy barrier to a cooler electrode, producing a useful electric power output. Caesium vapour is used to optimize the electrode work functions and provide an ion supply (by surface contact ionization or electron impact ionization in a plasma) to neutralize the electron charge.

The simplest type of thermionic converter consists of two electrodes separated by a vacuum gap (less than 5 mm). A typical simple caesium thermionic converter is shown in Figure 1.10. One electrode is the cathode, or emitter, and the other is the anode, or collector. The electrodes are made of refractory metals, usually molybdenum, rhenium, or tungsten. A heat source supplies enough thermal energy for appreciable thermionic emission to occur from the surface of the metal. After passing through the inter-electrode gap, which is a few tenths of a millimetre in size, the electrons impinge on the surface of the collector, where they create an excess of negative charge and increase the collector's negative potential. If heat is continuously supplied to the emitter and if the collector, which takes up heat from the electrons that reach it, is correspondingly cooled, then an electric current will be maintained in the external circuit, and work will thus be done. Since a thermionic converter is essentially a heat engine whose working fluid is an 'electron gas' (the electrons

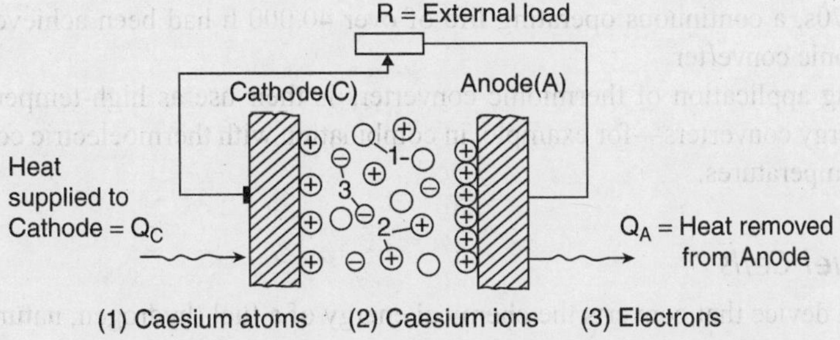

Figure 1.10 Schematic of a thermionic converter

'evaporate' from the heated emitter and 'condense' on the cooled collector), the efficiency of a thermionic converter cannot exceed that of the Carnot cycle. The thermionic converter produces a voltage of 0.5–1 V. This voltage is of the order of the contact potential difference, but differs from it by the value of the voltage drop across the inter-electrode gap and the voltage losses in the switching lines (see Fig. 1.10). The maximum density of the current generated by a thermionic converter is limited by the emission capacity of the emitter and may reach a few tens of amperes from 1 cm^2 of surface. To obtain optimal values of the work function of the emitter (2.5–2.8 electron volts [eV]) and collector (1.0–1.7 eV) and to compensate for the electron space charge formed near the electrodes, a slightly ionized caesium vapour is usually introduced into the interelectrode gap.

Positive caesium ions are formed when caesium atoms collide with electrons (1) at the hot cathode or (2) in the inter-electrode space. The first case is known as surface ionization. In the second case, a caesium atom may be ionized by a single electron impact or through stepwise ionization, in which the caesium atom is brought into an excited state by an initial collision with an electron and is ionized by subsequent collisions. In this second case, arcing occurs; this mode of operation is the one most widely used. Present-day thermionic converters operate at cathode temperatures of 1,700°–2,000°K and anode temperatures of 800°–1,100°K. At such temperatures, the power density at the cathode surface reaches tens of watts per cm^2, and the efficiency of the converters may exceed 20%.

According to the nature of the heat source, thermionic converters are classified as follows:

1. *Nuclear*: The heat for nuclear thermionic converters may come from a nuclear fission reaction (in reactor-powered converters) or the decay of a radioactive isotope (in isotope-powered converters). The first reactor-powered thermionic converter called Topaz in the world was built in the USSR in 1970. It has an electric power output of about 10 kW.
2. *Solar*: In solar thermionic converters, the emitter is heated by the thermal energy of solar radiation, which is collected with solar concentrators.
3. *Flame heated*: Flame-heated thermionic converters operate on the heat released in the combustion of organic fuel.

Thermionic converters have several advantages over traditional electromechanical converters:

1. Absence of moving parts
2. Compactness
3. High reliability
4. Possibility of operation without regular servicing.

As of mid-1970s, a continuous operating life of over 40,000 h had been achieved for an individual thermionic converter.

A promising application of thermionic converters is their use as high-temperature units of multistage energy converters—for example, in combination with thermoelectric converters operating at low temperatures.

1.11.3.4 Fuel Cells

A fuel cell is a device that converts the chemical energy of a fuel (hydrogen, natural gas, methanol, gasoline, etc.) and an oxidant (air or oxygen) into electricity. In principle, a fuel cell operates like a battery. However, unlike a battery, a fuel cell does not run down or require recharging. It will produce electricity and heat as long as fuel and an oxidizer are supplied. Fuel cells come in many varieties; however, they all work in the same general manner. They are made up of three segments that are sandwiched together:

1. anode
2. electrolyte
3. cathode

An electrolyte is any substance containing free *ions* that make the substance *electrically conductive*. The most typical electrolyte is an ionic solution, but molten electrolytes and *solid electrolytes* are also possible. Commonly, electrolytes are solutions of acids, bases, or salts. Furthermore, some gases may act as electrolytes under conditions of high temperature or low pressure. Electrolyte solutions can also result from the dissolution of some biological (e.g., DNA, polypeptides) and synthetic polymers (e.g., polystyrene sulfonate), termed polyelectrolytes, which contain charged functional groups.

Electrolyte solutions are normally formed when a salt is placed into a solvent such as water and the individual components dissociate due to the thermodynamic interactions between solvent and solute molecules in a process called *salvation*. For example, when Table salt, NaCl, is placed in water, the salt (a solid) dissolves into its component ions, according to the dissociation reaction

$$NaCl_{(s)} \rightarrow Na+_{(aq)} + Cl^{-}_{(aq)} \tag{1.8}$$

It is also possible for substances to react with water producing ions, for example, carbon dioxide gas dissolves in water to produce a solution that contains hydronium, carbonate, and hydrogen carbonate ions. Note that molten salts can be electrolytes as well. For instance, when sodium chloride is in a molten state, the liquid conducts electricity. An electrolyte in a solution may be described as concentrated, if it has a high concentration of ions, or dilute if it has a low concentration. If a high proportion of the solute dissociates to form free ions, the electrolyte is strong; if most of the solute does not dissociate, the electrolyte is weak. The properties of electrolytes may be exploited using *electrolysis* to extract constituent elements and compounds contained within the solution.

Two chemical reactions occur at the interfaces of the three different segments. The net result of the two reactions is that fuel is consumed, water or carbon dioxide is created, and an electric current is produced, which can be used to power electrical devices, normally referred to as the load (see Fig. 1.11).

At the anode, a catalyst oxidizes the fuel, usually hydrogen, turning the fuel into a positively charged ion and a negatively charged electron. The electrolyte is a substance specifically designed so ions can pass through it, but the electrons cannot. The freed electrons travel through a wire creating the electric current. The ions travel through the electrolyte to the cathode. Once reaching the cathode, the ions are reunited with the electrons and the two react with a third chemical, usually oxygen, to create water or carbon dioxide. The most important design features in a fuel cell are as follows:

1. The electrolyte substance usually defines the type of fuel cell.
2. The most common fuel is hydrogen.

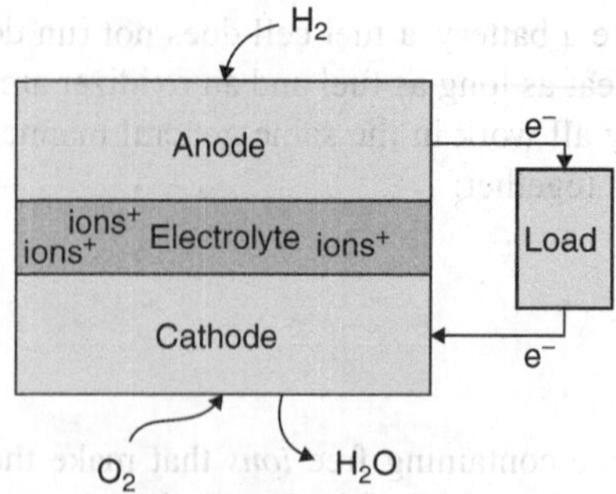

Figure 1.11 Block diagram of a fuel cell

3. The anode catalyst, which breaks down the fuel into electrons and ions. The anode catalyst is usually made up of very fine platinum powder.
4. The cathode catalyst, which turns the ions into the waste chemicals like water or carbon dioxide. The cathode catalyst is often made up of nickel.

A typical fuel cell produces a voltage from 0.6 V to 0.7 V at full-rated load. Voltage decreases as current increases, due to following factors:

1. Activation loss
2. Ohmic loss (voltage drop due to resistance of the cell components and interconnects)
3. Mass transport loss (depletion of reactants at catalyst sites under high loads, causing rapid loss of voltage).

To deliver the desired amount of energy, the fuel cells can be combined in series and parallel circuits, where the former yields higher voltage and the latter allows a higher current to be supplied. Such a design is called a fuel cell stack. The cell surface area can be increased to allow stronger current from each cell.

1.11.3.5 Thermoelectric Conversions

The Seebeck effect is the conversion of temperature differences directly into electricity and is named after the German physicist Thomas Johann Seebeck, who in 1821 discovered that a compass needle would be deflected by a closed loop formed by two metals joined in two places, with a temperature difference between the junctions. This was because the metals responded differently to the temperature differences, creating a current loop and a magnetic field.

The device diagram of the circuit on which Seebeck discovered the Seebeck effect is shown in Figure 1.12, and there are two different metals used. Seebeck did not recognize there was an electric current involved, so he called the phenomenon the thermomagnetic effect. Danish physicist Hans Christian Ørsted rectified the mistake and coined the term 'thermoelectricity'. The voltage created by this effect is of the order of several microvolts per Kelvin difference. One such combination, copper–constantan, has a Seebeck coefficient of 41 mV/K at room temperature. Using the principle known as 'the Seebeck Effect', electricity can be generated, if there is a temperature differential between the two sides of a thermoelectric module. Because this type of system

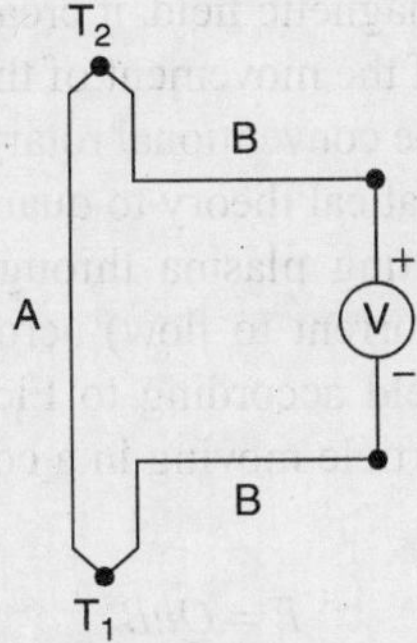

Figure 1.12 Seebeck effect

depends upon a consistent temperature differential to provide electricity, the modules are often combined with a known heat source such as natural gas or propane for remote power generation or waste heat recovery. They are often used in remote locations where power is required but solar energy is unreliable or insufficient, such as offshore engineering, oil pipelines, remote telemetry, and data collection. For power generation and waste heat recovery applications, thermoelectric modules have been optimized for efficient performance.

The term 'thermoelectric effect' encompasses three separately identified effects:

1. Seebeck effect
2. Peltier effect
3. Thomson effect

1.11.3.6 Magnetohydrodynamics

It provides a way of generating electricity directly from a fast moving stream of ionized gases without the need for any moving mechanical parts—no turbines and no rotary generators.

1.11.3.7 Working Principle

The magnetohydrodynamics (MHD) generator can be considered to be a fluid dynamo as shown in Figure 1.13. This is similar to a mechanical dynamo in which the motion of a metal conductor through a magnetic field creates a current in the conductor except that in the MHD generator, the metal conductor is replaced by conducting gas plasma.

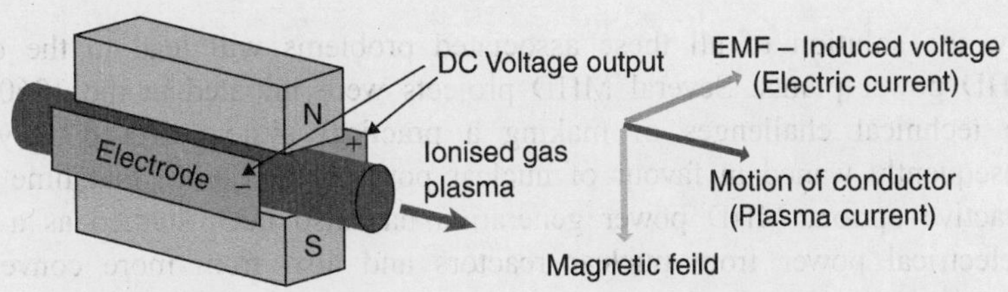

Figure 1.13 Principle of MHD generator

When a conductor moves through a magnetic field, it creates an electrical field perpendicular to the magnetic field and the direction of the movement of the conductor. This is the principle, discovered by Michael Faraday, behind the conventional rotary electricity generator. Dutch physicist Antoon Lorentz provided the mathematical theory to quantify its effects.

The flow (motion) of the conducting plasma through a magnetic field causes a voltage to be generated (and an associated current to flow) across the plasma, perpendicular to both the plasma flow and the magnetic field according to Fleming's right-hand rule. Lorentz Law describing the effects of a charged particle moving in a constant magnetic field can be stated as follows

$$F = QuB \tag{1.9}$$

where

F = force acting on the charged particle,
Q = charge of particle,
U = velocity of particle, and
B = magnetic field.

The calculation of open circuit voltage is straightforward. In a properly functioning MHD generator, open circuit voltage is given by

$$V = Bdu \tag{1.10}$$

where

B = flux density between the electrodes in Wb /m^2,
u = gas speed in m/s, and
d = electrode separation in m.

The relatively low ceiling of conversion efficiency of conventional generators prompted the scientists and engineers to look for new ways of converting heat energy directly into electrical energy. In the process, MHD shows promise as a way of generating electricity on a large scale.

MHD can be looked upon as a combination of fluid mechanics and electromagnetism; or in other words, behaviour of electrically conducting fluids in the presence of magnetic and electric fields. Operating principle of MHD is based on the fact that when as ionized (conductive) gas flows across a magnetic field an emf is induced in it, and if electrodes are placed in appropriate positions, this emf can deliver a current to an external load. However, MHD power generation is associated with many varying problems. Some important problems are as follows:

1. Production of very high temperature required for the plasma gases.
2. Increasing the working time of the plant by suitable design of plants.

Undoubtedly, the solution of all these associated problems will lead to the development of large MHD power plants. Several MHD projects were initiated in the 1960s, but overcoming the technical challenges of making a practical system proved very expensive. Interest consequently waned in favour of nuclear power which since that time has seemed a more attractive option. MHD power generation has also been studied as a method for extracting electrical power from nuclear reactors and also from more conventional fuel combustion systems.

1.12 RENEWABLE ENERGY

Renewable energy is the energy that comes from natural resources such as sunlight, wind, rain, tides, and geothermal heat, which are renewable (naturally replenished). The availability of the renewable energy resources is discussed in the following sections.

1.12.1 Worldwide Renewable Energy Availability

About 16% of global final energy consumption comes from renewable as shown in Figure 1.14, with 10% coming from traditional biomass, which is mainly used for heating, and 3.4% from hydroelectricity.

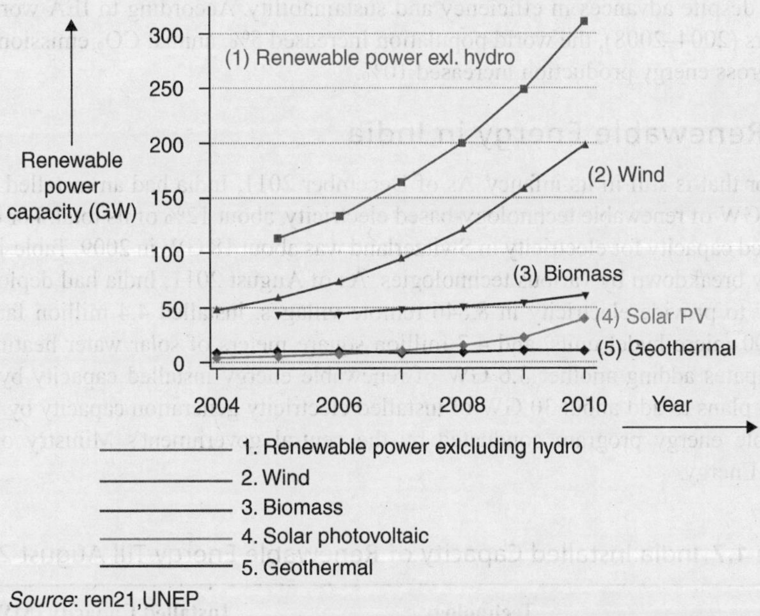

Source: ren21,UNEP

Figure 1.14 Worldwide renewable power capacity excluding hydro

New renewable energy (small hydro, modern biomass, wind, solar, geothermal, and biofuel) accounted for another 3% and were growing very rapidly. The share of renewable energy in electricity generation is around 19%, with 16% of global electricity coming from hydroelectricity, and 3% from new renewable energy.

Potential for renewable energy is given in Table 1.6.

Table 1.6 Potential for Worldwide Renewable Energy

Energy Resource	Energy Amount
Solar energy	1,600 EJ (444,000 TWh),
Wind power	600 EJ (167,000 TWh)

(Continued)

Table 1.6 Continued

Energy Resource	Energy Amount
Geothermal	500 EJ (139,000 TWh),
Biomass	250 EJ (70,000 TWh)
Mini hydropower	50 EJ (14,000 TWh)
Ocean energy	1 EJ (280 TWh)

More than half of the energy has been consumed in the last two decades since the industrial revolution, despite advances in efficiency and sustainability. According to IEA world statistics in four years (2004–2008), the world population increased 5%, annual CO_2 emissions increased 10%, and gross energy production increased 10%.

1.12.2 Renewable Energy in India

It is a sector that is still in its infancy. As of December 2011, India had an installed capacity of about 22.4 GW of renewable technology-based electricity, about 12% of its total. For context, the total installed capacity for electricity in Switzerland was about 18 GW in 2009. Table 1.7 provides the capacity breakdown by various technologies. As of August 2011, India had deployed renewable energy to provide electricity in 8,846 remote villages, installed 4.4 million family biogas plants, 1,800 micro-hydel units, and 4.7 million square meters of solar water heating capacity. India anticipates adding another 3.6 GW of renewable energy installed capacity by December 2012. India plans to add about 30 GW of installed electricity generation capacity by 2017 based on renewable energy program conducted by the central government's Ministry of New and Renewable Energy.

Table 1.7 India Installed Capacity of Renewable Energy Till August 2011

Type	Technology	Installed Capacity (MW)
Grid connected power system	Wind	14989
	Small Hydro	3154
	Biomass	1084
	Bagasse Cogeneration	1799
	Waste to energy	74
	Solar	46
Off-grid, Captive power	Biomass	141
	Biomass non-Bagasse cogeneration	328
	Waste to energy	76
	Solar	73
	Hybrid/Aerogen	01

1.12.3 Solar Energy

The annual amount of energy reaching the surface of the earth as solar radiation is about a billion kWh. An appropriate combination of solar thermal panels and photovoltaic cells could convert this to any desired combination of heat and electricity at an estimated efficiency of about 10%. Thus, the potential amount of energy that could be produced annually is 10^8 kWh; this would mean covering the entire surface of the globe with solar thermal panels and photovoltaic cells.

Solar heating and photovoltaic are potentially a very large source of energy in the forms, and the technology to use them already exists. Still, it has to be worked out how to integrate them with other technologies, and they need storage to be used effectively. Therefore, the contribution from this source must remain substantial, but uncertain at this stage.

The solar thermal power industry is growing rapidly with 1.2 GW under construction as of April 2009 and another 13.9 GW announced globally through 2014. Spain is the epicentre of solar thermal power development with 22 projects for 1,037 MW under construction, all of which are projected to come online by the end of 2010. In the United States, 5,600 MW of solar thermal power projects have been announced. In developing countries, three World Bank projects for integrated solar thermal or combined cycle gas turbine power plants in Egypt, Mexico, and Morocco have been approved.

Solar photovoltaic cells convert sunlight into electricity and photovoltaic production has been increasing by an average of more than 20% each year since 2002, making it a fast-growing energy technology. At the end of 2010, cumulative global photovoltaic (PV) installations surpassed 40 GW and PV power stations are popular in Germany and Spain.

Many solar photovoltaic power stations have been built, mainly in Europe.[9] As of December 2011, the largest photovoltaic (PV) power plants in the world are the Golmud Solar Park (China, 200 MW), Sarnia Photovoltaic Power Plant (Canada, 97 MW), Montalto di Castro Photovoltaic Power Station (Italy, 84.2 MW), Finsterwalde Solar Park (Germany, 80.7 MW), and Okhotnykovo Solar Park (Ukraine, 80 MW).

1.12.4 Wind Power

Wind energy is the kinetic energy of air in motion. Wind turbines extract the kinetic energy present in the wind, and convert it to rotary shaft motion. The shaft motion transmits power to generators by gearboxes, belts and pulleys, roller chains, or by hydraulic transmissions. The power in the wind is proportional to the cube of the wind velocity.

Let us assume

V is the wind speed (m/sec); ρ is the air density (approximately1.2 kg/m^2 at sea level); S is the cross-sectional area of air flow m^2; m is the mass of the wind passing per unit time through rotor area; t is the time period of wind flow (s).

Therefore, SVt is the volume of air passing through S (which is considered perpendicular to the direction of the wind) and

$$m = \rho SVt \tag{1.11}$$

Total wind energy flowing through an imaginary area A during the time t is

$$E = ½mV^2 = ½(SVt)V^2 = ½(St)V^3 \tag{1.12}$$

Power is the energy per unit time, and therefore, the wind power incident on A (e.g., equal to the rotor area of a wind turbine) is

$$P = E/t = \tfrac{1}{2}\rho SV^3 \tag{1.13}$$

Wind power in an open air stream is thus proportional to the third power of the wind speed; the available power increases eightfold when the wind speed doubles. Therefore, wind turbines for grid electricity need to be especially efficient at greater wind speeds. Thus, the general formula for available wind power is:

$$P = \tfrac{1}{2}\rho SV^3 \tag{1.14}$$

where P = power available in the wind in W. The available power of wind (P) is theoretically achievable; however, in practice, power extracted from the wind will theoretically be maximum and is given as

$$P = \tfrac{1}{2} C_p \rho SV^3 \tag{1.15}$$

where C_p is the performance coefficient of a wind machine. It is the ratio of the power extracted by the rotor to the power available in the wind stream. The maximum theoretical value of coefficient of performance is

$$C_{pmax} = 16/27 = 0.593$$

It is also known as Betz limit. Depending upon wind rotor design and wind speed, coefficient of performance (efficiency) of 30–50% can be achieved. It is established theoretically and experimentally as well. In all the cases, the power coefficient passes through a maximum at a particular value of tip speed ratio. Except the propeller type, the rise and fall of the power coefficient around the maximum value is quite rapid. The highest value of C_p is obtained with the propeller-type rotor.

Example 1.1

Power available in wind (in Watts) = ½ (air density)(swept area)(wind velocity^3)
where air density = 1.23 kg/m^3 at sea level.
Swept area is in m^2 and wind velocity is in m/s.
Let us work the formula for a 5-foot diameter turbine in a 10 miles/h wind:
Given
D = 5 feet = 1.524 m; swept area = $\pi D^2)/4$ = 1.8241m^2; wind speed = 10 mph = 4.4704 m/s.
Therefore, power available (watts) = ½ × 1.23 × 1.8241 × 4.4704^3 = 100.22 W.

Example 1.2

Consider that the wind speed for this 5-foot rotor is increased to 20 mph (8.9408 m/s).
Power available (Watts) = ½ (1.23) × (1.8241) × (8.9408)^3 = 802 W.

Example 1.3

Consider a 10-foot (3.048 m) diameter rotor for a 7.30 m^2 swept area in a 10 miles/h wind.
Thus, power available (watts) = 1/2 × (1.23) × (7.30) × (4.4704) = 401 W.

Example 1.4

If wind speed in Problem 1.3 is 20 miles/h. Power available (Watts) = ½ (1.23) × (7.3) × (8.9408) = 3,209 W.

Observations

1. From problem 1, it can be observed that there is very little power available in low winds.
2. The only way to increase the available power in low winds is by sweeping a large area with the blades.
3. The first key concept that this formula show is that when the wind speed doubles, the power available increases by a factor of 8
4. The second key concept from this formula is that the power available increases by a factor of 4 when the diameter of the blades doubles.

Total wind power capacity for top ten countries are given in Table 1.8.

Estimating the total energy carried by all the world's winds is possible, but not very useful. It does not seem sensible to include winds at one kilometre above the surface of the earth, in the middle of the larger oceans, or in the remote regions of the Arctic and Antarctic regions. Considering only sites on land or in coastal waters, the estimated total potential energy that can be provided is 20,000 TWh per year—about twice the 1987 global electricity consumption.

Electricity (25 TWh) was generated from wind in 1999. This is less than one hundredth of the electricity generated by hydro, but about 10 times that was generated by the wind in 1990. Wind generation capacity is growing very quickly, but because winds do not always blow when we need the electricity, there will be a limit to the use of this resource until there is some economical way to store the energy. Eighty-three countries around the world are using wind power on a commercial basis.

Table 1.8 Total Wind Power Capacity for Top 10 Countries

Country	Total Capacity End 2009 (MW)	Total Capacity June 2010 (MW)	Total Capacity End 2010 (MW)	Total Capacity June 2011 (MW)
Canada	2,550	3,319	4,008	4,611
China	26,010	33,800	44,733	52,800
France	4,521	5,000	5,660	6,060
Germany	25,777	26,400	27,215	27,981
India	10,925	12,100	13,066	14,550
Italy	4,850	5,300	5,797	6,200
Portugal	3,357	3,465	3,734	3,960
Spain	19,149	19,500	20,676	21,150
United Kingdom	4,092	4,600	5,204	5,707
United States	35,159	36,300	40,180	42,432
Rest of world	21,698	24,500	26,154	29,500
Total	1,59,213	1,75,000	1,96,630	2,15,000

1. Cambay Graben in Gujarat
2. Manikaran in Himachal Pradesh
3. Surajkund in Jharkhand
4. Chumathang in Jammu & Kashmir

1.12.5 Tidal Power

Tidal plants will probably be worthwhile only in places where the tidal range is particularly large, which means that estimating a 'world total' involves not a general assessment but careful site by site investigation. Using this approach, the potential output of the most promising sites is considered to be about 386 per year or around 10% of the world's total stations generating around 670 GHz (equal to 0.67) of electricity per year.

Tidal power is essentially a specific form of hydropower, and therefore, uses basically the same equipment as a regular tidal station. The difference is in the available power to extract from the tide. A reversible hydraulic turbine is used so the inflow and outflow of the tide can generate electricity.

The following is the diagram of a simple barrage: tidal barrages work like a hydro-electric scheme. However, the dam is much bigger. A huge dam or a barrage is built across a river estuary. When the tide goes in and out, the water flows through tunnels in the dam.

As the tide comes in, the dam allows the seawater to pass through into a holding basin. As soon as the tide is about to go down, the dam is closed. The water held back in this way will be used to feed the turbines at low tide. The ebb and flow of the tides can be used to turn a turbine, or it can be used to push air through a pipe, which then turns a turbine. Large lock gates, like the ones used on canals, allow ships to pass.

There would be a number of benefits, including protecting a large stretch of coastline against damage from high storm tides, and providing a ready-made road bridge. However, the drastic changes to the currents in the estuary could have huge effects on the ecosystem, and huge

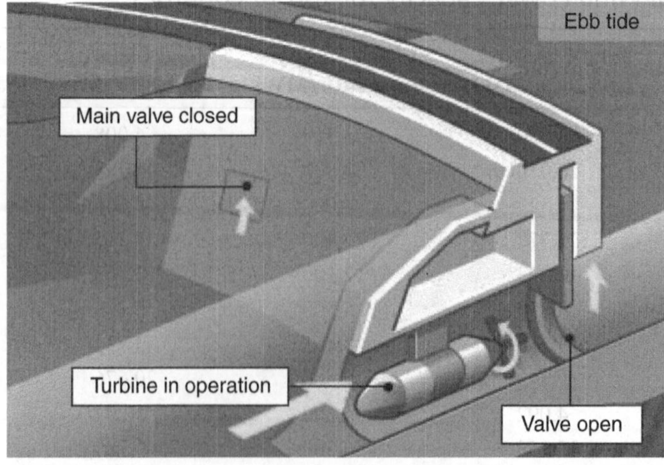

Source: www.worldcolleges.info/science-Tech/Tidal-barrages.php

Figure 1.15 Tidal barrage

numbers of birds that feed on the mud flats in the estuary when the tide goes out would have nowhere to feed.

A major drawback of tidal power stations is that they can only generate when the tide is flowing in or out—in other words, only for 10 h each day. However, tides are totally predictable, so we can plan to have other power stations generating at those times when the tidal station is out of action.

There are different types of turbines that are available for use in a tidal barrage. A bulb turbine is one in which water flows around the turbine. If maintenance is needed, then the water must be stopped. This is a time consuming process resulting in loss of generation. When rim turbines are used, the generator is mounted at right angles to the turbine blades.

A rough estimate of power output from a tidal barrage can be obtained from a simple energy balance method by considering the average change of potential energy during the draining process.

Let ρ is the density of water 1,025 kg/m^3 (seawater varies between 1,021 and 1,030 kg/m^3), g is the acceleration due to gravity (9.81 m/s^2), A is the horizontal area of the tidal pool (m^2), h is the vertical range of the tide (m), m is the total mass of water in the tidal basin above the low water level, T is the time interval between tides, i.e., the tidal period, 1/2 (factor half) is due to the fact that as the basin flows empty through the turbines, the hydraulic head over the dam reduces. The maximum head is only available at the moment of low water, assuming the water level is still present in the basin.

Therefore, $$m = \rho Ah \tag{1.16}$$

The height of the centre of gravity = $h/2$

Therefore, work done in raising the water

$$= 1/2mgh = 1/2\ \rho gAh^2 \tag{1.17}$$

= potential energy contained in a volume of water.

Average power for a tidal period is calculated by dividing the extracted energy by the tidal period (T).

Hence, the average power output, $$P_{ave} = 1/2\rho gAh^2/T \tag{1.18}$$

1.12.5.1 Calculation of Tidal Power Generation

Example 1.5

Let us assume that the tidal range of tide at a particular place is 32 feet = 10 m (approximately); the surface of the tidal energy harnessing plant is 9 km² (3 km × 3 km) = 3,000 m × 3,000 m. = 9 × 10^6 m^2; density of sea water = 1,025.18 kg/m^3

Mass of the sea water = volume of sea water × density of sea water

= (area × tidal range) of water × mass density

= (9 × 10^6 m^2 × 10 m) × 1,025.18 kg/m^3 = 92 × 10^9 kg (approximately)

Assuming factor half for obtaining average power, potential energy content of the water in the basin at high tide = ½ × area × density × gravitational acceleration × (tidal range)^2 = ½ × 9 × 10^6 m^2 × 1,025 kg/m^3 × 9.81 m/s^2 × (10 m)^2 = 4.5 × 10^{12} J (approximately)

Now, we have two high tides and two low tides everyday. At low tide, the potential energy is zero. Therefore, the total energy potential per day

$$= \text{energy for a single high tide} \times 2 = 4.5 \times 10^{12}\ \text{J} \times 2 = 9 \times 10^{12}\ \text{J}$$

Therefore, the mean power generation potential = energy generation potential/time in 1 day = 9×10^{12} J/86,400 s = 104 MW

Assuming the power conversion efficiency to be 30%, the daily-average power generated

$$= 104\ \text{MW} \times 30\%/100\%$$

$$= 31\ \text{MW (approximately)}$$

Because the available power varies with the square of the tidal range, a barrage is best placed in a location with very high-amplitude tides. Suitable locations are found in Russia, USA, Canada, Australia, Korea, and UK. Amplitudes of up to 17 m (56 feet) occur, for example, in the Bay of Fundy, where tidal resonance amplifies the tidal range

Example 1.6

Estimate the average power output of Severn Barrage tidal basin of 520 Km^2 area with a tidal range of 7 m and tidal period of 12.5 h. Assume the density of seawater as 1,025 kg/m^2.

Solution Tidal period (T) = 12.5 h =12.5 × 60 × 60 s = 45,000 s = 45 × 10^3 s.
Area of tidal basin (A) = 520 km^2 = 520 × 10^6 m^2
Tidal range or height (H) = 7 m
Density of seawater =1,025 kg/m^3
Acceleration due to gravity = 9.81 m/s^2
The average power output, $P_{ave} = 1/2 \rho g A h^2 / T$

$$= 1/2 \times 1{,}025 \times 9.81 \times 520 \times 10^6 \times (7)^2/45 \times 10^3$$

$$= 28.5 \times 10^8\ \text{W}$$

1.12.5.2 *Advantages of Tidal Power Generations*

Although the initial costs of tidal power generations are extremely high as compared to conventional hydroelectric power plants, they are associated with large number of the following advantages.

1. A firm output is guaranteed and the output can be predicted well in advance.
2. Tidal power is unaffected by the vagaries of nature like draught.
3. The impact of tidal power plants on ecology is minimum since no unhealthy waste such as exhaust gas, ash, atomic refuse, or atmospheric radiation are involved
4. Unlike hydroelectric power plant and atomic power plants, the tidal plant do not make demands on large areas of valuable lands.
5. Tidal power plants can also add to recreational facilities.
6. Tidal basins created can also increase the possibility of fish farming.

1.12.5.3 Development of Tidal Power Scheme in India

Nevertheless, the possibility of developing tidal power scheme in India may be examined in view of the following all aspects:

1. Economic aspects of tidal power schemes when compared to the conventional schemes.
2. Problems associated with the construction and operation of plant.
3. Problems related to the hydraulic balance of the system in order to minimize the fluctuation in the power output.

1.12.6 Wave Energy

Sea wave energy is now the vanguard in averting a world energy crisis not only on technical grounds but also from economic, environmental, and political considerations. Solar and wind energy fickle and widely diffused over the earth surface. However, nature has evolved a very elegant system of gathering solar energy by the windward shores, thus solving at a stroke the problems of collection storage and transmission of energy.

Technical system for the utilization of wave energy must always be developed with a view to one single or a few possible applications for the generated secondary energy. This approach promises maximum utilization of the generated secondary energy. Because of the technical problem in foresight, it may be assumed that wave energy converters will, in the beginning, attain only regional energy supplies.

1.12.6.1 Power Calculation for Ocean Waves

Typical sea wave propagation is shown in Figure 1.16.

The distance between two consecutive troughs defines wavelength λ. Wave height H (crest to trough) is proportional to wind intensity and its duration. The wave period T (crest to crest) is the time in seconds needed for the wave to travel with the wavelength λ and is proportional to sea depth. The frequency $f = 1/T$ indicates the number of waves that appears in a given position. Consequently, the wave speed is $v = \lambda/T = \lambda/f$. The ratio $\lambda/2H$ is called the wave declivity and when this value is greater than 1/7, it can be proved that the wave becomes unstable and vanishes.

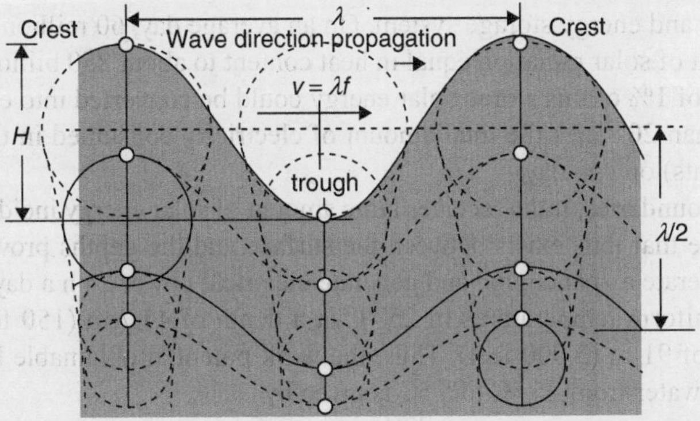

Figure 1.16 Sea wave propagation

Longer period waves have relatively longer wavelengths and move faster. Generally, large waves are more powerful. Ocean waves transport mechanical energy.
The power associated with a wave of wavelength λ, height H, and a front b is given by

$$P = \tfrac{1}{2}\, \rho g \lambda b H^2 \tag{1.19}$$

where

ρ = specific weight of water;
g = acceleration due to gravity;
H = wave height in meter;
and λ = the wave length in meter.

Then, power across each meter of wave front associated to uniform wave is given by

$$P_v = P/b = \tfrac{1}{2}\, \rho g \lambda H^2,\ \text{W/m} \tag{1.20}$$

For irregular waves of height H (in meter) and period T (in second), an expression for power per unit wave front can be derived as

$$P_v = 0.5\, H^2 T \text{ (kilowatt per meter, kW/m)} \tag{1.21}$$

Example 1.7

Consider moderate ocean swells, in deep water, a few kilometres off a coastline, with a wave height of 3 m and a wave period of 8 s. Calculate power potential per meter of coastline

$$P_v = 0.5(\text{kW/m}^3\text{s}) \times 3^2\ (\text{m}^2) \times 8(\text{s}) = 36\ (\text{kW/m})$$

It means that there are 36 kW of power potential per meter of coastline. In major storms, the largest waves offshore are about 15 m high and have a period of about 15 s. According to the abovementioned formula, such waves carry about 1.7 MW of power across each meter of wave front. An effective wave power device captures as much as possible of the wave energy flux. As a result, the waves will be of low height in the region behind the wave power device.

1.12.7 Ocean Thermal Energy

The oceans cover a little more than 70% of the earth's surface. This makes it the world's largest solar energy collector and energy storage system. On an average day, 60 million km^2 of tropical seas absorb an amount of solar radiation equal in heat content to about 250 billion barrels of oil. If less than one tenth of 1% of this stored solar energy could be converted into electric power, it would supply more than 20 times the total amount of electricity consumed in the United State (263 million inhabitants) on one day.

Within territorial boundaries, India receives large amount of solar energy incident upon it. The temperature difference that thus exists between the surface and the depths provides the source and sink needed to operate a heat engine and generate electrical power 24 h a day. Typically, this surface layer has a uniform temperatures of 25 °C at a depth of 45.7 m (150 feet) while 5 °C water lies at a depth of 91 m (3,000 feet). Thus, the work potential obtainable by lowering the temperature of 1 g of water from 25 °C to 5 °C is given by

$$_{278}\int^{298} \text{C}\,(T - 278)/T dt = 2.87\ \text{J/g} \tag{1.22}$$

This 2.87 J/g is just the work obtained per gram of flux by an ideal hydroelectric power plant with a 292.5 m (960 feet) head. Thus, the top isothermal layers of the tropical oceans have an effective head of 292.5 m. Further, there is an ample prospects for tapping this vast reservoir of work potential. Such tapping of energy is called ocean thermal energy conversion (OTEC).

1.12.7.1 Principles of Ocean Thermal Energy Conversion

The idea behind OTEC is the use of all natural collectors, the sea, instead of artificial collector. As shown in Figure 1.17, warm water is collected on the surface of the tropical ocean and pumped by a warm water pump.

The water is pumped through the boiler, where some of the water is used to heat the working fluid, usually propane or some similar material. If it is a cooler, you can use a material with a low boiling point like ammonia. The propane vapour expands through a turbine that is coupled to a generator generating electric power. Cold water from the bottom is pumped through the condensers, where the vapour returns to the liquid state. The fluid is pumped back into the boiler. Some small fraction of the power from the turbine is used to pump the water through the system and to power other internal operations, but most of it is available as net power.

Like other solar energy conversion system, the basic OTEC plant will be very attractive ecologically when compared to fossil fuel or nuclear power plant and will be virtually non-polluting. The production of ammonia at a tropical OTEC plant may prove attractive socially and economically. If interested to capitalize on the vast tropical ocean energy resource, hydrogen is likely to be the primary long term OTEC produced either directly or by the indirect route of using ammonia as a hydrogen carrier, which is then decomposed into N_2 and H_2 and fed in to H_2–O_2 fuel cells to produce electricity.

1.12.8 Biomass Energy

Plant matter created by the process of photosynthesis is called biomass. In plants, algae and certain types of bacteria, the photosynthetic process results in the release of molecular oxygen and

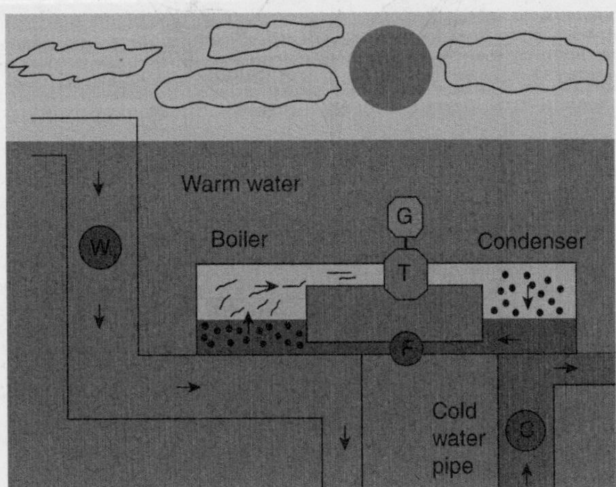

Figure 1.17 Principles of ocean thermal energy conversion

the removal of carbon dioxide from the atmosphere that is used to synthesize carbohydrates (oxygenic photosynthesis). Other types of bacteria use light energy to create organic compounds but do not produce oxygen (non-oxygenic photosynthesis). Photosynthesis provides the energy and reduced carbon required for the survival of virtually all living organisms on our planet, as well as the molecular oxygen necessary for the survival of oxygen-consuming organisms. In addition, the fossil fuels currently being burned to provide energy for human activity were produced by ancient photosynthetic organisms.

1.12.8.1 Photosynthesis in Plants

It is the process by which plants, some bacteria, and some protists use the energy from chlorophyll. Most of the time, the photosynthesis process uses water and releases the oxygen for human survival. Further, there will always be need for the food and energy.

In plants (see Fig. 1.18), photosynthesis occurs mainly within the leaves. Since photosynthesis requires carbon dioxide, water, and sunlight, all of these substances must be obtained by or transported to the leaves. Carbon dioxide is obtained through tiny pores in plant leaves called stomata. Oxygen is also released through the stomata. Water is obtained by the plant through the roots and delivered to the leaves through vascular plant tissue systems. Sunlight is absorbed by chlorophyll, a green pigment located in plant cell structures called chloroplasts. Chloroplasts are the sites of photosynthesis. Chloroplasts contain several structures, each having specific functions.

1.12.8.1.1 Oxygenic Photosynthesis The photosynthetic process in all plants and algae as well as in certain types of photosynthetic bacteria involves the reduction of CO_2 to carbohydrate and removal of electrons from H_2O, which results in the release of O_2. This process is known as oxygenic photosynthesis.

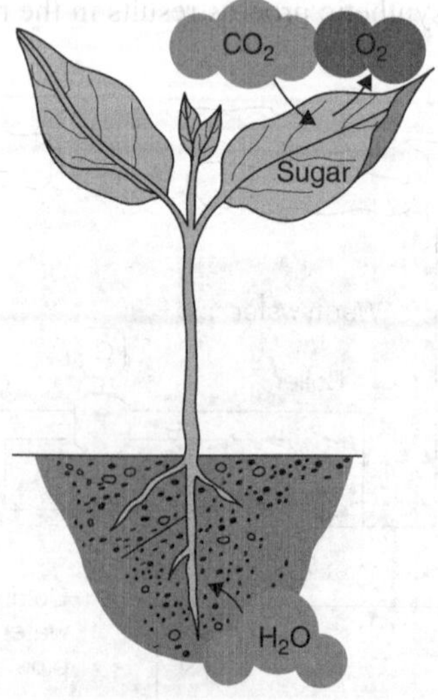

Figure 1.18 A typical plant

Some photosynthetic bacteria can use light energy to extract electrons from molecules other than water. These organisms are of ancient origin, presumed to have evolved before oxygenic photosynthetic organisms.

1.12.8.1.2 Anoxygenic Photosynthesis

These organisms occur in the domain bacteria and have representatives in four phyla—Purple Bacteria, Green Sulphur Bacteria, Green Gliding Bacteria, and Gram Positive Bacteria. By the middle of the 19th century, the key features of plant photosynthesis were known. Further, plants could use light energy to make carbohydrates from CO_2 and water.

In photosynthesis, solar energy is converted to chemical energy. The chemical energy is stored in the form of glucose (sugar). Carbon dioxide, water, and sunlight are used to produce glucose, oxygen, and water. The chemical equation for this process is

$$CO_2 + 2H_2O \xrightarrow{\text{Solar Light Energy}} [CH_2O] + O_2 + H_2O \quad (1.23)$$

where $[CH_2O]$ represents a carbohydrate (i.e., glucose, which is a six-carbon sugar). The synthesis of carbohydrate from carbon and water requires a large input of light energy. The standard free energy for the reduction of one mole of CO_2 to the level of glucose is +478 kJ/mol. Because glucose, a six-carbon sugar, is often an intermediate product of photosynthesis, the net equation of photosynthesis is frequently written as

$$6H_2O + 6CO_2 \xrightarrow{\text{Solar Light Energy}} C_6H_{12}O_6 + 6O_2 \quad (1.24)$$

The standard free energy for the synthesis of glucose is +2,870 kJ/mol. Most of the people do not understand chemically, so the abovementioned chemical equation is translated as

A combination of six molecules of water and six molecules of carbon dioxide produce one molecule of sugar and six molecules of oxygen.

Approximately 114 kilocalories of free energy are stored in plant biomass for every mole of CO_2 fixed during photosynthesis. Solar radiation striking the earth on an annual basis is equivalent to 1,78,000 TW, i.e. 15,000, times that of current global energy consumption. Although photosynthetic energy capture is estimated to be 10 times that of global annual energy consumption, only a small part of this solar radiation is used for photosynthesis. Approximately, two thirds of the net global photosynthetic productivity worldwide is of terrestrial origin, while the remainder is produced mainly by phytoplankton (microalgae) in the oceans that cover approximately 70% of the total surface area of the earth. Since biomass originates from plant and algal photosynthesis, both terrestrial plants and microalgae are appropriate targets for scientific studies relevant to biomass energy production.

Any analysis of biomass energy production must consider the potential efficiency of the processes involved. Although photosynthesis is fundamental to the conversion of solar radiation into stored biomass energy, its theoretically achievable efficiency is limited both by the limited wavelength range applicable to photosynthesis, and the quantum requirements of the photosynthetic process. Only light within the wavelength range of 4–7 μm is used by plants. This visible light wavelength range is known as photosynthetically active radiation (PAR).

For a 4-carbon sugar (C_4) plants, such as maize, sorghum, and sugarcane, which is the first product of photosynthesis? The photosynthetically active compounds captures photon energy of PAR is 80%. Photon energy of PAR lost (by absorption, reflection, and transportation) is 20%.

A minimum of eight photon energy of PAR are required to produce one mole of glucose and water each. Thus, energy of PAR conversion to glucose is 50%. Energy stored in the produced glucose from photon energy is 28%. Dark respiration is the reverse process of photosynthesis to sustain plants metabolic process. Photon energy consumed in this process from the stored energy of glucose is 40%.

Considering all the abovementioned facts, a maximum photosynthesis efficiency

$$= [0.5 \times 0.8 \times 0.28 \times (100 - 40)]100 = 6.72\%.$$

For a 3-carbons sugar (C_3) plants such as wheat, rice, soya bean, and tree account for 95% of global biomass. The photosynthetically active compounds captures photon energy of PAR is 74%. Photon energy of PAR lost (by absorption, reflection, and transportation) is 26%. A minimum of eight photon energy of PAR are required to produce one mole of glucose and water each.

Thus, energy of PAR conversion to glucose is 50%. Energy stored in the produced glucose from photon energy is 28%. Dark respiration is the reverse process of photosynthesis to sustain plants metabolic process. Photon energy consumed in this process from the stored energy of glucose is 70%.

Considering all the abovementioned facts, a maximum photosynthesis efficiency

$$= [0.5 \times 0.28 \times 0.74 \times (100 - 70)]100 = 3.12\%$$

Thus, the efficiency of biomass production of C_3 plants is lower than C_4 plants.

Biomass includes solid biomass (organic, non-fossil material of biological origins), biogas (principally methane and carbon dioxide produced by anaerobic digestion of biomass and combusted to produce heat and/or power), and liquid biofuel (bio-based liquid fuel from biomass transformation, mainly used in transportation applications), and municipal waste (wastes produced by the residential, commercial, and public services sectors and incinerated in specific installations to produce heat and/or power). The most successful forms of biomass are sugar cane bagasse in agriculture, pulp and paper residues in forestry, and manure in livestock residues. It is argued that biomass can directly substitute fossil fuels, as more effective in decreasing atmospheric CO_2 than carbon sequestration in trees. The Kyoto protocol encourages further use of biomass energy. Biomass may be used in a number of ways to produce energy. The most common methods are combustion, gasification, fermentation, and anaerobic digestion.

1.12.8.2 Biomass Energy in India

India is very rich in biomass. It has a potential of 19,500 MW (3,500 MW from Bagasse-based cogeneration and 16,000 MW from surplus biomass). Currently, India has 537 MW commissioned and 536 MW under construction. The facts reinforce the idea of a commitment by India to develop these resources of power production.

Table 1.9 is a list of some states with most potential for biomass production:

Table 1.9 Potential for Biomass Production in India

S. No.	State	Capacity (MW)
1	Andhra Pradesh	200
2	Bihar	200
3	Gujarat	200
4	Karnataka	300
5	Maharashtra	1,000
6	Punjab	150
7	Tamil Nadu	350
8	Uttar Pradesh	1,000

The potential available and the installed capacities for biomass and Bagasse are as follows:

Source	Potential	Installed
Biomass	16,000 MW	222 MW
Biogases (cogeneration) in existing sugar mills	3,500 MW	332 MW

The major part of waste obtainable after the energy utilisation are non-organic, having diversified nature and characteristics, and thus, their identification and separation from the main waste stream by improved techniques are an essential parameter of any energy recovery scheme. On site processing of waste for the reduction of in-home compactors and industrial shredders through improved technology should be employed that may be environmentally acceptable. Collection and transportation components of the waste energy conversion scheme are the most expensive components owing to the many varying social, technical, and other reasons. A careful cost analysis and implementation of this vital component will minimize the running cost of the scheme. The storage of waste for resource recovery and final disposal after suitable treatment is another component of scheme and selection of storage station and other associated problems invite careful attention. Normally, two types of energy recovery systems are used.

Separation of metals, paper, and glass from the rest through the process such as size reduction, screening, vibrating sorting, and electronic scanning; however, a truly homogeneous, inexpensive separation scheme will provide competitive input to waste energy utilization. The conversion of the remaining waste product to usable energy includes:

1. Generation of methane gas (biogas conversion) or other fuels (biological conversion).
2. Generation of electricity either from (step 1) or through thermo-mechanical process.
3. Compositing for fertilizers.

Treatment is meant for those processes designed to reduce waste to innocuous forms without or after energy recovery. The most familiar techniques are the burning of waste at high temperature in the presence of oxygen (known as incineration) and the breaking down of the complex compounds using heat in the absence of oxygen (known as pyrolysis). However, treatment techniques should be selected so as to be accepted socially, environmentally, and economically. The cheapest method for final disposal of waste before or after energy recovery is a systematic burial in ground.

1.12.8.3 Biological Hydrogen Production

Biological hydrogen production appears to be another promising conversion technique for Indian environment as compared to other hydrogen producing schemes because of the following points:

1. The only energy input is solar energy and a hydrogen donor (probably seawater).
2. Biological hydrogen schemes can be operated at relatively low temperature (20°–50° C).
3. Scheme is free from pollution.
4. The fuel produced, hydrogen gas, is a clear burning fuel (yielding water).
5. It can be used for multi-utility scheme, such as production of food for human and animal consumption, fuel as methane or alcohol, and commercially usable chemical products.

1.12.8.4 Biogas Conversion

Organic wastes, which, on natural anaerobic digestion produce methane gas is an essential by-product of civilization, often offensive in its raw state are available in plenty in both urban as well as rural areas. Anaerobic digestion not only provides valuable fuels and enhances the fertilizer value of the waste, but also provides a convenient, safe, aesthetic, and economical waste disposal method.

Anaerobic digestion is a complex biological conversion process during which organic matter is decomposed by anaerobic bacteria organism into stabilized minerals and gas. The organic substrate need not be pure. The entire substrate—carbohydrates, fats, and proteins—with the possible exception of a small amount of fibre is broken during the digestion process yielding methane and carbon dioxide.

Bio-energy conversion seems to be the most promising energy conversion techniques (specifically for India) in near future probably because of the following points:

1. The absence or the difficulties related to the installation of centralized power supply systems.
2. Increasing energy demand for energy even in remote rural areas or isolated parts of the country.
3. Basic need for large amounts of protein for food and feeding purpose, and inexpensive methods available for collecting and storing of energy.

Energy schemes utilizing plant (biomass) as source of liquid fuel (such as ethanol or methanol) are, therefore, worth attempting in addition to electrical power generation. The production of usable energy through algal and similar crops includes the following three important conversion steps.

1. Production of organic matters and photosynthesis.
2. Collection and processing of plant material.
3. Fermentation of organic matter to produce liquid and gaseous fuels and their storage.

Sugar crops, trees, grains, and grasses are various aquatic fuel sources and have relative potentials on each other and utilized in any biomass production schemes. Sugar crops and algal crops seem to be the most promising crops of importance suitable for the bio-energy conversion in India.

1.12.9 Decentralized and Dispersed Generation

Decentralized energy is the opposite of centralized energy. It is also referred to as distributed energy, distributed generation, or onsite power generation. It generates the power and energy that a residential, commercial, or industrial customer needs onsite. Examples of decentralized energy production are solar energy systems and solar refrigerator energy systems.

Decentralized generation is the production of electricity at or near the point of use, irrespective of size, fuel, or technology. Decentralized electric generation will reduce capital investment, lower the cost of electricity, reduce pollution, reduce production of greenhouse gas, and decrease vulnerability of the electric system to extreme weather conditions and terrorist attacks. Decentralized generation can be distributed or dispersed and can be powered by a wide variety of fossil fuels.

Distributed power generation is any small-scale power generation technology that provides electric power at a site closer to customers than central station generation. Dispersed generation is similar to decentralized energy, which is the opposite of centralized energy. It is defined as the efficient deployment of clean, efficient, and renewable power, located very near a load centre, that can be anywhere in size from 1 MW to 100 MW.

Distributed generation is used mainly for onsite power generation. Dispersed generation is strategically located on the transmission grid to overcome bottlenecks in the transmission and distribution system and to improve the stability of the system.

1.12.9.1 Features of Dispersed Generation

Dispersed generation reduces both power transfers between regions of the power system and power imbalance in each region. Dispersed generation also allows for a uniform distribution of the overall system by responding fast to demand variation. Dispersed generation offers more flexibility and can be dispatched in incremental blocks of power as needed. It provides reliability and stability to the system. Total failure can be avoided when the load centres are supported by dispersed generation. A major outage such as the one experienced in August 2003 could have been avoided with the help of dispersed generation powered by reciprocating engines and bringing power back online within 10 min.

1.12.9.2 Types of Distributed Energy Resources

Distributed energy resource (DER) systems are small-scale power generation technologies (typically in the range of 3 kW to 10,000 kW) used to provide an alternative to or an enhancement of the traditional electric power system. The usual problem with distributed generators is their high costs. Following may be considered as distributed energy resources:

1. One popular source is solar panels on the roofs of buildings. Unlike coal and nuclear, there are no fuel costs, pollution, mining safety, or operating safety issues. Solar power has a low capacity factor, producing peak power at local noon each day. Average capacity factor is typically 20%.
2. Another source is small wind turbines. These have low maintenance and low pollution. Construction costs are higher per watt than large power plants, except in very windy areas. Wind towers and generators have substantial insurable liabilities caused by high winds, but good operating safety.
3. Solid oxide fuel cells using natural gas, like the bloom energy server, have recently become a distributed energy resource.
4. Distributed cogeneration sources use natural gas-fired microturbines or reciprocating engines to turn generators. The hot exhaust is then used for space or water heating or to drive an absorptive chiller for air conditioning.

The clean fuel has only low pollution. Designs currently have uneven reliability, with some makes having excellent maintenance costs and others being unacceptable.

1.12.9.3 Advantages for Dispersed Generation

1. *Low, cost of electricity*: The fact that the consumer is benefited with low cost of electricity could well be the key driver for dispersed generation.
2. *Geographical factors*: Existence of transmission congestion and high price in major metropolitan areas provide ample potential for dispersed generation.
3. *Saving on outage cost*: The rising demand for premium power may force many industrial and commercial consumers to switch to dispersed generation to protect against the risk of power outages.
4. *Increasing demand in intermediate sector*: Flexibility to meet intermediate load accelerates the demand for dispersed generation.
5. *Low payback period*: As utility providers are worried about investing for long-term, dispersed generation calls for lesser investment and lower payback period.

1.12.9.4 Technological Options

Dispersed generation options can be classified either on the basis of the prime movers used or engines, turbines, fuel cells, or on the basis of fuel resources used (renewable and non-renewable).

Figure 1.19 illustrates the technology options for distributed generation. In India, many renewable energy technologies are being employed in a number of distributed generation projects. The technologies include biomass gasifier, solar thermal and photovoltaic systems, small wind turbines (aero-generators), and small hydro-power plants.

1.12.9.5 Relevance of Distributed Generation in India

In India, distributed generation has found three distinct markets:

1. Back-up small power generation systems including diesel generators that are being used in the domestic and small commercial sectors.

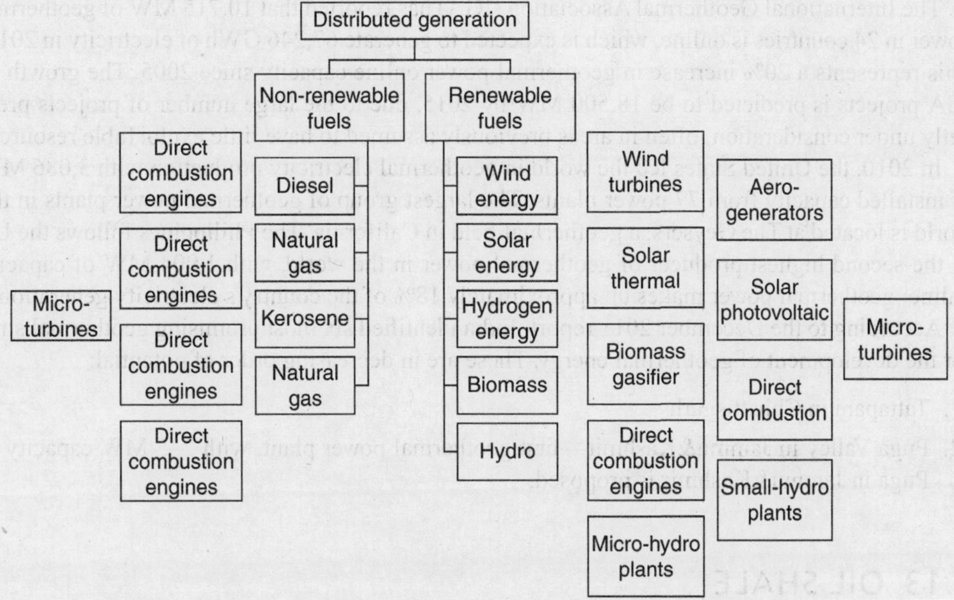

Figure 1.19 Technology options for distributed power generation

2. Stand-alone off-grid systems or mini grids for electrification of rural and remote areas.
3. Large captive power plants such as those installed by power intensive industries.

1.12.10 Geothermal Energy

There is a constant flow of heat from the hot interior of the earth, and a corresponding rise of temperature as one go deep underground. In most places, this is too little to be useful with present technology. However, some parts of the world are particularly favoured with very hot water or steam with only a few hundred meters below the surface, where the hot fluid breaks through naturally. Further, hot springs or geysers are found, and these have been used as heat source for hundreds of years.

As the resource varies so much from place to place, a 'world total' is obtained by detailed investigation of possible sites. The world total capacity for electricity production from geothermal resources is about 8,000 MW. The World Energy Council estimates that in 1999, about 68 TWh of energy came from geothermal sources, including both direct heat and electricity.

Geothermal power is cost effective, reliable, sustainable, and environmentally friendly, but has historically been limited to areas near tectonic plate boundaries. Recent technological advances have dramatically expanded the range and size of viable resources, especially for applications such as home heating and opening a potential for widespread exploitation. Geothermal wells release greenhouse gases trapped deep within the earth, but these emissions are much lower per energy unit than those of fossil fuels. As a result, geothermal power has the potential to help mitigate global warming, if widely deployed in place of fossil fuels.

The International Geothermal Association (IGA) has reported that 10,715 MW of geothermal power in 24 countries is online, which is expected to generate 67,246 GWh of electricity in 2010. This represents a 20% increase in geothermal power online capacity since 2005. The growth of IGA projects is predicted to be 18,500 MW by 2015, due to the large number of projects presently under consideration, often in areas previously assumed to have little exploiTable resource.

In 2010, the United States led the world in geothermal electricity production with 3,086 MW of installed capacity from 77 power plants. The largest group of geothermal power plants in the world is located at The Geysers, a geothermal field in California. The Philippines follows the US as the second highest producer of geothermal power in the world, with 1,904 MW of capacity online; geothermal power makes up approximately 18% of the country's electricity generation.

According to the December 2011 report, India identified six most promising geothermal sites for the development of geothermal energy. These are in decreasing order of potential:

1. Tattapani in Chhattisgarh
2. Puga Valley in Jammu&Kashmir—first geothermal power plant, with 2–5 MW capacity at Puga in Jammu&Kashmir is proposed.

1.13 OIL SHALE

The rapid depletion rates of global reserves of conventional fossil fuel (oil) have led to an increased focus on unconventional oil sources, many of which are associated with shale. Oil shale is a fine-grained sedimentary rock that contains solid bituminous materials (called kerogen, which is an organic matter) that release petroleum-like liquids (shale oil or gas) when the rock is heated from which oil or gas can be extracted. All types of kerogen consist mainly of hydrocarbons, smaller amounts of sulphur, oxygen, and nitrogen, and a variety of minerals.

They were formed millions of years ago by deposition of silt and organic debris on lake beds and sea bottoms. Heat and pressure then transformed the materials into oil shale in a process similar to that forms oil over long periods of time. Similar to traditional petroleum, natural gas, and coal, oil shale and kerogen are also fossil fuels. Fossil fuels are developed from the remains of algae, spores, plants, pollen, and a variety of other organisms that lived millions of years ago in ancient lakes, seas, and wetlands.

Dr Chudamani Ratnam, former chief of Oil India Ltd, from 1990 onwards, repeatedly claimed that India had a big treasure of shale oil in Arunachal Pradesh and other parts of the northeast. He said these deposits could produce 140 million tonnes per year for 100 years, making India a net oil exporter. However, the deposits looked uneconomic, so he was not taken seriously.

Shale formations are best known for shale oil—a type of unconventional oil. Deposits of oil shale are found in many areas around the world and large areas of the United States, Russia, Argentina, Libya, Israel, and China are known to have shale oil and gas reserves. In the United States, it is claimed that the Green River formation is an underground oil shale formation that contains as much as 1.8 trillion barrels of shale oil. Despite the fact that not all of this can be extracted, it is claimed that shale oil reserve in United States will be more than three times the proven petroleum reserves of Saudi Arabia.

Oil shale generally contains enough oil that it will burn without any additional processing, and it is known as the rock that burns.

There are several different types of shale found throughout the world including oil shale and bituminous shale. These two types are important sources of various grades of unconventional oil. Shale oil was considered a more secure source of oil during the World War II but commercial development started only in 1960s. However, difficulty of extracting and producing oil from shale and environmental damage made it a less attractive resource compared to oil from conventional wells.

Most processes of shale oil production use significant amount of water and the chemicals used may harm humans and animals. The process is energy intensive and can require the burning of more fossil fuels in order to provide the necessary power supply.

1.13.1 Extraction of Shale Oil

Oil shale is underground rock formations that contain trapped petroleum. The petroleum trapped within the rocks is known as tight oil and is difficult to extract.

Oil shale can be mined and processed to generate oil similar to oil pumped from conventional oil wells. However, extracting oil from oil shale is more complex than conventional oil recovery and currently it is more expensive.

The important extraction processes of shale oil are as follows:

1. *Ex situ retorting*: Since the oil substances in oil shale are solid and cannot be pumped directly out of the ground, the following steps must be involved:
   (a) The oil shale must be mined and brought to ground surface.
   (b) The mined oil shale is then heated at a high temperature (a process called retorting). It involves heating kerogen in a process called pyrolysis. Pyrolysis is a form of heating without oxygen. At about 60°C–160°C, kerogen reaches its natural 'oil window', and at 120°C–225°C, kerogen reaches its natural 'gas window'.
   (c) The resultant liquid must then be separated and collected.
2. *In situ retorting*: An alternative method of extracting shale oil under experimental investigation is referred to as *in situ* retorting. During the *in situ* process, oil shale is not mined or crushed. Instead, the rock is heated to its oil window while it is still underground.
   It involves the following steps:
   (a) Heating the oil shale while it is still underground
   (b) Pumping the resulting liquid to the surface
   However, improvements in drilling technology, such as the emergence of directional drilling, has made extraction of oil from shale less cost prohibitive. Production companies use a variety of methods to extract oil from shale.
3. *Hydraulic fracturing (fracking)*: It involves injecting pressured water and chemicals into a well in order to break into underground reservoirs. Steam can be injected underground in order to heat up oils in the surrounding shale formation, which then seep into the well. Acids can also be injected in order to increase the permeability of rock surrounding the well.
4. *Volumetric heating*: In this process, the rock is heated directly with an electric current. The heating element is injected either directly in a horizontal well or into a fractured area of the rock, until the oil shale begins producing shale oil. The oil could then be pumped directly from underground.

5. *Combined technologies*: Some methods are designed for both *in situ* and *ex situ* extraction. The internal combustion process uses a combination of gas, steam, and spent shale produced by *ex situ* processing. These compounds are burned for pyrolysis. The hot gas is continually cycled through die oil shale, pyrolyzing the rock and releasing oil.

1.13.2 Classification of Oil Shales

They are often classified as follows:

1. *Depositional history*: A sedimentary rock depositional history is the history of the type of environment in which the rock developed. The depositional history of an oil shale includes the organisms and sediments that were deposited, as well as how those deposits interacted with pressure and heat. The Van Krevelen Diagram is a method of classifying oil shales based on their depositional history. The diagram divides oil shales according to where they were deposited:
   (a) *In lakes (lacustrine)*: Oil shales from lacustrine environments are formed mostly from algae living in freshwater, saltwater, or brackish water. Lamosite and torbanite are the types of oil shales associated with lacustrine environments. Lamosite deposits make up some of the largest oil shale formations in the world. Torbanite deposits are found mainly in Scotland, Australia, Canada, and South Africa.
   (b) *In the ocean (marine)*: Oil shales from marine environments are formed mostly from deposits of algae and plankton. Kukersite, tasmanite, and marinite are the types of marine shales. Kukersite is found in the Baltic Oil Shale Basin in Estonia and Russia. Tasmanite is named after the region in which it was discovered, the island of Tasmania, Australia. Marinite, the most abundant of all oil shales, is found in environments that once held wide, shallow seas. Although marinite is abundant, it is often a thin layer and not economically practical to extract. The largest marinite deposits in the world are in the United States, stretching from the states of Indiana and Ohio through Kentucky and Tennessee.
   (c) *On land (terrestrial)*: Oil shales from terrestrial environments are formed in shallow bogs and swamps with low amounts of oxygen. The deposits were mostly the waxy or corky sterns of hardy plants. Cannel shale, also called cannel coal or 'candle coal', is probably the most familiar type of terrestrial oil shale. Cannel coal was used primarily as fuel for streetlights and other illumination in the 19th century.
2. *By their mineral content*: Oil shales are classified into three main types based on their mineral content:
   (a) *Carbonate-rich shale*: The deposits have high amounts of carbonate minerals. Carbonate minerals are made of various forms of the carbonate ion (a unique compound of carbon and oxygen). Calcite, for instance, is a carbonate mineral common in carbonate-rich shales. Calcite is a primary component of many marine organisms. Calcite helps form the shells and hard exteriors of oysters, sea stars, and sand dollars. Plankton, red algae, and sponges are also important sources of calcite.
   (b) *Siliceous shale*: It is rich in the mineral silica or silicon dioxide. Siliceous shales are formed from organisms such as algae, sponges, and microorganisms called radiolarians. Algae have a cell wall made of silica, whereas sponges and radiolarians have skeletons or spicules made of silica. Siliceous oil shale is sometimes not as hard as carbonate-rich shale and can more easily be mined.

(c) *Cannel shale*: It has terrestrial origins and is often classified as coal. It is formed from the remains of resin, spores, and corky materials from woody plants. It can contain the minerals inertinite and vitrinite. Cannel shale is rich in hydrogen and burns easily.

1.13.3 Use of Shale Oil (Tight Oil)

Shale oil has been used for more than thousands of years by mankind for meeting their energy requirements for road construction, caulking ship, and pipes leakage, and developing burning arrows (Agni band) for use during battle and decorative mosaic. It is only after World War II that shale oil got the newest attention.

1. Shale oil was used for a variety of products including paraffin wax.
2. R&D efforts proved that it can be used immediately as a fuel or upgraded as a refinery feedstock specification by adding hydrogen and removing sulphur and nitrogen impurities similar to crude oil.
3. It is burned to generate electricity.
4. Shale oil is similar to petroleum and can be refined into many different substances including diesel fuel, gasoline, and liquid petroleum gas (LPG).
5. Companies can also refine shale oil to produce other commercial products such as ammonia and sulphur. The spent rock can be used in cement production.

1.13.4 Problems Associated with Shale Oil Production

1. *High processing costs*: The high costs of heating and drilling wells made commercial oil shale production unprofitable, especially when the cheaper crude oil is available.
2. *Environmental concerns*: Mining for oil shale can have damaging effects on the environment, such as the following:
   (a) When shale oil is combusted (heated), it releases carbon dioxide into the atmosphere. Carbon dioxide is a greenhouse gas.
   (b) Substances in the oil shale, such as sulphides, react with water to form toxic compounds that are harmful to the environment and to human beings. Sulphides can cause effects from eye irritation to suffocation.
   (c) Water containing toxic substances is unusable and expensive to decontaminate.
   (d) The ash by-product can pollute ground, air, and water sources.
   (e) Another environmental disadvantage is that extraction of shale oil requires enormous amounts of freshwater. Water is necessary for drilling, mining, refining, and generating power.
   (f) It causes land and underground water degradation.

1.14 ENERGY STORAGE

Rapid fluctuations in the demand of electrical energy vary with the seasons, the days of the week, and even the hours of each day. However, the energy supply industry must be capable of generation, transmission, and distribution to meet the peak demand capacity. This results in the under-loaded or lowers the base load operation of the obtained energy. If this energy is stored and used during on-peak operation, there will be definite reduction in the capital cost of the generating systems and cause full use of transmission facilities in transmitting the stored energy to the appropriate locations without running the risk of environmental pollutions.

Further, the exploitation of new unconventional energy resources have indicated that all these resources suffer from the drawback of fluctuations in the energy output owing to many varying reasons. This necessitates the use of some form of energy storage techniques and is thus an essential parameter in the new unconventional energy resource development programme practicable at large scale.

Several energy storage techniques are available to handle large scale energy. Pumped hydropower, compressed air, thermal oil, thermal steam, and lead acid batteries had their own importance up to 1985, which were superseded by advanced batteries and hydrogen storage up to 2000 AD. Later, super-conducting magnetic energy storage may be the promising scheme for the future. All these energy storage schemes are briefly outlined in following subsections.

1.14.1 Hydro Pump Storage

Energy available during off-peak period is used for pumping water from a lower to a higher elevation where it is stored. The energy may then be recovered during on-peak periods by allowing the water down through a water turbine coupled to an electric generator. Natural bodies of water, existing hydro plants' reservoirs, especially constructed surface reservoirs or underground cavern or any possible combination of all are used as storage reservoirs. Pumping and generation may be obtained simultaneously by a reversible pump turbine connected to a motor generator.

1.14.2 Compressed Air Storage

Compressed air can be stored in a constant volume and variable pressure reservoirs or in a hydrostatically compensated and constant pressure reservoir, which may be either in natural cavern or a man-made cavity. A modified combustion turbine is used for energy recovery. Compressors and turbines are uncoupled so that they can be operated at different times. During power generation, the function of the combustion turbine compressor is replaced by air from storage.

1.14.3 Thermal Storage

Energy may be stored as sensible heat in fluids or as latent heat in materials that undergo a phase change that may assist in maintaining the steam supply constant while the electrical output of the plant is varied.

1.14.4 Electrochemical Storage or Battery Storage

Except all hydrogen storage, (which is treated as chemical storage), it includes both the conventional battery storage and the hybrid system suitable for the storage of the chemical reactants external to the electrochemical converters. Such a storage system is simple, reliable, and fairly compact system.

1.14.5 Inertial Storage

Energy during the off-peak period is stored in the rotating mass of a flywheel, which can be recovered during the on-peak period. dc Motor and converter, variable frequency field machine and similar equipment are used in such system.

Complete flywheel energy storage system needs a careful study of social, fatal, and other problems during its failure. However, a single motor–generator may be coupled to a number of flywheels.

1.14.6 Hydrogen Storage

Hydrogen storage seems to be the only well-defined, achievable, energy storage system that is sufficiently developed and is likely to outclass the other types of energy storage in the near future. Hydrogen storage may incorporate the well-established technology of gas storage in high pressure tanks or certain kind of metal hybrid storage. Hydrogen thus obtained during off-peak periods and stored can be used during on-peak period either directly or as electricity through fuel cells, etc.

1.14.7 Superconducting Magnetic Energy Storage

In this system, electrical energy is stored in a magnetic field produced by a circulating current in the winding of a magnet. Although this system seems to be very promising during the next century, a thorough study of energy loss during storage period, size etc., are required to be economically acceptable.

1.15 CONCLUSIONS

There is no doubt that various aspects of energy conversion and storage, which have been discussed in the previous sections, have individual and collective importance, but it is essential that attention must be diverted towards the development of reliable and cheap energy conversion and storage schemes capable of being well-appreciated by common people.

It may be mentioned that energy from the sun, biogas, and agricultural waste and their conversion through thermo-mechanical, photovoltaic, biological processes, and storage of energy as hydrogen energy seem to have much potential. While the need for the conservation of energy is recognized, it is too early for India to apply brakes on growing use of energy in various spheres of life, as India is still in a developing stage. However, necessary step must be taken to avoid excessive energy use for non-essential utility.

SUMMARY

- It may be difficult to find out exactly how people's senses of values have changed through history. Energy consumption pattern can be approximated as exponentially rising till today because of growing population except during the period between World War I and II.
- Energy system model is characterized by production and sustenance, inputs, outputs, feedback, and dissipation.

- System acceptability index (δ) can be taken as the measure of gross efficiency of the system. The numerical value of δ is a measure of the system's acceptability. A value of $\delta = 1$ can be approached by properly conducting any energy management program to avoid dissipation in the system.
- Energy is a key measure of techno-socio-economic development of a nation.
- Energy consumption and energy scarcity are the main cause of social inequalities between peoples and disparities between developed, developing, and underdeveloped nations.
- Energy resource is defined as any material objects that can be quantified (available in huge quantity) and transferrable (easily converted to useful form).
- Secondary energy resources are usable forms of energy generated by means of suitable plants to convert the primary energy. They are electrical energy, steam power, hot water power, hydrogen energy, and so on.
- The electrical energy is expected to remain as dominant usable form of energy (secondary energy) during centuries as its generation, transmission, distribution, and utilization methods are well established.
- Oil shale is a fine-grained sedimentary rock that contains solid bituminous materials (called kerogen, which is an organic matter) that release petroleum-like liquids (shale oil or gas) when the rock is heated from which oil or gas can be extracted.
- Difficulty of extracting and producing oil from shale and environmental damage caused made it a less attractive resource compared to oil from conventional wells.
- It is essential that attention must be diverted towards the development of reliable and cheap energy conversion and storage schemes that are capable of being well-appreciated by common people.

REVIEW QUESTIONS

1. Define the term energy. Explain its significance in context of techno-socio-economic development.
2. Draw and explain energy system model of any energy activity. Further, explain its parameters and significance.
3. Discuss causes of energy scarcity. Further, mention factors to be considered for solving energy crunch problems.
4. Define system acceptability index and explain its importance.
5. What is an energy system? Explain in brief.
6. Define and explain the term 'energy resources'. Discuss different ways of their classifications. Mention at least two energy resources in each category.
7. What are the conventional and unconventional energy sources? Describe briefly.
8. List various non-conventional energy resources. Give their availability, relative merits and demerits, and their classification.
9. Discuss the main features of non-conventional energy resources.
10. Briefly explain economic criteria for comparing non-conventional energy resources.
11. What are the conventional and unconventional energy sources? Describe briefly.

12. What are primary and secondary energy sources?
13. What are the advantages and limitations of renewable energy sources?
14. Explain why direct energy conversion processes are becoming more important as compared to conventional generation.
15. What is the various primary energy resources utilized in direct-energy conversion?
16. What are the practical difficulties in exploiting non-conventional energy resources?
17. Discuss the future prospects of solar energy use.
18. Explain the basic principle of Ocean Thermal Energy Conversion (OTEC).
19. How does biomass conversion take place?
20. What are the limitations of solar cells?
21. What is the meaning of biomass? Further, discuss its multipurpose utilization
22. Justify the statement 'the future fuel of the world will be hydrogen obtained by electrolysis of water with the energy'.
23. Describe the principle of solar photovoltaic energy conversion.
24. What are the different sources of geothermal energy?
25. Discuss the availability of geothermal energy.
26. Explain the basic reasoning that most of the direct energy conversion devices can be classified into 'cells' and 'heat' systems.
27. Explain and distinguish between tidal, wave, and ocean thermal energy.
28. Explain the basic principle of MHD generator. Further, discuss the practical problems associated with MHD power generation.
29. Discuss the principle and working of MHD power plant.
30. Discuss the performance and limitations of various fuel cells available.
31. Explain the theory of momentum with respect to wind power.
32. Discuss the theory and working principle of ocean thermal energy conversion systems.
33. Discuss the principle and working of sea wave and tidal energy conversion system.
34. Discuss the principle of a solar collector. How collector coating can be used to improve the performance of collector?
35. Discuss the principle of MHD generation.
36. Explain the meaning of decentralized power generation.
37. Describe dispersed generation and mention its applications.
38. State and explain different methods of energy storage.

Chapter 2

Energy from the Sun

KEY CONCEPTS

- The sun–earth geometric relationship and characteristics of the sun
- Earth–sun angles and their relationships
- The sunrise, sunset and day length equations
- Solar energy reaching the earth's surface and problems associated with harnessing full solar energy
- Extraterrestrial irradiation
- Multi-Purpose utilization of solar energy
- Solar thermoelectric conversion and applications
- Solar thermal energy applications
- Solar thermal energy storage

The sun's energy is the primary source of energy for all the surface phenomena and life on earth. Combined with the materials of the earth (including the molecules held close by the earth's gravitational force called the atmosphere), this energy is utilized for the survival of immense diversity of life forms that are found on the earth. The sun is a powerful source of energy and provides the earth with as much energy every hour as it is collectively used in a year worldwide. Solar energy is derived from the sun's radiation. It is important to continuously harness and increase the use of solar energy (and other clean, renewable energies) as fossil fuels are depleting at a rapid rate. As the global demand for energy grows and conventional energy resources becoming costly to extract, people have started utilizing the energy obtained from the sun.

The sun, which is our singular source of renewable energy, being at the centre of the solar system emits energy as electromagnetic radiation at an extremely large and relatively constant rate, that is, 24/7, throughout the year. The emission rate of this energy is equivalent to the energy produced in a furnace at a temperature of about 6,000 K. If we could harvest the energy coming from just 10 hectares (25 acres) of the surface of the sun, then we would have enough energy to supply the current energy demand of the world.

2.1 SUN–EARTH GEOMETRIC RELATIONSHIP

The term *earth rotation* refers to the spinning of the earth on its axis. One rotation takes exactly 24 h and is called a mean solar day. If one look down at the earth's North Pole from space, he or she would notice that the direction of rotation is counterclockwise. The opposite is true if the earth is viewed from the South Pole.

The orbit of the earth around the sun is called earth revolution. This celestial motion takes 365.25 days to complete one cycle. Furthermore, the earth's orbit around the sun is not circular, but elliptical (as shown in Fig. 2.1). An elliptical orbit causes the earth's distance from the sun to vary annually; however, this phenomenon does not cause the seasons. This annual variation in the distance from the sun does influence the amount of solar radiation intercepted by the earth by approximately 6%. On January 3rd, the earth comes closest to the sun (147.5 million kilometres) each year(Perehelion). The earth is farthest from the sun on July 4th, each year (or aphelion). The average distance of the earth from the sun over a one-year period is 150 million kilometres.

From Figure 2.1, the following conclusions are derived:

1. The earth's orbit around the sun is elliptical with a mean centre to centre distance from the sun is approximately 9.3×10^6 miles (1.5×10^8 Km).
2. While the earth makes its daily rotation and yearly revolution, the sun also rotates on its axis approximately once every month.
3. The earth's axis of rotation (the polar axis) is always inclined at an angle of 23.5° from the ecliptic axis.
4. This distance from the sun to the earth varies ±1.7% over the average distance. This causes the solar energy reaching the earth to vary ±3% during a year. The energy is received at its peak on 1st January and the lowest on 1st July.
5. The sun is 109 times larger in diameter than the earth.
6. The sun appears to move across the sky in an arc from east to west, owing to the rotation of the earth around its north-south axis.
7. Viewing the sun from the average miles, it subtends an arc of 0.53° (32 min).

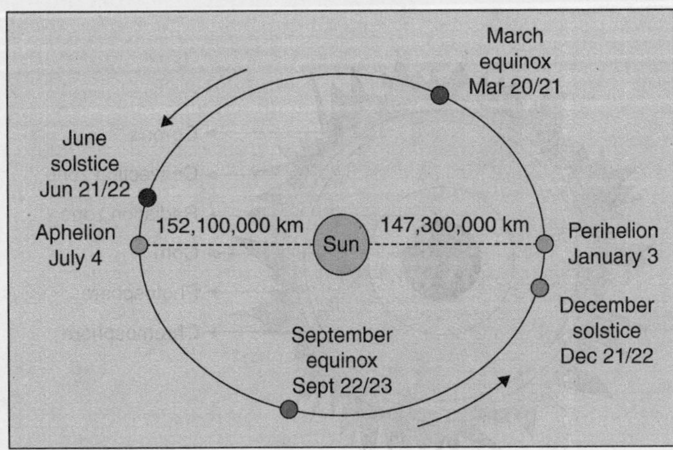

Source: http: //www.physicalgeography.net/fundamentals/6h.html

Figure 2.1 Sun–earth geometry

2.2 LAYER OF THE SUN

The sun can be divided into following six layers as shown in Figure 2.2:

1. Core
2. Radiative zone
3. Convection zone
4. Photosphere
5. Chromosphere
6. Corona

2.2.1 Core

The innermost layer of the sun is called the core. With a density of 160 g/cm^3, which is 10 times that of lead, the core might be expected to be solid. However, the core's temperature of 1,50,00,000°C keeps it in a gaseous state.

In the core, *fusion reactions* produce energy in the form of gamma rays and neutrinos. Gamma rays are photons with high energy and high frequency. The gamma rays are absorbed and re-emitted by many atoms on their journey from the envelope to the outside of the sun. When gamma rays leave atoms, their average energy is reduced. However, the first law of thermodynamics (which states that energy can neither be created nor be destroyed) plays an important role and the number of photons increases. Each high-energy gamma ray that leaves the solar envelope will eventually become one thousand low-energy photons.

The neutrinos are extremely nonreactive. Several experiments are being performed to measure the neutrino output from the sun. Chemicals containing elements with which neutrinos react are put in large pools in mines, and the neutrinos' passages through the pools can be measured by the rare changes they cause in the nuclei in the pools. For example, perchloroethane contains some isotopes of chlorine with 37 particles in the nucleus (17 protons and 20 neutrons).

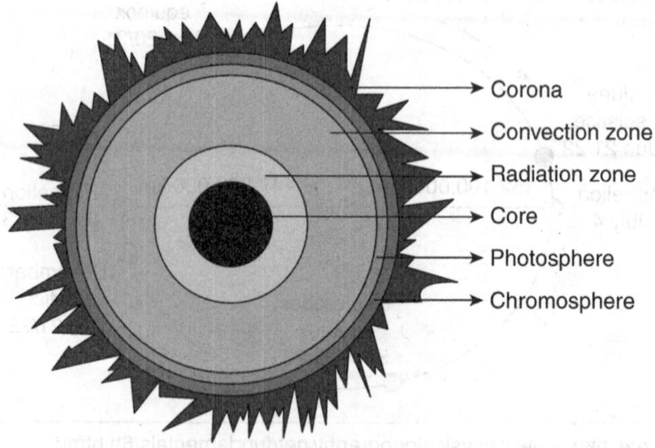

Figure 2.2 Interior of the sun

These Cl-37 molecules can take in neutrinos and become radioactive Ar-37 (18 protons and 19 neutrons). From the amount of argon present, the number of neutrinos can be calculated.

2.2.2 Solar Envelope

Outside of the core is the radiative envelope, which is surrounded by a convective envelope. The temperature is 4 million kelvin (7 million degrees F). The density of the solar envelope is much less than that of the core. The core contains 40% of the sun's mass in 10% of the volume, whereas the solar envelope has 60% of the mass in 90% of the volume. The solar envelope puts pressure on the core and maintains the core's temperature. The hotter a gas is, the more transparent it is.

The solar envelope is cooler and more opaque than the core. It becomes less efficient for energy to move by radiation, and as a result, heat energy starts to build up at the outside of the radioactive zone. The energy begins to move by convection in huge cells of circulating gas with several hundred kilometres in diameter. Convection cells nearer to the outside are smaller than the inner cells. The top of each cell is called a granule. These granules, when observed through a telescope, look like tiny specks of light. Variations in the velocity of particles in granules cause slight wavelength changes in the spectra emitted by the sun.

2.2.3 Photosphere

The photosphere is the zone from which the sunlight is both seen and emitted. The photosphere is a comparatively thin layer of low-pressure gasses surrounding the envelope. It is only a few hundred kilometres thick with a temperature of 6,000°C. The composition, temperature, and pressure of the photosphere are revealed by the spectrum of sunlight. When analysing the solar spectrum, William Ramsey discovered helium in 1896 and found that features of the gas did not belong to any gas known on earth. Hence, the newly discovered gas was named as helium in honour of Helios, the mythological Greek god of the sun.

2.2.4 Chromospheres

During an eclipse, a red circle can sometimes be seen outside the sun. This circle is called the chromospheres. Its red colouring is caused by the abundance of hydrogen. From the centre of the sun to the chromospheres, the temperature decreases proportionally as the distance from the core increases. The chromospheres' temperature, however, is 7,000 K, which is hotter than that of the photosphere. Temperatures continue to increase through the corona.

2.2.5 Corona

The outermost layer of the sun is called the corona or the crown. The corona is very thin and faint and is, therefore, very difficult to observe from the earth. Typically, we can observe the corona during a total solar eclipse or by using a coronagraph telescope, which simulates an eclipse by covering the bright solar disk. This outer layer is very dim—a million times dimmer than the photosphere and oddly enough, it is the hottest. At 10^6 K, it would seem that the heat would be unbearable for us, but remember in Physics, heat is a measure of molecular energy, that is, the movement of molecules within a space. Because the Corona extends several million kilometres into space, there is a lot of room for molecules to move. It is this movement that forms the source of the solar winds. The high temperature of the corona can force ions to move as fast as a million kilometres per hour.

Table 2.1 Characteristics of the Sun

Present Age	**4.5×10^9 years**	**Life Expectancy**	**10×10^9 years**
	Distance to Earth		
Mean	1.496×10^{11} m = 1.000 AU	variation	1.016735–0.98329 AU
Diameter (photosphere)	1.39×10^9 m	Angular diameter (from earth):	9.6×10^{-3} radians
Variation	±1.7%	Volume (photosphere):	1.41×10^{27} m^3
	Composition		
Hydrogen	73.46%	Helium	24.85%
Oxygen	0.77%	Carbon	0.29%
Iron	0.16%	Nitrogen, silicon, magnesium, sulphur, etc.	<0.1%
	Density		
Mean	14.1 kg/m^3	Centre	1,600 kg/m^3
	Solar Radiation		
Entire sun	3.83×10^{26} W	unit area of surface	6.33×10^7 W/m^2
At 1 AU (i.e., the solar constant)	1,367 W/m^2		
	Temperature		
Centre	1,50,00,000 K	Surface (photosphere)	6,050 K
Chromospheres	4,300–50,000 K	Corona	8,00,000–30,00,000 K
	Rotation		
Solar equator	26.8 days	30° latitude	28.3 days
60° latitude	30.8 days	75° latitude	31.8 days
Energy resource	$4\,H \rightarrow He + 2\,e^{-1} + 2\,\upsilon + \gamma$	Rate of mass loss:	4.1×10^9 kg/s

Source: Abridged from Eddy (1979); www.powerfromthesun.net/Book/chapter02/chapter02.html

2.3 EARTH-SUN ANGLES AND THEIR RELATIONSHIPS

In order to understand how to collect energy from the sun, one must first be able to predict the location of the sun relative to the collection device.

2.3.1 Hour Angle (ω)

The hour angle is the angular distance between the meridian of the observer and the meridian whose plane contains the sun.

To describe the earth's rotation about its polar axis, the concept of the hour angle (ω) is used. As shown in Figure 2.3, the hour angle is zero at solar noon (when the sun reaches its highest point in the sky). At this time, the sun is said to be 'due south' (or 'due north', in the Southern Hemisphere) since the meridian plane of the observer contains the sun. The hour angle increases by 15° every hour. An expression to calculate the hour angle from solar time is,

$$\omega = 15 \times (t_s - 12); \text{ (in degrees)} \tag{2.1}$$

Where, t_s is the solar time in hours.

Hour angle (ω) can be calculated simply as follows:

Since the earth makes one revolution on its axis in 24 h, then 15 minutes will be equal to 15/60 = 1/4 min

Therefore,

$$\omega = 1/4 \times t_m; \text{ (in degrees)} \tag{2.2}$$

Where, t_m is the time in minutes after local solar noon. ω will be +ve if solar time is after solar noon. However, ω will be –ve if solar time is before solar noon as shown in Figure 2.4.

Example 2.1

Calculate hour angle when it is 3 h after solar noon.

Solution Solar time = 12 + 3 = 15:00

Therefore, hour angle (w) = $15 \times (t_s - 12) = 15 \times (15 - 12) = 45°$

Example 2.2

Calculate hour angle when it is 2 h 20 min before solar noon.

Solution Solar time = $-1/4 \times (t_m) = -1/4 \times 140 = -35°$

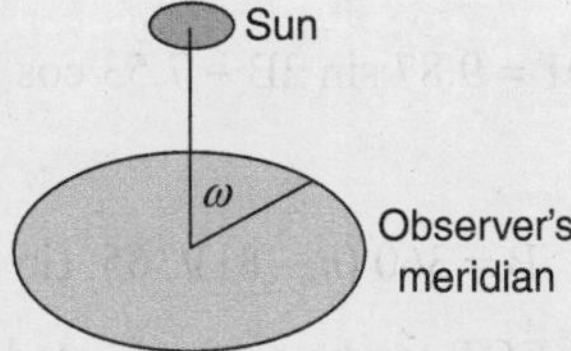

Source: http://www.brighton-webs.co.uk/energy/solar_earth_sun.aspx

Figure 2.3 Hour angle (ω)

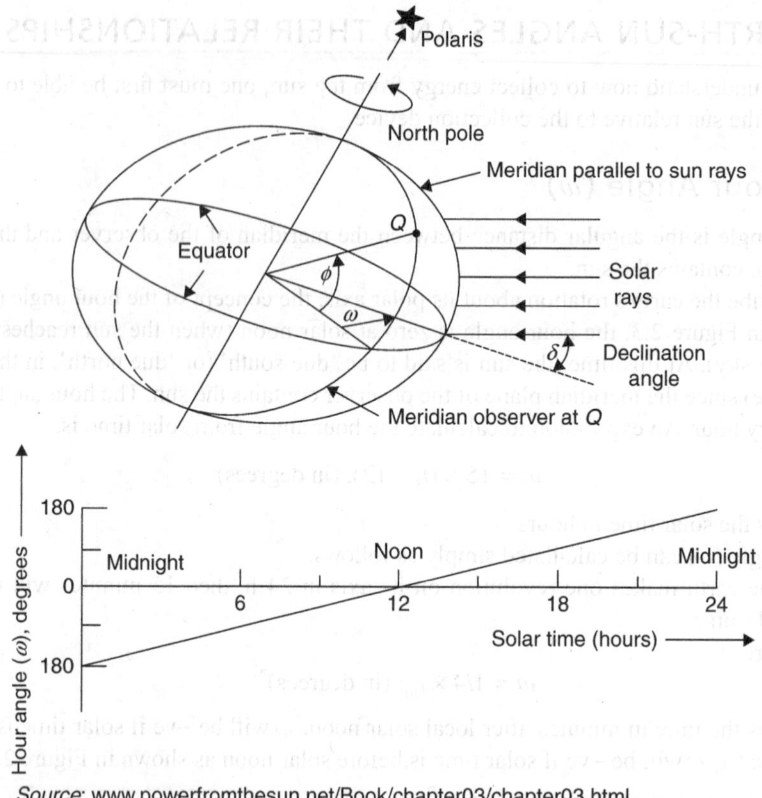

Source: www.powerfromthesun.net/Book/chapter03/chapter03.html

Figure 2.4 Variation of hour angle (ω) 24 h

2.3.2 Equation of Time

This is the difference between the local apparent solar time and the local mean solar time. The actual equation of time (EOT), which is mathematically defined as apparent solar time minus mean solar time, varies slightly from year to year due to variations in the earth's eccentricity and obliquity and in the time of the solstices and equinoxes. However, for a century, either side of the year 2000, it may be approximated (to an accuracy of better than 1%) by the formula:

$$\text{EOT} = 9.87 \times \sin(2\text{B}) - 7.67 \sin(\text{B} + 78.7°);\ \text{(in minutes)} \tag{2.3}$$

This can further be simplified as

$$\text{EOT} = 9.87 \sin 2\text{B} - 7.53 \cos \text{B} - 1.5 \sin \text{B} \tag{2.3a}$$

Where,

$$\text{B} = 360\,(n - 81)/365;\ \text{(in degrees)} \tag{2.4}$$

Another formula equation for EOT can be approximated as

$$\text{EOT} = 9.8 \times \sin(2\text{A}) + 7.6 \times \sin(\text{A} - 0.2) \tag{2.3b}$$

Where,

$$A = \text{K} \times (n + 10) + 0.033 \times \sin\,[\text{K}\,(n - 2)]$$

$$\text{K} = 2\pi/365$$

n is the total number of days of the year (e.g., $n = 1$ on Jan 1 and $n = 33$ on Feb 2)

Important to remember: During leap year, February month will have 29 days.

The level of accuracy required in determining the EOT will depend on whether the designer is doing system performance or developing tracking equations. An approximation for calculating the EOT in minutes is given by Woolf (1968) and is accurate to within about 30 s during daylight hours.

$$\text{EOT} = 0.258 \times \cos(\alpha_d) - 7.416 \times \sin(\alpha_d) - 3.648 \times \cos(2\,\alpha_d) - 9.228 \times \sin(2\,\alpha_d); \quad \text{(in minutes)}, \qquad (2.5)$$

Where the angle (α_d) is defined as a function of n.

$$\alpha_d = 360 \times (n - 1)/365.242; \text{ (in degrees)} \qquad (2.6)$$

n = the number of days counted from January 1

2.3.3 Declination Angle (δ)

The declination angle (δ) of the sun is the angle between the rays of the sun and the plane of the earth's equator. The earth's axial tilt (called the obliquity of the ecliptic by astronomers) is the angle between the earth's axis and a line perpendicular to the earth's orbit. The earth's axial tilt changes gradually over thousands of years, but its current value is about $\varepsilon = 23°26'$. Because this

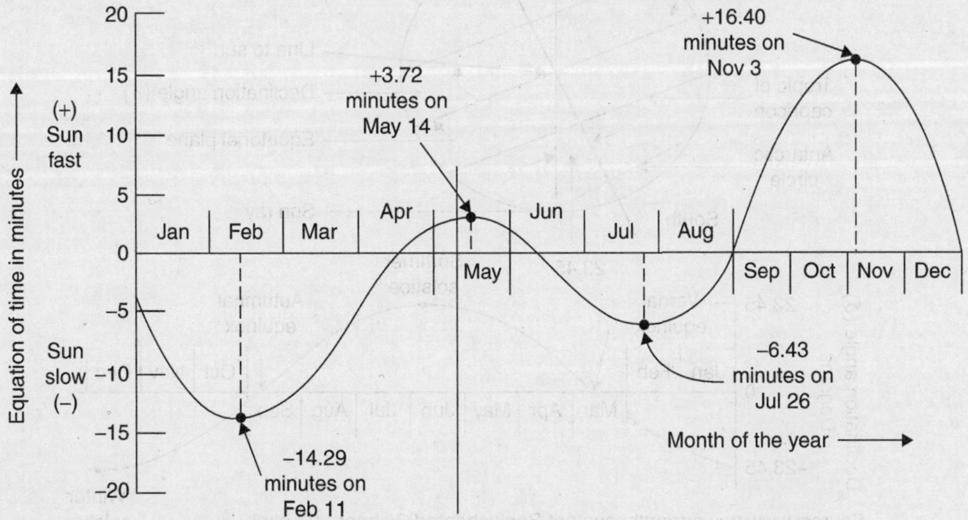

Source: www.powerfromthesun.net/Book/chapter03/chapter03.html

Figure 2.5 Equation of time

axial tilt is nearly constant, the solar declination angle (δ) varies with the seasons, and its period is one year. At the solstices, the angle between the rays of the sun and the plane of the earth's equator reaches its maximum value of 23°26'. Therefore, δ = +23°26' at the northern summer solstice and δ = −23°26' at the southern summer solstice.

At the moment of each equinox, the centre of the sun appears to pass through the celestial equator, and the declination angle (δ) is 0°.

The plane that includes the earth's equator is called equatorial plane. If a line is drawn between the centre of the earth and the sun, then the angle between this line and the earth's equatorial plane is called the declination angle (δ), as depicted in Figure 2.6.

At the time of year when the northern part of the earth's rotational axis is inclined towards the sun, the earth's equatorial plane is inclined 23.45° to the earth–sun line. At this time (about June 21), it is observed that the noon time sun is at its highest point in the sky and the declination angle (δ) = +23.45°. This condition is known as the summer solstice, and it marks the beginning of summer in the Northern Hemisphere.

As the earth continues its yearly orbit about the sun, a point is reached about 3 months later where a line from the earth to the sun lies on the equatorial plane. At this point, an observer on the equator would observe that the sun was directly overhead at noon time. This condition is called an equinox since anywhere on the earth, the time during which the sun is visible (daytime) is exactly 12 h and the time when it is not visible (night time) is 12 h. There are two such conditions during a year: the autumnal equinox on about September 23, which marks the start of the fall and the vernal equinox on about March 22, which marks the beginning of spring.

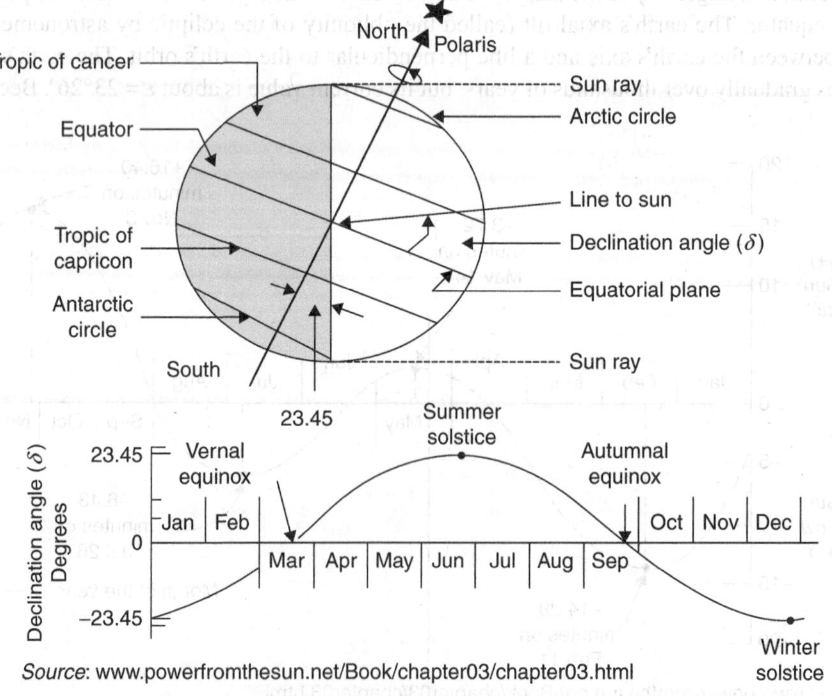

Source: www.powerfromthesun.net/Book/chapter03/chapter03.html

Figure 2.6 Declination angle (δ)

At the equinoxes, the declination angle (δ) is zero. The earth is shown in the summer solstice position when $\delta = +23.45°$.

Note the definition of the tropics as the intersection of the earth–sun line with the surface of the earth at the solstices and the definition of the Arctic and Antarctic circles by extreme parallel sun rays

The declination angle (δ) can be approximately obtained as,

$$\text{Sin}(\delta) \approx 0.39795 \times \cos[0.98563 \times (n - 173)] \quad (2.7)$$

Where, the argument of the cosine here is in degrees and n is the total number of days calculated from January 1. The annual variation of the declination angle is shown in Figure 2.6.

A formula that gives an approximation of declination in degrees based on the day number (e.g., 01 Jan = 1, 02 Jan = 2, etc.) is shown below.

$$\delta = 23.45 \times \sin[360 \times (284 + n)/365] \quad (2.8)$$

2.3.4 Latitude Angle (ϕ)

The latitude angle (ϕ) is the angle between a line drawn from a point on the earth's surface to the centre of the earth and the earth's equatorial plane. The intersection of the equatorial plane with the surface of the earth forms the equator and is designated as 0° latitude.

The earth's axis of rotation intersects the earth's surface at +90° S latitude (North Pole) and –90° latitude (South Pole). Any location on the surface of the earth can be then defined by the intersection of a longitude angle and a latitude angle.

Other latitude angles of interest are the Tropic of Cancer (+23.45° latitude) and the Tropic of Capricorn (–23.45° latitude). These represent the maximum tilts of the North and South Poles towards the sun. The other two latitudes of interest are the Arctic Circle (66.55° latitude) and Antarctic Circle (–66.5° latitude) representing the intersection of a perpendicular to the earth–sun line when the South and North Poles are at their maximum tilts towards the sun. The tropics represent the highest latitudes where the sun is directly overhead at solar noon, and the Arctic and Antarctic circles represent the lowest latitudes where there are 24 h of daylight or darkness. All of these events occur either at the summer or winter solstices.

2.3.5 Solar Altitude Angle (α)

It is defined as the angle between the central ray from the sun and a horizontal plane containing the observer, as shown in Figure 2.7. The earth's surface coordinates system for the observer at showing the Surface Azimuth angle (γ), the solar altitude angle (α), and the solar zenith angle (θ_Z) for a central sun ray along direction vector S. As an alternative, the sun's altitude may be described in terms of the solar zenith angle (θ_Z), which is simply the complement of the solar altitude angle (α) or

$$\theta_Z = 900 - \alpha; \text{ (in degrees)} \quad (2.9)$$

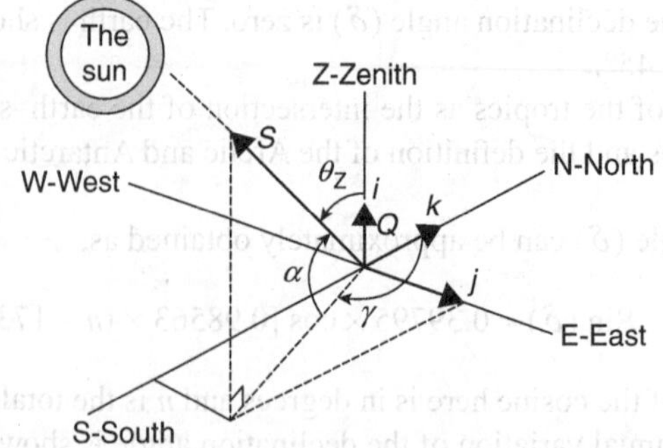

Source: www.powerfromthesun.net/Book/chapter03/chapter03.html

Figure 2.7 Solar altitude angle (α)

2.3.6 Solar Elevation Angle (α)

It is the elevation angle of the sun. That is, the angle between the direction of the geometric centre of the sun's apparent disk and the (idealized) horizon. It can be calculated, to a good approximation, using the following formula:

$$\sin \alpha = \cos \phi \cdot \cos \delta \cdot \cos \omega + \sin \phi \cdot \sin \delta \quad (2.10)$$

Where α = the solar elevation angle
ω = the hour angle in the local solar time
ϕ = the local latitude

2.3.7 Surface Azimuth Angle (γ)

$$\theta_z = 90° - \alpha \text{ (in degrees)}$$

The other angle defining the position of the sun is the surface azimuth angle (γ). It is the angle measured clockwise on the horizontal plane from the north-pointing coordinate axis to the projection of the sun's central ray.

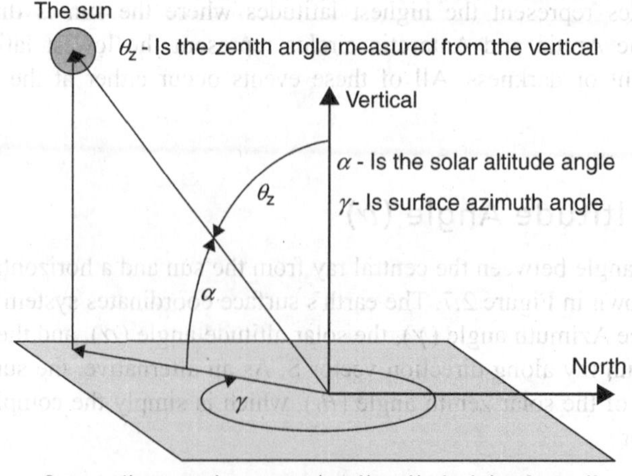

Source: //www.esrl.noaa.gov/gmd/grad/solcalc/azelzen.gif

Figure 2.8 Surface azimuth angle

The surface azimuth angle is the azimuth angle of the sun. It is most often defined as the angle from due north in a clockwise direction.

It can be calculated in various ways, and it has been explained in different ways during various periods. It can be calculated, to a good approximation, using the following formula; however, angles should be interpreted with care due to the inverse sign, i.e., $x = \sin^{-1}(y)$ has more than one solution, only one of which will be correct.

$$\sin(\gamma) = [-\sin(\omega) \times \cos(\delta) / \cos(\alpha)] \tag{2.11}$$

2.3.8 Relationship Between Different Sun–Earth Angles

Let us consider and define different earth–sun angles before establishing a relation between them as shown in Figure 2.9.

1. α = *Solar altitude angle*: It is defined as the angle between the central ray from the sun and a horizontal plane containing the observer.
2. β = *Slope angle*: It is defined as the angle between tilted and horizontal surfaces.
3. γ = *Surface azimuth angle*: It is the angle made in the horizontal plan between the line due south and the projection and of the normal to the surface on the horizontal plane.
4. γ_s = *Solar azimuth angle*: It is the angle made in horizontal plane between the line due south and the projection of line of site of the sun on the horizontal plane.

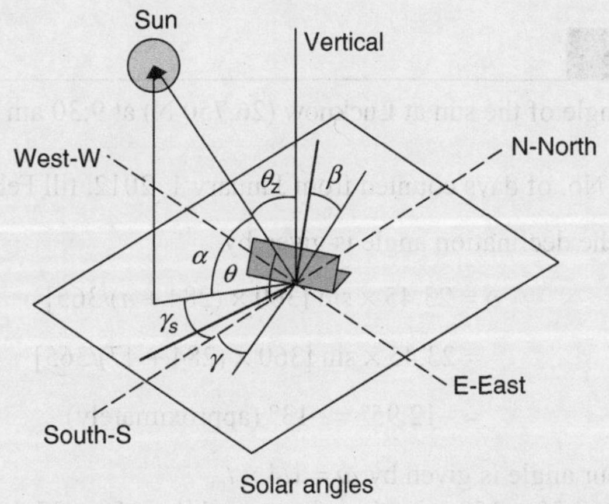

α - Is the solar altitude angle

β - Is the slope angle between the tilted and horizontal surface

γ - Is the surface azimuth angle

γ_s - Is the solar azimuth angle

δ - Is the declination angle

θ - Is the angle of incidence

θ_z - Is the zenith angle

Φ - Is the latitude angle

ω - Is the hour angle

Source: http://www.tboake.com/carbon-aia/strategies1a.html

Figure 2.9 Different solar angles

5. δ= *Declination angle*: It is the angle made by the line joining the centre of the sun and earth with its projection on the equatorial plane (Figs 2.6 and 2.7).
6. θ = *Angle of incidence*: The angle of incidence of a ray to a surface is measured as the difference in angle between the ray and the normal vector of the surface at the point of intersection.
7. θ_z = *Zenith angle*: It is simply the complement of the solar altitude angle (α).
8. ϕ= *Latitude angle*: The latitude angle (ϕ) is the angle between a line drawn from a point on the earth's surface to the centre of the earth and the earth's equatorial plane.
9. $\delta\omega$= *Hour angle*: The hour angle is the angular distance between the meridian of the observer and the meridian whose plane contains the sun.

It has been established that,

$$\cos\theta = \sin\phi \times (\sin\cdot\delta\cos\beta + \cos\delta\cdot\cos\gamma\cdot\cos\omega\cdot\sin\beta)$$
$$+ \cos\phi(\cos\delta\cdot\cos\omega\cdot\cos\beta - \sin\delta\cdot\cos\gamma\cdot\sin\beta)$$
$$+ \cos\delta\cdot\sin\gamma\cdot\sin\omega\cdot\sin\beta \tag{2.12}$$

The expressions for incidence angle (θ) can be further simplified as given below.

For horizontal surface, slope or tilt angle β= 0° and the angle of incidence θ becomes zenith angle θ_Z of the sun. Therefore,

$$\cos\theta_Z = \cos\phi\cdot\cos\delta\cdot\cos\omega + \sin\phi\cdot\sin\delta \tag{2.13}$$

Example 2.3

Calculate zenith angle of the sun at Lucknow (26.750 N) at 9:30 am on February 16, 2012.

Solution Total No. of days counted from January 1, 2012, till February 16, 2012, $n = 47$

From Eq. (2.8), the declination angle is given by

$$\delta = 23.45 \times \sin[360 \times (284 + n)/365]$$
$$= 23.45 \times \sin[360 \times (284 + 47)/365]$$
$$= -12.95° = -13° \text{ (approximately)}$$

From Eq. (2.2), hour angle is given by $\omega = 1/4 \times t_m$

Where t_m = 12:00 – 9:30 = 150 min; therefore, ω= 1/4 × 150 = 37.5 (since time is before solar noon, negative sign will be taken)

From Eq. (2.13), $\cos\theta_Z = \cos\phi\cdot\cos\delta\cdot\cos\omega + \sin\phi\cdot\sin\delta$

$$= \cos(26.75)\cdot\cos(-13)\cdot\cos(-37.5) + \sin(26.75)\cdot\sin(26.75)$$
$$= 0.589$$

Therefore, $\theta_Z = \cos^{-1}(0.589) = 53.914$

With vertical surface $\beta = 90°$, and then

$$\cos\theta = \sin\phi \cdot \cos\delta \cdot \cos\gamma \cdot \cos\omega - \cos\phi \cdot \sin\delta \cdot \cos\gamma \cdot + \cos\delta \cdot \sin\gamma \cdot \sin\omega \quad (2.14)$$

Horizontal surface $\beta = 0$,

$$\cos\theta = \sin\phi \cdot \sin\delta + \cos\phi \cdot \cos\delta \cdot \cos\omega \quad (2.14a)$$

The angle θ in this case is the zenith angle θ_Z. The complement of zenith angle is called the solar altitude angle.

For surface facing due south $\gamma = 0°$,

$$= \sin\delta \cdot \sin(\phi - \beta) + \cos\delta \cdot \cos\omega \cos(\phi - \beta) \quad (2.14b)$$

And for vertical surface facing due south $\beta = 90°$, $\gamma = 0°$

The solar azimuth angle γ_s is the angular displacement from south of the projection of the beam radiation on the horizontal plane.

Therefore, the solar azimuth γ_s can be written as

$$\sin\gamma_s = \cos\delta \cdot \sin\omega / \cos\alpha \quad (2.15)$$

Equation (2.15) can be solved for the sunset hour angle ω_{ss} when $\theta_Z = 90°$.

Therefore,

$$\cos\omega_{ss} = -\sin\phi \cdot \sin\delta / \cos\phi \cdot \cos\delta = -\tan\phi \times \tan\delta \quad (2.16)$$

2.3.9 Sunrise, Sunset, and Day Length Equations

The sunrise equation can be used to derive the time of sunrise and sunset for any solar declination and latitude in terms of local solar time (LST) when sunrise and sunset actually occur.

$$\cos\omega = -\tan\phi \times \tan\delta$$

Where

ω = the hour angle at either sunrise (when negative value is taken) or sunset (when positive value is taken)

ϕ = the latitude of the observer on the earth

δ = the sun declination angle

Since the hour angle at local solar noon is zero, with each 15° of longitude equivalent to 1 h, the sunrise and sunset from local solar noon is derived as follows:

$$T_H = 1/15 \times \omega_{ss} = 1/15 \cos^{-1}(-\tan\phi \times \tan\delta) \quad (2.17)$$

Therefore, daylight hour is given by $2T_H$.

Example 2.4

Find the solar altitude angle at 2 h after local solar noon on 1 June 2012 for a city, which is located at 26.75° N latitude. Moreover, find the sunrise and sunset hours and the day length.

Solution The declination on June 1 ($n = 153$) is

$$\delta = 23.45 \times \sin[360 \times (284 + 153)/365] = 22.17° \text{ approximately}$$

The hour angle at 2 h (120 min) after local solar noon is obtained by Eq. (2.2) as,

$$\omega = 1/4 \times (120) = 30°$$

The solar altitude angle is calculated as follows:

Since solar altitude angle $\theta_Z = 90° - \alpha$

$$\cos \theta_Z = \cos (90° - \alpha) = \sin \alpha$$

Therefore, $\sin \alpha = \cos \phi \cdot \cos \delta \cdot \cos \omega + \sin \phi \cdot \sin \delta$

$$= \cos 26.75° \cdot \cos 22.17° \cdot \cos 30° + \sin 26.75° \cdot \sin 22.17° = 0.953$$

$$\alpha = \sin^{-1} (0.953) = 72.364 = 72.4° \text{ approx.}$$

The daylight = sunset time + sunrise time (both measured from the solar noon)

$$= 2T_H = 2 \times 1/15 \cos^{-1} (-\tan \phi \times \tan \delta)$$

$$= 2/15 \cos^{-1} [-\tan (26.75°) \times \tan (22.17°)] = 10.43 \text{ h}$$

Therefore, the sun rises at 12:00 – 10.43/2 = 06:48 h = 06:48 am

Sunset time = 12:00 + 10.43/2 = 17:12 h

The hour angle corresponding to sunrise or sunset (ω_s) on a horizontal surface can be found from Eq. (2.12) if one substitutes the value of 90° for the zenith angle. We obtain

$$\cos \omega_s = -\tan \phi \tan \delta$$

$$\omega_s = \cos^{-1} (\tan \phi \tan \delta) \qquad (2.18)$$

Equation (2.18) yields positive and negative values for ω_s: the positive value corresponds to sunrise and the negative to sunset. Since 15° of the hour angle is equivalent to 1 h, the corresponding day length (in hours) is given as follows:

$$2T_H = 2/15 \cos^{-1} (\tan \phi \tan \delta) \qquad (2.19)$$

The hour angle at sunrise or sunset as seen by an observer on an inclined surface facing south ($\gamma = 0°$) is also given by Eq. (2.18); if the day under consideration lies between September 22 and March 21, in this period, the declination is negative, and the apparent plane of motion of the sun intersects the horizontal plane in an E–W line, which is an inclined plane. However, if the day under consideration lies between March 21 and September 22, the hour angle at sunrise or sunset (ω_{st}) would be smaller in magnitude than the value given by Eq. (2.18) and would be obtained by substituting in Eq. (2.14b). These yields

$$\omega_{st} = \cos^{-1} [-\tan (\phi - \beta) \tan \delta] \qquad (2.20)$$

Therefore, the magnitude of ω_{st} for an inclined surface facing south ($\gamma = 0°$) in the Northern Hemisphere is the smallest of the magnitudes of the values given by Eqs (2.18) and (2.20).

In general, for a plane surface not symmetrically oriented, hour angles at sunrise and sunset would be unequal in magnitude apart from having opposite's signs. The general procedure would be to calculate ω_{st} by substituting $\theta = 90°$ in Eq. (2.12) and by using Eq. (2.18), depending upon the day of the year and the orientation of the surface. Proper judgment would be needed in selecting the surface and in selecting the correct values from the solutions, thus obtained.

Example 2.5

Calculate the hour angle at sunrise and sunset on June 21 and December 21 for a surface inclined at an angle of 10° and facing due south ($\gamma = 0°$). The surface is located in Mumbai (19°07' N, 72°51' E).

Solution For December 21, using Eq. (2.20),

$$\omega_{st} = \cos^{-1}\,[-\tan\,(19.12° - 10°)\,\tan 23.45°] = 94.0°$$

For December 21, using Eq. (2.18),

$$\omega_{st} = \cos^{-1}\,[-\tan 19.12°\,\tan\,(-23.45°)] = 81.4°$$

2.3.10 Solar Time

It is the time based on the angular motion of the sun across the sky. With solar noon, the sun crosses the meridian of the observer. Solar time does not coincide with the local clock time. It is, therefore, necessary to convert standard time to solar time by applying the following correction. It is the time based on the 24-h clock, with 12:00 as the time that the sun is *exactly* due south. The concept of solar time is used in predicting the direction of sunrays relative to a point on the earth. Solar time is location (longitude) dependent and is generally different from local clock time, which is defined by politically defined time zones and other approximations. Solar time is used extensively to define the rotation of the earth relative to the sun. The time used for calculating the hour angle (ω) is the local apparent time. This LST can be obtained from the standard time (ST) by making two corrections as given below:

1. First correction arises because of the difference between the longitude of a location and the meridian on which the standard time is based.

   It has a correction of 4 min for every degree difference in longitude. The factor of 4 comes from the fact that earth rotates 1° for every 4 min.
2. The second correction is called the EOT.

It is due to the fact that the earth's orbit and rate of rotation are subject to small fluctuations.

Therefore, Time Correction Factor (T_C) in minutes can be obtained as

$$\text{LST} = \text{Standard Time (ST)} \pm T_C/60$$

Where T_C is in minutes

$$= \text{Standard time} \pm 4 \times (L_{STM} - \text{Longitude of location } (L_{LOL}) + \text{EOT} \qquad (2.21)$$

In India, standard time is based on 82.5°E. Twelve noon LST is defined as when the sun is the highest in sky.

L_{STM} = Local standard meridian time zone = 82.5°E (in India)

L_{LOL} = Longitude of location in degrees

EOT = EOT in minutes

Example 2.6

For a city located at 80.50 longitudes, calculate the solar time on March 15, 2011, at 10:30 am Indian Standard Time.

Solution The standard meridian for IST zone is 82.50 E.

Total Number of days on March 15 counted from January 1, 2011, $n = 74$
From Eq. (2.4), $B = 360\ (n - 81)/365$

$B = 360\ (74 - 81)/365 = -6.9$ (in degrees)

From Eq. (2.3a), EOT $= 9.87 \sin 2B - 7.53 \cos B - 1.5 \sin B$

$$= 9.87 \times \sin(-2 \times 6.9) - 7.53 \cos(-6.9) - 1.5 \sin(-6.9) = -9.65 \text{ min}$$

From Eq. (2.8), LST

$$= \text{Standard time} \pm 4 \times [L_{STM} - \text{Longitude of location } (L_{LOL})] + \text{EOT}$$

Given that, $L_{STM} = 82.5°$E; $L_{LOL} = 80.5°$; standard time = 10:30 am
Therefore, LST = 10:30 − 4 × (82.5 − 80.5) + (−9.65)
= 10:30 − 17.65
= 10:12.35 am

2.4 SOLAR ENERGY REACHING THE EARTH'S SURFACE

Solar radiation is electromagnetic radiation emitted by the sun. The sun is converting its mass into light particles called photons. The solar radiation that reaches on different locations of earth depends on several factors such as geographic location, time, season, local landscape, local weather, etc. The earth rotates around the sun in an elliptical orbit and is closer to the sun during a part of the year. When the sun is nearer to the earth, the earth's surface receives a little more solar energy. The rotation of the earth is responsible for hourly variations in sunlight.

When sunlight passes through the atmosphere, it is subjected to absorption, scattering, and reflection by air molecules, water vapour, clouds, dust, pollutants, forest fires, etc. When a photon is absorbed, its energy is changed into either electrical energy or heat energy. Scattering occurs when gas molecules and small particles diffuse from the incoming solar radiation in different directions without any alteration to the wavelength of electromagnetic energy. Reflection of solar radiation is a process where sunlight is redirected by 180° after it strikes an atmospheric particle, and mainly reflections are caused by clouds.

The radiation intensity on the surface of the sun is approximately 6.33×10^7 W/m^2. The solar energy reaching the periphery of the earth's atmosphere is considered to be constant for all practical purposes and is known as the *solar constant*. Because of the difficulty in achieving accurate measurements, the exact value of the solar constant is not known with certainty, but it is believed to be between 1,353 and 1,395 W/m^2. The solar constant value is estimated on the basis of the solar radiation received on a unit area exposed perpendicularly to the rays of the sun at an average distance between the sun and the earth.

Since radiation spreads out as the distance squared, by the time it travels to at a distance of one astronomical unit (AU) (roughly the mean distance from the sun to the earth), the radiant energy falling on the surface area is reduced to 1,367 W/m^2. There is a variation of solar intensity of about 1%, but this is a slow cycle. It is so small that it is negligible for the purpose of solar power.

It has already been established that huge amount of energy is stored on the earth each year by the radiation of hot solar rays at the average rate of 1.35 kW/m^2, which may be five thousand five hundred times more than the world energy requirement during the next century and is fifty thousand times more than the present-day energy availability from all the sources.

In India, the rate of solar radiation has been found to be 6–8 kWh/m^2, which may be considered as one of its important energy resource in the near future, especially for the rural areas.

2.4.1 Problems Associated with Harnessing Full Solar Energy

There are three important reasons why this cannot be done are as follows:

1. The earth is displaced from the sun, and since the sun's energy spreads out like light from a candle, only a small fraction of the energy leaving an area of the sun reaches an equal area on the earth.
2. The earth rotates about its polar axis, so that any collection device located on the earth's surface can receive the sun's radiant energy for only about one-half of each day.
3. The least predictable factor is the condition of the thin shell of atmosphere that surrounds the earth's surface.

At the best, the earth's atmosphere accounts for another 30% reduction in the sun's energy. As is widely known, however, the weather conditions can stop all but a minimal amount of solar radiation from reaching the earth's surface for many days in a row.

2.4.2 Extraterrestrial Irradiation

2.4.2.1 Solar Constant

It is a measure of flux density and is the amount of incoming solar electromagnetic radiation per unit area that would be incident on a plane perpendicular to the rays at a distance of one AU. It includes all types of solar radiation, not just the visible light.

The solar energy reaching the periphery of the earth's atmosphere is considered to be constant for all practical purposes and is known as the *solar constant*. Because of the difficulty in achieving accurate measurements, the exact value of the solar constant is not known with certainty but is believed to be between 1,353 and 1,395 W/m^2. The solar constant value is estimated on the basis of the solar radiation received on a unit area exposed perpendicularly to the rays of the sun at an average distance between the sun and the earth.

Its average value was thought to be approx. 1,366 W/m^2, varying slightly with solar activity, but recent recalibrations of the relevant satellite observations indicate a value closer to 1,361 W/m^2 is more realistic.

2.4.2.2 Solar Radiation Spectrum

Variations of solar irradiance with wavelength of solar radiation is called solar radiation spectrum as given in Figure 2.10. This spectrum of electromagnetic radiation striking the earth's

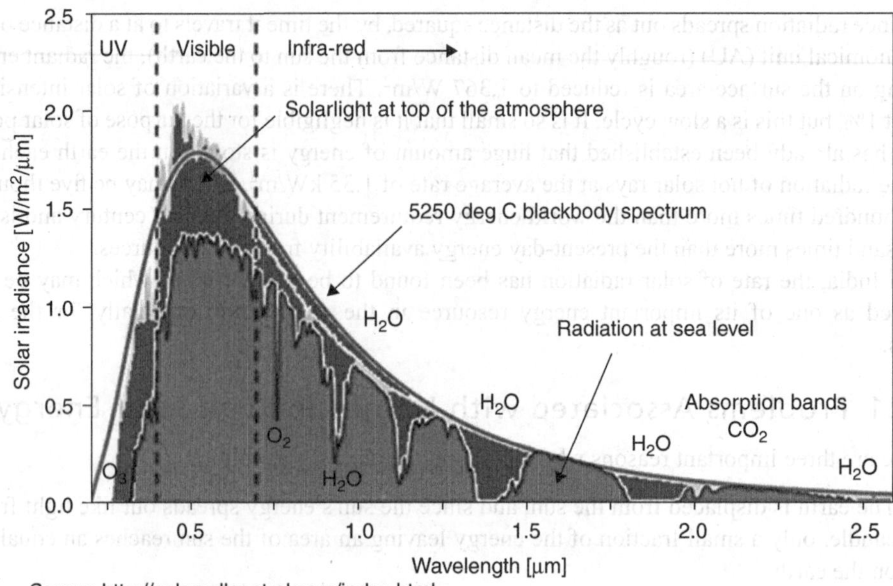

Source: http://solarcellcentral.com/index.html

Figure 2.10 Solar radiation spectrum

atmosphere spans a range of 0.1 μm to about 3 μm. This can be divided into five regions in increasing order of wavelengths as given below.

1. *Ultraviolet C or (UVC) range*: It spans a range of 0.1 μm to 0.28 μm. The term *ultraviolet* refers to the fact that the radiation is at higher frequency than violet light (and hence also invisible to the human eye). Owing to absorption by the atmosphere, only mere amount reaches the earth's surface (lithosphere). This spectrum of radiation has germicidal properties and is used in germicidal lamps.
2. *Ultraviolet B or (UVB) range*: It spans 0.28 μm to 0.315 μm. It is also greatly absorbed by the atmosphere, and along with UVC, it is responsible for the photochemical reaction leading to the production of the ozone layers.
3. *Ultraviolet A or (UVA) range*: It spans 0.315 μm to 0.4 μm. It has been traditionally held as less damaging to the DNA and hence used in tanning and PUVA (Photo-chemoUVA) therapy for psoriasis.
4. *Visible range or light*: It spans 0.38 μm to 0.78 μm. As the name suggests, it is this range that is visible to the naked eye.
5. *Infrared range*: It spans 0.7 μm to 1,000 μm. It is responsible for an important part of the electromagnetic radiation that reaches the earth. It is also divided into three types on the basis of wavelength:
   (a) Infrared-A: 0.7 μm to 1.4 μm
   (b) Infrared-B: 1.4 μm to 3.0 μm
   (c) Infrared-C: 3.0 μm to 100 μm

In practical terms, the biologically important output from the sun reaching the earth's surface can be divided into four wavelength regions such as ultraviolet B (UVB, 0.28 μm to

0.315 μm), ultraviolet A (UVA, 0.315 μm to 0.4 μm), visible light (0.38 μm to 0.78 μm), and infrared (0.7 μm to 1,000 μm).

Ultraviolet wavelengths shorter than 0.28 μm are heavily absorbed by molecular oxygen, ozone, and water vapour in the upper atmosphere and do not reach the surface of the earth in measurable amounts. All of the biologically important effects of sunlight are necessarily due to wavelengths in the range 0.28–100 μm.

Except in special circumstances, such as photo drug reactions and certain disease states, visible light does not appear to be harmful to normal individuals. Infrared is essentially heat, and although non-solar sources can cause skin tumours and cataracts, it is uncertain at present if the infrared in sunlight contributes significantly to the problem of skin cancer.

The major source of damaging effect of sunlight comes primarily from the ultraviolet portion of the spectrum between 0.29 μm and 0.4 μm (UVB and UVA). The different ultraviolet wavelengths penetrate the skin to different depths and have different biological consequences.

2.4.2.3 Solar Radiation Outside the Earth's Atmospheres

It has already been established by measurements that energy radiation received from the sun outside the earth's atmospheres is essentially constant. As already defined, the solar constant (I_{SC}) is the rate at which energy is received from the sun on a unit area perpendicular to the rays of the sun at a mean distance of the earth from the sun. The value of solar constant has been the subject of many investigations. Its standard value is taken from 1,353 to 1,367 W/m^2.

Because the earth's orbit is slightly elliptical, the intensity of solar radiation received outside the earth's atmosphere varies as the square of the earth–sun distance. Solar irradiance varies by ±3.4% with the maximum irradiance occurring at the perihelion, i.e., earth closest to the sun (January 3–5) and the minimum at the aphelion (i.e., July 5). This variation may be approximated by

$$I_0 = I_{SC}\,[1 + 0.034 \times \cos(360 \times n/365.25)]\ \text{W/m}^2 \tag{2.22}$$

Where I_0 is the extraterrestrial solar irradiance outside the earth's atmosphere and n is the total number of days counted from January 1.

An instructional concept, which is often used in solar irradiance models, is the extraterrestrial solar irradiance falling on a horizontal surface. Consider a flat surface just outside the earth's atmosphere and parallel to the earth's surface below. When this surface faces the sun (normal to a central ray), the solar irradiance falling on it will be the maximum possible solar irradiance (I_{0H}). If the surface is not normal to the sun, then the solar irradiance falling on it will be reduced by the cosine of the angle between the surface normal and the central ray from the sun. This concept is described pictorially in Figure 2.11. It can be seen that the rate of solar energy falling on both surfaces is the same. However, the surface area A is greater than its hypothetical projection area B, thus making the rate of solar energy per unit area (i.e. the solar irradiance), falling on surface A is less than on surface B.

The cosine effect relates to the concept of extraterrestrial horizontal irradiance. The extraterrestrial solar irradiance falling on a surface parallel to the ground is

$$I_{0H} = I_0 \times \cos\,\theta_z \tag{2.23}$$

Where I_0 is the extraterrestrial solar irradiance and θ_z is the angle between the two surfaces, which is the solar zenith angle.

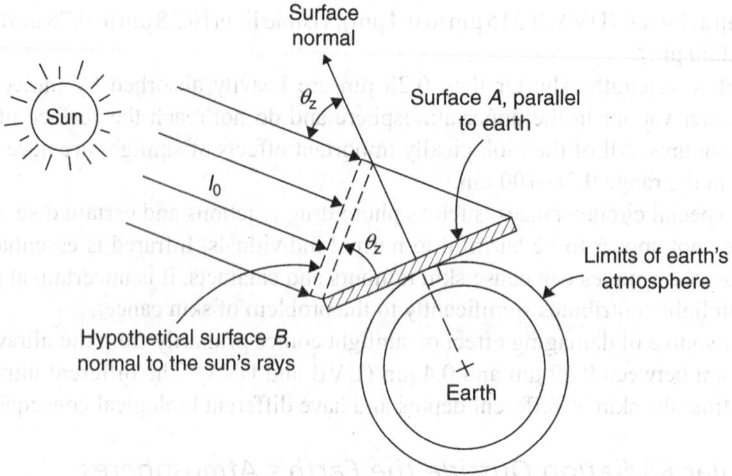

Figure 2.11 Cosine effect

2.4.2.3.1 Cosine Effect Reduction of radiation by the cosine of the angle between the solar radiation and a surface normal is called the cosine effect.

2.4.2.4 Solar Radiation on the Earth's Surface (Solar Insolation)

The rate at which solar energy reaches a unit area at the earth is called the solar irradiance or insolation. In other words, solar insolation means the sun's energy received over a horizontal surface. The units of measure for irradiance are watts per square metre (W/m^2). Solar irradiance is an instantaneous measure of rate and can vary over time. The maximum solar irradiance value is used in system design to determine the peak rate of energy input into the system. If storage is included in a system design, the designer also needs to know the variation of solar irradiance over time in order to optimize the system design.

It may be noted that total insolation on a horizontal surface on a clear day even for two different locations are not same. They show that a normal (perpendicular) surface to the sun's rays does increase energy received over a horizontal surface. Since locations are different and therefore not directly comparable but may be indicative of principles. For example, in the United States, a peak insolation power at noon of 1 KW/m^2 and 1.75 KW/m^2 in India can be assumed as standard. Even though it may not be reached in all areas at all times, it makes a nice design reference.

Since solar radiation before reaching the earth's surface is subjected to the mechanism of absorption and scattering while passing through several gases (e.g., water vapour, ozone, carbon dioxide, oxygen, etc.), maximum radiation reaches the earth's surface during clear sky (no clouds). In view of the absorption and scattering, solar radiation reaching the earth's surface is defined by the following terms:

1. *Beam radiation (direct solar radiation)*: The solar radiation received on the earth's surface without change of direction (without any attenuation) in line with sun.
2. *Diffuse radiation*: When solar radiation is subjected to attenuation and reaches the earth's surface from all parts of the sky hemisphere.

3. *Global radiation*: The sum of beam radiation and diffuse radiation is known as global radiation.

Air mass is a term normally used as a measure of the distance travelled by beam radiation through the earth's atmosphere before it reaches a location at the earth's surface. It is defined as the ratio of the mass of atmosphere through which the beam radiation passes to the mass of the atmosphere through which it will pass if the sun is directly overhead (i.e., at its Zenith). It has been proved that for location at sea level and zenith angles between 0° and 70°, air mass is obtained as,

$$\text{Air mass} = \sec \theta_z \tag{2.24}$$

The power incident on a tilted surface (Fig. 2.12) depends not only on the power contained in the sunlight but also on the angle between the tilted surface and the sun. When the absorbing surface and the sunlight are perpendicular to each other, the power density on the surface is equal to that of the sunlight. In other words, the power density will always be at its maximum when the solar-absorbing surface is perpendicular to the sun. However, as the angle between the sun and a fixed surface is continually changing, the power density on a fixed surface is less than that of the incident sunlight.

The amount of solar radiation incident on a tilted module surface is the component of the incident solar radiation that is perpendicular to the module surface. Figure 2.12 shows how to calculate the radiation incident on a titled surface (I_{TS}) if either the solar radiation measured on horizontal surface (I_{oh}) or the solar radiation measured perpendicular to the sun (I_O) is known.

The equations relating I_{TS}, I_{oh}, and I_O is

$$I_{oh} = I_O \sin \alpha \text{ and } I_{TS} = I_O \sin (\alpha + \beta) \tag{2.25}$$

where α is the elevation angle and β is the tilt angle of the surface measured from the horizontal.

$$\text{The elevation angle } \alpha = 90 - \phi + \delta \tag{2.26}$$

where ϕ is the latitude angle and δ is the declination angle given previously as:

$$\delta = 23.450 \times \sin [(360/365) \times (284 + n)] \tag{2.27}$$

where n is the number of days counted from January 1.

From the Eq. (2.22), (2.26) and (2.27)

$$I_{TS} = I_{oh} \times \text{Sin} (\alpha + \beta)/\text{Sin } \alpha \tag{2.28}$$

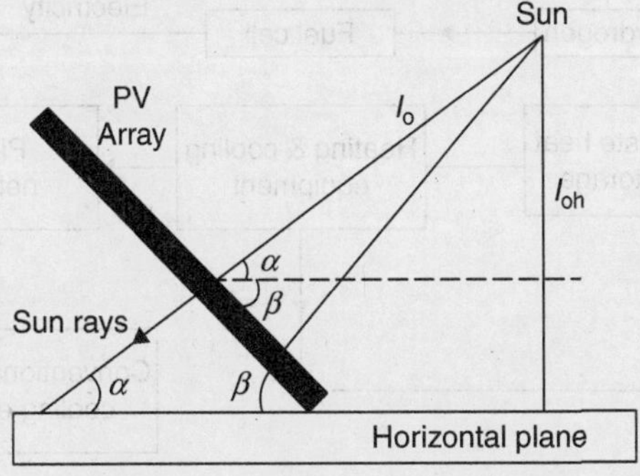

Figure 2.12 Tilted surface

2.5 SOLAR THERMAL ENERGY APPLICATIONS

Energy from the sun can be converted into usable form of energy for multi-purpose utilization as given in Figure 2.13 for the applications based on the controlled technology.

These technologies include passive and active systems.

2.5.1 Passive Systems

This system collects energy, without the need for pumps or motors, generally through the orientation, materials, and construction of a collector. These properties allow the collector to absorb, store, and use solar radiation. Passive systems are particularly suited to the design of buildings (where the building itself acts as the collector) and thermo siphoning solar hot water systems.

For new buildings, passive systems generally entail very low or no additional cost because they simply take advantage of the orientation and design of a building to capture and use solar radiation. In colder climates, a passive solar system can reduce heating costs by up to 40%, whereas in hotter climates, it can reduce the absorption of solar radiation and thus reduce cooling costs.

A passive solar system relies on natural sources to transfer heated water for domestic use, which is more prevalent in warmer climates with minor chance of freezing periods.

2.5.2 Active System

The most common active systems use pumps to circulate water or another heat absorbing fluid through solar collectors. These collectors are most commonly made of copper tubes bonded to

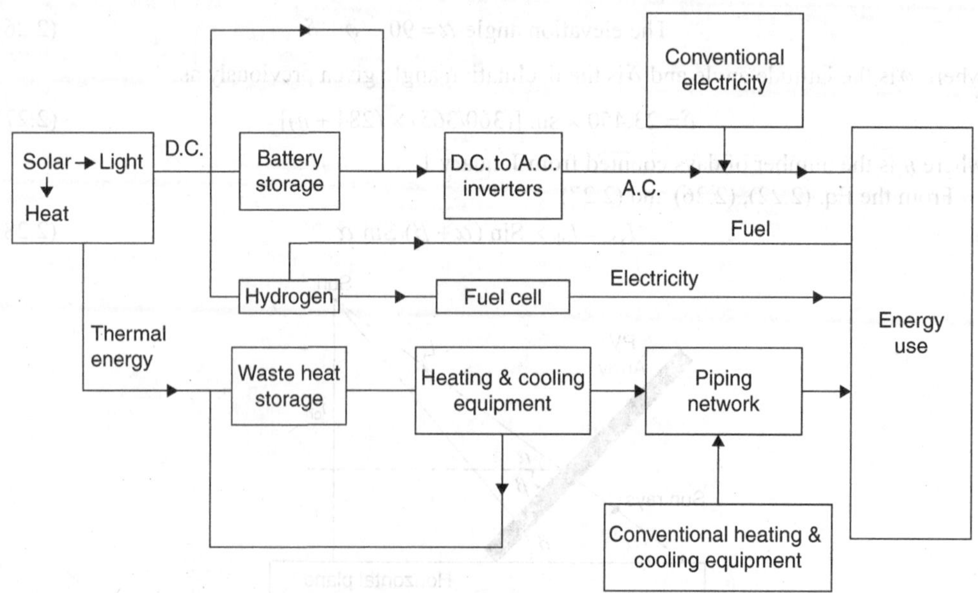

Figure 2.13 Multi-purpose utilization of solar energy

a metal plate, painted black, and encapsulated within an insulated box covered by a glass panel or 'glazing'. For pool heating and other applications where the desired temperature is less than 40°C, unglazed synthetic rubber materials are most commonly used.

An active pumped system can be either an open loop where the water is directly heated by the solar collector or closed loop where antifreeze or glycol mixture is heated before transferring its heat to the water by a heat exchanger. A popular design of the closed loop system is known as a drain back system. This freeze-proof design drains water back into a small holding tank when freezing temperatures occur.

2.5.3 Direct Thermal Applications

The sun's energy can be collected directly to create both high-temperature steam (greater than 100°C) and low-temperature heat (less than 100°C) for use in a variety of heat and power applications. Solar thermal collectors are, therefore, also classified as low-, medium-, and high-temperature collectors.

1. Low-temperature collectors are flat plates generally used to heat swimming pools.
2. Medium-temperature collectors are also usually flat plates but are used for heating water or air for residential and commercial use.
3. High-temperature collectors concentrate sunlight using mirrors or lenses and are generally used for electric power production.

These systems use mirrors and other reflective surfaces to concentrate solar radiation. Parabolic dish systems concentrate solar radiation to a single point to produce temperatures in excess of 1,000°C. Line-focus parabolic concentrators focus solar radiation along a single axis to generate temperatures of about 350°C. Central receiver systems use mirrors to focus solar radiation on a central boiler. The resulting high temperatures can be used to create steam either to drive electric turbine generators or to power chemical processes such as the production of hydrogen.

2.5.3.1 Low-temperature Solar Thermal Systems

It collects solar radiation to heat air and water for industrial applications including:

1. Space heating for homes, offices, and greenhouses
2. Domestic and industrial hot water
3. Pool heating
4. Desalination
5. Solar cooking
6. Crop drying

2.5.3.2 Domestic Water Heating

A solar domestic hot water system uses the sun's energy collected by a flat-plate solar collector and transfers the heat to water or another liquid flowing through the tubes. The system then draws upon this reservoir when hot water is needed. This system usually complements an existing electric or gas hot water system to reduce your utility bill and provide approximately 40%–70% of household's annual hot water needs. For domestic applications, the solar hot water system is a mature technology that can provide hot water to meet a significant (in some cases all) of the hot water needs in a domestic building.

In Europe, a solar heating system can generally meet between 50%–65% of domestic hot water requirements, while in subtropical climates, such as Asia and northern Australia, the percentage can be 80%–100% of needs.

2.5.3.3 Domestic Space Heating

A solar space heater collects the sun's energy by a solar collector and directs the energy into a thermal mass for storage later when the space is the coldest. A thermal mass can be a masonry wall, floor, or any storage drum used specifically to absorb and store the energy.

Many systems involve a distribution system and control devices to circulate the heat throughout the space and to prevent loss from the collector area. These systems may be combined with a solar hot water system and sized to accommodate both the uses, and also solar space heaters are more economical when it replaces electrical heating systems.

The same types of solar collectors used in a domestic solar heating system can also be used for space heating applications. In some countries, such as Sweden, large district heating systems have been built that heat large volumes of water during summer months for use in the winter heating season.

2.5.3.4 Solar Cooking

In developing countries, a solar cooker can provide basic cooking energy in areas of high solar radiation. In these areas, simple solar stills can also be used to purify water. These devices can be very simple and made using local materials and labour.

Solar cooking is a technology which has been given a lot of attention in recent years in developing countries. The basic design is that of a box with a glass cover. The box is lined with insulation and a reflective surface is applied to concentrate the heat onto the pots. The pots can be painted black to help with heat absorption. The solar radiation raises the temperature sufficiently to boil the contents in the pots. Cooking time is often a lot slower than conventional cooking stoves but there is no fuel cost.

Many variations have been developed on this theme but the main restriction has been of reducing the costs sufficiently to permit widespread dissemination. The cooker also has limitations in terms of only being effective during hours of strong sunlight. Another cooking stove is usually required for the periods when there is cloud or during the morning and evening hours. There have been large, subsidized solar cooking stove dissemination programs in India, Pakistan, and China.

2.5.3.5 Crop Drying

Controlled drying is required for various crops and products such as grain, coffee, tobacco, fruits, vegetables, and fish. Their quality can be enhanced if the drying is properly carried out. Solar thermal technology can be used to assist with the drying of such products. The main principle of operation is to raise the heat of the product, which is usually held within a compartment or box, while at the same time passing air through the compartment to remove moisture. The flow of air is often promoted using the 'stack' effect, which takes advantage of the fact that hot air rises and can, therefore, be drawn upwards through a chimney, while drawing in cooler air from below. Alternatively a fan can be used. The size and shape of the compartment varies depending on the product and the scale of the drying system. Large systems can use large barns, whereas smaller systems may have a few trays in a small wooden housing.

Solar crop drying technologies can help reduce environmental degradation caused by the use of fuel wood or fossil fuels for crop drying and can also help to reduce the costs associated with these fuels and hence the cost of the product. Helping to improve and protect crops also has beneficial effects on health and nutrition. Special features of solar drying systems are as follows:

1. Uses solar energy to heat the air. The hot air can be circulated for drying.
2. As a fuel source, it can be integrated with the existing systems as retrofit or merged with new construction.
3. It can be installed vertically or at any slope.
4. No need of glazing or insulation.
5. As a building material, it can form the roof or walls of the drier chamber.
6. One can walk on its surface, which is resistant to all types of weathering.
7. Increase in temperature, over ambient by 40°C to 60°C depending on insulation.
8. Improve quality of dried product due to uniform drying.

2.5.3.6 Space Cooling

The majority of the developing countries lies within the tropics and has little need of space heating. There is a demand for space cooling. The majority of the world warm climate cultures have again developed traditional, simple, elegant techniques for cooling their dwellings, often using effects promoted by passive solar phenomenon.

There are many methods for minimizing heat gain. These include constructing a building in shade or near water, using vegetation or landscaping to direct the wind into the building and good town planning to optimize the prevailing wind and available shade. Buildings can be designed for a given climate, i.e., domed roofs and thermally massive structures in hot arid climates, shuttered and shaded windows to prevent heat gain, and open structure bamboo housing in warm, humid areas. In some countries, dwellings are constructed underground and take advantage of the relatively low and stable temperature of the surrounding ground.

2.5.3.7 Daylighting

A simple and obvious use for solar energy is to provide light for use in buildings. Many modern buildings, office blocks, and commercial premises, for example, are designed in such a way that electric light has to be provided during the daytime to provide sufficient light for the activities taking place within. An obvious improvement would be to design buildings in such a way that that the light of the sun can be used for this purpose. The energy savings are significant and natural lighting is often preferred to artificial electric lighting.

2.5.3.8 Heating and Cooling System Design Considerations

For designing a heating or cooling system, the following points need special consideration:

1. Solar and weather conditions in the locality
2. Amount of solar radiation reaching the surface of a collector in a year
3. Development of improved collector materials with good resistance to degradation from sun light
4. Economic collector design
5. Impact of a hot dry climate on the solar system

2.5.4 Solar Electric Conversion and Applications

Energy from the sun is transformed into electricity through solar thermal, ocean thermal, photovoltaic, or wind conversion. A typical solar power conversion to electricity is shown in Figure 2.13a.

2.5.4.1 Solar Thermo–electro–mechanical Conversion (Heat to Power)

Solar thermal energy is concentrated and transferred to a working fluid for use either in a Rankine or in a Brayton cycle turbine generator. Most of the solar thermal energy development has been oriented towards obtaining heated fluid from a solar collector for direct heating applications. However, a more valuable form of energy as mechanical or electrical energy (both are equivalent in the thermodynamic sense) is sometimes desired either exclusively or in combination

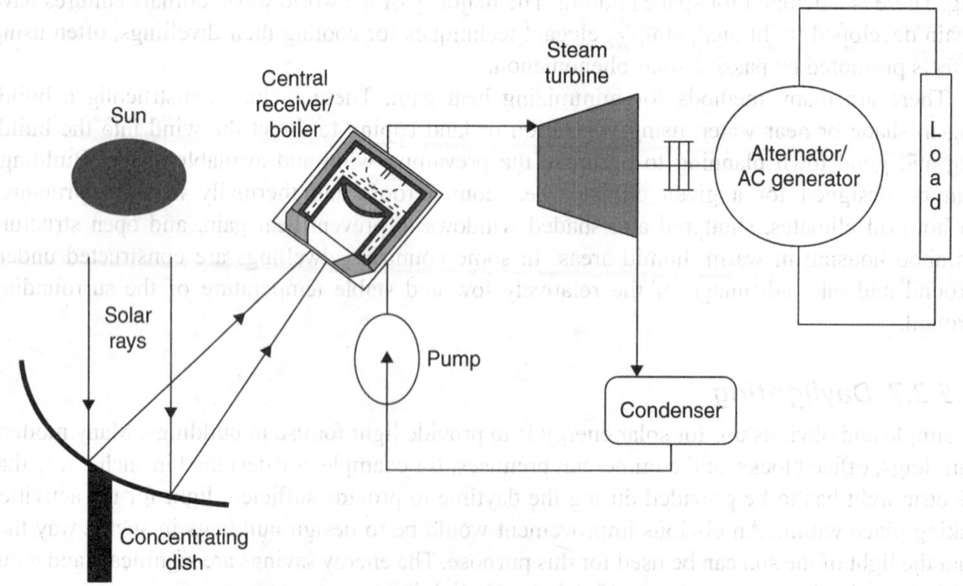

Figure 2.13a Thermo-electric conversion

with thermal energy. The device used to produce mechanical work or electricity (Fig. 2.13a) from solar generated heat is a power conversion cycle or heat engine referred to as Rankine Power Cycles.

2.5.4.1.1 Basic Rankine Cycle The Rankine cycle is a thermodynamic cycle which is used to produce electricity in many power stations, and it is the practical approach to the ideal Carnot cycle. Superheated steam is generated in a boiler and then expanded in a steam turbine. The turbine drives a generator to convert the work into electricity. The remaining steam is then condensed and recycled as feed water to the boiler. A disadvantage of using the water-steam mixture is that superheated steam has to be used; otherwise, the moisture content after expansion might be too high, which would erode the turbine blades. Instead of water, an organic fluid can be used. The major advantage is that these fluids can be used below 400°C and do not need to be overheated. In many cases, superheating is not necessary, resulting in a higher efficiency of the cycle. This is called an Organic Rankine Cycle. The most common power cycle used in solar power systems is, thus, the Rankine cycle, which combines constant-pressure heat addition and rejection processes with adiabatic reversible compression and expansion processes. It utilizes a working fluid that changes phase during the heat transfer processes to provide essentially isothermal heat addition and rejection. The working fluid is usually either water or organic liquids; however, liquid metals have also been used. The following description assumes water or steam as the working fluid.

The major components of a simple, ideal Rankine cycle are depicted in Figure 2.14 along with the thermodynamic states of the working fluid plotted on temperature–entropy coordinates.

Only ideal processes are depicted. The pressure of saturated liquid leaving the condenser at state 1 is raised in an adiabatic, reversible process by the (ideal) pump to state 2, where it enters the vapour generator (also called a boiler or steam generator). The compressed liquid is heated at

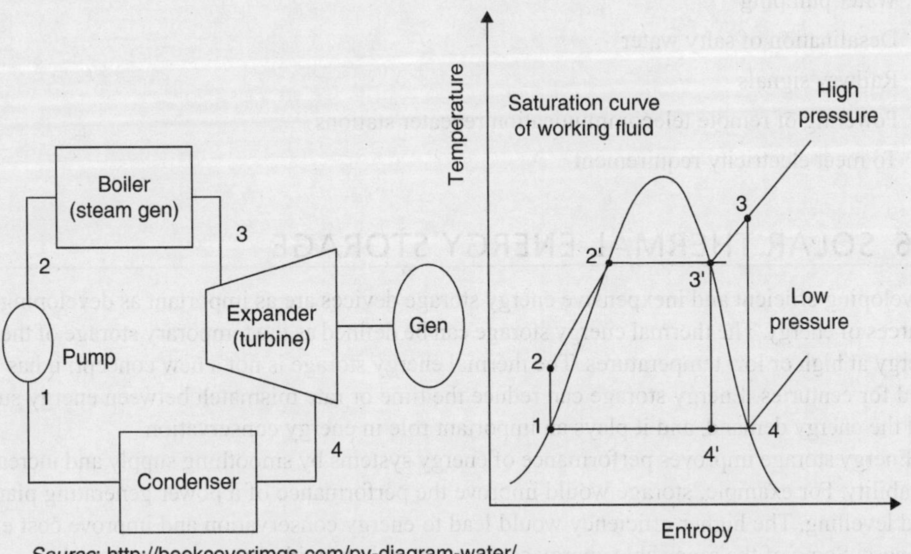

Source: http://bookcoverimgs.com/pv-diagram-water/

Figure 2.14 Basic Rankine cycle

constant pressure (often called preheat) until it reaches a saturated liquid state 2' and then at constant temperature (and pressure) until all the liquid has vapourized to become saturated vapour at 3'. More heat is added to superheat the saturated vapour at constant pressure, and its temperature rises to state 3'. The superheated vapour now enters an ideal expansion device (often a turbine) and expands in an adiabatic, reversible process to the low pressure maintained by the condenser indicated as state 4. The condenser converts the vapour leaving the turbine to liquid by extracting heat from it. Often during this expansion process, the vapour reaches saturation conditions and a mixture of saturated liquid and saturated vapour forms in the expander. The requirement to superheat the vapour from state 3' to 3 is defined by the amount of moisture that is permitted in the expander exhaust from state 4 to 4'. If the expander is a high-speed turbine, wet vapour produces destructive erosion of the blades. Some types of expanders such as piston and cylinder expanders permit some condensation during the expansion process. However, the amount of superheat is kept to a minimum so that the boiling temperature and the average heat-addition temperature can be maximized.

2.5.4.2 Photovoltaic Conversion (Light to Power)

Solar cells on which the sun shines deliver electric current. Although these devices have been used to provide energy in spacecraft with great success, their cost is so high (200%–500% of conventional generators) that large scale power plant applications are presently out of question. Nevertheless, the electricity is generated without boilers, turbines, generators, piping, or cooling tower. This electricity can either be used as it is or can be stored in the battery. This stored electrical energy later can be used for a number of applications such as the following:

1. Domestic lighting
2. Street lighting
3. Village electrification
4. Water pumping
5. Desalination of salty water
6. Railway signals
7. Powering of remote telecommunication repeater stations
8. To meet electricity requirement

2.6 SOLAR THERMAL ENERGY STORAGE

Developing efficient and inexpensive energy storage devices are as important as developing new sources of energy. The thermal energy storage can be defined as the temporary storage of thermal energy at high or low temperatures. The thermal energy storage is not a new concept; it has been used for centuries. Energy storage can reduce the time or rate mismatch between energy supply and the energy demand, and it plays an important role in energy conservation.

Energy storage improves performance of energy systems by smoothing supply and increasing reliability. For example, storage would improve the performance of a power generating plant by load levelling. The higher efficiency would lead to energy conservation and improve cost effectiveness. Some of the renewable energy sources can only provide energy intermittently.

Although the sun provides an abundant, clean, and safe source of energy, the supply of this energy is periodic following yearly and diurnal cycles; it is intermittent and often unpredictable and diffused. Its density is low compared with the energy flux densities found in conventional fossil energy devices such as coal or oil-fired furnaces. The demand for energy, on the other hand, is also unsteady following the yearly and diurnal cycles for both industrial and personal needs. Therefore, the need for the storage of solar energy cannot be avoided. Otherwise, solar energy has to be used as soon as it is received. In comparison, the present yield in energy gained by fossil fuels and waterpower amounts to about 70×10^{12} kWh. However, the technical use of solar energy presently poses problems primarily because of inefficient collection and storage. One of the important characteristics of a storage system is the length of time during which energy can be kept stored with acceptable losses. If solar energy is converted into a fuel such as hydrogen, then there will be no such time limit. Storage in the form of thermal energy may last only for shorter period because of losses by radiation, convection, and conduction. Another important characteristic of a storage system is its volumetric energy capacity or the amount of energy stored per unit volume. The smaller the volume, the better is the storage system. Therefore, a good system should have a long storage time and a small volume per unit of stored energy. Thermal mass systems can store solar energy in the form of heat at domestically useful temperatures for daily or seasonal durations. Thermal storage systems generally use readily available materials with high specific heat capacities such as water, earth, and stone. Well-designed systems can lower the peak demand, shift time-of-use to off-peak hours, and reduce the overall heating and cooling requirements.

2.6.1 Sensible Heat Storage

This means holding heat in a material without changing its phase when heat is added or removed. For examples, rocks and bricks become hot but remain solid. Oil may become hot but remains liquid. Much more heat can be stored in oil than in water since water can only be raised to 100°C without pressurizing it.

2.6.2 Latent Heat Storage

This is usually accomplished by using solar heat to melt a special material and then when the heat is needed, it is drawn from the said material. As it re-solidifies, it releases this heat. A very large amount of heat is stored in this way, and the temperature during melting or solidification remains constant. The material must melt at a 'reasonable' temperature and must be hot enough to begin and complete the cooking of food but also low enough to be attainable using solar energy. It must also be reasonably non-toxic, stable, easy to work with, and of course affordable. Salt, erythritol, and citric acid may be the candidates for this type of heat storage. Another latent heat storage approach is to chemically change a medium, usually under intense heat, and when it is reverted back to its original form, heat is released. Research is being done using quicklime (CaO) for this type of approach.

Phase change materials such as paraffin wax and Glauber's salt are another thermal storage media. These materials are inexpensive, readily available, and can deliver domestically useful temperatures (approximately 64°C).

SUMMARY

- The sun, which is our singular source of renewable energy, being at the centre of the solar system emits energy as electromagnetic radiation at an extremely large and relatively constant rate throughout the year.
- The emission rate of this energy is equivalent to the energy produced in a furnace at a temperature of about 6,000 K. If we could harvest the energy coming from just 10 hectares (25 acres) of the surface of the sun, then we would have enough energy to supply the current energy demand of the world.
- The atmosphere acts as a filter, which absorbs and reflects some portions of the electromagnetic spectrum (such as the ultraviolet region) that are harmful to humans and other life forms.
- The atmosphere provides a natural greenhouse effect, thereby maintaining the temperatures and climates in which humans and other living creatures on earth have evolved to survive.
- Sunlight drives plant life through photosynthesis, and animals survive by eating plants. Almost all microscopic forms of life such as bacteria, protozoa, and so on survive by using the energy of sunlight.
- The hour angle is the angular distance between the meridian of the observer and the meridian whose plane contains the sun.
- It is important to remember that during leap year, February month will be of 29 days.
- The declination angle (δ) of the sun is the angle between the rays of the sun and the plane of the earth's equator.
- The solar constant is a measure of flux density and it is the energy reaching the periphery of the earth's atmosphere. It is considered to be constant for all practical purposes with an average value of 1,361 W/m^2.
- The rate at which solar energy reaches a unit area at the earth is called the solar irradiance or insolation. The units of measure for irradiance are watts per square metre (W/m^2). The word *insolation* should not be confused with the word *insulation*.
- Energy storage improves performance of energy systems by smoothing supply and increasing reliability.
- Sensible heat storage means holding heat in a material without changing its phase when heat is added or removed.
- Latent heat storage is usually accomplished by using solar heat to melt a special material. It is drawn from the said material when heat is needed.

REVIEW QUESTIONS

1. Explain with a neat schematic diagram the binary cycle for thermo mechanical energy conversion and also mention its importance.
2. How much energy actually reaches the earth's surface from the sun? State and explain the terms *solar time* and *solar insolation*.

3. Explain the reason why solar energy system alone is not preferred for use in building services and also discuss the economic criterion for comparing investments on alternatives of solar and non-solar energy conversion systems.
4. Distinguish between global radiations and diffuse radiation. Which is applicable during cloudy atmosphere?
5. What are the reasons for variation in solar radiation reaching the earth than received at the outside of the atmosphere?
6. Write short notes on the following:
   (a) Beam and diffuse radiation
   (b) Solar constant
7. Define the following terms:
   (a) Altitude angle
   (b) Incident angle
   (c) Zenith angle
   (d) Solar azimuth angle
   (e) Latitude angle
   (f) Declination angle
   (g) Hour angle
8. What is angle of incidence of a solar beam radiation? What is its significance?
9. Explain basic Rankine cycle used with solar thermal power plant and derive the expression for thermal efficiency of Rankine cycle and also distinguish between high-temperature and low-temperature Rankine cycle.
10. Classify the methods of solar energy storage. Describe thermal energy storage system.
11. What is the principle of collection of solar energy used in a non-convective solar pond? Describe a non-convective solar pond for solar energy collection and storage.
12. What are the main applications of a solar pond? Describe briefly.
13. Write short notes on the following:
    (a) Maintenance of stable density gradient in a solar pond
    (b) Chemical energy storage method
    (c) Heat extraction method from a solar pond
    (d) Hydrogen storage
    (e) Electro-magnetic energy storage
14. State the principle of solar thermoelectric convertors.
15. Explain the following terms:
    (a) Solar insolation power curve
    (b) Thermal power collection curve
    (c) Energy storage
    (d) Energy supplied to load after solar hours
16. Why thermal storage is preferred in solar power plants?

Chapter 3

Solar Thermal Energy Collectors

KEY CONCEPTS

- Definition of solar collectors, applications, and classifications
- Types of solar collectors and comparison
- Configurations of certain practical solar thermal collectors
- Flat plate and concentrating solar collectors
- Material aspects of solar collectors
- Solar water and air heating systems
- Basic principles of solar cookers and its types
- Solar pond and its application

Sun's heat energy is a diffuse energy. It is always first collected and then concentrated. In residential systems, simple and cheap solar panels are used to collect the solar heat energy below 60°C. Residential panels for heat collection are referred to as flat plate collectors.

In utility scale systems, solar heat energy is required to be concentrated at high temperature level in the range 70°C–80°C at the collectors. The utility panels are, therefore, called concentrators.

Solar energy collectors are special kind of heat exchangers that transform solar radiation energy into internal energy of the transport medium. The major component of any solar system is the solar collector.

The solar collector absorbs the incoming solar radiation, converts it into heat, and then transfers this heat to a fluid (usually air, water, or oil) flowing through the collector. The solar energy, thus, collected is carried from the circulating fluid either directly to the hot water or space conditioning equipment or to a thermal energy storage tank, from which it can be drawn for use at night and/or cloudy days.

Solar collectors can be used in a large variety of applications. The following are the main areas of applications

1. *Solar water heating*: It includes thermosiphon, integrated collector storage systems, air systems, direct circulation, and indirect water heating systems.
2. *Solar space heating systems*: This includes both water and air systems.
3. *Solar refrigeration*: It includes both adsorption and absorption systems.

4. *Industrial process heat systems*: They include both low temperature (air and water based) applications and solar steam generation systems.
5. *Solar desalination systems*: They include both direct (solar stills) and indirect systems (conventional desalination equipment powered by solar collectors).
6. *Solar thermal power generation systems*: They include the parabolic trough systems, the power tower or central receiver systems, and the parabolic dish systems (dish/Stirling engine).

A Stirling engine is a heat engine operating by cyclic compression and expansion of air or other gas, the working fluid, at different temperature levels such that there is a net conversion of heat energy to mechanical work.

Similar to the steam engine, the Stirling engine is traditionally classified as an external combustion engine, as all heat transfers to and from the working fluid take place through the engine wall. This contrasts with an internal combustion engine where heat input is by combustion of a fuel within the body of the working fluid. Unlike a steam engine's (or more generally a Rankine cycle engine's) usage of a working fluid in both its liquid and gaseous phases, the Stirling engine encloses a fixed quantity of permanently gaseous fluid like air.

The amount of heat energy produced by a solar collector depends on the type of collector, its working surface direction towards the sun, meteorological conditions of the location, and many other factors.

3.1 TYPES OF SOLAR COLLECTORS

The collectors that are being marketed to utilize thermal energy from the sun can be subdivided into the following categories.

3.1.1 Flat Plate Collectors

Flat plate collectors are the most common type. They are also referred to as non- concentrating collectors and have the same area for intercepting and for absorbing solar radiation. A typical flat plate collector is shown in Figure 3.1.

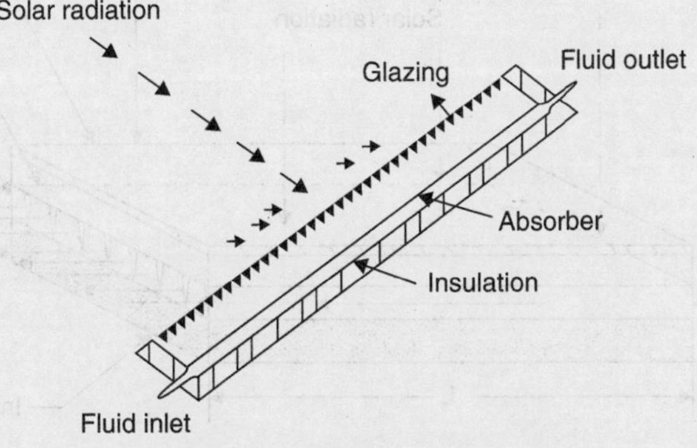

Figure 3.1 Flat plate collectors

It has five important parts:

1. *Dark flat plate absorber of solar energy*: The absorber consists of a thin absorber sheet (of thermally stable polymeric materials such as aluminium, steel, or copper to which a black or selective coating is applied) because of the fact that the metal is a good heat conductor. Copper is more expensive, but is a better conductor and less prone to corrosion than aluminium. In locations with average availability of solar energy, flat plate collectors are sized approximately 0.5 to 1 square foot per gallon of daily hot water use.
2. *Transparent cover*: This allows solar energy to pass through, but reduces heat losses.
3. *Heat-transport fluid (air, antifreeze, or water)*: To remove heat from the absorber, fluid is usually circulated through tubing to transfer heat from the absorber to an insulated water tank.
4. *Heat insulation backing*: Often backed by a grid or coil of fluid tubing.
5. *Insulated casing*: It is made of a glass or polycarbonate cover.

When solar radiation passes through a transparent cover and impinges on the blackened absorber surface of high absorptivity, a large portion of this energy is absorbed by the plate, and then transferred to the transport medium in the fluid tubes to be carried away for storage or use. The underside of the absorber plate and the side of casing are well insulated to reduce conduction losses. The transparent cover is used to reduce convection losses from the absorber plate through the restraint of the stagnant air layer between the absorber plate and the glass. It also reduces radiation losses from the collector as the glass is transparent to the short-wave radiation received by the sun, but it is nearly opaque to long-wave thermal radiation emitted by the absorber plate. For solar water heating systems in home and solar space heating flat plate collectors are the most common type of solar collector used.

3.1.1.1 Flat Plate Air Collectors

Schematic arrangement of a typical flat plate air collector is shown in Figure 3.2. It uses air as the heat transport medium. Air flat plate collectors are used mainly for solar space heating. The absorber plates can be made of metal sheets, layers of screen, or non-metallic materials. The air flows past the absorber by using natural convection or a fan. Since air does not conduct heat as easily as liquid, air collectors are typically less efficient than liquid collectors.

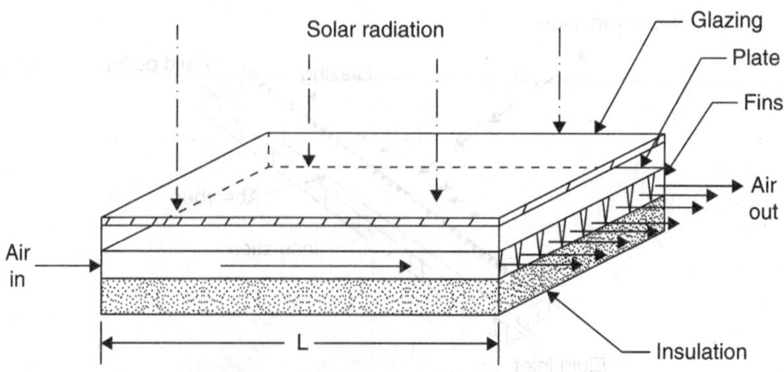

Figure 3.2 Flat plate air collectors

3.1.1.2 Flat Plate Liquid Collectors

These collectors use liquid as the heat transport medium. Liquid flat plate collectors heat liquid as it flows through tubes in or adjacent to the absorber plate as shown in Figure 3.3. The simplest liquid systems use household water that is heated as it passes directly through the solar collector and then flows to the house. Solar pool heating uses liquid flat plate technology, but the collectors are typically unglazed. The liquid tubes can be welded to the absorbing plate, or they can be an integral part of the plate. The liquid tubes are connected at both ends by large diameter header tubes.

3.1.2 Concentrating Collectors

By using reflectors to concentrate sunlight on the absorber of a solar collector, the size of the absorber can be dramatically reduced, which reduces heat losses and increases efficiency at high temperatures. Another advantage is that reflectors can cost substantially less per unit area than collectors. This class of collector is used for high-temperature applications such as steam production for the generation of electricity and thermal detoxification. These collectors are best suited to climates that have an abundance of clear sky days, and therefore, they are not so common in many regions. Stationary concentrating collectors may be liquid-based, air-based, or even an oven such as a solar cooker.

One such collector is a parabolic dish reflector, which is shown in Figure 3.4.

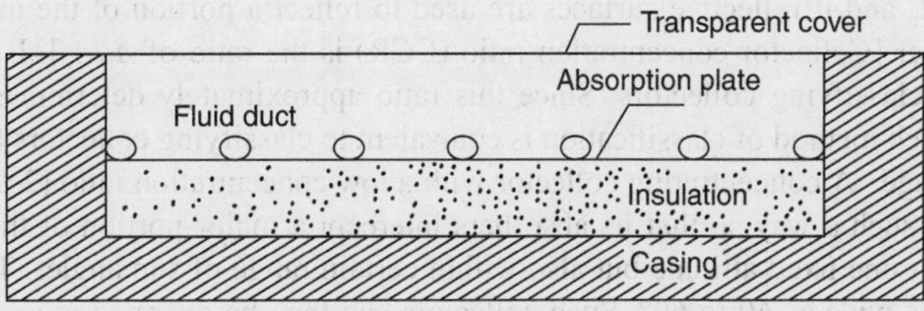

Figure 3.3 Flat plate liquid collectors

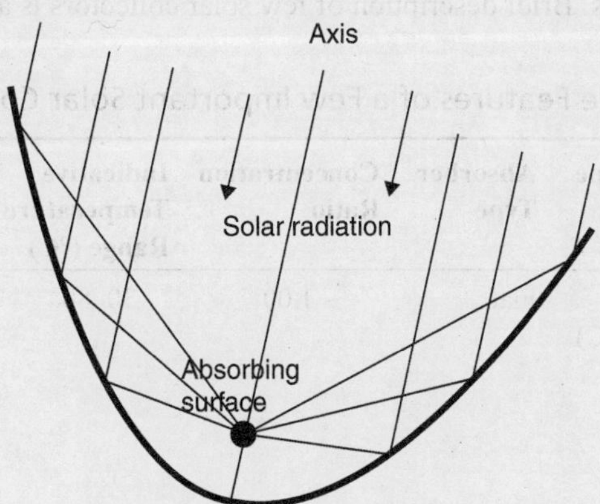

Figure 3.4 Parabolic reflector

3.1.2.1 Stationary Concentrating Collectors

These collectors are operated in a stationary mode for applications like air conditioning. Stationary concentrating collectors use compound parabolic reflectors and flat reflectors for directing solar energy to an accompanying absorber or aperture through a wide acceptance angle. The wide acceptance angle for these reflectors eliminates the need for a sun tracker. This class of collector includes parabolic trough flat plate collectors, flat plate collectors with parabolic boosting reflectors, and solar cookers. The development of the first two collectors has been done in Sweden. Solar cookers are used throughout the world, especially in the developing countries.

3.1.2.2 Tracking Concentrating Collectors

In the case of high temperature applications, like solar electric generation tracking, the sun is necessary. Heliostats are tracking mirrors that reflect solar energy onto a fixed target.

3.1.3 Comparison of Collectors

Two important performance parameters used for comparison of solar collectors are as follows:

1. Temperature range required for various range and
2. Collector concentration ratio

A collector is defined as concentrating collector if its absorber (fin) area A_d is smaller than the aperture area A_a and if reflective surfaces are used to reflect a portion of the incident sunlight into the absorber [Collector concentration ratio (CCR) is the ratio of A_a/A_d]. It is also used as a measure for classifying collectors. Since this ratio approximately determines the operating temperature, such method of classification is equivalent to classifying collectors by its operating temperature range. A concentrating collector with a low concentration ratio ($2 \le \text{CCR} \le 5$) can be designed in such a way so that its absorbers intercept a major portion of the incident sunlight not only at one fixed attitude but also within certain range of sun angles. This acceptance range may be as wide as 40 to 60°. Such collectors can then be operated in a stationary mode (i.e., tracking the sun is not required). The collector that have high concentration ratios (i.e., CCR >10), precise tracking is essentially required. Table 3.1 gives comparative features of a few important solar collectors. Brief description of few solar collectors is also presented.

Table 3.1 Comparative Features of a Few Important Solar Collectors

Motion	Collector Type	Absorber Type	Concentration Ratio	Indicative Temperature Range (°C)	Figure Nos.	Applications
Stationary	Flat plate collector (FPC)	Flat	1.00	30–80	Figure (3.1) Figure (3.2) Figure (3.3)	Space heating, space cooling, water heating, and for low temperature applications

(Continued)

Table 3.1 Continued

Motion	Collector Type	Absorber Type	Concentration Ratio	Indicative Temperature Range (°C)	Figure Nos.	Applications
	Evacuated tube collector (ETC)	Flat	1.00	50–200	Figure (3.4)	Space heating, space cooling, water heating, and for low temperature applications
	Compound parabolic collector (CPC)	Tubular	1–5	60–240	Figure (3.5)	For high temperature applications and air-conditioning
Single-axis tracking	Linear fresnel reflector (LFR)	Tubular	10–40	60–250	Figure (3.6)	For high-temperature applications and solar electricity generation
	Parabolic trough collector (PTC)	Tubular	15–45	60–300	Figure (3.7)	For high-temperature applications and solar electricity generation
	Cylindrical trough collector (CTC)	Tubular	10–50	60–300	Figure (3.8)	For high-temperature applications and solar electricity generation
Two-axis tracking	Parabolic dish reflector (PDR)	Point	100–1000	100–500	Figure (3.9)	For high-temperature applications and solar electricity generation
	Heliostat field collector (HFC)	Point	100–1500	150–2000	Figure (3.10)	For high-temperature applications and solar electricity generation

3.2 CONFIGURATIONS OF CERTAIN PRACTICAL SOLAR THERMAL COLLECTORS

Undoubtedly, there are many different ways that solar energy can be applied, but there are also many different methods for collecting the solar energy from incident radiation. The following are the list of some popular types of solar collectors.

3.2.1 Flat Plate Collectors

Flat plate collectors are the most common solar collector for solar water-heating systems in homes and solar space heating. A typical flat plate collector is an insulated metal box with a glass or plastic cover (called the glazing) and a dark-coloured absorber plate. These collectors heat liquid or air at temperatures less than 90°C.

Flat plate collectors are used for residential water heating and space heating installations.

3.2.1.1 Liquid Flat Plate Collectors

Liquid-based collectors use sunlight to heat a liquid that is circulating in a 'solar loop'. The fluid in the solar loop can be water, an antifreeze mixture, thermal oil, etc. The solar loop transfers the thermal energy from the collectors to a thermal storage tank. The simplest liquid systems use potable household water, which is heated as it passes directly through the collector and then flows to the house. The type of collector selection depends on how hot the water must be and the local climate. Unglazed collectors are typically used for swimming pool heating.

3.2.1.2 Air Flat Plate Collectors

These are used primarily for solar space heating. The absorber plates in air collectors can be metal sheets, layers of screen, or non-metallic materials. The air flows past the absorber by using natural convection or a fan. Because air conducts heat much less readily than liquid does, less heat is transferred from an air collector's absorber than from a liquid collector's absorber, and air collectors are typically less efficient than liquid collectors. The thermal energy collected from air-based solar collectors can be used for ventilation, air heating, space heating, and crop drying.

The most common air and liquid-based solar thermal collectors are as follows:

1. Glazed flat plate solar thermal collectors
2. Unglazed flat plate solar thermal collectors
3. Unglazed perforated flat plate solar thermal collectors
4. Back-pass flat plate solar thermal collectors
5. Batch flat plate solar thermal collectors
6. Solar cookers
7. Evacuated (vacuum tube) flat plate solar thermal collectors
8. Concentrating (flat plate collectors with flat reflectors)

3.2.2 Glazed Flat Plate Collectors

Glazed flat plate collectors are shown in Figure 3.1. They are very common and are available as liquid-based and air-based collectors. These collectors are better suited for moderate temperature applications where the demand temperature is 30°C–70°C and for applications that require heat during the winter months. The liquid-based collectors are most commonly used for the heating of domestic and commercial hot water, buildings, and indoor swimming pools. The air-based collectors are used for the heating of buildings, ventilation air, and crop drying

3.2.3 Unglazed Flat Plate Solar Collectors

Unglazed flat plate collectors account for the larger proportion of collector installed per year of any type of solar collector in many countries. Because they are not insulated, these collectors are best suited for low temperature applications where the temperature demand is below 30°C. By far, the primary market is for heating outdoor swimming pools, but other markets exist including heating seasonal indoor swimming pools, pre-heating water for car washes, and heating water used in fish farming operations. There is also a market potential for these collectors for water heating at remote and seasonal locations like summer camps. Unglazed flat plate collectors are usually made of black plastic that has been stabilized to withstand ultraviolet light. Since these collectors have no glazing, a large portion of the sun's energy is absorbed. However, because they are not insulated, a large portion of the heat absorbed is lost, particularly when it is windy and not warm outside. They transfer heat so well to air (and from air) that they can actually 'capture' heat during the night when it is hot and windy outside.

3.2.4 Unglazed Perforated Plate Collectors

The key to this type of collector is an industrial grade siding or cladding that is perforated with many small holes at a pitch of 2–4 cm. Air passes through the holes in the collector before it is drawn into the building to provide preheated fresh ventilation air. Efficiencies are typically high because the collector operates close to the outside air temperature. These systems can be very cost effective, especially when they replace conventional cladding on the building because only incremental costs need be compared to the energy savings. The most common application of this collector is for building ventilation air heating. Other possible components for this system are: a 20–30 cm air gap between the buildings, a canopy at the top of the wall that acts as a distribution manifold, and bypass dampers so that air will bypass the system during warm weather. Another application for this collector is crop drying. Systems have been installed in South America and Asia for drying of tea, coffee beans, and tobacco.

3.2.5 Back-pass Solar Collectors

Air-based collectors use solar energy to heat air. Their design is simple and they often weigh less than liquid-based collectors because they do not have pressurized piping. Air-based collectors do not have freezing or boiling problems. In these systems, a large solar absorber is used to heat the air. The simplest designs are single-pass open collectors. Collectors that are coated with a glaze can also be used to heat air for space heating.

3.2.6 Batch Flat Plate Solar Thermal Collectors

In ancient days, water tanks that were painted black were used as simple solar residential water heaters. Today, their primary market is for residential water heating in warm countries. Modern batch collectors have a glazing that is similar to the one used on flat plate collectors and/or a reflector to concentrate the solar energy on the tank surface. Because the storage tank and the solar absorber act as a single unit, there is no need for other components. On an area basis, batch collector systems are less costly than glazed flat plate collectors but also deliver less energy per year.

3.2.7 Flat Plate Collectors with Flat Reflectors

It is found that the addition of reflector on collector increases the solar yield on the collector and the overall thermal performance of the collector. The enhancement in the solar yield on the collector is about 44% in winter and 15% in summer conditions, which is consistent with more hot water demand in winter.

A variation of flat plate collector is shown in Figure 3.5.This simple reflector can markedly increase the amount of direct radiation reaching the collector. This is in fact a concentrate because the aperture is larger than the absorber plate.

3.2.8 Evacuated Tube Collectors

Conventional simple flat plate solar collectors were developed for use in sunny and warm climatic conditions. However, their benefits are greatly reduced when conditions become unfavourable during cold, cloudy, and windy days.

Furthermore, weathering influences such as condensation and moisture will cause early deterioration of internal materials resulting in reduced performance and system failure. Evacuated heat pipe solar collectors (tubes) operate differently than the other collectors available on the market. These solar collectors consist of a heat pipe inside a vacuum-sealed tube, as shown in Figures 3.6 and 3.7.

Evacuated tube collectors can achieve extremely high temperatures (75°C–180°C), making them more appropriate for cooling applications and for commercial and industrial applications.

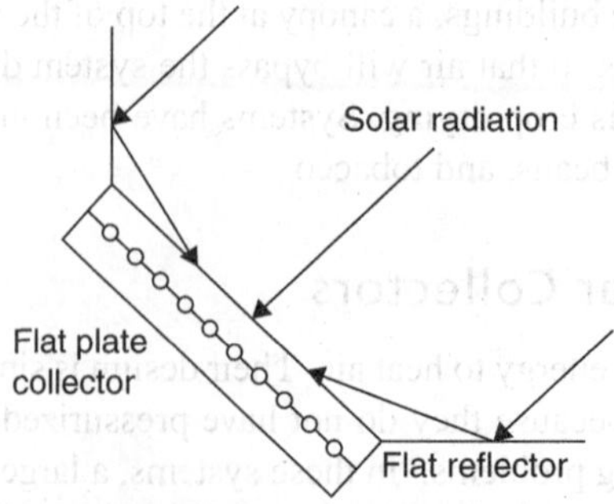

Figure 3.5 Flat plate collector with flat reflection

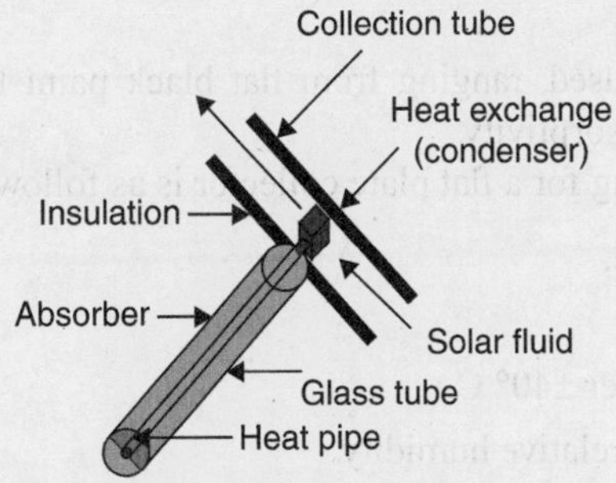

Figure 3.6 Typical evacuated tube solar collectors

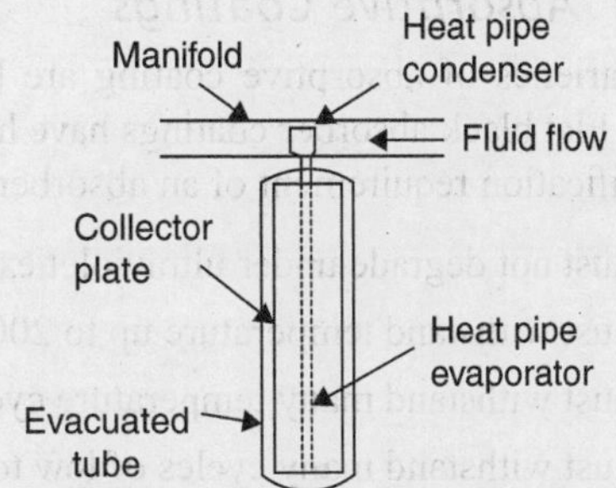

Figure 3.7 Schematic evacuated tube solar collectors

However, evacuated tube collectors are more expensive than flat plate collectors, as their unit area costs about twice than that of the latter. Evacuated tube collectors are efficient at high temperatures.

The collectors are usually made of parallel rows of transparent glass tubes. Each tube contains a glass outer tube and metal absorber tube attached to a fin. The fin is covered with a coating that absorbs solar energy well, but which inhibits radiation heat loss. Air is removed, or evacuated, from the space between the two glass tubes to form a vacuum, which eliminates conductive and convective heat loss.

Vacuum (also 'evacuated') tube solar collectors are amongst the most efficient and most costly types of solar collectors. These collectors are best suited for moderate temperature applications where the demand temperature is 50–95°C and also for very cold climates such as in the farthest northern part of Canada. Similar to the glazed flat plate solar collectors, applications of vacuum tube collectors include heating of domestic and commercial hot water, buildings, and indoor swimming pools. Due to their ability to deliver high temperatures efficiently, another potential application is for the cooling of buildings by regenerating refrigeration cycles. Vacuum tube solar collectors have a selective absorber for collecting sunlight that is in vacuum-sealed tube. Their thermal losses are very low even in cold climates.

3.3 MATERIAL ASPECTS OF SOLAR COLLECTORS

Flat, corrugated, or grooved plates, to which the tubes, fins, or passages are attached. The plate may be integrated with the tubes.

3.3.1 Absorber

The following are the types of solar flat plate absorbers that are most frequently used.

1. all copper plates are with integrated water passage (roll bond type). These plates can also be made of aluminium.
2. all copper (copper tube on copper sheet).
3. copper tube or aluminium fin
4. iron or steel
5. plastic (polymers)

3.3.1.1 *Absorptive Coatings*

Many varieties of absorptive coating are being used, ranging from flat black paint to baked enamel. Flat black absorber coatings have high absorptivity.

Specification requirement of an absorber coating for a flat plate collector is as follows:

1. It must not degrade under ultraviolet exposure.
2. It must withstand temperature up to 200° C.
3. It must withstand many temperature cycles over ±40° C.
4. It must withstand many cycles of low to high relative humidity.
5. It must not chalk, fade, or chip.
6. It must not be so thick that heat conduction through the paint to the metal absorber is impeded.

3.3.2 Glazing

One or more sheets of glass or other diathermanous (radiation transmitting) material is used as transparent covers. Following are its important functions:

1. It must reduce convective losses from the absorber plate.
2. It must suppress radiative heat losses from the absorber plate.
3. It must protect the absorber from the elements and from excessive UV exposures. A glazing material must be resistant to UV radiation.

Glass meets the entire abovementioned requirements and also compatible with the general requirement of longevity.

The following are the specification requirement of glazing materials:

1. They must be reasonably impact resistant.
2. Thin or no tempered glass panes are questionable because of the risk of damage from hail, birds, and vandalism.
3. Plastic materials of low tensile strength (i.e., Teflon) are not advisable.
4. They must be resistant to significant temperature shock.
5. Sudden rain will cause rapid overall limb changes. A leaf on a stagnant collector can cause high localized thermal stresses.

Thus, heat tempered glass is absolute necessity for outer collector glazing. Generally, plastic glazing can easily withstand the temperature shocks. Non-UV inhibited plastic materials are not acceptable.

Teflon (high transmittivity) and polyvinyl fluoride (PVF, Tedlar) are known to withstand UV radiation and is often used to protect other materials underneath from UV radiation.

3.3.2.1 *Areas of Practical Applications Attentions*

1. High cost
2. Longevity (is ultraviolet resistivity, thermal stability, corrosion resistance)
3. Black chrome coating is used in several high quality collectors. They are reasonable in cost and longevity. It can be used on copper, aluminium and steel absorbers.

4. Absorptivity is 92% to 95% in the visible spectrum and 10% to 20% in the infrared spectrum.
5. They must be able to withstand wind conditions.

3.3.2.2 Glazing Materials

1. Glass and fiberglass meet these requirements.
2. Tedlar used alone cannot serve the purpose.
3. Tedlar when bonded to the fiberglass, it acts as a good glazing material.
4. Optical rating must not change during its service life.
5. This requirement can hardly be met by any plastic glazing materials.
6. Fiberglass partially serves the purpose.

3.3.3 Insulation Shell

A solar flat plate collector must be insulated against excessive heat losses on its back side and on its edges as follows:

1. Back side – 3.5 inch of fiberglass insulation or 2 inch of foam insulation.
2. Side – 1 inch of fiberglass or 0.5 to 0.75 inch of foam insulation.

The following are the specifications to be met by insulating materials

1. It must withstand the maximum collector stagnation temperature rating (200°C) without damage. Foam materials shrink due to excessive heat.
2. The maximum stagnation temperature must not cause evaporation or sublimation of substances in the insulating materials such as the binder of the fiberglass.

Special fiberglass materials are available that have quite satisfactory outgassing rate.

3.4 CONCENTRATING COLLECTORS

As discussed in Section 3.1.2, they usually have concave reflecting surface to intercept and focus the sun's beam radiation to a smaller receiving area, thereby increasing the radiation flux. In other words, concentrating solar collectors use shaped mirrors or lenses to provide higher temperatures than the flat plate collectors.

However, each method of concentration has following drawbacks:

1. The mirror requires a clean smooth reflecting surface, because dust particles could scatter light away from the receiver or the light could be partly absorbed by a thin dirty film.
2. Smooth surface because contour error can also result in missing the receiver.

One criterion for the selection of a specific concentrator is the degree of concentration and hence temperature that is to be achieved. As a rule, concentrating energy on a point produces high to very high temperature; while on a line, it produces moderate to high temperature. Non-focusing concentrators produce low to moderate temperature.

Concentrating collectors are of various types and can be classified in many ways. They may be as follows:

1. Based on means of concentration: reflecting type use mirrors or refracting type use Fresnel lenses.
2. Based on reflecting surfaces used: parabolic, spherical, or flat.
3. Continuous or segmented.
4. Based on the formation of the image: imaging or non-imaging.
5. Imaging concentrator may focus on a line or at a point.
6. On the basis of collector concentration ratio or operating temperature range.
7. By the type of tracking.

The parabolic dish reflector (see Fig. 3.12) utilizes the point focus. The parabolic trough (see Fig. 3.10) is an example of line focus optics. Concentrating collectors are available in different configurations, which are discussed in the following sections.

3.4.1 Compound Parabolic Solar Collectors

These are non-imaging concentrators as shown in Figure 3.8.

3.4.2 Fresnel Solar Thermal Collectors

In order to deliver high temperatures with good efficiency, a high performance solar collector is required. Systems with light structures and low cost technology for process heat applications up to 400°C could be obtained with parabolic trough collectors. It can effectively produce heat at temperatures between 50°C and 400°C. They are made by bending a sheet of reflective material into a parabolic shape. A metal black tube, covered with a glass tube to reduce heat losses, is placed along the focal line of the receiver (see Fig. 3.9a). When the parabola is pointed towards the sun, parallel rays incident on the reflector are reflected onto the receiver tube. It is sufficient to use a single-axis tracking of the sun, and thus, long collector modules are produced.

Linear Fresnel reflector (LFR) technology relies on an array of linear mirror strips that concentrate light on to a fixed receiver mounted on a linear tower. It can be imagined as a broken-up parabolic trough reflector (see Fig. 3.9a), but unlike parabolic troughs, it does not have to be of parabolic shape, large absorbers can be constructed and the absorber does not have to move.

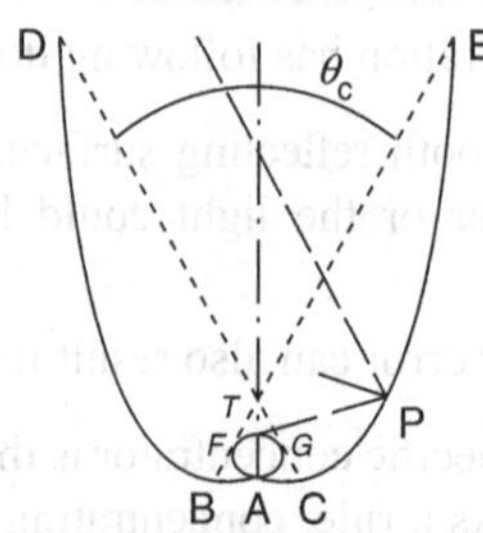

Figure 3.8 Compound parabolic solar collector

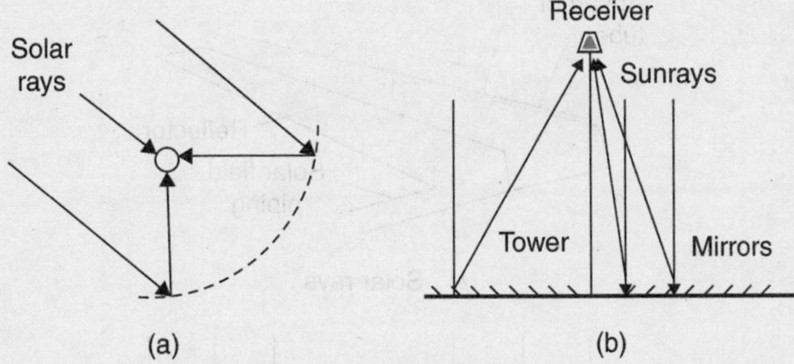

Figure 3.9 Fresnel trough collector (a) parabolic (b) linear

A representation of an element of an LFR collector field is shown in Figure (3.9b). The greatest advantage of this type of system is that it uses flat or elastically curved reflectors that are cheaper when compared to parabolic glass reflectors. Additionally, these are mounted close to the ground, thus minimizing structural requirements. One difficulty with the LFR technology is that the avoidance of shading and blocking between adjacent reflectors leads to increased spacing between reflectors. Blocking can be reduced by increasing the height of the absorber towers, but this increases the cost.

3.4.3 Parabolic Trough Solar Thermal Collectors

Parabolic troughs are devices that are shaped like the letter 'U', as shown in Figure 3.10.

Systems with light structures and low cost technology for process heat applications up to 400°C could be obtained with parabolic trough collectors. They can effectively produce heat at temperatures between 50°C and 400°C. They are made by bending a sheet of reflective material into a parabolic shape. A metal black tube, covered with a glass tube to reduce heat losses, is placed along the focal line of the receiver (see Fig. 3.10).

The troughs concentrate sunlight onto a receiver tube that is positioned along the focal line of the trough. Sometimes, a transparent glass tube envelops the receiver tube to reduce heat loss. Parabolic troughs often use single-axis or dual-axis tracking. In rare instances, they may be stationary. Temperatures at the receiver can reach 400°C and produce steam for generating electricity. In California, multi-megawatt power plants were built using parabolic troughs combined with gas turbines.

3.4.4 Cylindrical Trough Solar Collectors

Since linear translation does not introduce defocusing of the concentrated radiation, the aperture of a cylindrical trough need not track at all to maintain focus. However, as indicated in Figure 3.11, a high-rim angle cylindrical trough would have a focal plane not a focal line. To avoid a dispersed focus, cylindrical troughs would have to be designed with low rim angles in order to provide an approximate line focus. The advantage of cylindrical mirror geometry is that it need not track the sun in any direction as long as some means is provided to intercept the moving focus.

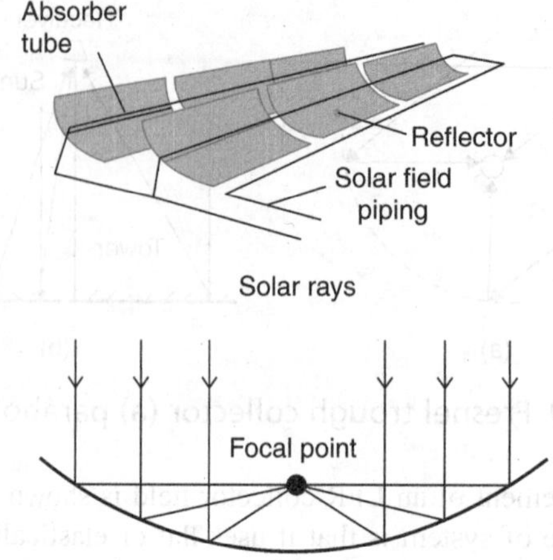

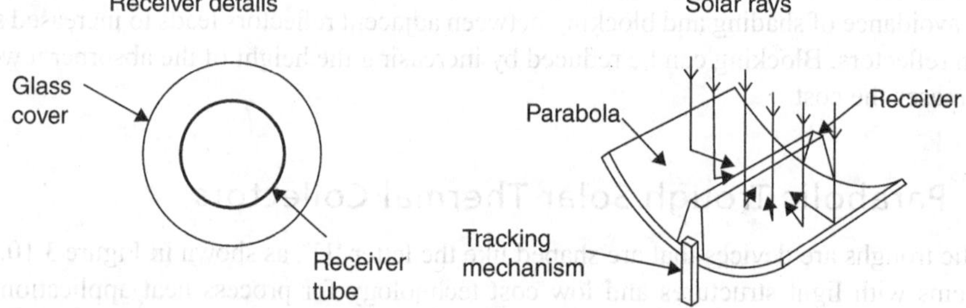

Figure 3.10 Parabolic trough reflector

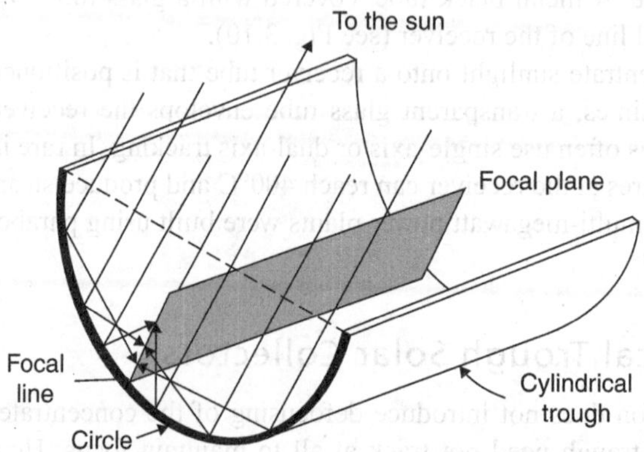

Figure 3.11 Cylindrical trough solar collectors

3.4.5 Parabolic Dish Systems

A parabolic dish collector is similar in appearance to a large satellite dish as shown in Figure 3.12, but has mirror-like reflectors and an absorber at the focal point. It uses a dual-axis sun tracker.

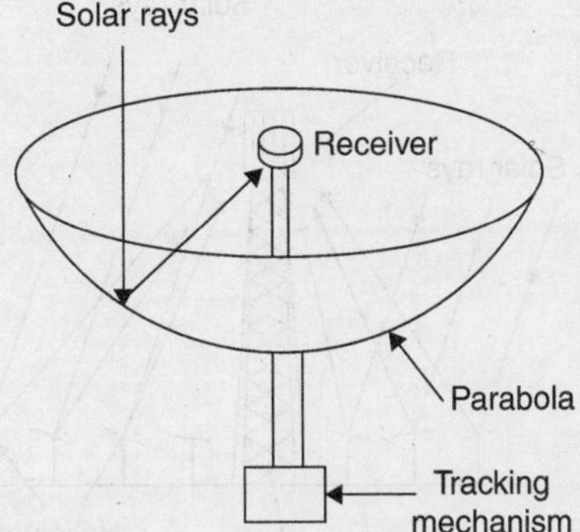

Figure 3.12 Parabolic dish solar thermal collector

It is a point-focus collector that tracks the sun in two axes

1. Concentrating solar energy onto a receiver located at the focal point of the dish.
2. The dish structure must track fully the sun to reflect the beam into the thermal receiver.

The receiver absorbs the radiant solar energy, converting it into thermal energy in a circulating fluid. The thermal energy can then either be converted into electricity using an engine–generator coupled directly to the receiver, or it can be transported through pipes to a central power conversion system. Parabolic-dish systems can achieve temperatures in excess of 1,500°C. Because the receivers are distributed throughout a collector field, like parabolic troughs, parabolic dishes are often called distributed receiver systems.

The following are the important advantages of parabolic dish reflectors:

1. They are the most efficient of all collector systems as they are always pointing the sun.
2. They typically have concentration ratio in the range of 500–2,000, and thus are highly efficient at thermal energy absorption and power conversion systems.
3. They have modular collector and receiver units that can either function independently or as part of a large system of dishes.

This type of solar thermal energy concentrator is primarily used with parabolic dish engine, which is an electric generator that uses sunlight instead of crude oil or coal to produce electricity.

3.4.6 Heliostat Field Solar Collectors

Heliostat is a mirror-based system that is used to continuously reflect sunlight onto a central receiver as shown in Figure 3.13. The collected solar energy is then converted into electrical power. Generally, it is a two-axis solar tracking flat mirror that reflects sunlight onto a fixed receiver or target. Furthermore, the geometry between the sun, mirror, and receiver are constantly changing throughout the day.

Heliostats have a potential of being the truly lowest cost solution to all the power one needs. These systems can be used for domestic heating, electricity, and lighting. The realization that all flat plate solar collectors have large amount of expensive collector area and still delivering only low

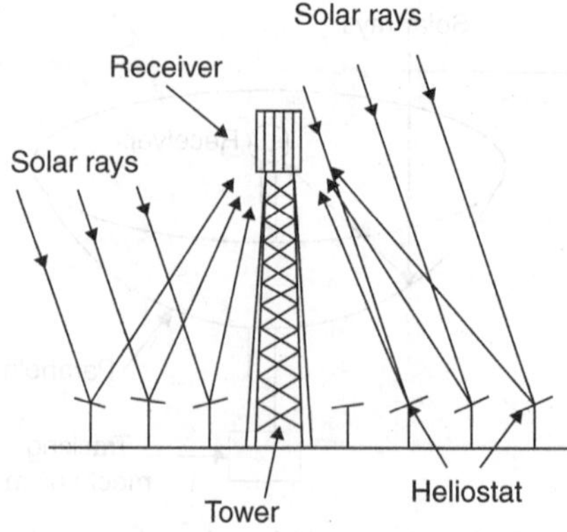

Figure 3.13 Heliostat filed solar collector

grade temperatures have compelled the researchers to look for a better solution. In this type of collector, a flat absorber efficiently transforms sunlight into heat. To minimize heat escaping, the plate is located between a glazing (glass pane or transparent material) and an insulating panel. The glazing is chosen so that a maximum amount of sunlight will pass through it and reach the absorber.

A heliostat uses a field of dual-axis sun trackers that direct solar energy to a large absorber located on a tower. To date, the only application for the heliostat collector is power generation in a system called the power tower. A power tower has a field of large mirrors that follow the sun's path across the sky. The mirrors concentrate sunlight onto a receiver on top of a high tower. A computer keeps the mirrors aligned so the reflected rays of the sun are always aimed at the receiver, where temperatures well above 1,000°C can be reached. High-pressure steam is generated to produce electricity.

3.4.6.1 Working of Practical Solar Heliostat

A practical solar heliostat is a mirror that makes precise movements up or down and left or right to reflect sunlight onto a fixed spot. As the sun marches across the sky, the heliostat adjusts its position, so the spot of reflected light remains stationary on the target. The relative spherical angular position of the thermal receiver to the heliostat is inputted to the computer. The computer solves the spherical trigonometry problem and commands the heliostat motor drive system such that the mirror is positioned angularly exactly halfway between the sun and the thermal receiver. Because every heliostat in an installation is in a unique position relative to the thermal receiver, each heliostat receives unique position commands.

When multiple practical solar heliostats reflect sunlight onto a single thermal receiver, the concentrated heat of the sunlight can be used to produce hot water or steam. Relatively cold water flows through the thermal receiver and is outputted as hot water. Although such a system can generate temperatures capable of melting steel, the temperature of the water is raised to just within a degree of boiling. For the purpose of heating and cooling a commercial building, this temperature uses the heat in sunlight at the highest efficiency and lowest cost. A higher temperature would mean more heat loss, as well as a more expensive system to withstand the higher temperature and pressure.

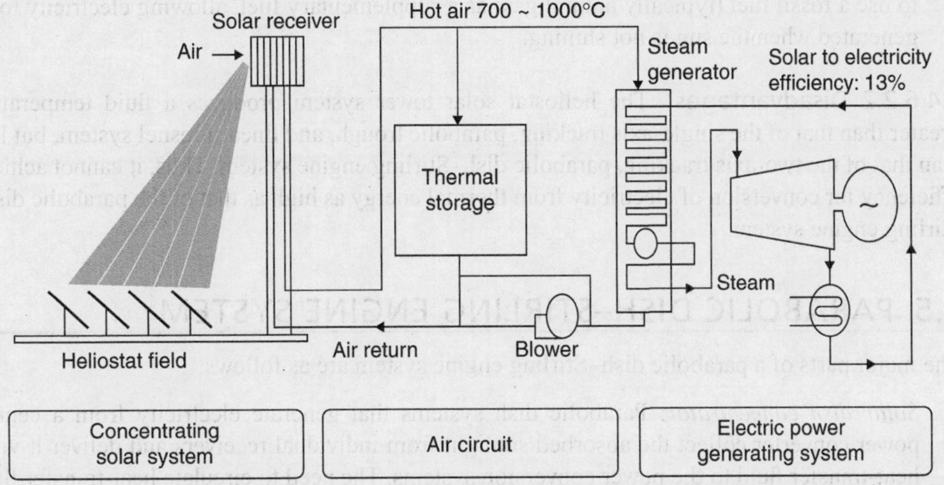

Source: http://energy.korea.com/archives/1117

Figure 3.14 Heliostat electric generating plant

The mirrors on practical solar heliostats have minimal reflection loss, so each heliostat reflects approximately its area in sunlight: about 1 kW of heat per square meter. As a frame of reference, a typical electric space heater produces 1.5 kW of heat. If 100 practical solar heliostats, each with 2.2 m^2 of mirror area, direct sunlight onto a single thermal receiver, the sunlight will be converted into 220 kw of heat.

With practical solar method of heat storage and heat distribution, 220 kw is more than enough energy to supply all of the heating needs of a 10,000 square foot commercial building. Using absorption chillers powered by hot water, it also has sufficient energy to supply the building's cooling needs.

However, the sun never shines at night, and often does not shine during the day. Energy storage is a fundamental requirement in any serious solar energy application. It is extremely expensive to store electrical energy, even though billions of dollars have been spent improving battery and fuel cell technology. The reverse is true for thermal energy. Hot water can be stored cheaply in a thermally insulated tank. As the volume of water and energy stored increases, the cost and losses of thermal energy storage drop rapidly.

3.4.6.2 Advantages and Disadvantages of the Heliostat Solar Tower Power Plant

A few advantages and disadvantages of the heliostat solar tower system in comparison with the other three concentrating solar power technologies under development are summarized here.

3.4.6.2.1 Advantages

1. Although the heliostat solar tower approach to solar power production is not as commercially developed as the solar parabolic trough system, it is more commercially developed than either the parabolic dish–Stirling engine or linear Fresnel systems.
2. Since the heliostat solar tower system produces steam to generate electricity with a conventional Rankine steam cycle, this system can be hybridized. In other words, it can be designed

to use a fossil fuel (typically natural gas) as a supplementary fuel, allowing electricity to be generated when the sun is not shining.

3.4.6.2.2 Disadvantages The heliostat solar tower system produces a fluid temperature greater than that of the single-axis tracking, parabolic trough, and linear Fresnel system, but less than that of the two-axis tracking, parabolic dish–Stirling engine system. Thus, it cannot achieve efficiency for conversion of electricity from thermal energy as high as that of the parabolic dish–Stirling engine system.

3.5 PARABOLIC DISH–STIRLING ENGINE SYSTEM

The major parts of a parabolic dish–Stirling engine system are as follows:

1. *Solar dish concentrator*: Parabolic dish systems that generate electricity from a central power converter collect the absorbed sunlight from individual receivers and deliver it via a heat-transfer fluid to the power conversion systems. The need to circulate heat-transfer fluid throughout the collector field raises design issues such as piping layout, pumping requirements, and thermal losses.
2. *Power conversion unit*: The power conversion unit includes the thermal receiver and the heat engine. The thermal receiver absorbs the concentrated beam of solar energy, converts it to heat, and transfers the heat to the heat engine. A thermal receiver can be a bank of tubes with a cooling fluid circulating through it. The heat transfer medium usually employed as the working fluid for an engine is hydrogen or helium. Alternate thermal receivers are heat pipes wherein the boiling and condensing of an intermediate fluid are used to transfer the heat to the engine. The heat engine system takes the heat from the thermal receiver and uses it to produce electricity. The engine–generators have several components; a receiver to absorb the concentrated sunlight to heat the working fluid of the engine, which then converts the thermal energy into mechanical work; an alternator attached to the engine to convert the work into electricity, a waste-heat exhaust system to vent excess heat to the atmosphere, and a control system to match the engine operation to the available solar energy. This distributed parabolic dish system lacks thermal storage capabilities, but can be hybridized to run on fossil fuel during periods without sunshine. The Stirling engine is the most common type of heat engine used in dish–engine systems.
3. *Tracking system*: A parabolic dish system uses a computer to track the sun and concentrate the sun's rays onto a receiver located at the focal point in front of the dish. In some systems, a heat engine, such as a Stirling engine, is linked to the receiver to generate electricity. Parabolic dish systems can reach 1,000 °C at the receiver, and achieve the highest efficiencies for converting solar energy to electricity in the small-power capacity range.

3.6 WORKING OF STIRLING OR BRAYTON HEAT ENGINE

After the array of mirrors focuses the sunlight, the concentrated sunlight then heats up the working fluid to temperatures of around 750°C within the receiver. The heated high temperature working fluid is then used in either a Stirling or Brayton heat engine cycle to produce mechanical power via

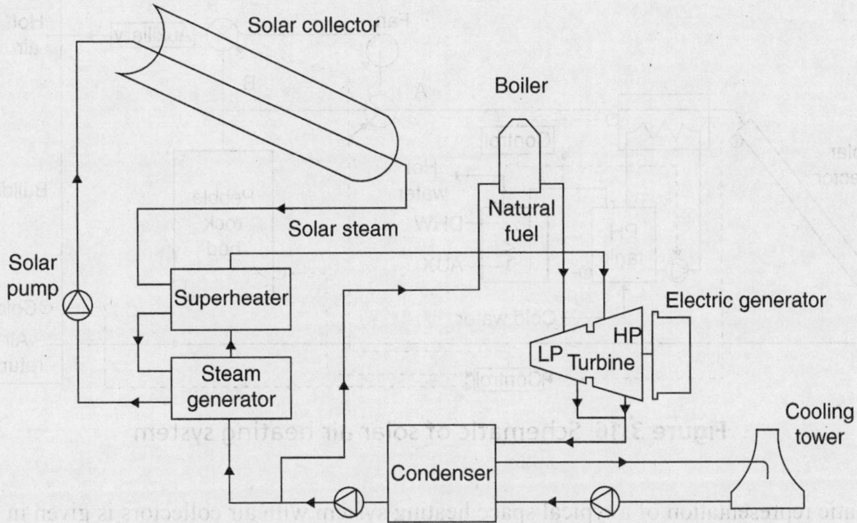

Figure 3.15 Schematic of solar electric generation

rotational kinetic energy and then electricity for utility use with an electric generator. An example of a Brayton cycle used to produce electricity for a parabolic dish power plant is shown in Figure 3.15. In the cycle, the concentrated sunlight focused on the solar fluid heats up the compressed working fluid of the cycle, i.e., air, replacing altogether or lowering the amount of fuel needed to heat up the air in the combustion chamber for power generation. As with all Brayton cycles, the hot compressed air is then expanded through a turbine to produce rotational kinetic energy, which is converted to electricity using the alternator. A recuperator is also utilized to capture waste heat from the turbine to preheat the compressed air and make the cycle more efficient.

3.7 SOLAR COLLECTOR SYSTEMS INTO BUILDING SERVICES

Although the operating costs of solar energy conversion system are generally low, the initial cost of purchasing and installing such system are high when compared to fossil fuel or other conventional energy system. Hence, the choice between solar energy systems against a conventional one must be cost effective on a long term basis (based on economic evaluation).

Since the availability of solar energy is intermittent and unpredictable it is rarely cost effective to have energy demands from solar energy alone. A solar system able to meet all the energy demands under the worst operating conditions for a long period would be greatly oversized.

The most economical way is to have

1. The solar system meets the basic energy demand while operating at its full capacity.
2. Let an auxiliary or backup system carry the peak load and unusual load.

The right proportion of solar versus auxiliary energy supply is to be determined by economics.

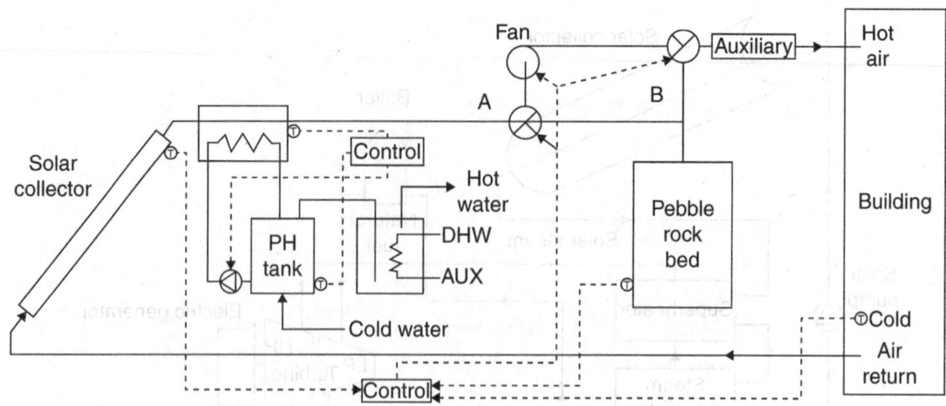

Figure 3.16 Schematic of solar air heating system

Schematic representation of a typical space heating system with air collectors is given in Figure 3.16. Dampers is indicated for the solar rock-bed charging mode. The main components of the building heating systems are as follows:

1. Air handling unit: a fan and two motor-driven dampers.
2. Heat storage unit (rock bed)
3. Temperature control system
4. Solar collectors

Depending on the position of dampers A and B, three modes of system operation can be achieved.

1. *Dampers A and B open*: This is the normal day time solar heating mode. The storage unit is bypassed. If the temperature sensor in the top of the collector array is below a necessary limit required for space heating, the auxiliary furnace is automatically turned on.
2. *Damper A open and damper B closed*: This mode is used whenever solar heat is collected but no space heating is required at the same time. The fan blows the solar heated air through the rock bed for thermal storage.
3. *Damper A closed and damper B open*: This mode is used during cloudy periods or during the night hours. The return air from the building is now pulled through the rock bed, where it picks up solar heat. The auxiliary furnace is activated automatically if the temperature is insufficient to meet the demand.

3.8 SOLAR WATER HEATING SYSTEMS

Most solar water heating systems have two main parts: a solar collector and a storage tank. The most common collector is called a flat plate collector. It consists of a thin, flat, rectangular box with a transparent cover that faces the sun mounted on the roof of building or home. Small tubes run through the box and carry the fluid – either water or other fluid, such as an antifreeze solution – to be heated. The tubes are attached to an absorber plate, which is painted with special coatings to absorb the heat. The heat builds up in the collector, which is passed to the fluid passing through the tubes.

An insulated storage tank holds the hot water. It is similar to water heater, but larger in size. In the case of systems that use fluids, heat is passed from hot fluid to the water stored in the tank through a coil of tubes. Solar water heating systems can be either active or passive systems.

1. The active systems, which are most common, rely on pumps to move the liquid between the collector and the storage tank.
2. The passive systems rely on gravity and the tendency for water to naturally circulate as it is heated.

3.8.1 Active Solar Water Heating Systems

The active water systems that can be used to heat domestic hot water are the same as the ones that provide space heat. A space heat application will require a larger system and additional connecting hardware to a space heat distribution system.

3.8.1.1 Parts of Water Heating Systems

There are five major components in active solar water heating systems:

1. Collector(s) to capture solar energy.
2. Circulation system to move a fluid between the collectors to a storage tank
3. Storage tank
4. Backup heating system
5. Control system to regulate the overall system operation

A typical active water heating system that exhibits effectiveness, reliability, and low maintenance is shown in Figure 3.17.

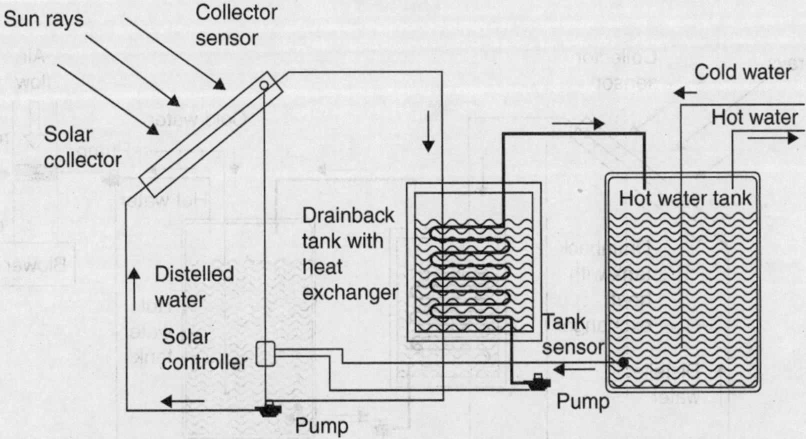

Source: http://biddecor.blogspot.in/2015/02/solar-water-heating.html

Figure 3.17 A typical hot water system

It uses distilled water as the collector circulating fluid. The collectors in this system will only have water in them when the pump is operating. This means that in the case of power failure as well as each night, there will be no fluid in the collector that could possibly freeze or cool down and delay the start-up of the system when the sun is shining. This system is very reliable and widely used. It requires that the collectors are mounted higher than the drain back tank or heat exchanger. This may be impossible to do in a situation where the collectors must be mounted on the ground. The fluids that are circulated into the collectors are separated from the heated water that will be used in the home by a double-walled heat exchanger. A heat exchanger is used to transfer the heat from the fluids circulating through the collectors to the water used in the home. The fluids that are used in the collectors can be water, oil, an antifreeze solution, or refrigerant. The heat exchangers should be double-walled to prevent contamination of the household water. The controller in these systems will activate the pumps to the collectors and heat exchanger when design temperature differences are reached. The heat exchanger may be separated from the storage tank or built into it. The systems that use antifreeze fluids need regular inspection (at least every 2 years) of the antifreeze solution to verify its viability. Oil or refrigerant circulating fluids are sealed into the system and will not require maintenance. A refrigerant system is generally more costly and must be handled with care to prevent leaking any refrigerant. This hot water system can be used for heating swimming pools and spas. Lower cost unglazed (no glass cover) collectors are available for this purpose.

3.8.2 Active Solar Space Heating

The active solar space heating system can use the same operational components as the domestic water heating systems as shown in Figure 3.18, but ties into a heating distribution system that can use heated fluids as a heat source.

The distribution system includes hydronic radiator, floor coil systems, and forced air systems.

The fluid that is heated and stored (typically water) and can be distributed into the house heating system in the following ways:

1. *Air distribution system*: The heated water in the storage tank is pumped into a coil located in the return air duct whenever the thermostat calls for heat. The controller for the solar

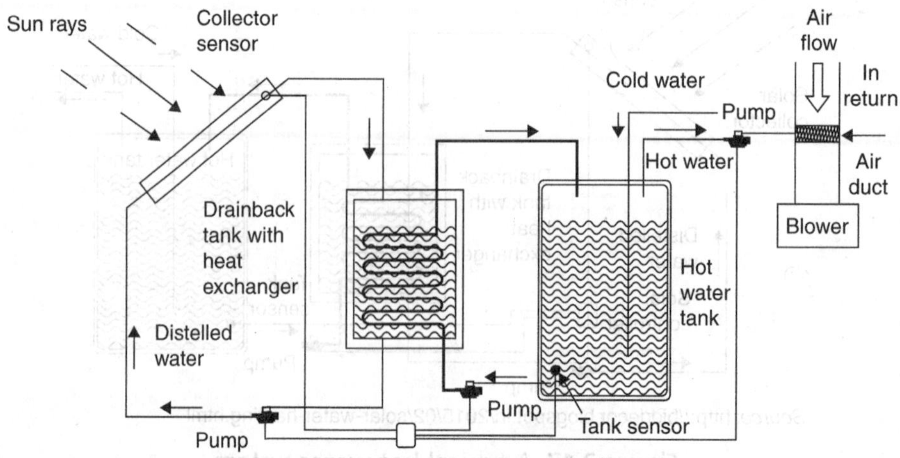

Figure 3.18 Typical space heating system

system will allow the pumping to occur if the temperature in the solar heated water is above a minimum amount needed to make a positive contribution to heating the home. An auxiliary heater can be used in two ways. It can add heat to the solar storage tank to maintain a minimum operating temperature in the storage tank at all times. In this case, the coil from the solar system will be located at the air handler supply plenum rather than in the return air duct. The auxiliary heater can also be a conventional furnace that will operate less often due to the warm air entering the air handler from the solar coil in the return duct.

2. *Hydronic system with radiators*: The heated water is circulated in series with a boiler into radiators located in the living spaces. Modern baseboard radiators operate effectively at 140°C. Solar heating systems can very often reach that temperature. Using the solar system's heated water as the source of water for the boiler will reduce the boiler's energy use, particularly if it senses the incoming temperature and will not operate when that temperature is above the required distribution temperature.
3. *Hydronic system with in-slab heat*: The solar heated water is pumped through distribution piping located in the floor of the home. Lower temperatures are used in this type of system (the slab is not heated above 80° in most cases). The auxiliary heat can be connected in series with the solar system's heated output water or it can be connected to the solar tank to provide a minimum temperature.

The space heating system, like the domestic water heating system, must be backed up by an auxiliary heating system. It is not practical to size a solar system to provide a home's entire heat requirement under the worst conditions. The system would become too large, too costly, and oversized for most of the time. The storage system should be sized to approximately 1.5 gallons of storage for each square foot of collector area.

3.9 PASSIVE SOLAR WATER HEATING SYSTEMS

A passive solar water heating system uses natural convection or household water pressure to circulate water through a solar collector to a storage tank or to the point of use. Active systems employ pumps and controllers to regulate and circulate water. Although passive system is generally less efficient than active systems, the passive approach is simple and economical.

Passive water heating systems must follow the same parameters for installations as that of active systems – south facing non-shaded location with the collector tilted at the angle of our latitude. Since the storage tank and collector are combined or in very close proximity, roof structural capacities must accommodate the extra weight of a passive system.

3.9.1 Types of Passive Water Heaters

Two types of passive water heaters are batch and thermosiphon systems.

3.9.1.1 Batch System

The batch system is the simplest of all solar water heating systems, as depicted in Figure 3.19. It consists of one or more metal water tanks painted with a heat absorbing black coating and placed in an insulating box or container with a glass or plastic cover that admits sunlight to strike the tank directly. The batch system's storage tank is the collector as well. These systems will use the existing house pressure to move water through the system. Each time a hot water tap

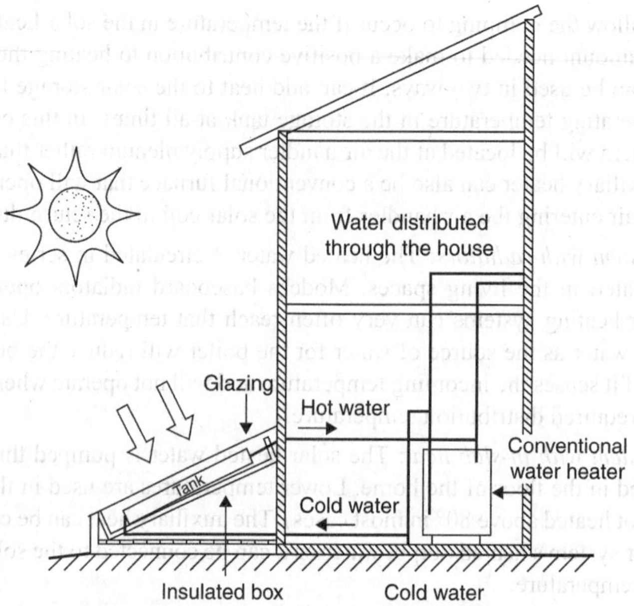

Figure 3.19 A typical schematic of batch domestic water heating system

is opened, heated water from the batch system tank is removed and replaced by incoming cold water. The piping that connects to and from the batch heater needs to be highly insulated. On a cold night, when no one is drawing hot water, the water in the pipes is standing still and vulnerable to freezing. In many applications, insulated polybutylene piping is used because the pipe can expand if frozen. The water in the batch heater itself will not freeze because there is adequate mass to keep it from freezing.

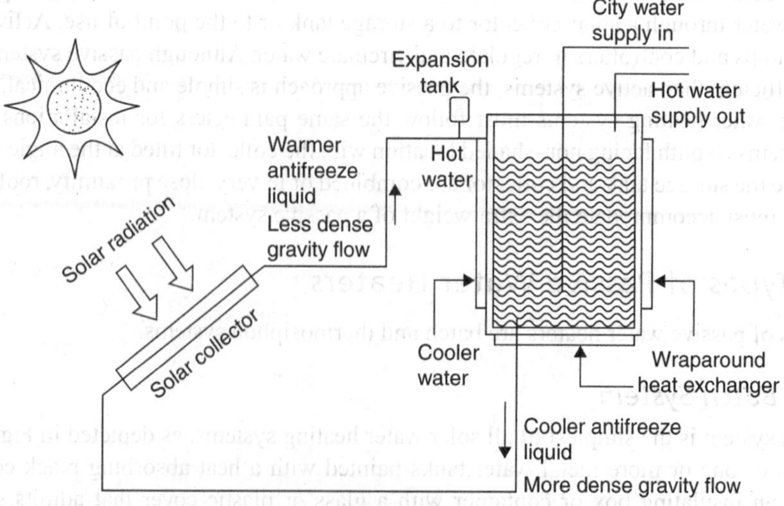

Source: http://solarheatcool.sustainablesources.com/

Figure 3.20 Thermosiphon system

Since the tank that is storing the heated water is sitting outside, there will be heat loss from the tank during the night. This can be minimized by an insulating cover placed on the heater in the evening. The most effective use of a batch water heater is to use hot water predominantly in the afternoon and evenings when the temperature in the tank will be the highest.

3.9.1.2 Thermosiphon Systems

The thermosiphon system uses a flat plate collector and a separate storage tank that must be located higher than the collector as shown in Figure 3.20. The collector is similar to those used in active systems.

The storage tank located above the collector receives heated water coming from the top of the collector into the top of the storage tank. Colder water from the bottom of the storage tank will be drawn into the lower entry of the solar collector to replace the heated water that was thermosiphoned upward. The storage tank may or may not use a heat exchanger. The thermosiphon system is more costly and complex than the batch system. In our area, it is best to use an indirect system (one that employs a heat exchanger). In that case, antifreeze can be used in the system eliminating freeze ups.

3.10 APPLICATIONS OF SOLAR WATER HEATING SYSTEMS

The following are a few industrial applications of solar water heaters.

1. *Hotels*: bathing, kitchen, washing, laundry applications
2. *Dairies*: ghee (clarified butter) production, cleaning and sterilizing, pasteurization
3. *Textiles*: bleaching, boiling, printing, dyeing, curing, ageing, and finishing
4. *Breweries and distilleries*: bottle washing, work preparation, boiler feed heating
5. *Chemical/bulk drugs units*: fermentation of mixes, boiler feed applications
6. *Electroplating or galvanizing units*: heating of plating baths, cleaning, degreasing applications
7. *Pulp and paper industries*: boiler feed applications, soaking of pulp.

3.11 ACTIVE SOLAR SPACE COOLING

Solar space cooling is quite costly to implement. It is best to use a solar system that serves more than just the cooling needs of a house to maximize the return on investment and not leave the system idle when cooling is not required. Significant space heating and/or water heating can be accomplished with the same equipment used for the solar cooling system. Active solar absorption cooling system is presented in Figure 3.21 in which (T) represents the sequence of flow.

Heat from solar collectors separates a low boiling refrigerant in a generator that receives the pressurized refrigerant from an absorber. Solar heat can also be used in the evaporation stage of the cycle.

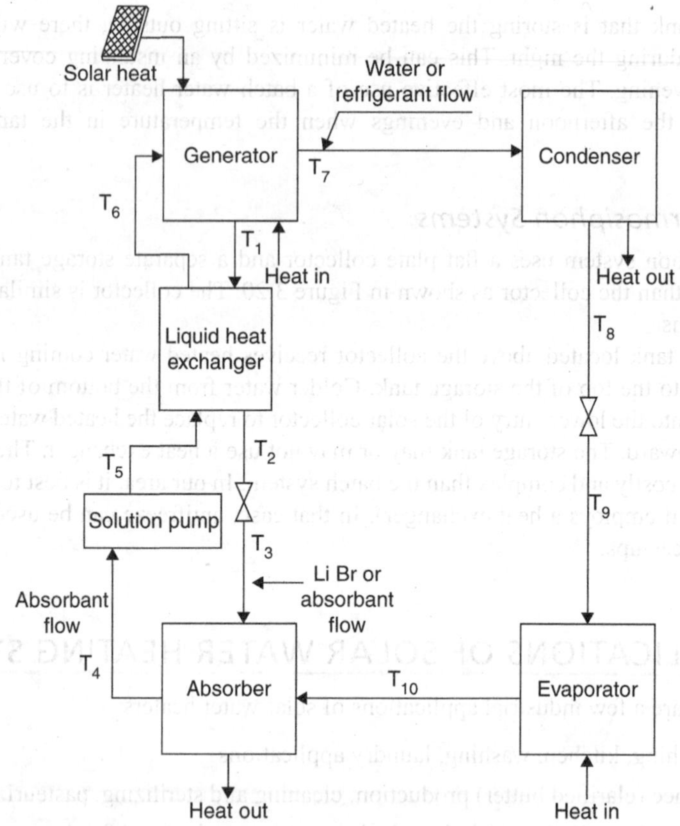

Figure 3.21 Schematic of solar absorption cooling system (T represents the sequence of flow)

3.12 SOLAR AIR HEATING

Solar-heated air can be used for drying most crops that require warm air. Solar heated air is ideal for drying delicate foods since it will not burn or risk potential damage from high temperature steam heat. Solar heat is non-polluting and best of all, it incurs no fuel costs.

Existing commercial drying operations can be converted to utilize solar heat by installing our system to remove heat from the building's metal roof or wall. We remove heat from under the metal panels, add the duct, and connect the ducts to the intake of the drier fans. The system then removes the heated air from the underside of the panels and passes the air to the drying chamber.

Simple sensors are installed in the air flow and use thermostatic controls to turn off the incoming air flow when the temperature is not high enough for solar heating. The existing system then operates, as it always has, burning high-cost fuel but serving the drying process.

For some in-field applications, one can use a ground-mounted polymer system that is low cost and very transportable. In new building, metal roofs and walls are integrated with the building's structure. By trapping air into a confined space which has sunlight hitting it, the trapped air is heated up naturally by the solar power of the sun. This type of natural heat transfer is also used for solar water heating. The hot air rises, and this is a key component to solar heating with convection. Inside houses, the coldest air is closest to the floor and the warmest is up high along the ceiling. As the warm air at the top of the room cools, it drops lower, and as the cooler air is warmed up, it rises. In order to use this process naturally in houses and particularly to make use of the warming power of solar energy, means has to be found or create a way to have the cool air go into a space that is warmed up by the sun. That same space usually allows the warm air to escape once it has reached a certain point.

In many cases, all you need is some sort of confined area to direct and control airflow. For example, if you have a sunny window, you could put a piece of black fabric, wood, plastic, or metal against that window frame to trap the air in for a short period of time. There needs to be a space between the black material you choose to use, and the glass window pane itself. Usually, this space is at least a few inches, but there can be space as much as 5–6 inches between your window glass and the black material.

It is a solar thermal technology in which the energy from the sun, solar insolation, is captured by an absorbing medium and used to heat air. Solar air heating is a renewable energy heating technology used to heat or condition air for buildings or process heat applications. It is typically the most cost effective out of all the solar technologies, especially in commercial and industrial applications, and it addresses the largest usage of building energy in heating climates, which is space heating and industrial process heating.

Solar air collectors can be commonly divided into two categories:

1. Unglazed air collectors or transpired solar collector (used primarily to heat ambient air in commercial, industrial, agricultural and process applications).
2. Glazed solar collectors (recirculation types that are usually used for space heating).

3.13 SOLAR DRYERS

Solar dryers can be utilized for various domestic purposes. They also find numerous applications in industries such as textiles, wood, fruit and food processing, paper, pharmaceutical, and agro-industries.

3.13.1 Advantages

Solar dryers are more economical when compared to dryers that run on conventional fuel or electricity. The drying process is completed in the most hygienic and eco-friendly way. Solar drying systems have low operation and maintenance costs.

Solar dryers last longer. A typical dryer can last 15–20 years with minimum maintenance.

3.13.2 Limitations

1. Drying can be performed only during sunny days, unless the system is integrated with a conventional energy-based system.
2. Due to limitations of solar energy collection, the solar drying process is slow in comparison with dryers that use conventional fuels.
3. Normally, solar dryers can be utilized only for drying at 40°C–50°C.

One well-known type of solar dryer is shown in Figure 3.22.

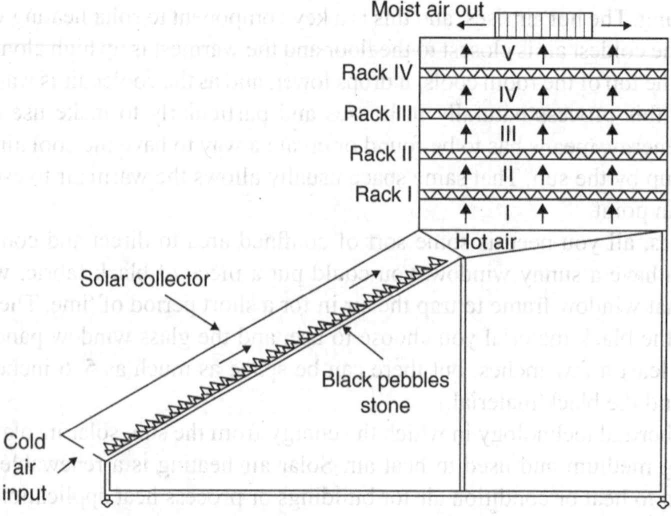

Figure 3.22 A rice solar dryer

It was designed for the particular requirements of rice but the principles hold for other products and design types, since the basic need to remove water is the same. Air is drawn through the dryer by natural convection. It is heated as it passes through the collector and then partially cooled as it picks up moisture from the rice. The rice is heated both by the air and directly by the sun. Warm air can hold more moisture than cold air so the amount required depends on the temperature to which it is heated in the collector as well as the amount held (absolute humidity) when it entered the collector.

A rock-bed dryer is shown in Figure 3.23. In this dryer, air drawn by natural convection through an air inlet (A), circulates the heat collected by the primary solar energy collector (B), throughout the drying chamber (C), which is packed with limestone rocks of relatively uniform diameter.

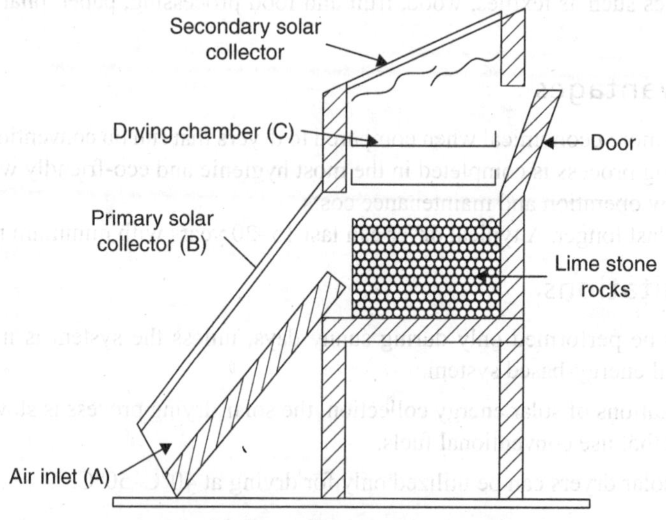

Figure 3.23 Rock-bed solar dryer

The heat would then stratify across the rock bed but, since rocks are poor thermal conductors, temperature differences would slowly disappear when air is not moving through the rock bed. Thus, samples positioned above the rock bed can continue drying during the night. This type of a solar dyer requires very little maintenance. Solar heated air can be used for drying most crops that require warm air. This air is ideal for drying delicate foods since it will not burn or risk potential damage from high temperature steam heat. Solar heat is non-polluting and best of all, it incurs no fuel costs.

Existing commercial drying operations can be converted to utilize solar heat by installing system to remove heat from the building's metal roof or wall. Heat is removed from under the metal panels, add the duct, and connect the ducts to the intake of the dryer fans. The system then removes heated air from the underside of the panels and passes the air to the drying chamber.

Sensors are installed in the air flow and use thermostatic controls to turn off the incoming air flow when the temperature is not high enough for solar heating. The existing system then operates on, as it always has, and auxiliary standby conventional fuel system. Solar-heated air can be used to dry

1. Crops, timber, distillers grains, and textiles
2. Tea, coffee, beans, tobacco, etc.
3. Food for dehydration or processing
4. Sludge, manure, and compost

3.14 CROP DRYING

Controlled drying is required for various crops and products, such as grain, coffee, tobacco, fruits, vegetables, and fish. Their quality can be enhanced if the drying is properly carried out. Solar thermal technology can be used to assist with the drying of such products. The main principle of operation is to raise the heat of the product, which is usually held within a compartment or box; while, at the same time, passing air through the compartment to remove moisture. The flow of air is often promoted using the 'stack' effect that takes advantage of the fact that hot air raises, and therefore, it can be drawn upwards through a chimney, while drawing in cooler air from below. Alternatively, a fan can be used. The size and shape of the compartment varies depending on the product and the scale of the drying system. Large systems can use large barns, while smaller systems may have a few trays in a small wooden housing.

Solar crop drying technologies can help reduce environmental degradation caused by the use of fuel wood or fossil fuels for crop drying and can also help to reduce the costs associated with these fuels and hence the cost of the product. Improving and protecting crops also have beneficial effects on health and nutrition.

3.15 SPACE COOLING

The majority of the developing countries lies within the tropics and have little need of space heating. However, there is a demand for space cooling. The majority of the world warm climate cultures have again developed traditional, simple, elegant techniques for cooling their dwellings, often using effects promoted by passive solar phenomenon.

There are many methods for minimizing the heat gain. These include situating a building in shade or near water, using vegetation or landscaping to direct wind into the building, good town planning to optimize the prevailing wind and available shade. Buildings can be designed for a given climate; domed roofs and thermally massive structures in hot arid climates, shuttered and shaded windows to prevent heat gain, open structure bamboo housing in warm and humid areas. In some countries, dwellings are constructed underground and take advantage of the relatively low and stable temperature of the surrounding ground. There are as many options, as there are people.

Solar heating by convection is a natural process that involves trapping air and letting it warm up before releasing it back into a given space. Convection heating is often used as a solar heating source because the two naturally go hand in hand (see Fig. 3.24). Various types of solar air heating or ventilation systems available in the market are shown in Figure 3.25.

Type 1 is a very simple construction: ambient air passes from a glazed or unglazed collector directly into the room to provide ventilation and heating. Applications include vacation cottages (dehumidification) and large industrial buildings requiring adequate ventilation.

Type 2 circulates room air to the collector. The heated air rises to a thermal storage ceiling from which it is conveyed back into the room. This system uses natural convection and is well suited for apartment buildings.

Type 3 is particularly suited for retrofitting poorly insulated buildings. Collector heated air passes through a cavity between an outer insulated wall and an inner facade. This creates a buffer that considerably reduces heat loss via the facade of the building.

Type 4 is the classical solar air heating system and is commonly used. Collector heated air is circulated through channels in the floor or in the wall. Heat is radiated into the room with a time delay of 4 to 6 h. The advantage of this system consists in the large radiating surfaces, which provide for a comfortable climate. Systems with forced ventilation (fans) provide the best efficiency and thermal output. They may be used in buildings with large surfaces, which serve as radiation sources.

Type 5 is an advanced version of type 4; room air is circulated through separate channels of the storage. Thus, heat can be stored for a long period of time and released when it is needed. However, this type is rarely used as investment costs are rather high.

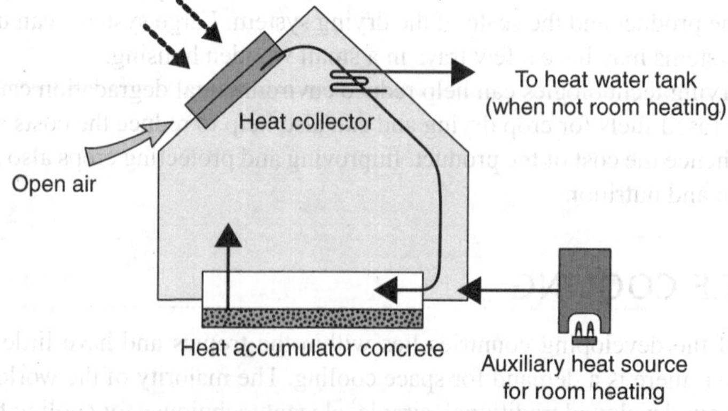

Figure 3.24 A typical air heating system

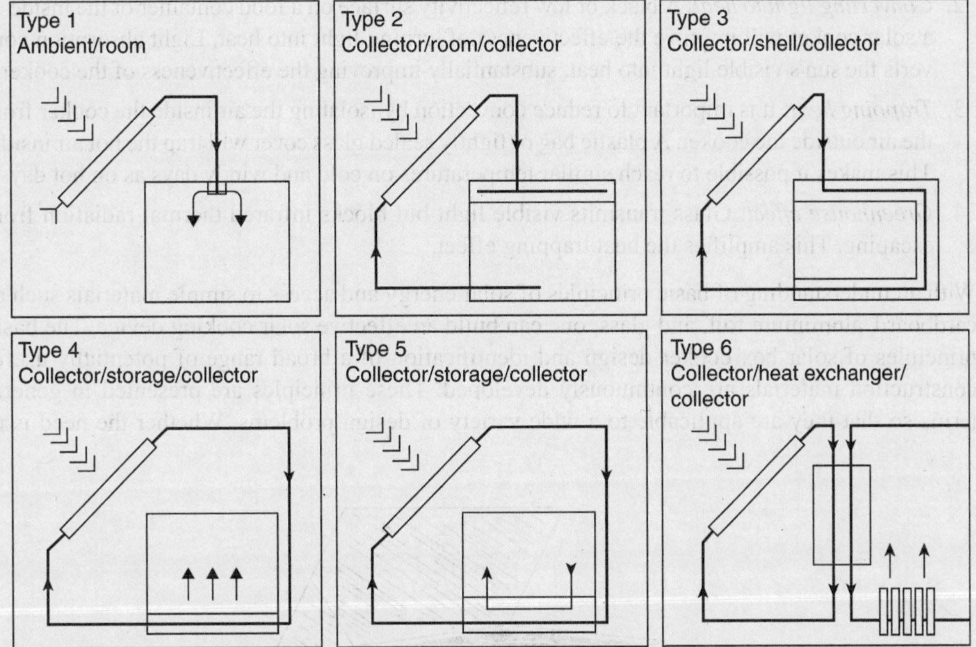

Figure 3.25 Various types of solar air heating systems

Type 6 combines a solar air collector and, via a heat exchanger, a conventional heating system. Thus, common radiators and floor or wall heating components may be used. This system can also provide domestic hot water and is particularly suited for retrofitting and for buildings in which heat has to be transported over long distances.

3.16 SOLAR COOKERS

Solar cooking is a technology that has been given a lot of attention in recent years in developing countries. The basic design is that of a box with a glass cover. The box is lined with insulation and a reflective surface is applied to concentrate the heat onto the pots. The pots can be painted black to help with the heat absorption. The solar radiation raises the temperature sufficiently to boil the contents in the pots. Cooking time is often considerably slower than conventional cooking stoves, but there is no fuel cost.

People use solar cookers primarily to cook food and pasteurize water, although additional uses are continually being developed. Numerous factors including access to materials, availability of traditional cooking fuels, climate, food preferences, cultural factors, and technical capabilities affect people's approach to solar cooking.

Many types of cookers exist. Simple solar cookers use the following basic principles:

1. *Concentrating sunlight*: A reflective mirror of polished glass, metal, or metalized film is used to concentrate light and heat from the sun into a small cooking area, making the energy more concentrated and increasing its heating power.

2. *Converting light to heat*: A black or low reflectivity surface on a food container or the inside of a solar cooker will improve the effectiveness of turning light into heat. Light absorption converts the sun's visible light into heat, substantially improving the effectiveness of the cooker.
3. *Trapping heat*: It is important to reduce convection by isolating the air inside the cooker from the air outside the cooker. A plastic bag or tightly sealed glass cover will trap the hot air inside. This makes it possible to reach similar temperatures on cold and windy days as on hot days.
4. *Greenhouse effect*: Glass transmits visible light but blocks infrared thermal radiation from escaping. This amplifies the heat trapping effect.

With an understanding of basic principles of solar energy and access to simple materials such as cardboard, aluminium foil, and glass, one can build an effective solar cooking device. The basic principles of solar box cooker design and identification of a broad range of potentially useful construction materials are continuously developed. These principles are presented in general terms so that they are applicable to a wide variety of design problems. Whether the need is to

Figure 3.26 Box-type solar cooker

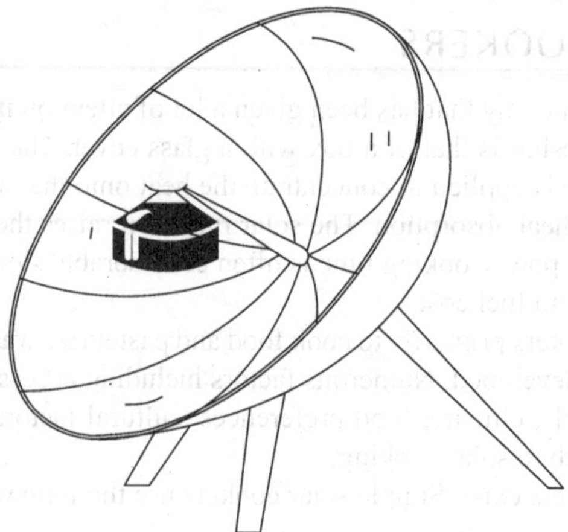

Figure 3.27 Reflector-type solar cooker

cook food, pasteurize water, or dry fish or grain, the basic principles of solar, heat transfer, and materials apply. The application of a wide variety of materials and techniques as people make direct use of the sun's energy is continuously under development.

3.16.1 Types of Solar Cookers

All the solar cookers are subdivided into four configurations:

1. Solar cooking boxes (see Fig. 3.26) are well insulated boxes with a double glass lid and a cover with a reflector on the inside. Solar cooking boxes keep the food warm in the afternoon and evening. They are presently the most successful type of solar cookers in the world.
2. Reflector cookers concentrate the sun's radiation by a more or less parabolic reflector into a focal region, where the cooking vessel is fixed. Success in disseminating solar reflector cookers has only been reported from China, as shown in Figure 3.27.

   Reflective materials are used to concentrate light and heat from the sun into a small cooking area, making the sun's energy more concentrated and, therefore, more powerful, resulting in the fastest cooking times of all cooker designs.

   Parabolic cookers require more precision to focus the sunlight on the cooking vessel and are, therefore, the most complex design to build. If the sunlight is not focused exactly on the cooking vessel, the food will not cook efficiently.

3. Solar steam and convection cookers use vapour or hot air as heat transfer medium. Water is evaporated or air is heated up mostly in flat plate or vacuum collectors and then led in a piping system to the cooking vessel. Collector and cooking place can be separated and thus cooking in the shadow is possible. Most steam and convection cookers have a low efficiency and a high price, and further, they require relatively much effort in manufacturing.

Heat storage solar cookers do overcome the most important disadvantage of solar cooking in general. They allow cooking after sunset and some of them even in the morning. They collect the solar energy by high efficiency flat plate or vacuum tube collectors or concentrators and store the heat in a solid heat storage block or a liquid storage medium in a tank. They always are heavy and bulky; further, they are complex constructions, and generally, very expensive.

3.16.2 Advantages

Solar energy cooking has a variety of advantages, out of which the most important are as follows:

1. Cooking with solar energy saves fuel wood and/or chemical fuels.
2. Cooking with solar energy is clean and healthy and reduces health problems related to kitchen smoke.
3. Solar cooking enables individual families to do without commercial fuels, and thus, money can be saved.
4. Solar cooking saves time and effort that would otherwise be spent in collecting fuel wood.
5. Food cooked in box-type solar cookers cannot burn and does not have to be stirred or watched.
6. Food cooked in box-type solar cooker is cooked gently so that more of the nutrients and flavour of the food are conserved than when cooking on the fire.

3.16.3 Disadvantages

The following are some disadvantages related to the principle of solar cooking.

1. Solar cooking requires good weather with relatively steady sunshine.
2. Solar cooking cannot completely replace the conventional wood, gas, or kerosene fire.
3. Solar cooking is only possible during the daytime and not in the mornings and evenings (except with storage-type solar cookers).
4. Most types of solar cookers require industrially manufactured components. These can easily be destroyed, and it is difficult or impossible to repair or replace them with local material.
5. Some solar cooking boxes do not attain high temperatures. This requires long cooking time.
6. Boiling, roasting, and grilling require high temperatures, and thus, it is only possible in a few types of solar cookers
7. Some reflector-type solar cookers demand understanding, skill, and almost constant attention when handling and cooking with them.
8. The person doing the cooking has to stay out in the sun to avoid the risks of being dazzled or burnt.
9. Generally, families that need solar cookers mostly cannot afford them.

3.17 SOLAR POND

One of the best ways of harnessing solar energy is through solar ponds. It is basically a pool of water that collects and also stores solar energy. The peculiarity of the solar pond is that it has layers of salt solutions of differing concentrations, and thus, different densities to a certain depth. Once this depth is reached, then water with uniform, high salt concentration is obtained. The solar pond is a relatively low technology and low cost approach for harvesting solar energy. To develop a solar pond, pond is filled with three layers of water as shown in Figure 3.28.

1. The top layer is cold and has relatively little salt content.
2. Next is the intermediate insulating layer that has a salt gradient that maintains a density gradient. It is this density gradient that helps in preventing heat exchange with the natural convection of water.
3. The bottom layer is hot up to 100°C and has a high salt content.

It is because of these different salt contents in the different layers of water that the different layers have different densities. With the different densities in the water, the development of convection currents is prevented, which would have transferred heat to the surface of the pond, and then to the air above. Without these convection currents, heat is trapped in the salty bottom layer of the solar pond, which is used for heating of buildings, industrial processes, generation of electricity, and other purposes. In addition to the abovementioned uses, solar ponds can also be used in water desalination and for storage of thermal energy.

In this system, a large salty lake is used as a plate collector. With the right salt concentration in the water, the solar energy can be absorbed at the bottom of the lake. The heat is insulated by different densities of the water, and at the bottom, the heat can reach 90°C, which is high enough to run a vapour cycle engine; at the top of the pond, the temperature can reach 30°C. There are

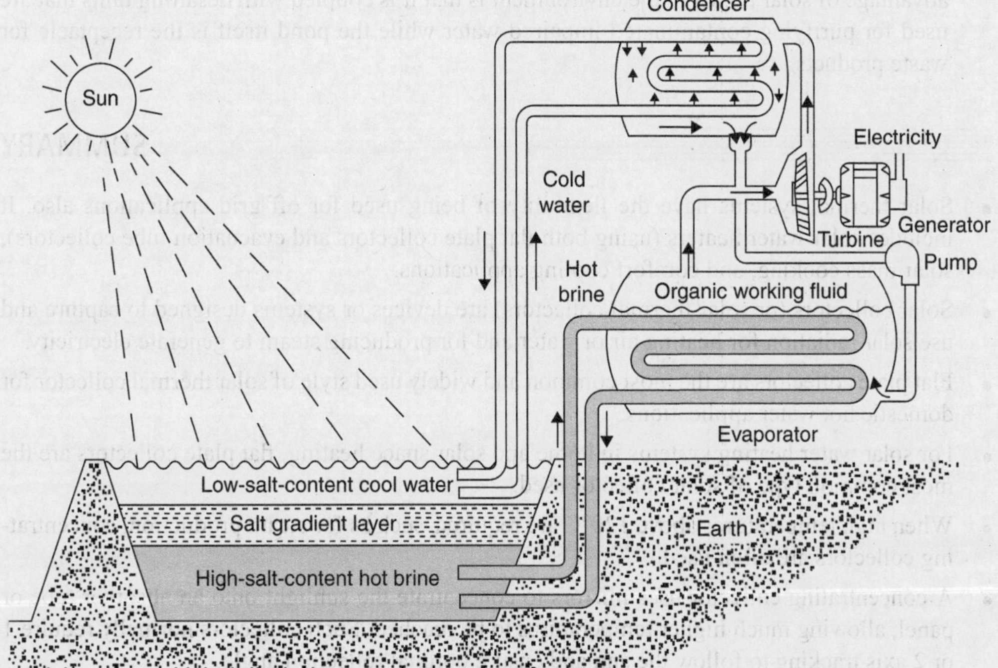

Source: http://www.powerfromthesun.net/Book/chapter06/chapter06.html

Figure 3.28 Solar pond

three different layers of water in a solar pond: the top layer has less concentration of salt, the intermediate layer acts as a thermal insulator, and finally, the bottom layer has a high concentration of salt. These systems have a low solar to electricity conversion efficiency, less than 15% (having an ambient temperature of 20°C and storage heat of 80°C). One advantage of this system is that because the heat is stored, it can run day and night if required. Further, due to its simplicity, it can be constructed in rural areas in developing countries.

3.17.1 Advantages of Solar Pond

There are many advantages of using a solar pond to meet the energy requirements of a place.

1. The greatest advantage lies in the fact that it has a low cost per unit area of collection and also an inherent capacity for storage purposes. In addition to this, it is possible to easily construct solar ponds over large areas with which it is possible for the diffusion of solar resources to get concentrated on a grand scale.
2. Not only is a solar pond a great source of generation of electricity, it produces many environmental advantages when compared to the use of other fossil fuels for producing electricity. With a solar pond, the greatest advantage to the environment is that the heat energy is provided without the burning of any fuel, which reduces pollution.
3. Another advantage is that because there is no use of conventional energy resources for creating electricity in solar ponds, conventional energy resources are conserved. Further, the third

advantage of solar ponds to the environment is that it is coupled with desalting units that are used for purifying contaminated impaired water while the pond itself is the receptacle for waste products.

SUMMARY

- Solar thermal systems have the flexibility of being used for off-grid applications also. It includes solar water heaters (using both flat plate collectors and evacuation tube collectors), solar mass cooking, and comfort cooling applications.
- Solar collectors (or solar thermal collectors) are devices or systems designed to capture and use solar radiation for heating air or water and for producing steam to generate electricity.
- Flat plate collectors are the most common and widely used style of solar thermal collector for domestic hot water applications.
- For solar water heating systems in home and solar space heating, flat plate collectors are the most common type of solar collector used.
- When high temperatures above 120°C are required, such as for steam production, concentrating collectors are often used.
- A concentrating collector uses mirrors to concentrate the sunlight onto an absorber tube or panel, allowing much higher temperatures to be reached. Such collectors normally require 1 or 2 axis tracking to follow the sun and ensure optimal reflection angle.
- Collector Concentration Ratio (CCR) is the ratio of Aa//Ad. A collector is defined as concentrating if its absorber (fin) area Ad is smaller than the aperture area Aa. It is also used as a measure for classifying collectors. Since this ratio approximately determines the operating temperature, such method of classification is equivalent to classifying collectors by its operating temperature range.
- Heliostat is a mirror-based system used to continuously reflect sunlight onto a central receiver.
- Solar cooking technology has been given a lot of attention in recent years in developing countries. Solar cookers are primarily used to cook food and pasteurize water, although additional uses are continually being developed.
- Solar pond is basically a pool of water that collects and also stores solar energy. The peculiarity of the solar pond is that it has layers of salt solutions of differing concentration and thus, different densities to a certain depth. Once this depth is reached, then water with uniform, high salt concentration is obtained. Solar pond is a relatively low cost technology for harvesting solar energy.

REVIEW QUESTIONS

1. What are solar collectors? Give their classification and compare them based on construction and area of applications.
2. With neat sketches, discuss important parts of any flat plate solar collector. Further, discuss material aspects of individual parts.

3. Give schematic representation of an air-type flat plate solar collector system integrated with an auxiliary system for space heating and explain its working.
4. What are the main components of a flat plate solar collector? Explain.
5. Compare flat plate, parabolic dish of collectors to be used for a solar thermal plant with respect to (i) temperature, (ii) concentration ratio (iii) suitability, and (iv) cost.
6. State clearly the difference between the distributed collector system and central receiver system in solar thermal applications.
7. How are solar air collectors classified?
8. What are the main applications of a dryer?
9. Why orientation is needed in concentrating-type collectors? Describe the different methods of sun tracking.
10. What are the advantage and disadvantages of concentrating collectors over a flat plate collector?
11. Explain the considerations in installing a flat plate collector system with reference to the geographical location and the angle of tilt.
12. State the function of the following
    1. Solar thermal collector
    2. Central receiver
    3. Draw a diagram of a distributed central receiver solar thermal power plant and explain the arrangements of solar heliostats.

Chapter 4

Solar Cells

A solar cell or photovoltaic cell is an electrical device that converts the energy of light directly into electricity by photovoltaic effect. They are made of special materials that are semiconductors. These semiconductors produce electricity when sun light falls onto its surface. It is 5 to 11 times more expensive to produce electricity from the sun than it is from fossil and nuclear fuels. The first problem is with the cost of the solar cell technology that uses expensive semiconductor materials to convert sunlight directly into DC electricity. Moreover, the solar cell efficiency is not very encouraging yet. Practical efficiency of solar cells lies somewhere in between 10% and 25%. Reduction of solar cell material costs and increasing the practical efficiency of solar cells are lively topics of current research that will necessitate huge investment and time consuming R&D activities.

The other problem with solar cell power that has stifled its use is the fact that energy production only takes place when the sun is shining. Large storage systems, therefore, are required to be developed to provide a constant and reliable source of electricity when the sun is not shining at night or during cloudy weather.

The sun provides approximately 10^5 TW of energy to the earth that is about 10^4 times more than the present rate of the world's energy consumption. Photovoltaic cells are being increasingly used to tap this huge resource and will play key role in future sustainable energy systems. So far, solid-state junction devices, usually made of silicon, crystalline, or amorphous, and profiting from the experience and material availability resulting from the semiconductor industry, have dominated photovoltaic solar energy converters. These systems have, by now, attained a mature state

KEY CONCEPTS

- Definition and need of solar cell
- Basic elements and important requirements of silicon solar cell
- Solar cell materials and their comparison
- Practical solar cells and their comparative features
- The main components of photovoltaic system and their functions
- Theory of solar cells
- Equivalent circuit and current–voltage characteristic of a solar cell
- Performance parameters (efficiency and fill factor) of solar cell
- Photovoltaic panels (series and parallel arrays)
- Application of photovoltaic systems

serving a rapidly growing market, which is expected to rise to 3,000 GW by 2030. However, the cost of photovoltaic electricity production is still too high to be competitive with nuclear or fossil energy.

Photovoltaic conversion is the process of direct conversion of light into electricity at the atomic level. Some materials exhibit a property known as the photoelectric effect that causes them to absorb photons of light and release electrons. When free electrons are captured, electricity is generated. The photoelectric effect was first noted by a French physicist, Edmund Becquerel, in 1839, who found that certain materials would produce small amounts of electric current when exposed to light. In 1905, Albert Einstein described the nature of light and the photoelectric effect on which photovoltaic technology is based, for which he later won a Nobel Prize in physics.

In simple terms, It is a method of generating electrical power by converting solar radiation into direct current electricity using semiconductors that exhibit the photovoltaic effect. Solar cells are devices that convert solar energy directly into electricity via the photovoltaic effect. Photovoltaic converter is the other name for solar cells. They are responsible for producing energy out of sunlight it receives. Photovoltaic or solar cells are made of special materials that are semiconductors. These semiconductors produce electricity when sunlight falls onto its surface. Solar electric cells are simple cells to use, they do not require anything but sunlight to operate; they are long lasting, reliable, and easy to maintain. Normally, solar panels' lifetime is 25 years or more.

4.1 NEED FOR SOLAR CELLS

The development of solar cell use has been advocated because of the following needs

1. For low maintenance, long lasting sources of electricity suitable for places remote from both the main electricity grid and the people, that is, satellites, remote site water pumping, outback telecommunications stations, and lighthouses;
2. For cost effective power supplies for people living in the remote areas that have no access to the main electricity grid; for example, aboriginal settlements, outback sheep and cattle stations, and some home sites in grid-connected areas.
3. For non-polluting and silent sources of electricity, i.e., tourist sites, caravans, and campers
4. For a convenient and flexible source of small amounts of power, that is, calculators, watches, light meters, and cameras;
5. For renewable and sustainable power, as a means of reducing global warming.

Together these needs have produced a growing market for photovoltaic (solar cells) that has stimulated innovation. As the market has grown, the cost of cells and systems has declined, and new applications have been discovered.

4.1.1 Components of a Solar Cell System

These include solar cell panels, one or more batteries, a charge regulator or controller for a stand-alone system, an inverter for a utility grid-connected system, requirement of alternating current (AC) rather than direct current (DC), wiring, and mounting hardware or a framework.

4.1.2 Key Elements of Silicon Solar Cell

The basic elements of a solar cell is described in Figure 4.1.

The following are the basic elements:

1. *Substrate*: It is an unpolished p-type wafer referred to as p-region base material. The important parameters to be kept in mind while choosing a wafer for solar cells are its orientation, resistivity, thickness, and doping. Typical thickness of wafers used for solar cells is 180–300 μm. The typical resistivity values are in 1–2 Ωcm. The doping should be close to $5 \times 10^{15}/cm^3$ to $1 \times 10^{16}/cm^3$. The wafer can be single crystalline or multi-crystalline.
2. *Emitter*: The emitter formation involves the doping of silicon with pentavalent impurities such as phosphorus, arsenic, and antimony. However, for solar cell applications, phosphorus is the widely used impurity. The doping is done by the process of diffusion. The basic idea is to introduce the wafer in an environment rich in phosphorus at high temperatures. The phosphorus diffuses in, due to the concentration gradient, and it can be controlled by varying the time and temperature of the process. Commonly used diffusion technique makes use of POCl3 as the phosphorus source. The process is done at temperatures of 850°C to 1,000°C. The typical doping concentration will be of the order of $1 \times 10^{19}/cm^3$. The junction depths are in the range of 0.2–1 μm. This is also commonly known as *n*-region diffused layers
3. *Electrical contacts*: These are essential to a photovoltaic cell since they bridge the connection between the semiconductor material and the external electrical load. It includes

   (a) *Back contact*: It is a metallic conductor completely covering back. The back contact of a cell is located on the side away from the incoming sunlight and is relatively simple. It usually consists of a layer of aluminium or molybdenum metal.

   (b) *Front contact*: Current collection grid of metallic finger type is arranged in such a way that photon energy falls on n-region diffused layers. The front contact is located on the side facing the light source and is more complicated. When light falls on the solar cell,

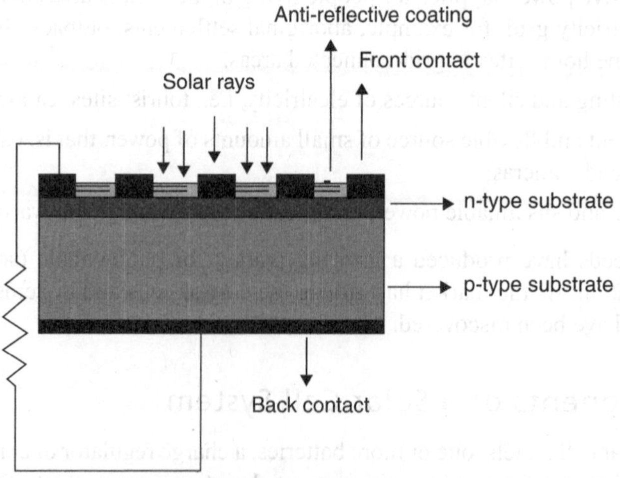

Figure 4.1 Basic elements of a photovoltaic cell

a current of electrons flow over the surface. If contacts are attached at the edges of the cell, it will not work well due to the great electrical resistance of the top semiconductor layer; only a small number of electrons will make it to the contact. To collect the maximum current, the contacts must be placed across the entire surface of a solar cell. This is done with a grid of metal stripes or fingers. However, placing a large grid, which is opaque, on the top of the cell shades active parts of the cell from the light source; as a result, this significantly reduces the conversion efficiency. To improve the conversion efficiency, the shading effect must be minimized.

(c) *Anti-reflective coatings*: Anti-reflective coatings are applied to reduce surface reflection and maximize cell efficiency in solar glass and silicon solar cell manufacturing. It helps to reduce the reflection of desirable wavelengths from the cell, allowing more light to reach the semiconductor film layer, increasing solar cell efficiency. When a thin-film nano-coating of anti-reflection coating of silicon dioxide (SiO_2) and titanium dioxide (TiO_2) is applied, there seems to be an increase in cell efficiencies by 3–4%.

4.1.2.1 Important Requirements of Solar Cell (Photovoltaic Cell)

Following are the important factors that need careful attention in the design of solar cells:

1. *Photon energy of solar light*: Photons with a certain level of energy can free electrons in the semiconductor material from their atomic bonds to produce an electric current. The energy of a photon must be larger than the bandgap energy to free an electron. However, photons with more energy than the bandgap energy will expand that extra amount as heat when freeing electrons.
2. *Bandgap energy of semiconductors*: When light falls on crystalline silicon, electrons within the crystal lattice can be freed. This level of energy is known as the bandgap energy. It is defined as the amount of energy required to dislodge an electron from its covalent bond and allow it to become part of an electrical circuit.

   Thus, it is important for a solar cell to be tuned through slight modifications to the silicon's molecular structure to optimize the photon energy. An efficient solar cell converts as much sunlight as possible into electricity.
3. *Photon energy absorption*: Photons are absorbed in the p-layer of solar cell. It is important that this p-layer absorbs maximum possible photons and also releases many electrons as possible up to maximum extent.
4. *Electron conduction*: The material design to free electrons as close to the junction as possible allows the electric field to send the free electrons through the conduction layer and into the electrical circuit, which is the another challenge of fabrication process. By optimizing these characteristics, the solar cell conversion efficiency is improved.
5. *Electric contact resistance*: Another important aspect in the solar cell design is to minimize the electrical resistance losses when applying grid contacts to the solar cell material. These losses relate to:

   (a) The material properties of the solar cell that oppose the flow of an electric current and results in heating.

(b) The shading effects must be balanced against electrical resistance losses in designing grid contacts.

The usual approach is to design grids with many thin, conductive fingers spreading to every part of the cell surface. The fingers of the grid must be thick enough to conduct well (with low resistance), but thin enough not to block the incoming light. This type of grid maintains low resistance losses while shading only ~3–5% of the cell surface.

6. *Antireflective Coating*: Silicon is a highly reflective material behaving as a mirror, thereby reflecting more than one-third of the solar light falling on it. Minimization of the amount of solar light reflected is also an important factor to improve the conversion efficiency of a solar cell.

4.1.3 Creating P-type and N-type Semiconductors

In a crystalline silicon solar cell, p-type silicon must contact n-type silicon to create the built-in electrical field. The process of doping, which is used to create these materials, introduces an atom of another element into silicon crystal to alter its electrical properties. The dopant, which is the introduced element, has either three or five valence electrons, which is one less or one more than silicon that have four valence electrons.

4.1.3.1 N-type Semiconductors

Phosphorus atoms, which have five valence electrons, are used to dope n-type silicon because phosphorus provides its fifth free electron. A phosphorus atom occupies the same place in the crystal lattice that was formerly occupied by the silicon atom it replaced. Four of its valence electrons take over the bonding responsibilities of the four silicon valence electrons that they had replaced. However, the fifth valence electron remains free, having no bonding responsibilities. When phosphorus atoms are substituted for silicon in a crystal, many free electrons become available.

The most common method of doping is to coat a layer of silicon material with phosphorus, and then, heat the surface. This allows the phosphorus atoms to diffuse into the silicon. The temperature is then reduced so the rate of diffusion drops to zero. Other methods of introducing phosphorus into silicon include gaseous diffusion, a liquid dopant spray-on process, and a technique in which phosphorus ions are precisely driven into the surface of the silicon.

However, the n-type silicon cannot form an electric field by itself. It also needs p-type silicon. Boron, which has only three valence electrons, is used for doping p-type silicon. Boron is introduced during silicon processing, when the silicon is purified for use in photovoltaic devices. When a boron atom takes a position in the crystal lattice formerly occupied by a silicon atom, a bond will be missing an electron. In other words, there is an extra positively charged hole.

4.1.3.2 P-type Semiconductors

In a photovoltaic cell, photons are absorbed in the p-layer. Therefore, it is important that this layer be 'tuned' to the properties of incoming photons so it can absorb as many as possible and, thus, free up as many electrons as possible. The design of the p-layer must also keep the electrons from meeting up with holes and recombining with them before they can escape from the PV cell. To accomplish these goals, p-layers are designed to free electrons as close to the junction as possible, so that the electric field can help to send the free electrons through the conduction layer (the *n*-layer) and in the electrical circuit.

By optimizing these characteristics, the PV cell's conversion efficiency (how much light energy is converted into electrical energy) is improved.

4.1.3.3 Fabrication of Silicon Solar Cell

It includes

1. Pure silicon is placed in an induction furnace where it melts.
2. Boron is then added to the melt from which p-type crystals are withdrawn.
3. The p-type base material is then placed in a diffusion furnace containing a gaseous n-type dopant like phosphorous.
4. n-type dopant is allowed to diffuse on to the surface, thus forming p–n junction.
5. Metal conductor grids are added as back and front contact for current collection.

4.2 SOLAR CELL MATERIALS

Many combinations of materials and methods of fabrication of photovoltaic cells are now either in practical use or in various developmental stage. However, silicon is the most widely used basic material because of its suitability and its availability in abundance. More than 80% of solar cells currently produced are crystalline silicon solar cells. Nearly all of the other 20% are developed as amorphous silicon solar cells. Silicon wafers have long been the primary base for assembly.

The absorption coefficient of a material indicates how far light with a specific wavelength (or energy) can penetrate the material before being absorbed. A small absorption coefficient means that light is not readily absorbed by the material. Again, the absorption coefficient of a solar cell depends on two factors: the material making up the cell, and the wavelength or energy of the light being absorbed.

The bandgap of a semiconductor material is the minimum energy needed to move an electron from its bound state within an atom to a free state. This free state is where the electron can be involved in conduction. The lower energy level of a semiconductor is called the 'valence band.' The higher energy level where an electron is free to roam is called the 'conduction band.' The bandgap (often symbolized by Eg) is the energy difference between the conduction band and the valence band.

Solar cell material has an abrupt edge in its absorption coefficient, because light with energy lesser than the material's bandgap cannot free an electron, it is not absorbed. A solar cell consists of semiconductor materials.

4.2.1 Silicon

This remains the most popular material for solar cells, including these types:

1. Monocrystalline or single crystal silicon
2. Multi-crystalline silicon

3. Polycrystalline silicon
4. Amorphous silicon

Polycrystalline wafers are made by a casting process in which molten silicon is poured into a mould and allowed to set. Then, it is sliced into wafers. As polycrystalline wafers are made by casting, they are significantly cheaper to produce, but not as efficient as monocrystalline cells. The lower efficiency is due to imperfections in the crystal structure resulting from the casting process.

Amorphous silicon, one of the thin-film technologies, is made by depositing silicon onto a glass substrate from a reactive gas like silane (SiH_4). This type of solar cell can be applied as a film to low cost substrates such as glass or plastic. Other thin-film technologies include thin multi-crystalline silicon, copper indium diselenide or cadmium sulphide cells, cadmium telluride or cadmium sulphide cells and gallium arsenide cells. There are many advantages of thin-film cells including easier deposition and assembly, the ability to be deposited on inexpensive substrates or building materials, the ease of mass production, and the high suitability to large applications.

Other types of PV materials that show commercial potential include copper indium diselenide ($CuInSe_2$), cadmium telluride (CdTe), and amorphous silicon as the basic material.

4.2.2 Thin Film

Thin-film solar cells use layers of semiconductor materials only a few micrometres thick. Thin-film technology has made it possible for solar cells to now double as these materials:

1. Rooftop or solar shingles
2. Roof tiles
3. Building facades
4. Glazing for skylights or atria

Thin-film photovoltaic cells made of CuInSe or CdTe that are being increasingly employed along with amorphous silicon. The recently discovered cells based on mesoscopic inorganic or organic semiconductors commonly referred to as 'bulk' junctions due to their three-dimensional structure. These junctions are very attractive alternatives that offer the prospect of very low cost fabrication. The prototype of this family of devices is the dye-sensitized solar cell (DSC), which accomplishes the optical absorption and the charge separation processes by the association of a sensitizer as light-absorbing material with a wide bandgap semiconductor of mesoporous or nano-crystalline morphology. Further, research is booming in the area of third generation photovoltaic cells where multi-junction devices and a recent breakthrough concerning multiple carrier generation in quantum-dot absorbers offer promising perspectives.

4.3 PRACTICAL SOLAR CELLS

Solar cells are now manufactured from a number of different semiconductors that are summarized in the following points. In addition, there is considerable activity to commercially manufacture the dye-sensitized solar cells.

Table 4.1 Efficiency of Different Types of Solar Cells

Cell Material	Theoretical Efficiency (%)	Practical Efficiency (%)	Technology
Mono or multi-crystalline silicon	20–26	12–18	Ingot or wafer
Amorphous silicon	12–14	5 – 10	Thin film
Copper indium diselenide	16–18	8–10	Thin film
Cadmium telluride	15–16	5–8	Thin film
Gallium arsenide	26–32	18–25	Ingot or wafer
Ribbon-grown silicon	10–16	10–12	Ingot or wafer
Ovshinsky silicon	8–10	≤10	Thin film

1. *Crystalline silicon cells*: They dominate the photovoltaic market. To reduce the cost, these cells are now often made from multi-crystalline material, rather than from the more expensive single crystals. Crystalline silicon cell technology is well established. The modules have long lifetime (20 years or more) and their best production efficiency is approaching 18%.
2. *Amorphous silicon solar cells*: They are cheaper (but also less efficient) type of silicon cells made in the form of amorphous thin films that are used to power a variety of consumer products; however, larger amorphous silicon solar modules are also becoming available.
3. *Cadmium telluride and copper indium diselenide*: Thin-film modules are now beginning to appear on the market and hold the promise of combining low cost with acceptable conversion efficiencies.
4. *High-efficiency solar cells*: From gallium arsenide, indium phosphide, or their derivatives are used in specialized applications, for example, to power satellites or in systems that operate under high-intensity concentrated sunlight.

The principal characteristics of different types of cell in or near commercial production are summarized in Table 4.1.

4.4 FUNCTIONS OF A SOLAR CELL

The three important functions of a solar cell are as follows:

1. *Carrier generation*: The generation of carriers occurs in the bulk of the material.
2. *Carrier separation*: The separation of carriers would require a p–n junction.
3. *Carrier collection*: For effective carrier collection, a proper contacting method is required.

The starting material for a silicon solar cell is silicon where the carrier generation takes place. The critical element of a silicon solar cell is a p–n junction. The electron–hole pairs generated across the device, and these carriers are separated out by the junction leading to the formation of a potential, and hence a current. To create an electric field within a crystalline silicon photovoltaic (PV) cell, two silicon semiconductor layers are sandwiched together. p-type (or positive) semiconductors have an abundance of positively charged holes, and n-type (or negative) semiconductors have an abundance of negatively charged electrons. When n- and p-type silicon layers contact, excess electrons move from the n-type side to the p-type side. The result is a build-up of positive charge along the n-type side of the interface and a build-up of negative charge along the p-type side.

Because of the flow of electrons and holes, the two semiconductors behave like a battery, creating an electric field at the surface where they meet and this surface is called the p–n junction. The electrical field causes the electrons to move from the semiconductor and move towards the negative surface, making them available for the electrical circuit. At the same time, the holes move in the opposite direction, that is, toward the positive surface, where they await incoming electrons.

4.4.1 Main Components of Photovoltaic System

1. *Photovoltaic cell*: Thin squares, discs, or films of semiconductor material that generate voltage and current when exposed to sunlight.
2. *Module*: Photovoltaic cells wired together and laminated between a clear strait glazing and encapsulating substrate.
3. *Array*: One or more modules with mounting hardware and wired together at specific voltage.
4. *Charge controller*: Power-conditioning equipment to regulate battery voltage.
5. *Battery storage*: A medium that stores direct current (DC) electrical energy.
6. *Inverter*: An electrical device that changes direct current to alternating current (AC) to operate loads that require alternating current.
7. *DC loads*: Appliances, motors, and equipment powered by DC.
8. *AC loads*: Appliances, motors, and equipment powered by AC.

4.5 THEORY OF SOLAR CELL (PHOTOVOLTAIC CELL)

The basic mechanism behind a solar cell is based on the photoelectric effect and semiconductor physics. A photon with energy greater than the bandgap energy ($h\nu >$ Egap) incident on a semiconductor can excite electrons from the valence band to the conduction band, allowing for current flow. The maximum current density is then given by the flux of photons with this energy. Excess energy is lost to thermalization ($h\nu -$ Egap). The excitation of the electron to the conduction band results in a hole in the valence band. In the case of organic polymer solar cells, the electron and its corresponding hole exist in a bound state due to Coulomb attraction. This state, known as an exciton, has a lower energy than an unbound electron and hole.

The photovoltaic device is technically very easy to convert light into electricity by purely electronic process. The principle of operation depends on the fact that a semiconductor will absorb radiation of high frequency to excite an electron across the energy bandgap and that electron excited in this way returns to the valence band and combines only after comparatively long time (10–100 μs).

4.5.1 Process of Photovoltaic Potential Development

A junction of p-type and n-type semiconductors is extremely useful in making photovoltaic cells. A p–n junction (called as junction diode) can be formed in a single semiconductor crystal having both p-type and n-type regions, it is nevertheless instructive to treat it that way in the discussion.

4.5.1.1 *Energy Band Diagrams of the p-type and n-type Crystals Before the Contact*

Figure 4.2 illustrates the energy band diagrams of the p-type and n-type crystals before the contact with the Fermi level.

In p-type semiconductor, the Fermi level is close to valance band, and in n-type semiconductor, the Fermi level is close to conduction band.

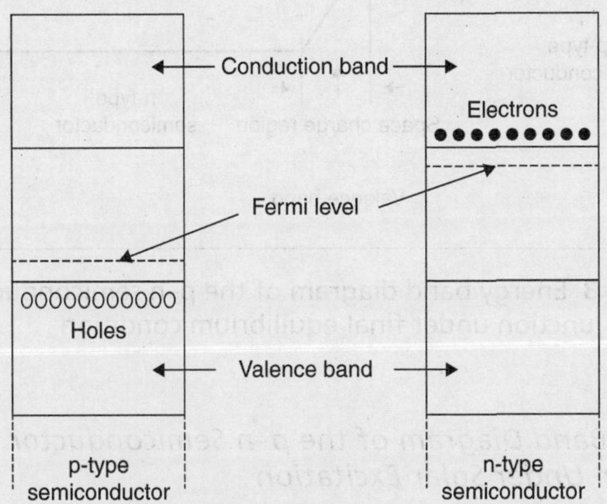

Figure 4.2 Energy band diagram of p-type and n-type semiconductor before the contacts

4.5.1.2 *Energy Band Diagram of the p–n Semiconductor Junction at the Final Equilibrium Condition*

Figure 4.3 shows the energy band diagram of the diode under final equilibrium condition in the absence of any externally applied voltage or energy source.

When two crystals are pushed together, electrons diffuse from the n-side to the p-side, where they recombine with free holes. At the same time, positive holes diffuse from the p-side to the n-side, where they recombine with free electrons.

The diffusion of free electrons leaves a net positive charge in the n-side, while the reverse diffusion of free holes leaves a net negative charge in the p-side. The charge distribution so created gives rise to an electric field and hence a potential difference across junction. A potential difference at the contact having an energy ΔE is soon developed of such magnitude as just to stop any further diffusion of charge carriers.

This contact potential causes the energy level of the p-side to be displaced upwards and those of the n-side downwards so that finally the Fermi levels of the two sides are lined up and the system achieves equilibrium.

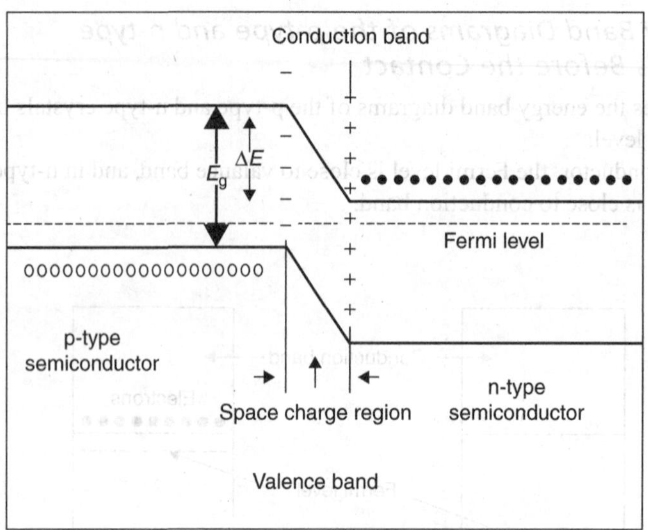

Figure 4.3 Energy band diagram of the p–n semiconductor junction under final equilibrium condition

4.5.1.3 Energy Band Diagram of the p–n Semiconductor Junction Under Solar Excitation

Figure 4.4 describes a p–n junction under solar excitation.

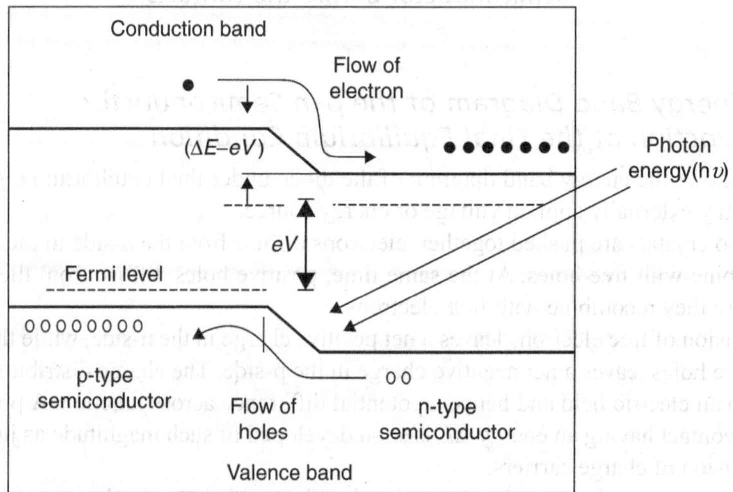

Figure 4.4 Energy band diagram of the p–n semiconductor junction under solar excitation

Since in this case, the external resistance of the circuit is external load, a suitable potential difference (V) is required for the passage of the current. The required potential difference with an energy equivalence of (eV) must be obtained by a difference in Fermi level on the two sides of the junction. The potential hill across the junction is now reduced to ($\Delta E - eV$), where (ΔE) is the short-circuit high potential hill.

It should be noted that since the potential hill has been reduced, the junction is now not so effective in separating the solar-generated electron–hole pairs as in the short-circuit case. This means that some of the charge carriers can now manage to cross the junction in the wrong direction, resulting in a leakage current at the junction (I_0). This causes reduction in the load current (I), which is delivered to the load as compared to the short-circuit case.

When solar electromagnetic radiation with photon energy hν(where h is Planck's constant and ν is frequency) greater than the bandgap energy (E_g) strikes a p–n junction, electrons in valence band can acquire enough energy to jump to the conduction band, thus producing electron–hole pairs. The action of the junction is to separate the pairs; electron pass down the potential hill and into the n-side, whereas hole goes up the hill and into the p-side, resulting in the height of the potential hill. A new equilibrium condition would finally be reached across the junction with the appearance of a potential difference. This is the so called open-circuit voltage (V_{OC}), which is a function of the intensity of the incident radiation.

Now, if the junction is connected by an external wire of negligible resistance, the Fermi levels are equal on both sides. The solar-generated electron–hole pairs are forced to separate and go across the full potential hill (ΔE). The short-circuit current (I_{SC}) that prevails in this situation is also a function of the intensity of incident radiation.

To make use of the solar-generated electron–hole pairs, one has to lead the electrons from the n-side, take them through an external circuit to do useful work, and then return them to the p-side, where they can meet and recombine with the holes coming from the opposite direction.

4.5.2 Junction Current (I_J)

It is the net current flow from p-side to the n-side due to all charge carriers. The minority carriers (electrons in the p-side and hole in the n-side) can cross the junction easily because electron prefers to go downhill and holes prefer to go uphill.

On the other hand, the majority carriers (holes in the p-side and electrons in the n-side) cannot cross the junction unless they possess energies in excess of the barrier ($\Delta E - eV$).

Since the carriers tend to have Maxwellian distribution, the fraction of the majority carriers succeeding in making the trip is

$$\text{Exp}\,[-(\Delta E - eV)/kT]$$

where k is the Boltzmann's constant and T is the absolute temperature.

Let n_1 = density of electrons in the p-side; n_2 = density of holes in the p-side; n_3 = density of electrons in the n-side; n_4 = density of holes in the n-side; I_1 = electron current from the p-side to the n-side; I_2 = hole current from the p-side to the n-side; I_3 = electron current from the n-side to the p-side; I_4 = hole current from the n-side to the p-side.

It follows that the net current flowing from the p-side to the n-side is given by.

$$I_J = -I_1 + I_2 + I_3 - I_4 \tag{4.1}$$

where

$I_1 = K_1 \times n_1$; $I_2 = K_2 \times n_2 \times \text{Exp}\,[-(\Delta E - eV)/kT]$; $I_3 = K_3 \times n_3\ \text{Exp}\,[-(\Delta E - eV)/kT]$; $I_4 = K_4 \times n_4$

where K_1, K_2, K_3, and K_4 are constants.

Substituting the above mentioned expressions in Equation (4.1), gives,

$$I_J = -(K_1n_1 + K_4n_4) + (K_2n_2 + K_3n_3) \times \text{Exp}\,[-(\Delta E - eV)/kT] \quad (4.2)$$

When the system is in an equilibrium condition with no illumination (i.e., when $I_J = 0$ and $V = 0$), then Equation (4.2) can be written as

$$K_1n_1 + K_4n_4 = (K_2n_2 + K_3n_3) \times \text{Exp}\,(-\Delta E/kT) \quad (4.3)$$

The junction current equation must also be true for negative values of V. When the negative V is sufficiently large, the second term of Equation (4.2) is negligible.

Thus, $$I_J = -(K_1n_1 + K_4n_4) = -I_0 \quad (4.4)$$

This means that with a high value of negative V, a p–n junction will stop all majority carriers' flow and allows only minority carriers to cross the junction, producing a reverse current (I_0), which is usually called the reversed saturation or dark current.

Substituting Equations (4.3) and (4.4) in Equation (4.2) results in

$$I_J = I_0\,[\text{Exp}\,(eV/kT) - 1] \quad (4.5)$$

Equation (4.5) is the current–voltage (i–v) relation for a p–n junction diode.

4.5.3 Solar Cell Performance (Equivalent Circuit of a Solar Cell)

In analysing the cell performance, the photovoltaic process of solar cell can be modelled as a macroscopic equivalent circuit as shown in Figure 4.5.

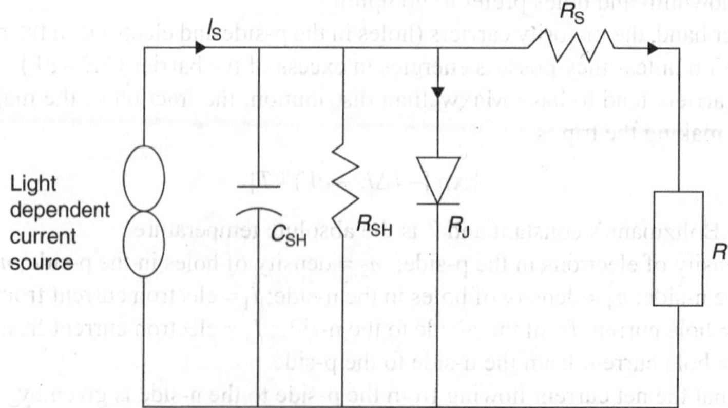

Figure 4.5 Equivalent circuit of photovoltaic cell

The circuit consists of light-dependent current source supplying current I_S to a network of resistances including

1. Junction resistance, R_J
2. Internal shunt resistance, R_{SH}
3. Internal series resistance, R_S
4. Internal shunt capacitance
5. External load resistance, R

The internal shunt resistance (R_{SH}) is usually much larger than the external load resistance so that most of the available current can be delivered to the load. The internal series resistance (R_S) is usually much less than the external load resistance (R) so that less power is dissipated internally within the cell.

A more realistic model of solar cell can be derived from Figure 4.5 without introducing considerable errors in the analysis. Resulting equivalent circuit is shown in Figure 4.6.

Based on this simplified circuit, load current is readily obtained as,

$$I = I_S - I_J \tag{4.6}$$

Junction current from the current–voltage relationship of a p–n junction diode is obtained as

$$I_J = I_0 \,[\text{Exp}\,(eV/kT) - 1] \tag{4.7}$$

where

I_0 = Reverse saturation current; V = Voltage developed or applied across the junction;
e = the electron charge; k = the Boltzmann's constant; and T = the absolute temperature

The light flux causes the flow of load current that may be taken as the difference of the generated short-circuit current (slightly less than photovoltaic current and junction current).

The short-circuit current is always less than the idealized value because of the properties of the materials and procedure used for the fabrication of device. However, for the ideal case, Equation (4.6) is valid.

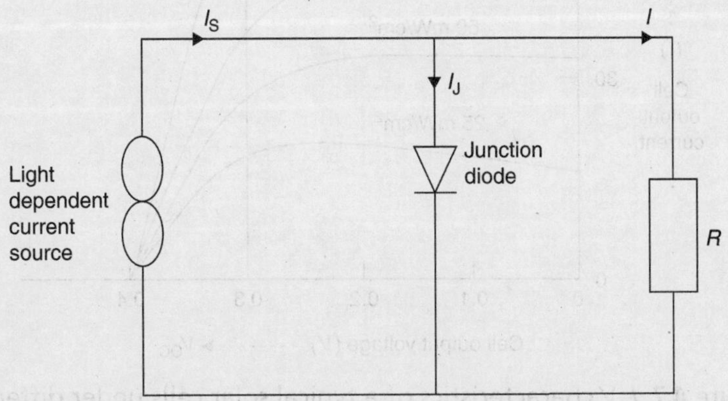

Figure 4.6 Simplified equivalent circuit of photovoltaic cell

Under short-circuit condition $V = 0$, $I = I_S$, which is the short-circuit current generated by the light-dependent current source.

Under open-circuit condition $I = 0$, maximum open-circuit voltage becomes V_{OC}

Substitution of Equation (4.5) in Equation (4.6) gives

$$I = I_S - I_0\,[\text{Exp}\,(eV/kT) - 1] \tag{4.8}$$

Applying open-circuit conditions, $I = 0$, $V = V_{OC}$

$$\text{Exp}\,(eV_{OC}/kT) - 1 = I_S/I_0 \tag{4.9}$$

or

$$V_{OC} = (kT/e)\,[\ln\{(I_S/I_0) + 1\}] \tag{4.10}$$

4.5.4 *I–V* Characteristics of Solar Cells

The voltage output of the cell (V), in general, can be obtained as

$$V = (kT/e)\,\log_e\,[1 + (I_S - I)/I_0] \tag{4.11}$$

Equation (4.11) represents the *I–V* characteristic of solar cell and it is shown in Figure 4.7 under different illumination levels.

On an *I–V* characteristic, the vertical axis refers to the current (I) and the horizontal axis refers to voltage (V). The actual *I–V* curve typically passes through two significant points:

1. The short-circuit current (I_{SC}) is the current produced when the positive and negative terminals of the cell are short-circuited and the voltage between the terminals is zero, which corresponds to a load resistance of zero.

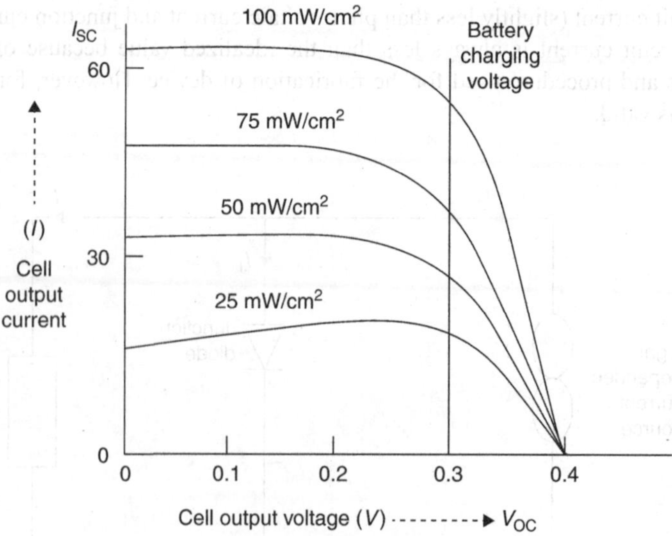

Figure 4.7 *I–V* characteristics of a typical solar cells under different illumination levels

2. The open-circuit voltage (V_{OC}) is the voltage across the positive and negative terminals under open-circuit conditions when the current is zero, which corresponds to a load resistance of infinity.

The cell may be operated over a range of voltages and currents.

4.5.4.1 Output Power

The cell output power $P = I \times V$

The output power depends on the value of load resistance for a given light intensity.

$$P = [I_S - I_0 \{\text{Exp}\,(eV/kT) - 1\}] \times V \tag{4.12}$$

Since $P = I \times V$, the maximum power occurs when the product of IV has it maximum value.

4.5.4.2 Maximum Output Power of the Cell

Differentiating Equation (4.12) with respect to V and setting the derivative equal to zero yields the value of external load voltage (V_{MP}) that gives the maximum output power.

Thus, from $dP/dV = 0$ and substituting $V = V_{MP}$ in the resulting equation yields,

$$[I_S - I_0 \{\text{Exp}\,(eV_{MP}/kT) - 1\}] + V_{MP}\,[-I_0 \{(e/kT)\,\text{Exp}\,(eV_{MP}/kT)\}] = 0 \tag{4.13}$$

This can be rearranged as

$$[\text{Exp}\,(eV_{MP}/kT]\,[1 + eV_{MP}/kT] = (1 + I_S/I_0) \tag{4.14}$$

Equation (4.13) is an implicit equation for V_{MP} that maximizes the power in terms of short-circuit current (I_S), the reverse saturation current (I_0), and the absolute temperature (T). If all these parameters are known, the value of V_{MP} can be evaluated from Equation (4.12) or (4.14) by either trial and error method or numerical method or graphical method.

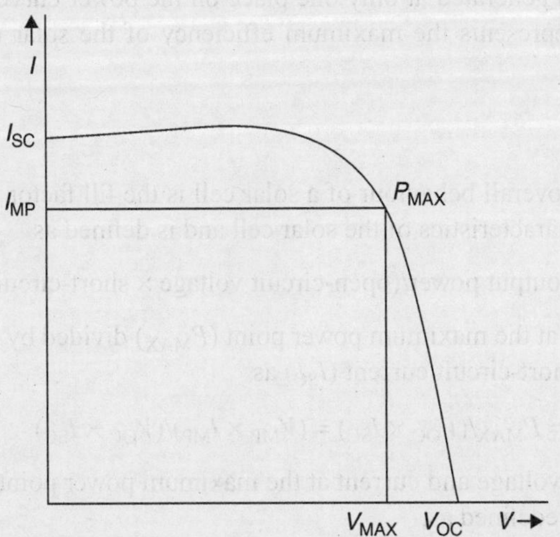

Figure 4.8 The *I–V* characteristic of an ideal solar cell

The load current (I_{MP}) that maximizes the output power can be found by substituting Equation (4.14) in Equation (4.8) as follows:

$$I_{MP} = I_S - I_0\,[(eV_{MP}/kT) - 1] = I_S - I_0\,[\{(1 + I_S/I_0)/(1 + eV_{MP}/kT)\} - 1]$$

$$= \{(eV_{MP}/kT)/(1 + eV_{MP}/kT\}\,(I_S - I_0) \quad (4.15)$$

The maximum power P_{MAX} produced by the conversion device is reached at a point on the characteristic where the product IV is maximum. This is shown graphically in Figure 4.8, where the position of the maximum power point represents the largest area of the rectangle shown.

4.6 EFFICIENCY OF SOLAR CELLS

Energy conversion efficiency (η) is defined as the ratio of power output of cell (in watts) at its maximum power point (P_{MAX}) and the product of input light power (E, in W/m^2) and the surface area of the solar cell (S in m^2) under standard conditions

$$\eta = \text{maximum output power/(irradiance} \times \text{area)} = P_{MAX}/(E \times S) \quad (4.16)$$

The performance of a photovoltaic device defines the prediction of the power that the cell will produce. Current–voltage (I–V) relationships, which measure the electrical characteristics of solar cell devices, are represented by I–V curves (see Figs. 4.7 and 4.8). These I–V curves are obtained by exposing the cell to a constant level of light while maintaining a constant cell temperature, varying the resistance of the load, and measuring the current that is produced.

By varying the load resistance from zero (a short circuit) to infinity (an open circuit), researchers can determine the highest efficiency as the point at which the cell delivers maximum power. The power is the product of voltage and current. Therefore, on the I–V curve, the maximum-power point (P_{MAX}) occurs where the product of current and voltage is a maximum. No power is produced at the short-circuit current with no voltage or at open-circuit voltage with no current. Therefore, the maximum power generated is expected to be somewhere between these two points. Maximum power is generated at only one place on the power curve, at about the 'knee' of the curve. This point represents the maximum efficiency of the solar device at converting sunlight into electricity.

4.6.1 Fill Factor

Another term defining the overall behaviour of a solar cell is the fill factor (FF). It is a measure of squareness of the I–V characteristics of the solar cell and is defined as

$$\text{FF} = \text{Maximum output power/(open-circuit voltage} \times \text{short-circuit current)}$$

It is the available power at the maximum power point (P_{MAX}) divided by the product of open-circuit voltage (V_{OC}) and short-circuit current (I_{SC}) as

$$\text{FF} = P_{MAX}/(V_{OC} \times I_{SC}) = (V_{MP} \times I_{MP})/(V_{OC} \times I_{SC}) \quad (4.17)$$

where V_{MP} and I_{MP} are the voltage and current at the maximum power point.

Equation (4.17) can be redefined as,

$$\text{FF} = (\eta \times S \times E)/(V_{OC} \times I_{SC}) \quad (4.18)$$

The fill factor is directly affected by the values of the cell's series and shunt resistances. Increasing the shunt resistance (R_{SH}) and decreasing the series resistance (R_S) lead to a higher fill factor, thus resulting in greater efficiency, and bringing the cell's output power closer to its theoretical maximum.

Example 4.1

For a typical photovoltaic cell, the following performance parameters are obtained from the *I–V* characteristics.

Open-circuit voltage (V_{OC}) = 0.611
Short-circuit current (I_{SC}) = 2.75
Voltage corresponding to cell maximum power output (V_{MP}) = 0.5
Current corresponding to cell maximum power output (I_{MP}) = 2.59
Calculate fill factor of the cell.

Solution Fill factor $= (V_{MP} \times I_{MP})/(V_{OC} \times I_{SC})$

$= (0.5 \times 2.59)/(0.611 \times 2.75) = 0.7707$

4.6.2 Factors Limiting the Efficiency of the Cell

1. *Wavelength of solar spectrum*: Cell response to only a portion of wavelength available in the solar spectrum. Photon with wavelength > 1.1 μm does not have sufficient energy to create electron–hole pair in silicon cell.
2. *Temperature*: Normal operating temperature of silicon cells can reach 60°C in peak sunlight and these temperature decreases the efficiency of the cells. Therefore, it is important to provide heat sinks of the best quality available. Gallium arsenide cells are capable of operating at high temperature where focused energy can be used.
3. *Mounting of the cells*: It should be to a heat sink (usually an aluminium plate) either heat conductive but electrically insulated. This will reduce operating temperatures and make the cell more efficient. In case free water source is available, heat sinks can be water cooled.
4. *Arrangement and maintenance of solar cell*: The negative side of the cells usually faces the sun and has antireflection coatings. These coatings should be protected from dust, bird dropping, by a clear plastic or glass cover. Accumulated dust on the cover will reduce the output power by about 10%.
5. *Position of the cell*: The cell or panel should be positioned either facing south in the north of equator or facing north in the south of equator for maximum power output and fixed panel applications. The angle off the ground should be equal to the latitude of the place for year around average or can be changed monthly to face the sun at noon for more efficiency.

4.7 PHOTOVOLTAIC PANELS (SERIES AND PARALLEL ARRAYS)

As single solar cell has a working voltage and current of about 0.5 V and 50 mA, respectively, they are usually connected together in series (positive to negative) to provide larger voltages.

Parallel connection of several strings of cells will give rise to higher current output when compared with single series string of cells.

Photovoltaic panels (as shown in Fig. 4.9) are made in a wide range of sizes for different purposes. They generally fall into one of three basic categories:

Low voltage or low power panels are made by connecting between 3 and 12 small segments of amorphous silicon photovoltaic with a total area of a few square centimetres for voltages between 1.5 and 6 V and outputs of a few milliwatts. Although each of these panels is very small, the total production is large. They are used mainly in watches, clocks and calculators, cameras, and devices for sensing light and dark, such as night lights.

Small panels of 1–10 W and 3–12 V, with areas from 100 cm^2 to 1,000 cm^2 are made by either cutting 100 cm^2 single or polycrystalline cells into pieces and joining them in series, or by using amorphous silicon panels. The main uses are for radios, toys, small pumps, electric fences, and trickle charging of batteries.

Large panels, ranging from 10 to 60 W, and generally either 6 or 12 V, with areas of 1,000 cm^2 to 5,000 cm^2 are usually made by connecting from 10 to 36 full-sized cells in series. They are used either separately for small pumps and caravan power (lights and refrigeration) or in arrays to provide power for houses, communications, pumping, and remote area power supplies (RAPS).

If the load resistance is very low, the cell acts as if it is shortened at the output of light falling on it. If the load resistance is very high, the cell acts as if it is open-circuited and the voltage rises very rapidly to maximum voltage. The current at a voltage is limited by the amount of sunlight and load resistance. This characteristic is ideal for charging battery.

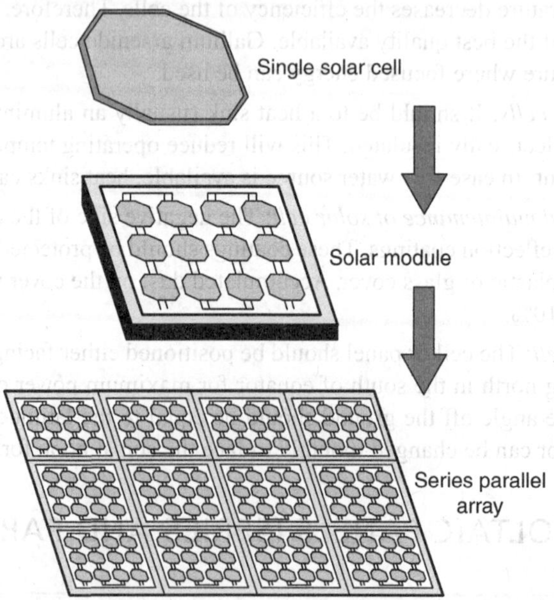

Figure 4.9 Assembly of series–parallel array

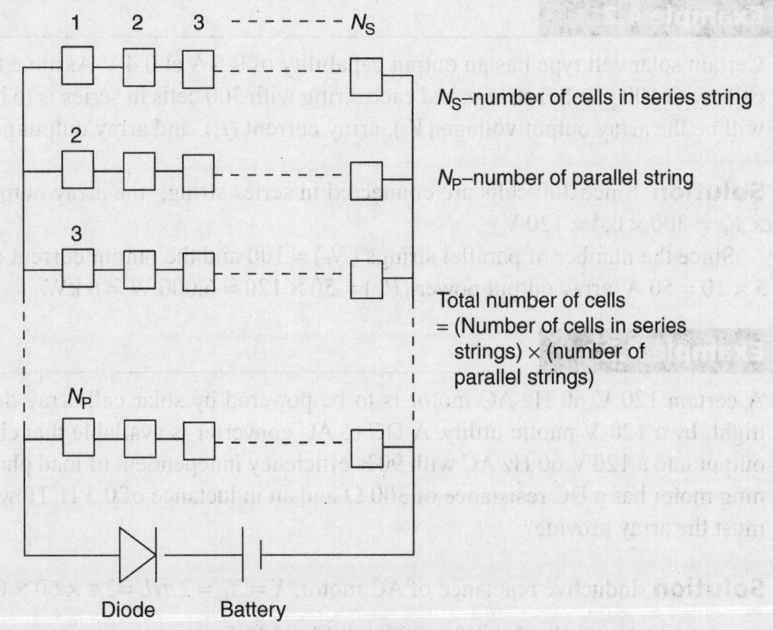

Figure 4.10 Series–parallel array with diode and battery

For charging, a 12 V battery by a 2 cm × 2 cm (0.3 V battery charging voltage), silicon cells required = 12/0.3 = 40 cells in series string.

A number of optimal solar array designs are available. However, the arrangement of series–parallel array has been most preferable, as it results in optimal performance characteristic under many conditions including shading, cell failure, non-uniform illumination, and unequal I–V characteristics.

A diode is placed in series with the positive terminal of battery as shown in Figure 4.10. This will prevent reverse current flow (a small battery drain) when the cells are not receiving sufficient light to chary battery.

4.7.1 Number of Solar Cell Required in Series

Solar cells must be electrically connected in series to provide the bus voltage (V_B) to the space craft load or batteries) and any voltage drops in the blocking diodes (V_D) and in the wiring (V_w).

The required number of cells (N_S) in series is calculated as

$$N_S = (V_B + V_D + V_w)/V_{MP} \tag{4.19}$$

V_{MP} = solar cell voltage at maximum power (or battery charging voltage) under operating temperature and intensity. For silicon diode, $V_D = 0.7$ V.

4.7.2 Number of Solar Cell in Parallel Strings

Let N_P be the number of parallel strings; I_L is the load current; and I_{MP} is the current corresponding to maximum power point on I–V plot. Therefore,

N_P = (the load current)/(current corresponding to maximum power point on I–V plot.

$$= I_L/I_{MP} \tag{4.20}$$

Example 4.2

Certain solar cell type has an output capability of 0.5 A at 0.4 V. Assume that an array of such cells with 100 parallel strings and each string with 300 cells in series is to be building up. What will be the array output voltage (V_a), array current (I_a), and array output power (P_a)?

Solution Since 300 cells are connected in series string, the array output voltage (V_a) = $N_S \times V_C = 300 \times 0.4 = 120$ V.

Since the number of parallel strings (N_P) = 100 and the output current capability of array = $5 \times 10 = 50$ A, array output power (P_a) = $50 \times 120 = 6{,}000$ W = 6 kW.

Example 4.3

A certain 120 V, 60 Hz AC motor is to be powered by solar cell array during the day and at night, by a 120-V public utility. A DC to AC converter is available that changes the array DC output into a 120 V, 60 Hz AC with 90% efficiency independent of load phase angle, while running motor has a DC resistance of 300 Ω and an inductance of 0.3 H. How much power output must the array provide?

Solution Inductive reactance of AC motor, $X = X_L = 2\pi fL = 2\pi \times 60 \times 0.3 = 113\ \Omega$

Motor impedance, $Z = R + jX_L = 300 + j113 = 320\ \Omega$
Motor current, $I = V/Z = 120/320 = 0.375$ A
Power drawn by the motor $P_m = I^2R = (0.375)^2 \times 300 = 42.2$ W.
The power could also be calculated using phase angle as

$$P_m = VIR/Z = 0.375 \times 120 \times 300/320 = 42.2 \text{ W.}$$

Hence, the array power motor = input power of motor = P_m/η_m

$$= 42.2/0.9 = 46.9 \text{ W}$$

Example 4.4

A solar cell array is required to deliver 100 W peak output at 120 V DC bus voltage. The solar cells to be used are rated for 0.1 W peak output at 0.4 V. Assuming that there are no assembly losses, define the array.

Solution Maximum power rating of each cell (P_C) = 0.1 W

Let N_T is the total number of cells. Total output power of array (P_{MAX}) = 100 W
Hence, $N_T = P_{MAX}/P_C = 100/0.1 = 1{,}000$ cells
Further, number of cells in series = N_S = array output voltage (V_a)/V_{MP}

$$= 120/0.4 = 300 \text{ cells}$$

Since $N_T = N_S \times N_P$
$N_P = 1{,}000/300 = 3.33$ parallel strings
Therefore, a decision must be taken to use either 3 or 4 parallel strings.
With $N_P = 4$, $N_T = 1{,}200$ and array power = $0.1 \times 1{,}200 = 120$W
With $N_P = 3$, $N_T = 900$ and array power = $0.1 \times 90 = 90$ W

Example 4.5

Determine the load profile for the power system shown in Figure P4.5. Assume that the battery has normally 25 V and the loads are as follows:

Load 1 is a constant power load transponder that draws 50 W continuously, day and night. Load 2 is an electric-motor driven water pump that operates three times a day for 1 h: once before sunrise, once near noon, and once after sunset, and draws a starting current of 20 A for 5 s and running current of 4 A. Load 3 is a scientific instrument that operates approximately every 2 h for 6 min, day and night, and draws a current of 3 A.

Solution

For load 1:

The current drawn by load 1 is, $I_1 = 50/25 = 2$ A

The other load current can be plotted arbitrarily, with time phasing of loads 2 and 3 on the assumption that all the worst case loads can be 'on' simultaneously.

Thus, peak current drain from the battery at night.

$$= 2 + 20 + 3 = 25 \text{ A}$$

The average current drain defined by the area under the curve of combined load [in units of ampere hours (Ah) and ampere second (As)], as shown in Figure P4.5.

Load 1: 2 A × 24 h = 48 Ah.

Load 2: At starting, 3 times × 20 A × 5 s = 300 As = 300 As/3,600 s = 0.08 Ah
For pump running time, 3 × 1 h × 4 A = 12 Ah.

Load 3: The instrument is operated for 12 times in 24-h day and night

$$= 12 \times 0.1 \text{ h} \times 3 \text{ A} = 3.6 \text{ Ah}.$$

Combined loads = 48 + 0.08 + 12 + 3.6 = 64 Ah
Average combined load current = 64 Ah /24 h = 2.7A
Hence, load = 2.7 A × 25 V = 67 W.

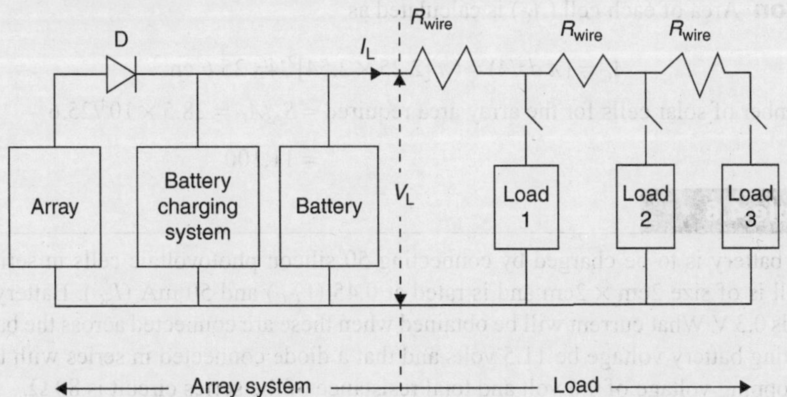

Figure P4.5 A typical solar power system supplying electrical loads

Example 4.6

Establish the preliminary solar array area and battery size for the average load of 67 W for 24 h. Solar cell efficiency is 10% and sum total of all array design and degrade array factor is 0.5. Battery charging efficiency is 60%. The load is to be supported for seven continuous days of cloudy weather (no sunshine) and the battery is to be fully recharged in 3 days.

Average monthly insolation is 181 kWh/m^2.

Solution The required array output is composed of the load and the battery recharge current as given.

The average load given is 67 W for 24 h. For continuous seven days under cloudy weather condition, we get $67 \times 24 \times 7 = 11.3$ kWh. Therefore, input power required for battery charging is $11.3/0.6 = 18.8$ kWh.

To recharge the battery in 3 days, each day having 9.7 h of sunshine in winter, the array must provide the following output.

Thus, the array output capability (P_a) is

$$P_a = 18.8/(9.7 \times 3) \text{ W (for the battery recharging)} + 67 \text{ W (for the load during 9.7 h)}$$

$$+ 67\,(24 - 9.7)/(9.7 \times 0.6) = 165 \text{ W (to carry the daily load through night)}$$

$$= 650 + 67 + 165 = 882 \text{ W} = 0.882 \text{ kW}$$

$$\text{Daily average insolation } D_I = 181/(9.7 \times 30) = 0.62 \text{ kW/m}^2$$

The required area of solar array (S_A) is,

$$S_A = P_a/(D_I \eta F) = 0.882 \times 10^3/(0.62 \times 103 \times 0.10 \times 0.5)$$

$$= 28.5 \text{ m}^2$$

Example 4.7

Find the number of solar cells for the array area of 28.5 m^2 if each cell has a diameter of 2.25 inches.

Solution Area of each cell (A_C) is calculated as

$$A_C = (\pi d^2/4) = (\pi [2.25 \times 2.54]^2/4 = 25.6 \text{ cm}^2$$

The number of solar cells for the array area required $= S_A/A_C = 28.5 \times 10^4/25.6$

$$= 11{,}100$$

Example 4.8

A 12 V battery is to be charged by connecting 50 silicon photovoltaic cells in series with it. Each cell is of size 2cm × 2cm and is rated at 0.45 (V_{OC}) and 50 mA (I_{SC}). Battery charging voltage is 0.3 V. What current will be obtained when these are connected across the battery? Let the existing battery voltage be 11.5 volts and that a diode connected in series with the battery has a dropping voltage of 0.6 volt and total resistance of the series circuit is 80 Ω.

Solution Schematic arrangement of battery charging circuit is shown in Figure P4.8.

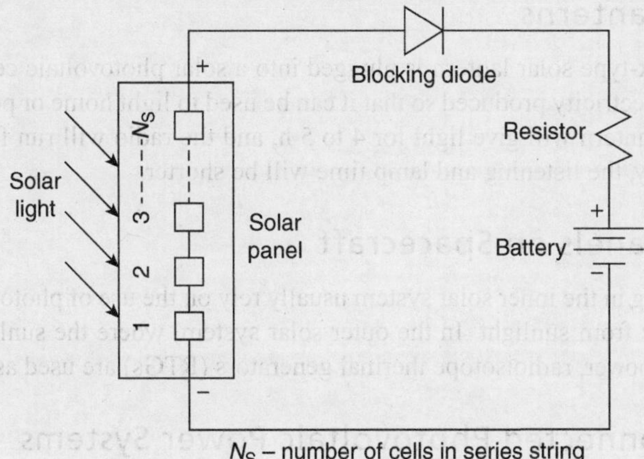

Figure P4.8 Battery charging by solar panel

Battery charging voltage output of solar panel = 50 × 0.3 = 15 V
Existing battery voltage + diode voltage = 11.5 + 0.6 = 12.1 V
Battery current = (15 – 12.1)/80 = 36.25 mA

4.8 APPLICATION OF SOLAR CELL SYSTEMS

In late 1980s and early 1990s, the major markets for solar panels were remote area power supplies and consumer products (watches, toys, and calculators). However, in the mid-1990s, a major effort was launched to develop building-integrated solar panels for grid-connected applications.

4.8.1 Solar Water Pumps

There are more than 10,000 solar powered water pumps in use in the world today. They are widely used on farms to supply water to livestock. In developing countries, they are used extensively to pump water from wells and rivers to villages for domestic consumption and irrigation of crops. In solar water pumping system, the pump is driven by motor run by solar electricity instead of conventional electricity drawn from utility grid. A solar photovoltaic water pumping system consists of a photovoltaic array mounted on a stand and a motor-pump set compatible with the photovoltaic array. It converts the solar energy into electricity, which is used for running the motor pump set. The pumping system draws water from the open well, bore well, stream, pond, canal, etc.

4.8.2 Solar Vehicle

It is an electric vehicle powered completely or significantly by direct solar energy. Usually, photovoltaic (PV) cells contained in solar panels convert the sun's energy directly into electric energy. The term 'solar vehicle' usually implies that solar energy is used to power all or part of a vehicle's propulsion. Solar power may be also used to provide power for communications or controls or other auxiliary functions.

4.8.3 Solar Lanterns

When the Petromax-type solar lantern is plugged into a solar photovoltaic cell, its rechargeable battery stores the electricity produced so that it can be used to light home or power a radio. When fully charged, the lantern will give light for 4 to 5 h, and the radio will run for 15 h. If both are used simultaneously, the listening and lamp time will be shorter.

4.8.4 Solar Panels on Spacecraft

Spacecraft operating in the inner solar system usually rely on the use of photovoltaic solar panels to derive electricity from sunlight. In the outer solar system, where the sunlight is too weak to produce sufficient power, radioisotope thermal generators (RTGs) are used as a power source.

4.8.5 Grid-connected Photovoltaic Power Systems

These are power systems energized by photovoltaic panels that are connected to the utility grid. Grid-connected photovoltaic power systems comprise photovoltaic panels, battery charging regulators, solar inverters, power conditioning units, and grid-connected equipments. When conditions are right, the grid-connected PV system supplies the excess power, beyond consumption by the connected load, to the utility grid. Residential grid-connected photovoltaic power systems that have a capacity less than 10 kW can meet the load of most consumers. It can feed excess power to the grid, which, in this case, acts as a battery for the system.

4.8.6 Cathodic Protection Systems

Cathodic protection is a method of protecting metal structures from corrosion. It is applicable to bridges, pipelines, buildings, tanks, wells, and railway lines. To achieve cathodic protection, a small negative voltage is applied to the metal structure and this prevents it from oxidizing or rusting. The positive terminal of the source is connected to a sacrificial anode that is generally a piece of scrap metal, which corrodes instead of the structure. Photovoltaic solar cells are often used in remote locations to provide this voltage.

4.8.7 Electric Fences

Electric fences are widely used in agriculture to prevent stock or predators from entering or leaving an enclosed field. These fences usually have one or two 'live' wires that are maintained at about 500 V DC. These give a painful, but harmless shock to any animal that touches them. This is generally sufficient to prevent stock from pushing them over. These fences are also used in wildlife enclosures and secure areas. They require a high voltage, but very little current and they are often located in remote areas where the cost of electric power is high. These requirements can be met by a photovoltaic system involving solar cells, a power conditioner, and a battery.

4.8.8 Remote Lighting Systems

Lighting is often required at remote locations where the cost of power is too high to use the grid. Such applications include security lighting, navigation aids, (e.g., buoys and beacons), illuminated road signs, railway crossing signs, and village lighting. Solar cells are suited to such applications, although a storage battery is always required in such systems. They usually consist of a solar photovoltaic panel, a battery charging regulator, a storage battery, power conditioner, and a low voltage, high-efficiency DC fluorescent lamp. These systems are very popular in remote areas, especially in developing countries and this is one of the major applications of solar cells.

4.8.9 Telecommunications and Remote Monitoring Systems

Good communications are essential for improving the quality of life in remote areas. However, the cost of electric power to drive these systems and the high cost of maintaining conventional systems has limited their use. Solar panel has provided a cost-effective solution to this problem through the development of remote area telecommunications repeater stations. These stations typically consist of a receiver, a transmitter, and a solar cell-based power supply system. Thousands of these systems have been installed around the world and they have an excellent reputation for reliability and relatively low costs for operation and maintenance.

Similar principles apply to solar powered radios and television sets, emergency telephones, and monitoring systems. Remote monitoring systems may be used for collecting weather data or other environmental information and for transmitting it automatically via radio to the home base.

4.8.10 Rural Electrification

Storage batteries are widely used in remote areas to provide low-voltage electrical power for lighting and communications as well as for vehicles. A photovoltaic-powered battery charging system usually consists of a small solar cell array and a charge controller. These systems are widely used in rural electrification projects in developing countries.

4.8.11 Water Treatment Systems

In remote areas, electric power is often used to disinfect or purify drinking water. Photovoltaic cells are used to power a strong ultraviolet light that can be used to kill bacteria in drinking water. This can be combined with a solar-powered water pumping system. Desalination of brackish water can be achieved via PV-powered reverse osmosis systems.

SUMMARY

- Photovoltaic conversion is the process of direct conversion of light into electricity at the atomic level. Some materials exhibit a property known as the photoelectric effect that causes them to absorb photons of light and release electrons. When free electrons are captured, electricity is generated.
- Single junction silicon solar cells produce approximately 0.5–0.6 V_{OC} (open circuit voltage), so they are usually connected together in series to provide larger voltages.
- An important feature of solar cells is that the voltage of the cell does not depend on its size and remains fairly constant with changing light intensity. However, the current in a device is almost directly proportional to light intensity and size. The output current varies depending on the size of the cell. In general, a typical commercially available silicon cell produces a current between 28 and 35 milliamps per square centimetre.

- Energy conversion efficiency (η) is defined as the ratio of power output of a cell (in watts) at its maximum power point (P_{MAX}) and the product of input light power (E, in W/m^2) and the surface area of the solar cell (S in m^2) under standard conditions as,

$$\eta = \text{maximum output power/(irradiance} \times \text{area)} = P_{MAX}/(E \times S)$$

- Fill Factor (FF) is another term defining the overall behaviour of a solar cell. It is a measure of squareness of I–V characteristic of solar cell and is given as,

$$\text{FF} = \text{Maximum output power/(open circuit voltage} \times \text{short circuit current)}$$

- Solar cells exhibiting closed to square current – voltage characteristic have high efficiency.
- A solar photovoltaic module is used for converting sun light into electrical energy. They are usually manufactured as a sealed unit with a given output voltage and wattage rating.
- To produce a larger voltage, a number of pre-wired cells in series, all encased in tough, weather-resistant package, are used to form a module. A typical module may have 36, 54, 72, or 96 cells in series.
- Multiple modules can be wired in series to increase the voltage and in parallel to increase the current. Such combinations of modules are referred to as an array. They are used to provide any level of power requirements from mere watts (W) to kilowatt (kW) and megawatt (MW) sizes.
- In a larger photovoltaic array, individual photovoltaic modules are connected in both series and parallel.
- A series-connected set of solar cells or modules is called a string.
- Photovoltaic (PV) panels convert sunlight to electricity that can be used to supplement or replace the electricity supplied by the utility grid. PV panels are most commonly installed on rooftops and are most effective with a southerly exposure that provides full sun. Other possible installations include a ground mount, a pole mount, etc.

REVIEW QUESTIONS

1. Describe principle of solar PV conversion.
2. State applications of solar PV systems.
3. Draw a schematic of a solar PV electric plant.
4. Describe the layout of a typical solar PV array.
5. What is the principle of solar photovoltaic power generation?
6. What are the main elements of PV systems?
7. What are the advantages and disadvantages of photovoltaic solar energy conversion?
8. Write remarks on
    1. Solar cell arrays.
    2. Limitations of solar cells

9. Clearly, explain the construction of a p–n junction and its use to convert sunlight directly into electricity. What distinguishes a solar cell from a conventional p–n junction diode?
10. Establish the following relationship:

$$N_p/n_n = p_n/p_p = \Sigma \text{Exp}\,(-eV_0/kT)$$

where n_n and p_p are majority carriers and p_n and n_p are minority carriers in n and p semiconductor layers, respectively; e is electronic charge; V_0 is the potential barrier; k is the Boltzmann's constant; and T is the absolute temperature.
11. Give the detailed equivalent circuit for a photovoltaic cell. Discuss the effect of series resistance R_s, shunt resistance R_{sh} and minority carrier lifetime on the performance of the solar cell. Explain how, from test data, for I–V characteristic, the parameter R_s and diode factor can be determined.
12. A solar cell at a given insulation level has short-circuit current density as 180 A/m^2, and reverse saturation current density equal to 8×10^{-9} A/m^2. At an operating temperature of 27°C and condition of maximum power, find the effective surface area needed to produce an output of 100 W and also estimate the conversion efficiency, if the radiation intensity is 930 W/m^2.
13. Draw and explain electrical equivalent circuit model and current voltage characteristics of solar cells.
14. Explain the term fill factor and its importance as a performance parameter for a solar cell.
15. Solar cell array is required to deliver 100 W peak output at 120 V DC bus voltage. The solar cell to be used is rated for 0.1 W peak output at 0.4 V. Assuming that there are no assembly losses, calculate and design solar cell array.
16. Draw and explain electrical equivalent circuit model of a solar cell. Starting with the assumptions made, if any, derive and draw I–V characteristic of solar cell. Further, obtain an expression for the external load voltage that gives maximum cell output power.

Chapter 5

Hydrogen Energy

KEY CONCEPTS

- Hydrogen energy and its benefits
- Introduction to methods of hydrogen production technologies
- Associated problems with hydrogen energy
- Use of hydrogen energy
- Hydrogen energy storage-applications

Hydrogen is the simplest element and an atom of hydrogen consists of only one proton and one electron. Despite its simplicity and also availability in abundance in the universe, it does not occur naturally as a gas on the earth. It is always combined with other elements. Water H_2O, for example, is a combination of hydrogen and oxygen. It is also available in hydrocarbons used as fuels. Hydrogen can be separated from hydrocarbons through the application of heat—a process known as reforming. Another method of hydrogen production is known as electrolysis of water. In this process, a direct electrical current passes through water to separate water into its components of oxygen and hydrogen. Using sunlight as their energy source, some algae and bacteria give off hydrogen under certain conditions.

Hydrogen energy can be thought of usable form (secondary energy resource) next to electrical energy in future, if technology for hydrogen transportation, distribution, and utilization is fully established and accepted by human society.

Hydrogen, as an energy vector, is considered as one of the most interesting element of the future energy system. Many energy experts have started believing that of all the new usable forms of energy, hydrogen may be preferred next to electrical energy with several advantages.

Hydrogen is the simplest element. An atom of hydrogen consists of only one proton and one electron. It is also the most plentiful element in the universe. Despite its simplicity and its availability in abundance, hydrogen does not occur naturally as a gas on the earth—it is always combined with other elements. Water, for example, is a combination of hydrogen and oxygen (H_2O). It is also found in many organic compounds, notably the hydrocarbons that make up many of our fuels, such as gasoline, natural gas, methanol, and propane. Hydrogen can be separated from hydrocarbons through the application of heat—a process known as reforming. Currently, most hydrogen is made this way from natural gas. An electrical current can also be used to separate water into its components of oxygen and hydrogen. This process is known as electrolysis. Using sunlight as their energy source, some algae and bacteria give off hydrogen under certain conditions.

It is a clean burning fuel that can be produced in virtually unlimited quantities at a cheap cost. Further, it can be readily converted into electricity when needed either through mechanical energy or directly through electrochemical fuel cells and, thus, an excellent medium for storing off-peak power. Another advantage of hydrogen as a fuel is that it can become an industrial raw material for ammonia synthesis, petroleum reforming, fat hardening, and be used in a number of other chemical industries.

Based on the assumption, it can be said that the most promising methods for producing hydrogen is the one that provides the products at the lowest cost. The following methods utilizing solar energy as the primary source (avoiding fossil fuels and other sources) may be found suitable for many countries and India, in particular:

1. Electrolysis of water by electrical current generated by a thermal solar plant in a conventional steam cycle.
2. Thermochemical decomposition of water using solar energy to heat the vessel.
3. Direct photocatalytic decomposition of water by solar radiation.
4. Electrolysis of water by electrical current generated by photovoltaic conversion.

Hydrogen can also be produced through coal gasification and steam–iron reaction.

Indian experts in the field have recommended that the following R&D work should be undertaken without delay in this country:

1. Technoeconomic assessment of the use of hydrogen in the country's energy system.
2. Development of advanced technology for the production of hydrogen through water electrolysis, coal gasification, and steam–iron reaction.
3. Utilization of solar energy for the production and microbiological generation from agriculture wastes and sewage.
4. Development of materials for optimum storage of hydrogen as solid hydrides.
5. Production, storage, transportation, and transfer of liquid hydrogen.
6. Optimum utilization as a liquid fuel.
7. Hydrogen–oxygen fuel cell.

5.1 BENEFITS OF HYDROGEN ENERGY

The three basic benefits of hydrogen energy are as follows:

1. *Use of hydrogen greatly reduces pollution*: When hydrogen is combined with oxygen in a fuel cell, energy in the form of electricity is produced. This electricity can be used to power

vehicles, as a heat source, and for many other uses. The advantage of using hydrogen as an energy carrier is that when it combines with oxygen, the only by-products are water and heat. No greenhouse gasses or other particulates are produced by the use of hydrogen fuel cells.

2. *Hydrogen can be produced locally from numerous sources*: Hydrogen can be produced either centrally, and then distributed, or onsite where it will be used. Hydrogen gas can be produced from methane, gasoline, biomass, coal, or water. Each of these sources brings with it different amounts of pollution, technical challenges, and energy requirements.
3. *A sustainable production system if hydrogen is produced from electrolysis of water*: Electrolysis is the method of separating water into hydrogen and oxygen. Renewable energy can be used to power electrolysers to produce hydrogen from water. Using renewable energy provides a sustainable system that is independent of petroleum products and is non-polluting. Some of the renewable sources used to power electrolyses are wind, hydro, solar, and tidal energy. After the hydrogen is produced in an electrolyser, it can be used in a fuel cell to produce electricity. The by-products of the fuel cell process are water and heat. If fuel cells operate at high temperatures, the system can be set up as a co-generator, with the waste energy used for heating.

5.2 HYDROGEN PRODUCTION TECHNOLOGIES

The choice of production methods will vary depending on the availability of feedstock or resource, the quantity of hydrogen required, and the required purity of hydrogen. Researchers are developing a wide range of processes for producing hydrogen economically and in an environmentally friendly way. These processes can be divided into three major research areas:

1. Thermochemical production technologies
2. Electrolytic production technologies
3. Photolytic production technologies

5.2.1 Thermochemical Production Technologies

Hydrogen bound in organic matter and in water makes up 70% of the earth's surface. Breaking up these bonds in water allows us produce hydrogen, and then, to use it as a fuel. There are numerous processes that can be used to break these bonds. Following sections discuss a few methods for producing hydrogen that are currently used or are under research and development. Most of the hydrogen now produced on an industrial scale by the process of steam reforming, or as a by-product of petroleum refining and chemical production.

5.2.1.1 Steam Reforming

Steam reforming uses thermal energy to separate hydrogen from the carbon components in methane and methanol and involves the reaction of these fuels with steam on catalytic surfaces. The first step of the reaction decomposes the fuel into hydrogen and carbon monoxide. Then, a 'shift reaction' changes the carbon monoxide and water to carbon dioxide and hydrogen. These reactions occur at temperatures of 200°C or greater.

Steam reforming of natural gas is currently the least expensive method and is responsible for more than 90% of hydrogen production worldwide. Natural gas is first cleared from sulphur compounds. It is then mixed with steam and send over a nickel–alumina catalyst inside a tubular reactor heated externally, where carbon monoxide (CO) and hydrogen (H_2) are generated. This step is followed by a catalytic water-gas shift reaction that converts the CO and water to hydrogen and carbon dioxide (CO_2). The hydrogen gas is then purified.

The endothermic reforming reaction is:

$$CH_4 + H_2O + 206\ (kJ/kg) \Rightarrow CO + 3H_2 \tag{5.1}$$

It is usually followed by the exothermic shift reaction:

$$CO + H_2O \Rightarrow CO_2 + H_2 + 41(kJ/kg) \tag{5.2}$$

The overall reaction is:

$$CH_4 + 2H_2O + 165\ (kJ/kg) \Rightarrow CO_2 + 4H_2 \tag{5.3}$$

The residual stream from the initial purification step is part of the fuel gas burned in the reformer in order to supply the required heat. Hence, the CO_2 contained in this gas is currently vented with the flue gas. If CO_2 were to be captured, an additional separation step would be needed.

The technology is suitable for large reformers (e.g., 100,000 tons per year), where yields higher than 80% can be achieved. Small-scale reformers especially designed for feeding small fuel cells show low efficiencies.

The production of hydrogen from natural gas is an integral part of the strategy to introduce hydrogen into the transportation and utility energy sectors, by reducing the cost of conventional and developing innovative hydrogen production processes that rely on cheap fossil feedstocks. Today, nearly all hydrogen production is based on fossil raw materials. Worldwide, 48% of hydrogen is produced from natural gas, 30% from oil (mostly consumed in refineries), 18% from coal, and the remaining 4% via water electrolysis.

Modification of the conventional steam methane reforming (SMR) process to incorporate an adsorbent in the reformer to remove CO_2 from the product stream may offer a number of advantages over conventional processes. Disturbing the reaction equilibrium in this way drives the reaction to produce additional hydrogen at lower temperatures than conventional SMR reactors.

Although still in the research stage, the cost of hydrogen from this modified process is expected to be 25%–30% lower, primarily because of reduced capital and operating costs. In addition, the adsorption of the CO_2 in the reforming stage results in a high-purity CO_2 stream from the adsorbent regeneration step. This has interesting implications in a carbon-constrained world.

5.2.1.2 Partial Oxidation or Ceramic Membrane Reactor

Scientists are developing a ceramic membrane reactor for the simultaneous separation of oxygen from air and the partial oxidation of methane. If successful, this process could result in improved production of hydrogen and/or synthesis gas when compared to conventional reformers.

In the partial oxidation process, natural gas (or other liquid or gaseous hydrocarbons) and oxygen are injected into a high-pressure reactor. The oxygen to carbon ratio is optimally set for maximizing the yield of CO and H_2 and avoiding the formation of soot. Further steps and equipment remove the large amount of heat generated by the oxidation reaction, shift the CO with water to CO_2 and H_2, and remove the CO_2, which can then be captured, and purify the hydrogen produced. This process needs oxygen, which is usually provided by an air distillation plant. Partial oxidation can also be helped by an oxidation catalyst. It is then called catalytic partial oxidation.

The partial oxidation reaction for natural gas is:

$$CH_4 + \tfrac{1}{2}\,O_2 \Rightarrow CO + 2\,H_2 \qquad (5.4)$$

After the partial oxidation reaction, the process gas is similar to that of the steam reforming process. Since the reaction is exothermic, a heating system is not required, which is a major advantage resulting in size and capital cost reduction. However, partial oxidation is typically less energy efficient than steam reforming.

5.2.1.3 Biomass Gasification and Pyrolysis

The thermal processing techniques for plant material (biomass) and fossil fuels are similar, with a number of the downstream unit operations being essentially the same for both feedstocks. Using agricultural residues and wastes, or biomass specifically grown for energy uses, hydrogen can be produced via pyrolysis or gasification.

Biomass pyrolysis produces a liquid product (bio-oil) that, like petroleum, contains a wide spectrum of components that can be separated into valuable chemicals and fuels.

Unlike petroleum, bio-oil contains a significant number of highly reactive oxygenated components derived mainly from constitutive carbohydrates and lignin. These components can be transformed into products, including hydrogen. Co-product strategies are designed to produce high value chemicals, like phenolic resins, in conjunction with hydrogen.

5.2.2 Electrolytic Production Technologies

Another way to produce hydrogen is by electrolysis. Electrolysis separates the elements of water—H_2 and oxygen (O)—by charging water with an electrical current. Adding an electrolyte like salt improves the conductivity of the water and increases the efficiency of the process. The charge breaks the chemical bond between the hydrogen and the oxygen and splits apart the atomic components, creating charged particles called ions. The ions form at two poles: the anode, which is positively charged, and the cathode, which is negatively charged. Hydrogen gathers at the cathode and the anode attracts oxygen.

Electrolysis is the process of producing hydrogen and oxygen from water in an electrochemical cell. Two types of electrochemical methods, alkaline or proton exchange membrane (PEM), are used in commercially available equipment commonly referred to as electrolysers.

An alkaline electrolyser immerses the two electrodes, the cathode and the anode, into an aqueous alkaline electrolyte, typically a solution of sodium or potassium hydroxide, and a voltage is applied across the electrodes. The resulting migration of ions in solution results in the production of hydrogen at the cathode and oxygen at the anode according to the following equation:

Cathode reaction $\quad 4H_2O + 4e^- \text{—}\ 2H_2 + 4OH^- \qquad (5.5)$

Anode reaction $\quad 4OH^- \text{—}\ O_2 + 2H_2O + 4e^- \qquad (5.6)$

In a PEM electrolyzer, the mobile ion is a proton in an electrolyte that is a proton-conducting polymer memebarane. In this case, the reactions at the electrodes are as follows:

Cathode reaction $$4H^+ + 4e^- — 2H_2 \quad (5.7)$$

anode reaction $$2H_2O — O_2 + 4H^+ + 4e^- \quad (5.8)$$

Currently, the best conversion efficiency (i.e., overall system efficiency for converting electrical power to power stored as hydrogen) for commercial electrolysers is approximately 70%.

5.2.2.1 Water Electrolysis

Until the 1950s, water electrolysers were in widespread use for hydrogen (or oxygen) production. Currently, electrolysis provides only a small percentage of the world's hydrogen, most of which is supplied to applications requiring small volumes of high purity hydrogen (or oxygen, such as for breathing atmospheres for submarines). There is significant renewed interest in the use of electrolysers to produce hydrogen as a fuel for automotive applications, with a number of refuelling stations installed around the world. In addition, research continues in the integration of intermittent renewable resources (PV and wind) with electrolysers for producing hydrogen that has to be used as a fuel or for energy storage.

5.2.2.2 Steam Electrolysis

Steam electrolysis is a variation of the conventional electrolysis process. Some of the energy needed to split the water is added as heat instead of electricity, making the process more efficient than conventional electrolysis. At 2,500°C, water decomposes into hydrogen and oxygen. This heat could be provided by a solar energy concentrating device to supply the heat. The problem here is to prevent the hydrogen and oxygen from recombining at the high temperatures used in the process.

5.2.2.3 Photoelectrolysis

Multi-junction cell technology developed by the PV industry is being used for photoelectrochemical (PEC) light harvesting systems that generate sufficient voltage to split water and are stable in a water or electrolyte environment. Theoretical efficiency for tandem junction systems is 42%; practical systems could achieve 18% –24% efficiency; low-cost multi-junction amorphous silicon (a-Si) systems could achieve 7% –12% efficiency. This is one of the advantages of a direct conversion hydrogen generation system. Not only does it eliminate most of the costs of the electrolyser, but it also has the possibility of increasing the overall efficiency of the process. Research results for the development of PEC water splitting systems have shown a solar-to-hydrogen efficiency of 12.4% for the lower heating value (LHV) of hydrogen using concentrated light. Low-cost a-Si tandem designs with appropriate stability and performance are also being developed. An outdoor test of the a-Si cells resulted in a solar-to-hydrogen efficiency of 7.8% LHV under natural sunlight.

5.2.2.4 Thermochemical Water Splitting

Thermochemical water splitting uses chemicals such as bromine or iodine assisted by heat. This causes the water molecule to split. It takes several steps—usually three—to accomplish this entire process.

Several high-temperature thermochemical reactions are under study, which have high efficiency and practical applicability with nuclear heat sources.

One of the most promising may be the sulphur–iodine (SI) cycle, where three chemical reactions achieve the dissociation of water:

$$I_2 + SO_2 + 2H_2O \Rightarrow 2HI + H_2SO_4(120°C) \quad (5.9)$$

$$H_2SO_4 \Rightarrow SO_2 + H_2O + ½\,O_2(830–900°C) \quad (5.10)$$

$$2HI \Rightarrow I_2 + H_2(300–450°C) \quad (5.11)$$

The overall reation being: $$H_2O \Rightarrow H_2 + ½\,O_2 \quad (5.12)$$

The efficiency of the sulphur–iodine process increases from 30% at 750°C to 60% at 1,000°C. Other thermochemical cycles show efficiencies of 40%–50% at typical temperatures of 700°C.

Several high-temperature nuclear reactors have been developed that could produce heat at the required temperature. The high-temperature helium reactor and the molten-salt reactor appear to offer the best perspectives for hydrogen production. Other reactor types may be used if efficient hydrogen production processes can be developed at temperatures of 500°C.

5.2.2.5 By-product of Sodium or Potassium Chloride Electrolysis

Hydrogen is a by-product of sodium or potassium chloride electrolysis that produces chlorine and caustic soda or potash:

$$NaCl + H_2O + electricity \Rightarrow ½\,Cl_2 + NaOH + ½\,H_2 \quad (5.13)$$

$$KCl + H_2O + electricity \Rightarrow ½\,Cl_2 + KOH + ½\,H_2 \quad (5.14)$$

Chlorine is one of the most common chemicals in the world. It is produced in huge quantities.

5.2.2.6 Reversible Fuel Cells or Electrolysers

Operating the proton exchange membrane (PEM) fuel cell 'in reverse' as an electrolyser is possible, but optimum operating conditions for the power production mode and for the hydrogen production mode are significantly different. Design issues for the reversible fuel cell system include thermal management, humidification, and catalyst type and loading.

5.2.3 Photolytic Production Technologies

Solar energy can be used to convert water to hydrogen and oxygen directly. Electricity need not be produced by photovoltaic cell. Hydrogen production can be achieved by using either photo-electrochemical or photo-biological methods.

5.2.3.1 Photoelectrochemical Processes

Photoelectrochemical processes use two types of electrochemical systems to produce hydrogen. One uses soluble metal complexes as a catalyst, while the other uses semiconductor surfaces. When the soluble metal complex dissolves, the complex absorbs solar energy and produces an electrical charge that drives the water-splitting reaction. This process mimics photosynthesis.

The other method uses semiconducting electrodes in a photochemical cell to convert optical energy into chemical energy. The semiconductor surface serves two functions: to absorb solar energy and to act as an electrode. Light-induced corrosion limits the useful life of the semiconductor.

5.2.3.2 Biological and Photobiological Processes

Certain photosynthetic microbes produce hydrogen in their metabolic activities using light energy. By employing catalysts and engineered systems, hydrogen production efficiency could reach 24%. Photo-biological technology holds great promise but because oxygen is produced along with the hydrogen, the technology must overcome the limitation of oxygen sensitivity of the hydrogen-evolving enzyme systems. Researchers are addressing this issue by screening for naturally occurring organisms that are more tolerant of oxygen, and by creating new genetic forms of the organisms that can sustain hydrogen production in the presence of oxygen. A new system is also being developed that uses a metabolic switch (sulphur deprivation) to cycle algal cells between the photosynthetic growth phase and the hydrogen production phase.

Unlike cyanobacteria or algae, photosynthetic bacteria do not oxidize water. However, they do evolve hydrogen from biomass (previously generated from sunlight, water, and carbon dioxide). These bacteria use several different enzymatic mechanisms with near-term commercial potential for biological hydrogen production from biomass. One mechanism, in particular, looks promising for applications as a biological conditioning agent for upgrading thermally generated fuel gases to a level where they can be directly injected into hydrogen fuel cells. This same system has the potential to subsequently evolve into a second-generation photo-biological method to produce hydrogen from water.

From the abovementioned points, it is evident that biological and photo-biological processes use algae and bacteria to produce hydrogen. Under specific conditions, the pigments in certain types of algae absorb solar energy. The enzyme in the cell acts as a catalyst to split the water molecules. Some bacteria are also capable of producing hydrogen, but unlike algae, they require a substrate to grow on. The organisms not only produce hydrogen, but also can clean up pollution as well.

When considering the production process, the cost of electricity required for the electrolysis process is one of the barriers to sustainable energy. Besides electrolysis, the production of hydrogen has been accomplished by a catalytic reaction of waste aluminium. The end products are hydrogen and alumina that can be reused to make aluminium.

With increasing use of hydrogen and technical advances, the costs of production, distribution, and product manufacturing will become increasingly affordable. By continuing to build partnerships between business, government, universities, and non-profit organizations, hydrogen will be the foundation of a sustainable energy economy.

5.3 HYDROGEN ENERGY STORAGE

When compared to the electrical energy, the development of safe, reliable, compact, and cost-effective hydrogen storage technologies is one of the most technically challenging barriers to the widespread use of hydrogen as a usable form of energy. To be competitive with conventional vehicles, hydrogen-powered cars must be able to travel more than 450 km between fills. This is a challenging goal because hydrogen has physical characteristics that make it difficult to store in large quantities without taking up a significant amount of space.

5.3.1 Compressed Gas and Liquid Hydrogen Storage Tanks

Hydrogen has a very high energy content by weight (about three times more than gasoline), but it has a very low energy content by volume (liquid hydrogen is about four times less than gasoline). This makes hydrogen a challenge to store. Liquefied hydrogen is denser than gaseous hydrogen, and thus, it contains more energy in a given volume. Similar sized liquid hydrogen tanks can store more hydrogen than compressed gas tanks, but it takes energy to liquefy hydrogen. However, the tank insulation required to prevent hydrogen loss adds to the weight, volume, and costs of liquid hydrogen tanks.

5.3.2 Materials-based Storage

Hydrogen can be stored in materials by following different processes. It can be stored on the surfaces of solids (by adsorption process) or within solids (by absorption process).

In adsorption process, hydrogen attaches to the surface of a material either as hydrogen molecules (H_2) or hydrogen atoms (H). This is also referred to as surface adsorption storage.

In absorption process, hydrogen molecules dissociate into hydrogen atoms that are incorporated into the solid lattice framework. This is also known as intermetallic hydride storage. This method may make it possible to store larger quantities of hydrogen in smaller volumes at low pressure and at temperatures close to room temperature. Finally, hydrogen can be strongly bound within molecular structures, as chemical compounds containing hydrogen atoms in the form of compressed gas or cryogenic liquid.

5.3.3 Methods of Hydrogen Energy Storage

Based on the abovementioned processes, methods of hydrogen energy storage may be classified as follows:

5.3.3.1 Compression

The hydrogen can be compressed into containers or underground reservoirs. This is a relatively simple technology, but the energy density and efficiency (65%–70%) are low. Further, problems have occurred with the mechanical compression.

However, this is, at present, the most common form of hydrogen storage for the transport industry, with the hydrogen compressed to approximately 700 bar (the higher the storage pressure, the higher the energy density). However, the energy required for the compression is a major drawback.

5.3.3.2 Liquefied Hydrogen

The hydrogen can be liquefied by pressurising and cooling. Although the energy density is improved, it is still four times less than conventional petrol. Further, keeping the hydrogen liquefied is very energy intensive, as it must be kept below 20.27K

5.3.3.3 Metal Hydrides

Certain materials absorb molecular hydrogen such as nanostructured carbons and clathrate hydrate. By absorbing the hydrogen in these materials, it can be easily transported and stored. Once required, the hydrogen is removed from the parent material. The energy density is similar to that obtained for liquefied hydrogen. The extra material required to store the hydrogen is a major problem with this technique, as it creates extra costs and mass. This is still a relatively new technology, and therefore, with extra development, it could be a viable option, especially if the mass of material is reduced. Carbon-based absorption can achieve higher energy densities but it has higher costs and even lesser demonstrations. Both the metal-hydride and carbon-based absorption use thermal energy. This thermal heat could be got from the waste heat of other processes with HESS, such as the electrolyser or fuel cell, to improve overall efficiency.

Each storage technique is in the early stages of development, and hence, there is no optimum method, at present, with research being carried out in each area. Scientists are investigating several different kinds of materials, including metal hydrides, adsorbent materials, and chemical hydrides, in addition to identifying new materials with potential for favourable hydrogen storage attributes.

Hydrogen storage in materials offers great promise, but additional research is required to better understand the mechanism of hydrogen storage in materials under practical operating conditions and to overcome critical challenges related to it.

5.4 USE OF HYDROGEN ENERGY

Aside from the production of hydrogen, the everyday use and acceptance of hydrogen must be carefully introduced. Today, hydrogen is being used to power commercial buses both by internal combustion engines burning a combination of hydrogen and other fuels and solely by hydrogen used in fuel cells. Hydrogen is used in many commercial applications from welding metal to dying fabrics for making electronics, plastics, and fertilizers. When a renewable economically viable production process of hydrogen can be achieved, the advantages will be spread out to many industries. Some of the proving grounds for various production methods can be locally developed to provide hydrogen for these industries.

Hydrogen can be used as a mobile source of power for transportation by being compressed and stored in small tanks for applications similar to gasoline or propane.

The following are the two superior ways of using hydrogen energy:

1. *Internal combustion engine (ICE)*: It is expected that the ICE will act as a transition technology while fuel cells are improving, because the modifications required to convert an ICE to operate on hydrogen are not very significant.
2. *Fuel cell (FC)*: It is expected to be the generator of choice for future hydrogen-powered energy applications owing to its virtually emission-free, efficient, and reliable characteristics. A fuel cell converts stored chemical energy, in this case hydrogen, directly into electrical energy.

5.5 APPLICATIONS OF HYDROGEN ENERGY

Hydrogen and fuel cells have a wide range of applications for use almost anytime and anywhere.

5.5.1 At Home Sector

Fuel cells are ideal for residential zones. They are virtually silent with no moving parts and provide reliable power 24/7. In addition, a fuel cell, which is large enough to power an entire home, is about the size of a traditional AC unit. Fuel cells already power thousands of homes in Japan and are beginning to power similarly in the United States.

5.5.2 At Work Sector

Fuel cells can be produced in stacks large enough to power the large office buildings, but only occupy the area of couple of parking spaces. Again, fuel cells are a great fit in this situation, as they are noiseless, environmentally friendly, and efficient. Distributed power from fuel cells does not rely on transmission lines, and thus eliminates the need for backup power generators.

5.5.3 At Transport and Industrial Sectors

Fuel cells are just as mobile as human beings. Fuel cells can power cars, buses, airplanes, cell phones, laptops, and more. With nearly 10 times the lifespan of batteries on a single charge, fuel cells can keep powered no matter where the road takes the transport vehicles.

Hydrogen is an ideal replacement for fossil fuels such as coal, oil and natural gas in furnaces, internal combustion engines, turbines, and jet engines. Today, environmental pressures are concentrating on the hydrogen research and the development efforts to utilize hydrogen as an alternative fuel to power our mobility and transportation needs. In electrified vehicles, for example, it is used to run fuel cells that convert hydrogen efficiently (back) to electricity. The application spectrum of fuel cells is vast. They have the potential to replace conventional power generators such as combustion engines or even large batteries in cars, buses, forklift trucks (FLTs), submarines, and backup and power plants.

5.6 ADVANTAGES OF HYDROGEN ENERGY

1. Uncoupling of primary energy sources and utilization.
2. Hydrogen is a gas; thus, it is easier to store than to store electricity.
3. Hydrogen can be obtained from any primary energy source, including renewable energy source.
4. Decentralized production is possible. Hydrogen is viewed as capable of providing services where electricity is not available, in particular as a fuel for vehicles and energy storage in remote areas.
5. Very efficient when used in fuel cells.
6. Very good experience of hydrogen as a chemical reactant (ammonia, methanol, and oil refining).
7. Very good safety records (for a specific range of applications).

5.7 DISADVANTAGES OF HYDROGEN ENERGY

1. Poor overall energy efficiency when produced from electricity made with fossil fuels.
2. Very low density and poor specific volume energy density.
3. Need for high pressures and very low temperatures if stored in the liquid phase.
4. Specific safety problems and poor public acceptance (Hindenburg syndrome and Apollo Challenger space shuttle).
5. No existing infrastructures for transport, distribution, and storage.
6. Rather high cost (till today).

5.8 PROBLEMS ASSOCIATED WITH HYDROGEN ENERGY

The serious problems that are affecting the development of hydrogen for household and transport applications are as follows:

1. *Hydrogen storage*: The concerns surrounding the storage of hydrogen are a major issue. It must be stored at extremely low temperatures and high pressure. A container capable of withstanding these specifications is larger than a standard gas tank. Hydrogen storage could be viewed as a problem by consumers.
2. *High reactivity of hydrogen*: Hydrogen is extremely reactive. It is combustible and flammable. The Hindenburg disaster, where a hydrogen-filled blimp exploded and many people died, has caused a fear of hydrogen
3. *Cost and methods of hydrogen fuel production*: Current production of hydrogen takes a lot of energy. If one has to burn fossil fuels to make hydrogen, what has really been gained? New, clean energy technology or hydrogen production methods will need to be developed for hydrogen vehicles to make sense.
4. *Consumer demand*: Another problem for hydrogen fuel is consumer demand and the cost to change all gasoline filling stations and vehicle production lines into hydrogen. The major transport companies will not start to produce hydrogen vehicles until there is consumer demand. Why would a person pay for an expensive hydrogen vehicle?
5. *Cost of changing the infrastructure*: To accommodate hydrogen equipment and appliances.

SUMMARY

- Hydrogen is an energy carrier but not a source of energy. Therefore, it must be produced.
- Hydrogen is a clean burning fuel that can be produced in virtually unlimited quantities at a very low cost. It can be readily converted into electricity when needed either through mechanical energy or directly through electrochemical fuel cells.
- Hydrogen is an industrial raw material for ammonia synthesis, petroleum reforming, fat hardening, and a number of other chemical industries.
- Hydrogen is an extremely environment friendly fuel; when it burns, it releases only water vapour into the atmosphere, but the problem is that it is not easy to store.

- Hydrogen is a by-product of sodium or potassium chloride electrolysis that produces chlorine and caustic soda or potash.
- Hydrogen is highly explosive in gas form, and sometimes, it can be hazardous to work around and use.

REVIEW QUESTIONS

1. What is hydrogen energy?
2. Discuss the benefits of hydrogen energy.
3. State and explain methods of hydrogen production technologies.
4. Discuss the applications, advantages, and disadvantages of hydrogen energy.
5. Mention the problems associated with the development and application of hydrogen energy.

Chapter 6

Wind Energy

Sun is the main source of wind, and hence, wind is considered a form of solar energy. Winds are caused by the uneven heating of the atmosphere by the sun, the irregularities of the earth's surface, and rotation of the earth. Wind flow patterns are modified by the earth's terrain, water bodies, and vegetative cover. The wind flow or motion energy is 'harvested' by modern wind turbines. Wind power has been utilized for several centuries. The invention of sail boats are the first and most important example of driving them by using wind energy. The earliest known wind-powered grain mills and water pumps were used by the Persians, the Indians, and the Chinese. Wind turbines convert the kinetic energy of the wind into mechanical power. This mechanical power can be used for specific tasks (such as grinding grain or pumping water) or imparting motion to an electric generator that converts mechanical power into electricity.

Windmills have been used for many centuries for pumping water and milling grain. The discovery of the internal combustion engine and the development of electrical grids caused many windmills to disappear in the early part of this century. However, in recent years, there has been a revival of interest in wind energy and attempts are underway all over the world to introduce cost-effective wind energy conversion systems for this renewable and environmentally friendly energy source. In developing countries, wind power can play a useful role for water supply and irrigation (wind pumps) and electrical generation (wind generators). These two variants of windmill technology are discussed in separate technical briefs. This brief gives a general overview of the resource and of the technology for extracting energy from the wind.

KEY CONCEPTS

- Energy availability in the wind
- Wind energy scenario in india and worldwide
- Wind turbine site selection
- Wind turbine power output variation with steady wind speed
- Different parts of wind turbines
- Classification and description of wind machines
- Vertical axis and horizontal axis wind rotors
- Principles of wind energy conversion (aerodynamics)
- Simple wind turbine theory (betz theory of momentum)
- Characteristics of windmill rotors (rotor design)
- Types of electrical generators used with wind turbines
- Applications of wind turbines

6.1 WINDMILLS

If the mechanical energy is used directly by machinery, such as for a pump or grinding stones, the machine is usually called a windmill.

Wind mills

Wind turbines

6.2 WIND TURBINES

Wind turbines deliver their power through a revealing shaft, and in this respect, they are similar to other prime movers such as diesel engine and stream turbines. A generator can be coupled to this shaft and the electrical power delivered can be used to serve the multitude of different purposes for which electricity is required today.

In practice, an important difference between the wind turbine and the power delivered by engines and stream turbines is that the power delivered by wind turbines to the same extent uncontrolled and unpredictable over very short periods of time. In most applications for electricity, power is normally required on demand and not whenever available. Therefore, it is important to have some storage or reserve supply. This requirement has been one of the main limitations. If the mechanical energy is then converted to electricity, the machine is called a wind generator.

6.3 ENERGY AVAILABILITY IN THE WIND

The power in the wind is proportional to the cube of wind velocity. The general formula for wind power is

$$P = ½ (\rho S V^3) \tag{6.1}$$

where P = power in watts (W); ρ = the density of air in kilograms per cubic metre (kg/m^3); The density of air at sea level and room temperature is approximately 1.3 kg/m^3; A = cross-section area of the wind passing through in square metre (m^2); and V = velocity of the wind in metres per second (m/s).

Wind speed increases with height above the ground because of the earth's boundary layer. This effect is modelled using a power law relation

$$V_z = V_{10} (Z/10)^{\alpha} \tag{6.2}$$

where V_Z = wind speed at some height z (in m); V_{10} is the wind speed at 10 m (the height often used for meteorological reporting of wind speed); and α = power law exponent or index and it

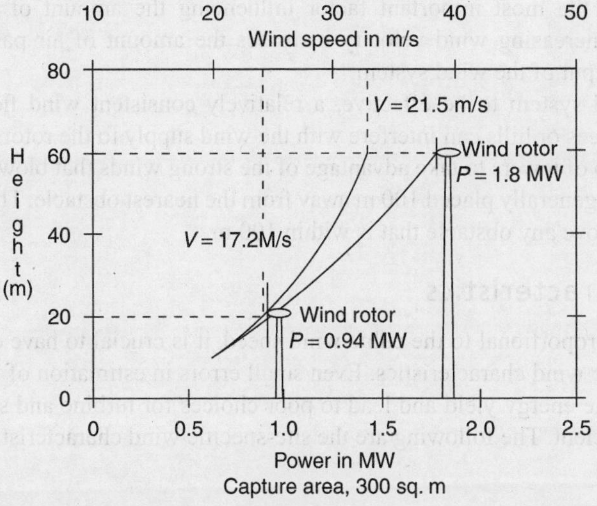

Figure 6.1 Relationship between wind speed, power, and height

varies over a wide range from 0.1 to 0.6 depending on the atmospheric conditions and the terrain near the wind turbine, but a value of 0.2 is common for wind turbine analysis.

For the example shown in Figure 6.1, the wind speed at 10 m height is 15 m/s wind speed.

At 20 m height, the wind speed is 17.2 m/s. Further, there is 0.94 MW of power available in the wind, for a 300 m^2 capture area (1 MW = 1 megawatts = 10^6 watts).

At 60 m, the wind speed is about 25% higher, but the power is almost doubled (≈1.8 MW). Due to physical limits (e.g., Betz limit) as well as inefficiencies in the rotor, generator, and gearboxes, not all of this power can be captured by a wind turbine.

6.3.1 Wind Potential

In order for a wind energy system to be feasible, there must be an adequate wind supply. A wind energy system usually requires an average annual wind speed of at least 15 km/h. Table 6.1 represents a guideline of different wind speeds and their potential in producing electricity.

A wind generator will produce lesser power in summer than in winter at the same wind speed as air has lower density in summer than in winter. Similarly, a wind generator will produce lesser power in higher altitudes—as air pressure as well as density is lower—than at lower altitudes.

Table 6.1 Wind Turbine Performance with Wind Speed

Average Wind Speed Suitability (km/h)	Wind Turbine Performance
Up to 15	Extremely poor
18	Poor
22	Moderate
25	Good
29	Excellent

The wind speed is the most important factor influencing the amount of energy a wind turbine can produce. Increasing wind velocity increases the amount of air passing the rotor, which increases the output of the wind system.

In order for a wind system to be effective, a relatively consistent wind flow is required. Obstructions such as trees or hills can interfere with the wind supply to the rotors. To avoid this, rotors are placed on top of towers to take advantage of the strong winds that blow high above the ground. The towers are generally placed 100 m away from the nearest obstacle. The middle of the rotor is placed 10 m above any obstacle that is within 100 m.

6.3.2 Wind Characteristics

As the wind power is proportional to the cubic wind speed, it is crucial to have detailed knowledge of the site-specific wind characteristics. Even small errors in estimation of wind speed can have large effects on the energy yield and lead to poor choices for turbine and site. An average wind speed is not sufficient. The following are the site-specific wind characteristics that are pertinent to wind turbines:

1. Mean wind speed: only interesting as a headline figure, but does not tell how often high wind speeds occur.
2. Wind speed distribution: diurnal, seasonal, annual patterns
3. Turbulence: short-term fluctuations
4. Long-term fluctuations
5. Distribution of wind direction
6. Wind shear (profile)

Information on those dimensions and tools for basic yield calculations is required. However, due to the sensitivities, no calculation can replace on-site wind measuring campaigns. The following are the main causes of high turbulence:

1. Inhomogeneous landscapes
2. Steep cliffs or mountain tops
3. Regions with many obstacles—buildings and others

Turbulence adversely affects the performance of wind machines. It includes

1. Reduced production of energy
2. Increased wear and tear shorten lifetime of the turbine
3. Increased dynamic loads on the blades

6.3.3 Wind into Electricity

Although the equation (6.1) gives the power in the wind, the actual power that can be extracted from the wind is significantly lesser than this figure suggested as above. The actual power will depend on the following several factors:

1. The type of machine and rotor used the sophistication of blade design
2. Friction losses

3. The losses in the pump or other equipment connected to the wind machine.
4. There are also physical limits to the amount of power that can be extracted realistically from the wind.

It is shown theoretically in succeeding section that any windmill can only possibly extract a maximum of 59.3% of the power from the wind (this is known as the Betz limit). In reality, this figure is usually around 45% (maximum) for a large electricity producing turbine and around 30%–40% for a wind pump. Therefore, modifying the formula for 'power in the wind', it can be said that the power that is produced by the wind machine can be given by the following formula:

$$P_M = ½C_p \rho SV^3 \tag{6.3}$$

where P_M = power (in watts) available from the machine and C_p = coefficient of performance of the wind machine.

It is also worth bearing in mind that a wind machine will only operate at its maximum efficiency for a fraction of the time it is running, due to the variations in the wind speed. A rough estimate of the output from a wind machine can be obtained using the following equation:

$$P_A = 0.2\, SV^3 \tag{6.4}$$

where P_A = the average power output in watts over the year and V = the mean annual wind speed in m/s.

6.4 WIND RESOURCES

Unfortunately, the general availability and reliability of wind speed data is extremely poor in many regions of the world. Large areas of the world appear to have average annual wind speeds below 3 m/s and are unsuitable for wind power systems; further, almost equally large areas have wind speeds in the intermediate range (3–4.5 m/s), where wind power may or may not be an attractive option. In addition, significant land areas have mean annual wind speeds exceeding 4.5 m/s, where wind power would most certainly be economically competitive.

6.4.1 Worldwide Wind Energy Scenario in 2010

As per the World Wind Energy Report 2010, wind energy scenario in 2010 is summarized as follows:

1. Worldwide capacity reached 196,630 MW, out of which 37,642 MW were added in 2010, slightly less than the capacity in 2009.
2. Wind power showed a growth rate of 23.6%, the lowest growth since 2004 and the second lowest growth of the past decade. All wind turbines installed by the end of 2010 worldwide can generate 430 TWh per annum; this wind power is more than the total electricity demand of the United Kingdom, the sixth largest economy of the world, and equalling 2,5% of the global electricity consumption.
3. China became number one in total installed capacity and the centre of the international wind industry, and it added 18,928 MW within one year, accounting for more than 50% of the world's market for new wind turbines.

4. Major decrease in new installations can be observed in North America and USA lost its number one position in total capacity to China.
5. Many Western European countries are showing stagnation, whereas there is strong growth in the number of Eastern European countries.
6. Germany keeps its number one position in Europe with 27,215 MW, followed by Spain with 20,676 MW.
7. The highest shares of wind power can be found in three European countries: Denmark (21%), Portugal (18%), and Spain (16%).
8. Asia accounted for the largest share of new installations (54,6%), followed by Europe (27,0%) and North America (16,7%).
9. Latin America (1,2%) and Africa (0,4%) still played only a marginal role in new installations.
10. Africa: North Africa represents still lion share of installed capacity, while wind energy plays hardly a role yet in Sub-Saharan Africa.
11. Nuclear disaster in Japan and oil spill in Gulf of Mexico will have long-term impact on the prospects of wind energy. Governments need to urgently reinforce their wind energy policies.
12. WWEA sees a global capacity of 600,000 MW as possible by 2015 and more than 1,500,000 MW by 2020.

6.4.2 Wind Energy in India

The Indian wind energy sector has an installed capacity of 14,158.00 MW (as on March 31, 2011). In terms of wind power installed capacity, India is ranked fifth in the world. Today, India is a major player in the global wind energy market.

The potential is far from exhausted. Indian Wind Energy Association has estimated that with the current level of technology, the 'on-shore' potential for utilization of wind energy for electricity generation is of the order of 65,000 MW. The unexploited resource availability has the potential to sustain the growth of wind energy sector in India in the years to come.

Wind in India are influenced by the strong south-west summer monsoon, which starts in May–June, when cool, humid air moves towards the land; further, the weak north-east winter monsoon, which starts in October, when cool, dry air moves towards the ocean. During March–August, the winds are uniformly strong over the whole Indian Peninsula, except the eastern peninsular coast. Wind speeds during November–March are relatively weak, although high winds are available during a part of the period on the Tamil Nadu coastline. A notable feature of the Indian programme has been the interest among private investors or developers in setting up of commercial wind power projects. The gross potential is 48,561 MW (source C-wet) and a total of about 14,158.00 MW of commercial projects have been established until March 31, 2011. The break-up of projects implemented in prominent wind potential states (as on March 31, 2011) is as given in Table 6.2.

Wind power potential has been assessed assuming 1% of land availability for wind farms requiring at 12 hectare/MW in sites having wind power density in excess of 200 W/m^2 at 50 m hub-height.

Table 6.2 State-wise Wind Power Installed Capacity in India

State	Gross Potential (MW)	Total Capacity (MW) Till 31.03.2011
Andra Pradesh	8,968	200.2
Gujarat	10,645	2,175.6
Karnataka	11,531	1,730.1
Kerala	1,171	32.8
Madhya Pradesh	1,019	275.5
Maharashtra	4,584	2310.7
Orissa	255	–
Rajasthan	4,858	1,524.7
Tamil Nadu	5,530	5,904.4
Others		4
Total(All India)	48,561	14,158

6.5 WIND TURBINE SITE SELECTION

The selection of a wind farm site is complex and time consuming, and also it involves multiple disciplines working on parallel paths. Financing, government permits, meteorological studies, land use restrictions, and design have to be completed well along before a site is approved and before the construction can begin. However, it is imperative in all of the above-referenced steps that construction expertise be involved and consulted to achieve maximum use of the approved site. Generally, there are three principle sources of construction expertise participating in wind farm projects. They are the design team responsible for conceptual and eventual site design, the developer or construction manager of the project, and the wind turbine generator contractor.

Wind is the energy resource that drives a wind turbine. A windmill needs to be placed on a high tower located in wind area. Not just any wind will do, a wind turbine needs air that moves uniformly in the same direction. Eddies and swirls, 'turbulence' in short, does not make good resource for a wind turbine. The rotor cannot extract energy from turbulent wind, and the constantly changing wind direction due to turbulence causes excessive wear and premature failure of turbine. This means that turbine must be placed high enough to catch strong winds, and above turbulent air. Since the tower price goes up quickly with height, there is a limit to what is practical and affordable.

6.5.1 Turbine Height

In general, wind turbines should be sited well above trees, buildings, and other obstacles. When the wind flows over an obstacle like a building or a tree, the wind is slowed down and turbulent

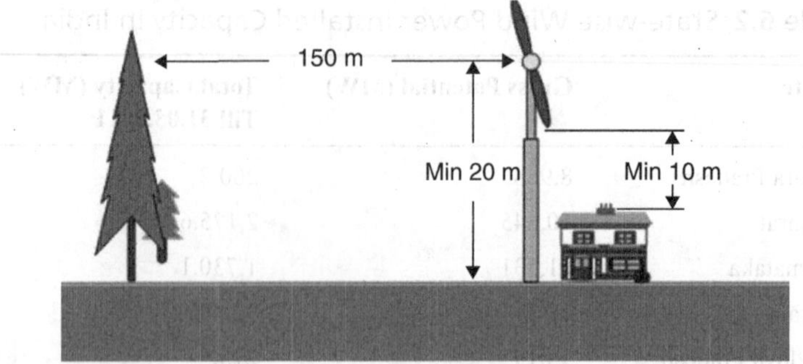

Figure 6.2 Installation of wind turbine (simple rule of thumb)

air is created, and if a wind turbine is located in this zone of turbulence, the result will be poor energy production and increased wear and tear on the turbine. One way to get above the zone of turbulence is to put the wind turbine on a tall tower.

Figure 6.2 is an illustration of a simple rule of thumb that is often used to specify a minimum tower height for a residential-sized wind turbine. The rule of thumb is to make sure that the tower is tall enough so that the entire turbine rotor is at least 10 m above the tallest obstacle within 150 m of the tower. Because trees grow and towers do not, the growth of trees over the lifetime of the wind turbine (typically 20–30 years between major rebuilds) should be considered in installation.

This should really be regarded as an absolute minimum for a wind turbine; at 10 m above an obstacle, there will still be some amount of turbulence and additional clearance is highly desirable. Changes in height of obstacles should be kept in mind as well. For example, if the obstacle like trees that are expected to grow up to 20 m high, it is advisable to use a 33-m tower.

Likewise, a 20-m tower should only be used when the terrain is very flat with no obstacles in a wide area around; for example, at the edge of the sea, or on top of a cliff with a clear area around it, or in the tundra. For most situations, a 20-m tower will only save a little money up front, while short selling energy production in the long run. To go beyond the rule of thumb, the airflow over any blunt obstruction, including a tree, tends to create a 'bubble' of turbulent air of twice the height of the obstacle, extending 20 times the height of the obstacle behind it. Therefore, your 10-m high house disturbs the air up to 200 m away. The tree line with 33 m trees disturbs the air up to 70 m high at a distance of 300 m away (see Fig. 6.3). Wind turbine may be located either upwind of the obstructions, or far enough downwind. Notice from Figure (6.3) that preference should be given to a site upwind of obstructions, but keep in mind that tall features downwind of the turbine can also influence the wind going through the blades, as shown in Figure (6.3).

Upwind and downwind are relative to the prevailing wind direction, where the wind blows from most of the time. A wind atlas can sometimes tell about prevailing wind direction, and if there is one at all. Some sites have winds that did not read the rule book, and there it is equally likely to blow from more than one direction.

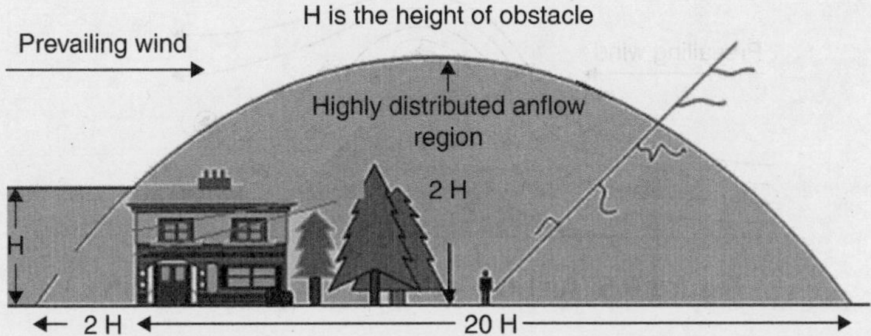

Figure 6.3 Installation of wind turbines

The Danish Wind Power Association made a very nice, interactive, calculator that allows one to plug in various obstacles (for example, a row of trees), set their height and distance to the wind turbine (see Fig. 6.4), and visually show what effect this will have on wind speed and energy. The calculator shows the percentage of the wind speed at various distances and heights behind the obstacle. It is necessary to remember that although the effect of obstacles is not just to diminish wind speeds, but they also make the air swirl, creating turbulence. Turbulence is an energy reducers when it comes to wind turbines.

Another aspect of proper windmill installation is the distance from occupied buildings. All wind turbines produce some amount of sound. Even though there are the most quiet wind turbines in the market (Scirocco), they also produce sound. Some people find its sound soothing, since it tells them they are making energy, while it drives others absolutely bonkers. For that reason, it is a good idea to place wind turbine some distance away from house, 100 feet is a good number for minimum separation. There is also such a thing as too much distance, since the length and gauge of the wiring that are needed will increase. With the ever-increasing price of copper, this makes it more expensive to install your turbine. This effect can be used as an advantage. For obstructions that are not smooth, like a cliff (i.e., a sudden rise in the landscape), it gets trickier. Sharp edges create turbulence, as illustrated in Figure 6.5.

The airflow at the top of the cliff can be stronger than the average wind speed in the area, but close to the cliff's edge, it may also be very turbulent, making it a poor site for a turbine. If there is a cliff edge on land and want to use it for installing a wind turbine, a 20 m high tower should be used to get above the turbulent air. Even if it seems that the wind is always blowing hard at the cliff's edge, the lee side (downwind of the prevailing winds) of a bluff object makes for a very

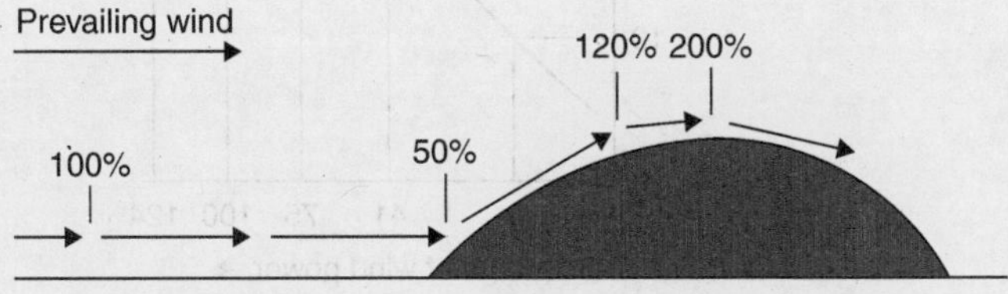

Figure 6.4 Wind speed around a cliff

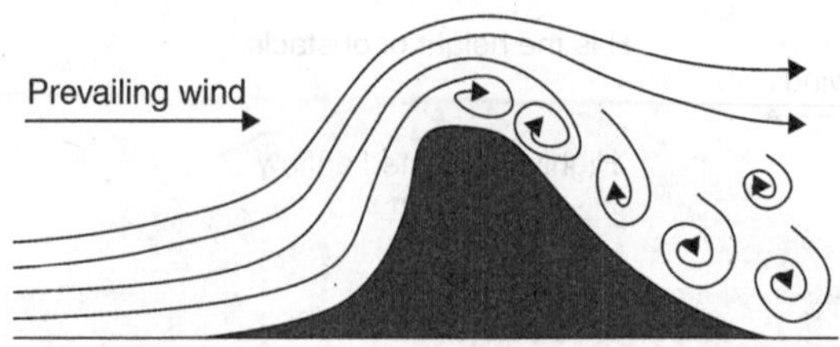

Figure 6.5 Airflow at different levels of cliff

poor wind turbine site. The bluff object will create large turbulence on its downwind side, and the average wind speed will drop off precipitously as well. This leaves no energy for the wind turbine to harvest.

Weather data usually reports wind speeds at 10 m above the ground level, and the spreadsheet can take care of translating that to a wind speed at turbine height. In order to understand how changes in tower height affect the power in the wind for an unobstructed site, the process is shown in Figure 6.5.

The energy in the wind increases with the cube of the wind speed ($P \alpha V^3$), and wind speed increases with height. An increase of just 26% in wind speed means twice as much power available in the wind, and your wind turbine will produce almost twice as much. By doubling the wind speed, almost eight times as much power can be harvested. Therefore, a small additional investment in tower height may be well worth it due to the increased energy production (see Fig. 6.6).

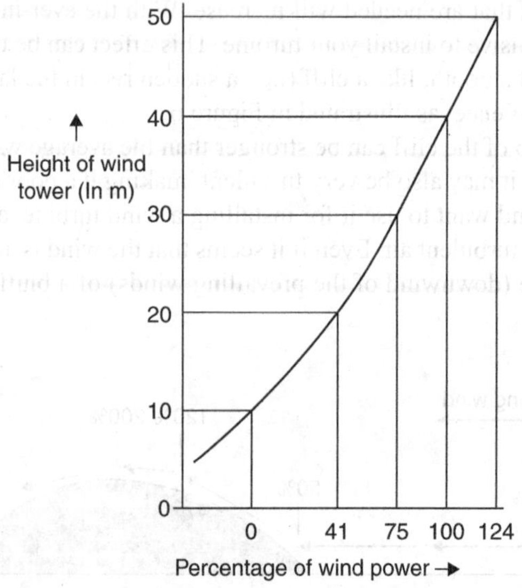

Figure 6.6 Wind power versus tower height

The airflow that close to the building is generally very turbulent, leading to premature failure and poor power production. It is usually noisy too. Every wind turbine has some amount of vibration associated with it, and this too will be transmitted inside the house. We know, the thought of bolting a little turbine to the house, just over the roof line, to offset your electricity use (as that salesman put it) is appealing. The harsh reality is that it does not work. Several studies were done, involving dozens of roof-top turbines. They all concluded that those turbines do not work. Their energy production is negligible and some were even net users of electricity (because their inverters draw power, even when nothing is going into the grid). It is necessary to say 'no' to building-mounted turbines.

6.5.2 Considerations and Guidelines for Site Selection

When looking for a place for a wind turbine, engineers consider factors such as wind hazards, characteristics of the land that affect wind speed, and the effects of one turbine on nearby turbines in wind farms. The following important factors need careful considerations:

1. *Hill effect*: When it approaches a hill, wind encounters high pressure because of the wind that has already built up against the hill. This compressed air rises and gains speed as it approaches the crest, or top of the hill. The installation of wind turbines on hilltops takes advantage of this increase in speed.
2. *Roughness or the amount of friction that earth's surface exerts on wind*: Oceans have very little roughness. A city or a forest has a great deal of roughness, which slows the wind.
3. *Tunnel effect*: The increase in air pressure undergoes when it encounters a solid obstacle. The increased air pressure causes the wind to gain speed as it passes between, for example, rows of buildings in a city or between two mountains. Placing a wind turbine in a mountain pass can be a good way to take advantage of wind speeds that are higher than those of the surrounding air.
4. *Turbulence*: Rapid changes in the speed and direction of the wind, often caused by the wind blowing over natural or artificial barriers are called turbulence. Turbulence causes not only fluctuations in the speed of the wind but also wear and tear on the turbine. Turbines are mounted on tall towers to avoid turbulence caused by ground obstacles.
5. *Variations in wind speed*: During the day, winds usually blow faster than they do at the night because the sun heats the air, setting air currents in motion. In addition, wind speed can differ depending on the season of the year. This difference is a function of the sun, which heats different air masses around earth at different rates, depending on the tilt of the earth towards or away from the sun.
6. *Wake*: Energy can neither be created nor destroyed. As wind passes over the blades of a turbine, the turbine seizes much of the energy and converts it into mechanical energy. The air coming out of the blade sweep has less energy because it has been slowed. The abrupt change in the speed makes the wind turbulent, a phenomenon called wake. Because of wake, wind turbines in a wind farm are generally placed about three rotor diameters away from one another in the direction of the wind, so that the wake from one turbine does not interfere with the operation of the one behind it.
7. *Wind obstacles*: Trees, buildings, and rock formations are the main obstacles in the installation of wind turbines. Any of these obstacles can reduce wind speed considerably and

increase turbulence. Wind obstacles like tall buildings cause wind shade, which can considerably reduce the speed of the wind, and therefore, the power output of a turbine.

8. *Wind shear*: It is the differences in wind speeds at different heights. When a turbine blade is pointed straight upward, the speed of the wind hitting its tip can be, for example, 9 miles (14 km) per hour, but when the blade is pointing straight downward, the speed of the wind hitting its tip can be 7 miles (11 km) per hour. This difference places stress on the blades. Further, too much wind shear can cause the turbine to fail.

Choosing the right site for wind turbine is the most important decision. Further, the location plays a vital part in the performance and efficiency of a wind turbine. The following guidelines can be followed to evaluate site for the installation of wind turbines:

1. Turbines work best when on high and exposed sites. Coastal sites are especially good.
2. Town centres and highly populated residential areas are usually not suitable sites for wind turbines.
3. Avoid roof-mounted turbines as there is no guarantee that these devices will not damage property through vibration.
4. The farther the distance between the turbine and the power requirement, the more power will be lost in the cable. The distance of the cabling will also impact the overall cost of the installation.
5. Turbulence disrupts the air flow that can wear down the blades and reduces the lifecycle of the turbine. It is recommended that installing a turbine may be considered only when the distance between the turbine and the nearest obstacle is more than twice the height of the turbine, or when the height of the turbine is more than twice the height of the nearest obstacle.
6. Small turbines require an average wind speed of over 4.5 m/s to produce an efficient level of electricity.
7. If site is in a remote location, connecting wind turbine to the national grid will be very expensive and it may be worth considering an off-grid connection instead using battery storage.

6.5.3 Wind Turbine Power Output Variation with Steady Wind Speed

Figure 6.7 gives the power output from a wind turbine variation with steady wind speed.

There are five important characteristic wind speeds and they are as follows:

1. Start-up speed is the speed at which the rotor and blade assembly begin to rotate.
2. The cut-in wind speed is the speed when the machine begins to produce power.
3. The design wind speed is the speed when the windmill reaches its maximum efficiency.
4. The rated wind speed is the speed when the machine reaches its maximum output power.
5. The furling wind speed is the speed when the machine furls to prevent damage at high wind speeds.

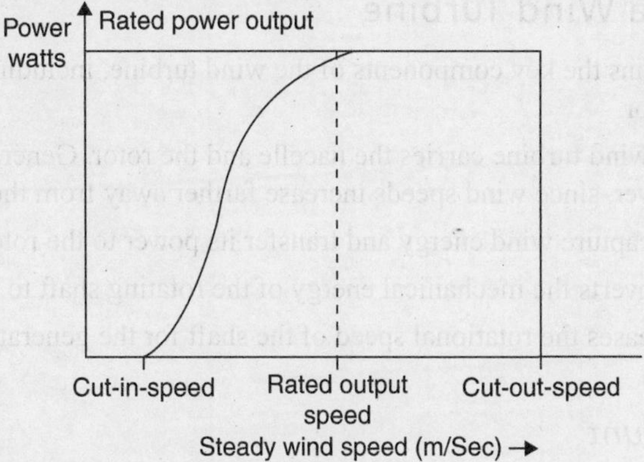

Figure 6.7 Wind turbine power output with steady wind speed

Cut-in speed is the minimum wind speed at which the wind turbine will generate usable power. This wind speed is typically between 7 and 10 mph for most turbines. At very low wind speeds, there is insufficient torque exerted by the wind on the turbine blades to make them rotate. However, as the speed increases, the wind turbine will begin to rotate and generate electrical power. The speed at which the turbine first starts to rotate and generate power is called the cut-in speed and is typically between 3 and 4 m/s.

Rated speed At wind speeds between cut-in and rated, the power output from a wind turbine increases as the wind increases. The output of most machines levels off above the rated speed. Most manufacturers provide graphs, called 'power curves,' showing how their wind turbine output varies with wind speed. As the wind speed rises above the cut-in speed, the level of electrical output power rises rapidly as shown. However, typically somewhere between 12 m/s and 17 m/s, the power output reaches the limit that the electrical generator is capable of. This limit to the generator output is called the rated power output and the wind speed at which it is reached is called the rated output wind speed. At high wind speeds, the design of the turbine is arranged to limit the power to this maximum level and there is no further rise in the output power. How this is done varies from design to design; however, typically, with large turbines, it is done by adjusting the blade angles so as to keep the power at the constant level.

Cut-out speed At very high wind speeds, usually around 25 m/s, most wind turbines cease power generation and shutdown. The wind speed at which the shutdown occurs is called the cut-out speed, or sometimes the furling speed. Having a cut-out speed is a safety feature that protects the wind turbine from damage. Shutdown may occur in one of the several ways. In some machines, an automatic brake is activated by a wind speed sensor. Some machines twist or 'pitch' the blades to spill the wind. Still others use 'spoilers,' drag flaps mounted on the blades or the hub that are automatically activated by high rotor rpms, or mechanically activated by a spring-loaded device, which turns the machine sideways to the wind stream. Normal wind turbine operation usually resumes when the wind drops back to a safe level. As the speed increases above the rate output wind speed, the forces on the turbine structure continue to rise and, at some point, there is a risk of damage to the rotor. As a result, a braking system is employed to bring the rotor to a standstill.

6.5.4 Parts of a Wind Turbine

1. The nacelle contains the key components of the wind turbine, including the gearbox and the electrical generator.
2. The tower of the wind turbine carries the nacelle and the rotor. Generally, it is an advantage to have a high tower, since wind speeds increase farther away from the ground.
3. The rotor blades capture wind energy and transfer its power to the rotor hub.
4. The generator converts the mechanical energy of the rotating shaft to electrical energy.
5. The gearbox increases the rotational speed of the shaft for the generator.

6.5.4.1 Blade Count

The determination of the number of blades involves design considerations of aerodynamic efficiency, component costs, system reliability, and aesthetics. Noise emissions are affected by the location of the blades upwind or downwind of the tower and the speed of the rotor. Given that the noise emissions from the blades' trailing edges and tips vary by the fifth power of blade speed, a small increase in the tip speed can make a large difference. Wind turbines developed over the last 50 years have almost universally used either two or three blades. Aerodynamic efficiency increases with number of blades but with diminishing return. Increasing the number of blades from one to two yields a 6% increase in aerodynamic efficiency, whereas increasing the blade count from two to three yields only an additional 3% in efficiency. Further increasing the blade count yields minimal improvements in aerodynamic efficiency and sacrifices too much in blade stiffness as the blades become thinner.

Component costs that are affected by blade count are primarily for materials and manufacturing of the turbine rotor and drive train. Generally, the fewer the number of blades, the lower the material and manufacturing costs will be. In addition, the fewer the number of blades, the higher the rotational speed can be. This is because blade stiffness requirements to avoid interference with the tower limit how thin the blades can be manufactured, but only for upwind machines; deflection of blades in a downwind machine results in increased tower clearance. Fewer blades with higher rotational speeds reduce peak torques in the drive train, resulting in lower gearbox and generator costs.

The limitation on the available power in the wind means that the more the blades are, the lesser the power each can extract. A consequence of this is that each blade must also be narrower to maintain aerodynamic efficiency. The total blade area as a fraction of the total swept disc area is called the solidity, and aerodynamically, there is an optimum solidity for a given tip speed; the higher the number of blades, the narrower each one must be. In practice, the optimum solidity is low (only a few percentage) and this means that even with only three blades, each one must be very narrow. To slip through the air easily, the blades must be thin relative to their width, so the limited solidity also limits the thickness of the blades. Furthermore, it becomes difficult to build the blades strong enough if they are too thin, or the cost per blade increases significantly as more expensive materials are required.

For this reason, most large machines do not have more than three blades. The other factor influencing the number of blades is aesthetics: it is generally accepted that three-bladed turbines are less visually disturbing than one- or two-bladed designs.

6.5.4.2 *Blade Materials*

Wood and canvas sails were used on early windmills due to their low price, availability, and ease of manufacture. Small blades can be made from light metals such as aluminium. However, these materials require frequent maintenance. Wood and canvas construction limits the airfoil shape to a flat plate, which has a relatively high ratio of drag to force captured (low aerodynamic efficiency) when compared to solid airfoils. The constructions of solid airfoil designs require inflexible materials such as metals or composites. Some blades also have incorporated lightning conductors.

New wind turbine designs push power generation from the single megawatt range to upwards of 10 MW using very large blades. A large area effectively increases the tip-speed ratio of a turbine at a given wind speed, thus increasing its energy extraction. Computer-aided engineering software such as HyperSizer (originally developed for spacecraft design) can be used to improve blade design. The current production of wind turbine blades is as large as 100 m in diameter with prototypes in the range of 110–120 m. In 2001, an estimated 50,000,000 kg of fiberglass laminate were used in wind turbine blades.

An important goal of large blade systems is to control blade weight. Since blade mass scales as the cube of the turbine radius, loading due to gravity constrains systems with large blades. Proven fiberglass composite fabrication techniques are used in manufacturing blades in the 40–50 m range. Each technique use a glass fibre reinforced polymer composite materials constructed with different complexities. Perhaps the largest issue with more simplistic, open-mould, wet systems are the emissions associated with the volatile organics released. Pre-impregnated materials and resin infusion techniques avoid the release of volatiles by containing all reaction gases. However, these contained processes have their own challenges, namely the production of thick laminates necessary for structural components becomes more difficult. As the preform resin permeability dictates the maximum laminate thickness, bleeding is required to eliminate voids and insure proper resin distribution. One solution to resin distribution is a partially pre-impregnated fiberglass. During the evacuation, the dry fabric provides a path for airflow and, once heat and pressure are applied, resin may flow into the dry region resulting in a thoroughly impregnated laminate structure.

Epoxy-based composites have environmental, production, and cost advantages over other resin systems. Carbon fibre-reinforced spars for load bearing can reduce weight and increase stiffness. Using carbon fibres in 60 m turbine blades is estimated to reduce total blade mass by 38% and decrease the cost by 14% when compared to 100% fiberglass. Carbon fibres have the added benefit of reducing the thickness of fiberglass laminate sections and also solves the problems associated with resin wetting of thick lay-up sections. Wind turbines may also benefit from the general trend of increasing use and decreasing cost of carbon fibre materials.

6.6 CLASSIFICATION AND DESCRIPTION OF WIND MACHINES

The basic wind energy conversion device is the wind turbine. Although various designs and configurations exist, these turbines are generally grouped into two types depending on the position of the rotor axis. Figure 6.8 illustrates the two types of turbines and typical subsystems for an electricity generation application.

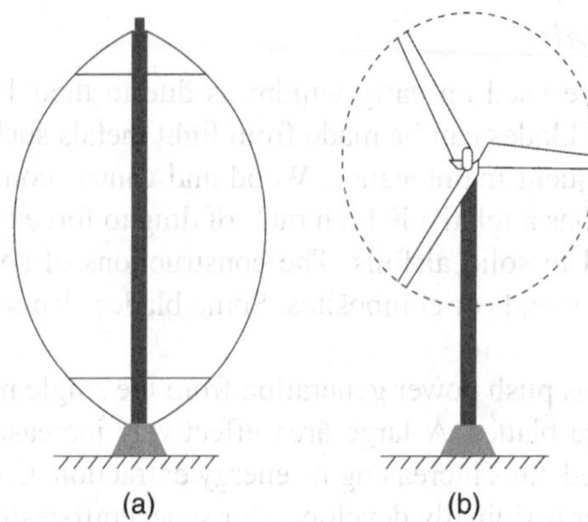

Figure 6.8 Wind rotor configurations (a) Vertical axis (b) Horizontal axis

Two important wind rotor configurations are as follows:

1. In vertical-axis wind turbines (VAWT), the axis of rotation is vertical with respect to the ground (and roughly perpendicular to the wind stream), as shown in Figure 6.8(a). The following are the two main types of VAWT:
   (a) Darrieus (which uses lift forces generated by aerofoils)
   (b) Savonius (which uses drag forces)
2. Horizontal-axis turbines, in which the axis of rotation is horizontal with respect to the ground (and roughly parallel to the wind stream), as represented in Figure 6.8(b). Horizontal-axis wind turbines (HAWT) can be further divided into three types:
   (a) Dutch windmills
   (b) Multi-blade water-pumping windmills
   (c) High-speed propeller-type wind machines

6.6.1 Savonius Drag-type Vertical-axis Wind Turbines

The Savonius wind generator was developed by a Finnish engineer known as Sigurd J. Savonius in 1920s. During that time, horizontal axis windmills were already widely used. However, early conventional windmill designs were quite sophisticated and costly to arrange and maintain. The Savonius

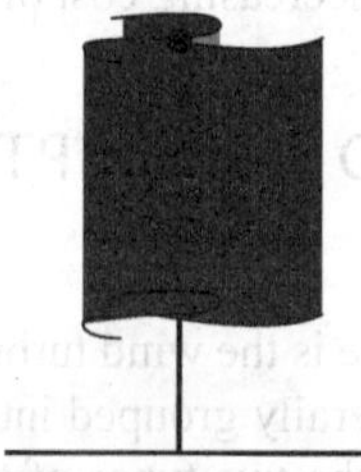

Figure 6.9 Savonius drag-type VAWT

wind turbine became a perfect alternative. The system was adapted for purposes such as water pumping and grain grinding. The Savonius turbine, as shown in Figure 6.9, is one of the simplest turbines.

Aerodynamically, it is a drag-type device consisting of two or three scoops. Figure 6.9 shows the plan view the Savonius drag-type VWAT; if one views the rotor from the top, a two-scoop machine would look like an 'S' shape in cross-section . Because of the curvature, the scoops experience less drag when moving against the wind than with the wind. The differential drag causes the Savonius turbine to spin. Because they are drag-type devices, Savonius turbines extract much less of the wind's power than other similarly sized lift-type turbines. Much of the swept area of a Savonius rotor may be near the ground, if it has a small mount without an extended post, making the overall energy extraction less effective due to the lower wind speeds found at lower heights.

They can be made in many different ways with buckets, paddles, sails, and oil drums. The Savonius rotor is S-shaped (when viewed from above). All of these designs turn relatively slow, but yield a high torque. They can be useful for grinding grain, pumping water, and many other tasks. They, however, are not good for generating large amounts of electricity. The drag-based VAWTs usually turn below 100 RPM. One might use a gearbox, but then efficiency suffers and the machine may not start at all easily.

One of the most important advantages of VAWT is their easy maintenance. This refers to the fact that all the working parts are at ground level. Hence, there is no need to climb tall towers to reach the blades. Another advantage involves placement. No calculations for wind direction and speed are required. For example, place the Savonius in your yard in the most suitable and inconspicuous location. It will capture the wind from any placement. Its small rotation translates into quiet operation that will produce small but steady electricity. However, one drawback to the Savonius' use on any large scale is its slow turning rotation, which requires a manual start. Thus, this limits the amount of electricity it can generate. They are used in electricity generation with the benefit that they continue to generate electricity in the strongest winds without being damaged, they are very quiet, and they are relatively easy to make. Savonius turbines do not scale well to kilowatt sizes; however, they are useful for small-scale domestic electricity generation. They are ideally suited to applications such as pumping water and grinding grain for which slow rotation and high torque are essential. Because of the torque yield of a Savonius wind turbine, the bearings used must be very sturdy and may require servicing every couple of years.

6.6.1.1 Advantages of Savonius

1. always self-starting, if there are at least three scoops
2. relatively easy to make

6.6.1.2 Disadvantages of Savonius

1. low efficiency: around 15%.

6.6.2 Darrieus Lift-type Vertical-axis Machines

The Darrieus wind turbine is a type of VAWT used to generate electricity from the energy carried in the wind. The turbine consists of a number of aerofoil usually but not always vertically mounted on a rotating shaft or framework. This design of wind turbine was patented by Georges Jean Marie Darrieus, a French aeronautical engineer in 1931.

In the original versions shown in Figure 6.10(a) of the Darrieus design, the aerofoils are arranged so that they are symmetrical and have zero rigging angles, that is, the angle that the aerofoils are set relative to the structure on which they are mounted. This arrangement is equally effective no matter which direction the wind is blowing—in contrast to the conventional type, which must be rotated to face into the wind.

A special type of the Darrieus turbine is the so-called 'Giromill'. It uses the same principle as a Darrieus wind turbine to capture wind energy but uses two or three straight blades individually attached to a vertical axis instead of curved blades (see Fig. 6.10b). Darrieus 1927 patent also covered practically for any possible arrangement using vertical airfoils. One of the more common types is the Giromill or H-bar design, in which the long 'egg beater' blades of the common Darrieus design are replaced with straight vertical blade sections that are attached to the central tower with horizontal supports.

It is also possible to use helical blades wrapped around a vertical axis in order to minimize the pulsating torque (see Fig. 6.10c). The blades of a Darrieus turbine can be canted into a helix, for example, three blades and a helical twist of 60°, similar to Gorlov's water turbines. Since the wind pulls each blade around both the windward and leeward sides of the turbine, this feature spreads the torque evenly over the entire revolution, thus preventing destructive pulsations.This design is used by the Turby, an Urban Green Energy and Quiet Revolution brand of wind turbine.

The most successful VAWT is the Darrieus rotor illustrated in Figure 6.10. The most attractive feature of this type of turbine is that the generator and transmission devices are located at ground level. Additionally, they are able to capture the wind from any direction without the need to yaw. However, these advantages are counteracted by a reduced energy capture. Wind turbines are mechanical devices specifically designed to convert part of the kinetic energy of the wind into useful mechanical energy. Several designs have been devised throughout the times. Most of them comprise a rotor that turns round propelled by lift or drag forces, which result from its interaction with the wind. Furthermore, despite having the generator and transmission at ground level, maintenance is not simple since it usually requires rotor removal. In addition, these rotors are

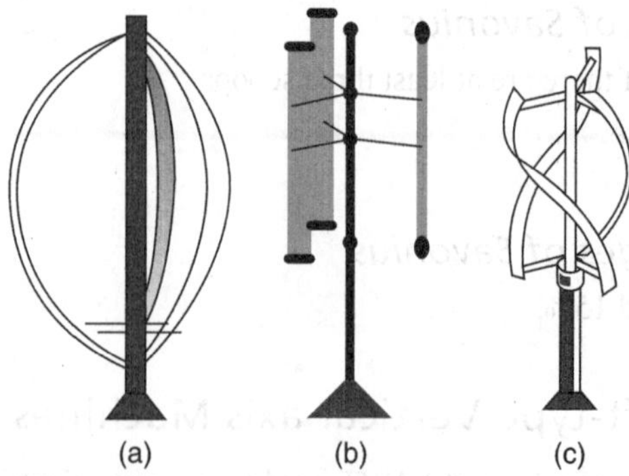

Figure 6.10 Darrieus wind turbines

supported by guy ropes taking up large land extensions. Due to these reasons, the use of VAWT has considerably declined during the last decades.

When the Darrieus rotor is spinning, the aerofoils are moving forward through the air in a circular path. Relative to the blade, this oncoming airflow is added vectorially to the wind, so that the resultant airflow creates a varying small positive angle of attack to the blade. This generates a net force pointing obliquely forwards along a certain 'line-of-action'. This force can be projected inwards past the turbine axis at a certain distance, giving a positive torque to the shaft, thus helping it to rotate in the direction it is already travelling in. The aerodynamic principles that rotate the rotor are equivalent to that in autogiros and normal helicopters in autorotation.

When the rotor is stationary, no net rotational force arises, even if the wind speed rises quite high—the rotor must already be spinning to generate torque. Thus, the design is not normally self-starting. Under rare conditions, Darrieus rotors can self-start, so some form of brake is required to hold it when stopped.

One problem with the design is that the angle of attack changes as the turbine spins, so each blade generates its maximum torque at two points on its cycle (front and back of the turbine). This leads to a sinusoidal (pulsing) power cycle that complicates design. In particular, almost all Darrieus turbines have resonant modes where, at a particular rotational speed, the pulsing is at a natural frequency of the blades that can cause them to (eventually) break. For this reason, most Darrieus turbines have mechanical brakes or other speed control devices to keep the turbine from spinning at these speeds for any lengthy period of time.

Another problem arises because the majority of the mass of the rotating mechanism is at the periphery rather than at the hub, as it is with a propeller. This leads to very high centrifugal stresses on the mechanism, which must be stronger and heavier than otherwise to withstand them.

The lift based VAWT shaped like an 'eggbeater' is the Darrieus VAWT, as shown in Figure 6.10. The long VAWT blades have many natural frequencies of vibration that must be avoided during operation.

6.6.2.1 Advantages of Darrieus

1. The equipment (gear box and generator) can be placed close to the ground.
2. There is no need of a mechanism to turn the rotor against the wind.

6.6.2.2 Disadvantages of Darrieus

1. The efficiency is not very remarkable
2. The Darrieus is not a self-starting turbine, the starting torque is very low but it can be reduced by using three or more blades that result in a high solidity for the rotor.
3. Because wind speeds are close to the ground level, there is very low wind speed on the lower part of the rotor.
4. They are very difficult to mount high on a tower to capture the high level winds. Because of this, they are usually forced to accept the low, more turbulent winds, and they produce less in possibly more damaging winds.

5. Guy cables are usually used to keep the turbine erect. They also impose a large thrust loading on the main turbine bearings and bearing selection is critical. Like all types of turbines, replacing main bearings requires that the turbine be taken down.
6. They have not performed well in the commercial wind turbine market although a number of small, household-sized units have recently hit the market.

6.6.3 Advantages of Vertical-axis Wind Turbines (VAWT)

1. The turbine generator and gearbox can be placed lower to the ground, thus facilitating easy maintenance and low construction costs.
2. The main advantage of VAWT is it does not need to be pointed towards the wind to be effective. In other words, they can be used on the sites with high variable wind direction.
3. Since VAWT are mounted close to the ground, they are more bird friendly and do not destroy the wildlife.
4. VAWT is quiet, efficient, economical, and perfect for residential energy production, especially in urban environments.

6.6.4 Disadvantages of Vertical-axis Wind Turbines (VAWT)

Despite the abovementioned advantages, VAWT suffer from the following serious drawbacks.

1. As the VAWT are mounted close to the ground, less wind speed is available to harness, which means less production of electricity.
2. VAWT are very difficult to erect on towers, which means they are installed on base, such as ground or building.
3. Another disadvantage of VAWT is the inefficiency of dragging each blade back through the wind.

There are advantages and disadvantages to VAWT. The most desirable attributes include nearly silent operation and an avian-friendly design. However, most VAWT are difficult to mount on a high tower to capture the high speed and less turbulent winds found at high elevations. As such, they must utilize the generally low speed winds found near ground level with a consequent reduction in power output. The silver lining is that having the generator mounted at or near ground level makes maintenance significantly less challenging. Further, VAWT are omnidirectional and accept wind from any direction. This somewhat reduces the penalty of utilizing more turbulent air. VAWT tend to be significantly more expensive to produce, and this, above all else, is likely responsible for the small number of models available for purchase. In terms of polar applications, there are really only two VAWT manufacturers to consider: the Finish Windside and the Italian Ropatec design. They are extremely robust and heavy machines are capable of withstanding the most extreme environmental conditions. They are also quite expensive with a very poor power-to-weight ratio.

6.6.5 Horizontal-axis Wind Turbines

Horizontal-axis wind turbines, also abbreviated as HAWT, are the common style that most of us think of when we think of a wind turbine. An HAWT has a similar design to a windmill; it has blades that look like a propeller that spin on the horizontal axis. They have the main rotor shaft and electrical generator at the top of a tower, and they must be pointed into the wind. Small turbines are pointed by a simple wind vane placed square with the rotor (blades), while large turbines generally use a wind sensor coupled with a servomotor. Most large wind turbines have a gearbox, which turns the slow rotation of the rotor into a faster rotation that is more suitable to drive an electrical generator.

Since a tower produces turbulence behind it, the turbine is usually pointed upwind of the tower. Wind turbine blades are made stiff to prevent the blades from being pushed into the tower by high winds. Additionally, the blades are placed in a considerable distance in front of the tower and are sometimes tilted up a small amount.

Downwind machines have been built, despite the problem of turbulence, because they do not need an additional mechanism for keeping them in line with the wind, and because in high winds, the blades can be allowed to bend, which reduces their swept area, and thus their wind resistance. Since turbulence leads to fatigue failures, and reliability is so important, most HAWTs are upwind machines. Nowadays, almost all commercial wind turbines connected to grid have horizontal-axis two-bladed or three-bladed rotors, as shown in Figure 6.11. The rotor is located at the top of a tower where the winds have more energy and are less turbulent. The tower also holds up a nacelle. The gearbox and the generator are assembled inside. There is also a yaw mechanism that turns the rotor and nacelle. In normal operation, the rotor is yawed to face the wind in order to capture as much energy as possible. Although it may be very simple in low power applications, the yaw system is likely to be one of the more complicated devices in high power wind turbines. Finally, the power electronics are arranged at ground level.

6.6.5.1 Giromill Wind Turbine

The giromill is typically powered by two or three vertical aerofoil (as shown in Fig. 6.12) attached to the central mast by horizontal supports. Giromill turbines work well under turbulent wind conditions and are the affordable option where a standard horizontal axis windmill-type turbine is unsuitable.

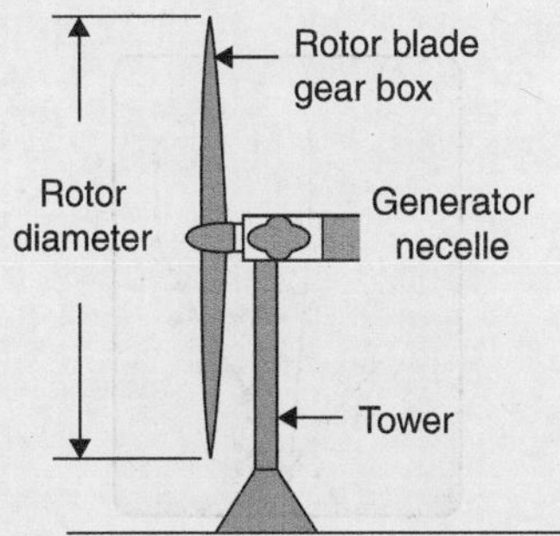

Figure 6.11 Horizontal-axis wind turbine

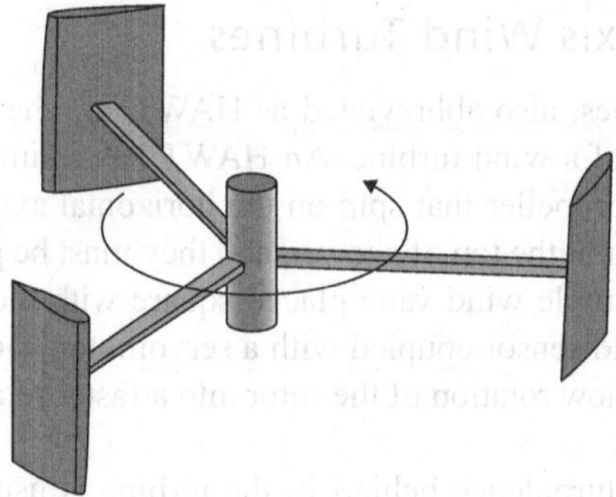

Figure 6.12 Giromill

6.6.5.2 Dutch Windmills

Dutch windmills (see Fig. 6.13) were the frontrunners to the windmills widely used across Europe for grinding grains. They operated on the thrust exerted by the wind. Their blades were inclined at an angle to the wind to result in rotation. Sails or wooden slats were used to manufacture these blades (wind turbine).

A Dutch windmill (multi-blade water-pumping windmills) have a large number of blades. The blades are generally made of wooden or metallic slats. This is used to rotate the shaft of a water pump. The location of the mill is not governed by the availability of wind, but by the availability of water. Hence, it is designed to be able to operate at low wind speeds. Their remote locations and purpose of use make their efficiency to take the backseat. Reliability, sturdiness, and low cost are the prime criteria for the design of these windmills. A tail vane is generally mounted on the turbine to orient it to face the wind

Horizontal-axis wind turbines dominate the majority of the wind industry. Horizontal axis means the rotating axis of the wind turbine is horizontal or parallel with the ground. In high wind applications, HAWT are used. However, in small wind and residential wind applications, vertical axis turbines have their place. The advantage of horizontal wind is that it is able to produce more electricity from a given amount of wind. Therefore, if one is trying to produce as much wind as possible at all times, horizontal axis is likely the choice for the same. However, their

Figure 6.13 Dutch windmills

disadvantage is that it is generally heavy and it does not produce well in turbulent winds. Using VAWT, the rotational axis of the turbine stands vertical or perpendicular to the ground.

Horizontal-axis wind turbines are the most common type used. All of the components (blades, shaft, and generator) are on top of a tall tower, and the blades face into the wind. The shaft is horizontal to the ground. The wind hits the blades of the turbine that are connected to a shaft causing rotation. The shaft has a gear on the end that turns a generator. The generator produces electricity and sends the electricity into the power grid. The wind turbine also has some key elements that add to efficiency. Inside the Nacelle (or head) is an anemometer, wind vane, and controller that read the speed and direction of the wind. As the wind changes direction, a motor (yaw motor) turns the nacelle so the blades are always facing the wind. The power source also comes with a safety feature. In the case of extreme winds, the turbine has a break that can slow the shaft speed. This is to inhibit any damage to the turbine under extreme conditions.

6.6.5.3 Advantages of Horizontal-axis Wind Turbines

1. *Stability*: Blades are to the side of the turbines' centre of gravity, helping stability.
2. *Ability to wing warp*: This gives the turbine blades the best angle of attack.
3. *Ability to pitch the rotor blades in a storm to minimize damage*: Variable blade pitch, which gives the turbine blades the optimum angle of attack. Allowing the angle of attack to be remotely adjusted gives great control, so the turbine collects the maximum amount of wind energy for the time of day and season.
4. *Tall tower allows access to strong wind in sites with wind shear*: In some wind shear sites, every 10 m up, the wind speed can increase by 20% and the power output by 34%.
5. *High efficiency*: Since the blades always move perpendicularly to the wind, and receives power through the whole rotation. In contrast, all VAWT, and the most proposed airborne wind turbine designs involve various types of reciprocating actions, requiring airfoil surfaces to backtrack against the wind for part of the cycle. Backtracking against the wind leads to inherently low efficiency.
6. Tall tower allows placement on uneven land or in offshore locations.
7. Can be sited in forest above tree-line.
8. Most are self-starting.

6.6.5.4 Disadvantages of Horizontal-axis Wind Turbines

1. It is difficult to transport (20% of equipment costs) and install. Tall masts and blades are more difficult to transport and install. Transportation and installation can now cost 20% of equipment costs. Further, it requires tall cranes and skilled operators.
2. Strong tower construction is required to support the heavy blades, gearbox, and generator.
3. Effect radar in proximity reflections from tall HAWTs may affect side lobes of radar installations creating signal clutter, although filtering can suppress it.

4. Local opposition to aesthetics mast height can make them obtrusively visible across large areas, disrupting the appearance of the landscape, and sometimes creating local opposition.
5. Fatigue and structural failure caused by turbulence downwind variants suffer from fatigue and structural failure caused by turbulence when a blade passes through the tower's wind shadow (for this reason, the majority of HAWTs use an upwind design, with the rotor facing the wind in front of the tower).
6. Difficult to maintain.
7. They require an additional yaw control mechanism to turn the blades toward the wind.

6.7 PRINCIPLES OF WIND ENERGY CONVERSION (AERODYNAMICS)

The VAWT and HAWT use either lift or drag forces to harness the wind. Out of these types, the horizontal-axis lift device is the most commonly used. In fact, other than a few experimental machines, virtually all windmills come under this category.

There are two primary physical principles by which energy can be extracted from the wind. These are through the creation of either lift or drag force (or through a combination of the two), as shown in Figure 6.14.

Air flow over a stationary airfoil produces two forces, a lift force perpendicular to the air flow and a drag force in the direction of air flow, as shown in Figure 6.14. The existence of the lift force depends upon laminar flow over the airfoil, which means that the air flows smoothly over both sides of the airfoil. If turbulent flow exists rather than laminar flow, there will be little or no lift force. The air flowing over the top of the airfoil has to speed up because of a greater distance to travel and this increase in speed causes a slight decrease in pressure. This pressure difference across the airfoil yields the lift force, which is perpendicular to the direction of air flow. The air moving over the airfoil also produces a drag force in the direction of the air flow. This is a loss term and is minimized as much as possible in high-performance wind turbines

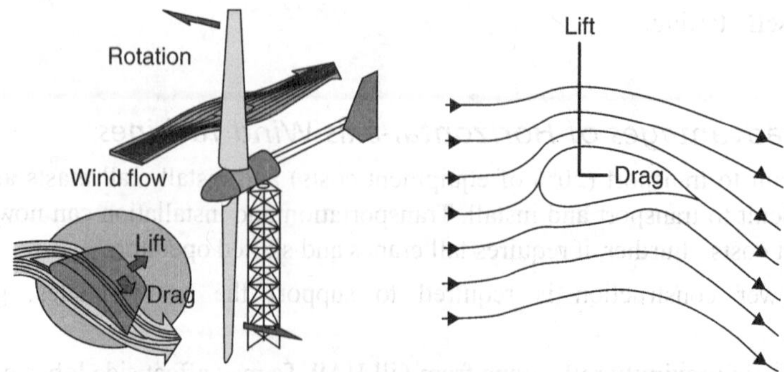

Figure 6.14 Principles of wind turbine aerodynamics

The basic features that characterize lift and drag are as follows:

1. Drag is in the direction of air flow.
2. Lift is perpendicular to the direction of air flow.
3. Generation of lift always causes a certain amount of drag to be developed.
4. With a good aerofoil, the lift produced can be 30 times greater than the drag
5. Lift devices are generally more efficient than drag devices.

The difference between the drag and the lift is illustrated by the difference between using a spinnaker sail, which fills like a parachute and pulls a sailing boat with the wind, and a Bermuda rig, the familiar triangular sail that deflects with wind and allows a sailing boat to travel across the wind or slightly into the wind.

Drag forces provide the most obvious means of propulsion and these being the forces felt by a person (or object) exposed to the wind. Lift forces are the most efficient means of propulsion, but being more subtle than drag forces are not so well understood.

6.7.1 Lift Force

The lift force (F_L) arises in a direction that is perpendicular to the airstream caused by Bernoulli's effect that lowers the pressure on the top of the airfoil when compared with the pressure on its bottom. The curvature on the top leads to a higher stream velocity than at the bottom and hence a lower pressure. Let (F_L) is the lift force in Newton, (S_L) is the cross-sectional area of airfoil in m^2, ρ is the air density in kg/m^2, and V is the wind speed in m/s^2. Then, lift coefficient (C_L) is defined as follows:

$$C_L = [F_L/S_L] / [(1/2)\rho V^2]$$

6.7.2 Drag Force

Similarly, drag force (F_D) is described as

$$C_D = [F_D/S_D] / [(1/2)\rho V^2]$$

where C_D = drag coefficient and S_D = effective area of airfoil in the direction of drag force.

The lift and drag force vary with the angle that rotor blade makes with the direction of wind stream. This angle is called as angle of attack. The resultant of drag and lift forces constitute the thrust force that effectively rotate the blade.

6.7.3 Capturing Wind Power

Just like an aeroplane wing, wind turbine blades work by generating lift due to their shape. The more curved side generates low air pressures, while high pressure air pushes on the other side of the aerofoil. The net result is a lift force perpendicular to the direction of the flow of the air.

The lift force increases as the blade is turned to present itself at a greater angle to the wind. This is called the angle of attack (see Fig. 6.15a). At very large angles of attack, the blade 'stalls' and the lift decreases again. Therefore, there is an optimum angle of attack to generate the maximum lift.

There is, unfortunately, also a retarding force on the blade, that is, the drag. This is the force parallel to the wind flow, and also increases with angle of attack. If the aerofoil shape is good, the lift force is much bigger than the drag, but at very high angles of attack, especially when the blade stalls, the drag increases dramatically. Therefore, at an angle slightly less than the maximum lift angle, the blade reaches its maximum lift or drag ratio. The best operating point will be between these two angles.

Both the lift and the drag are proportional to the air density, the area of the airfoil, and the square of the wind speed. Suppose now that we allow the airfoil to move in the direction of the lift force. This motion or translation will combine with the motion of the air to produce a relative wind direction, as shown in Figure 6.15(a). The airfoil has been reoriented to maintain a good lift to drag ratio. The lift is perpendicular to the relative wind, but it is not in the direction of airfoil translation.

The lift and drag forces can be split into components parallel and perpendicular to the direction of the undisturbed wind, and these components combined to form net force F_1 in the direction of translation and net force F_2 in the direction of the undisturbed wind. Force F_1 is available to do useful work. Force F_2 must be used in the design of the airfoil supports to assure structural integrity.

A practical way of using F_1 is to connect two such airfoils or blades to a central hub and allow them to rotate around a horizontal axis, as shown in Figure 6.15(b). Force F_1 causes a torque that drives some load connected to the propeller. The tower must be strong enough to withstand force F_2.

These forces and the overall performance of a wind turbine depend on the construction and orientation of the blades. One important parameter of a blade is the pitch angle, which is the angle between the chord line of the blade and the plane of rotation, as shown in Figure 6.15(c).

The chord line is the straight line connecting the leading and trailing edges of an airfoil. The plane of rotation is the plane in which the blade tips lie as they rotate. The blade tips actually

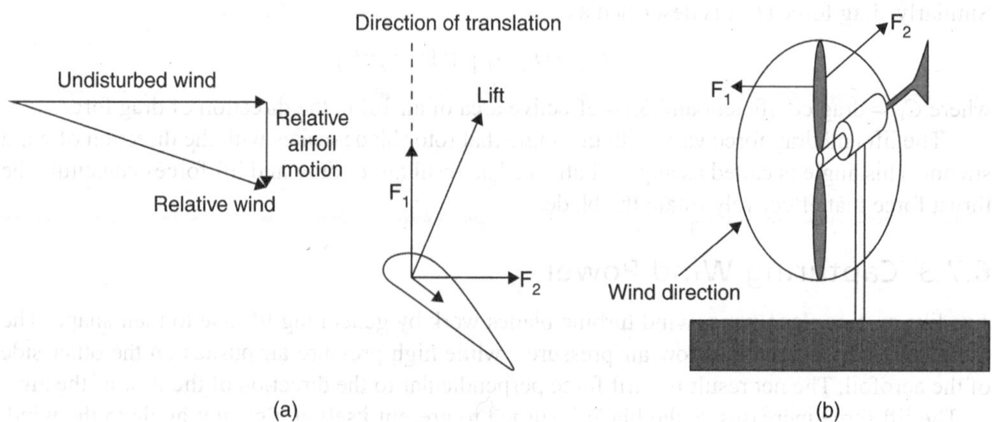

Figure 6.15 (a) Lift and drag on a translating airfoil (b) aerodynamic forces on a turbine blade

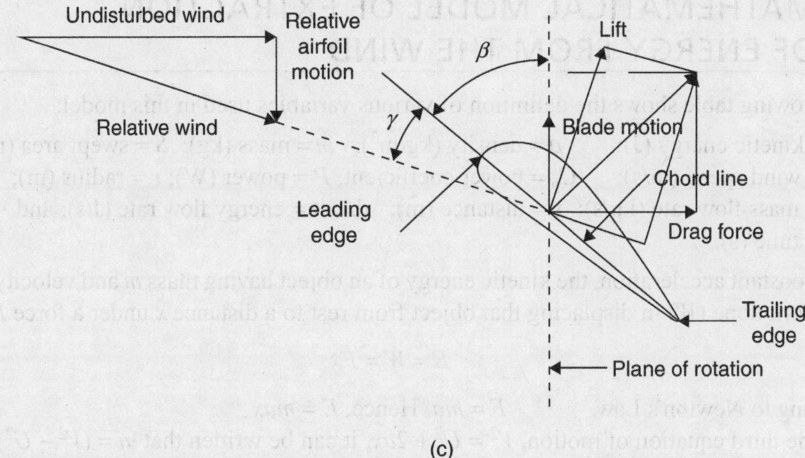

Figure 6.15 (c) Definition of pitch angle β and angle of attack γ

trace out a circle that lies on the plane of rotation. Full power output would normally be obtained when the wind direction is perpendicular to the plane of rotation. The pitch angle is a static angle, depending only on the orientation of the blade.

Another rtant blade parameter is the angle of attack, which is the angle γ between the chord line of the blade and the relative wind or the effective direction of air flow. It is a dynamic angle, depending on both the speed of the blade and the speed of the wind. The blade speed at a distance r from the hub and an angular velocity ω_m is $r\omega_m$.

A blade with twist will have a variation in angle of attack from hub to tip because of the variation of $r\omega_m$ with distance from the hub. The lift and drag have optimum values for a single angle of attack, so a blade without twist is less efficient than a blade with the proper twist to maintain a nearly constant angle of attack from hub to tip. Even the blades of the Old Dutch windmills were twisted to improve the efficiency. Most modern blades are twisted, but some are not for cost reasons. A straight blade is easy and cheap to build and the cost reduction may more than offset the loss in performance.

When the blade is twisted, the pitch angle will change from hub to tip. In this situation, the pitch angle measured three-fourths of the distance out from the hub is selected as the reference. Since the drag is in the downwind direction, it may seem that it would not matter for a wind turbine as the drag would be parallel to the turbine axis, so would not slow the rotor down. It would just create 'thrust', the force that acts parallel to the turbine axis; hence, it has no tendency to speed up or slow down the rotor. When the rotor is stationary (e.g. just before start-up), this is indeed the case. However, the blade's own movement through the air means that, as far as the blade is concerned, the wind is blowing from a different angle. This is called apparent wind. The apparent wind is stronger than the true wind but its angle is less favourable: it rotates the angles of the lift and drag to reduce the effect of lift force pulling the blade round and increase the effect of drag slowing it down. It also means that the lift force contributes to the thrust on the rotor. The result of this is that, to maintain a good angle of attack, the blade must be turned further from the true wind angle.

6.8 MATHEMATICAL MODEL OF EXTRACTION OF ENERGY FROM THE WIND

The following table shows the definition of various variables used in this model:

E = kinetic energy (J); ρ = density (kg/m^3); m = mass (kg); S = swept area (m^2);
v = wind speed (m/s); C_p = power coefficient; P = power (W); r = radius (m);
dm/dt = mass flow rate (kg/s); x = distance (m); dE/dt = energy flow rate (J/s); and
t = time (s).

Under constant acceleration, the kinetic energy of an object having mass m and velocity v is equal to the work done (W) in displacing that object from rest to a distance x under a force F, that is,

$$E = W = Fx$$

According to Newton's Law, $F = ma$. Hence, $E = max$ (6.5)

Using the third equation of motion, $V^2 = U^2 + 2ax$, it can be written that $a = (V^2 - U^2)/2x$.
Since the initial velocity of the object is zero, $u = 0$, and hence, $a = V^2/2x$
Substituting it in Equation (6.5), the kinetic energy of a mass in motions is

$$E = ½\, mV^2 \tag{6.6}$$

The power in the wind is given by the rate of change of energy:

$$P = \mathrm{d}E/\mathrm{d}t = ½\, \dot{m}\, V^2 \tag{6.7}$$

The mass flow rate is given by, $= \rho S\mathrm{d}x/\mathrm{d}t$ and the rate of change of distance is given by

$$V = \mathrm{d}x/\mathrm{d}t.\text{Thus,} = \rho SV.$$

Substituting the abovementioned expression in Equation (6.7) yields

$$P = 1/2\, \rho SV^3 \tag{6.8}$$

A German physicist Albert Betz concluded in 1919 that no wind turbine can convert more than 16/27 (59.3%) of the kinetic energy of the wind into mechanical energy turning a rotor. At present, this is known as the Betz limit or Betz' law. The theoretical maximum power efficiency of any design of wind turbine is 0.59 (i.e., no more than 59% of the energy carried by the wind can be extracted by a wind turbine). This is called the 'power coefficient' and is defined as:

C_{pmax} = 0.59 and is shown in Figure 6.16. Further, wind turbines cannot operate at this maximum limit. The C_p value is unique to each turbine type and is a function of wind speed that the turbine is operating in. Once we incorporate various engineering requirements of a wind turbine—strength and durability, in particular—the real world limit is well below the Betz limit with values of 0.35–0.45 common even in the best designed wind turbines.

By the time one take into account the other factors in a complete wind turbine system—for example, the gearbox, bearings, generator, etc.—only 10%–30% of the power of the wind is ever actually converted into usable electricity.

Hence, the power coefficient needs to be factored in Equation (6.8) and the extractable power from the wind is given by

$$P_{available} = C_p\, (1/2\rho SV^3) \tag{6.9}$$

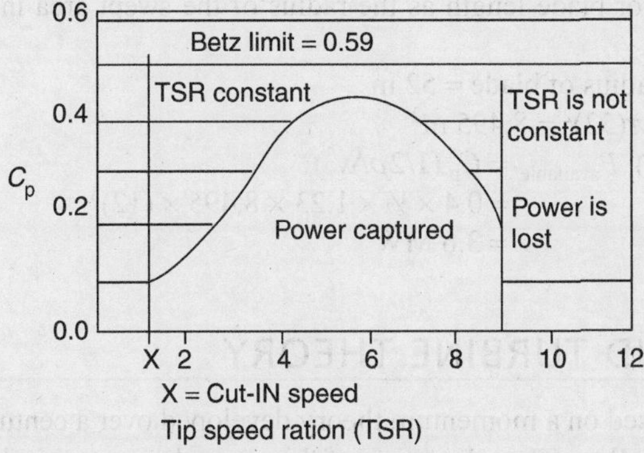

Figure 6.16 Variation of C_p with tip-speed ratio (TSR)

The swept area of the turbine can be calculated from the length of the turbine blades using the equation for the area of a circle:

$$S = \pi r^2 \tag{6.10}$$

where the radius is equal to the blade length as shown in Figure P6.1.

Example 6.1

A three-bladed wind rotor with blade length of 52 m is operating in a wind stream having wind velocity of 12 m/s. Air density is 1.23 kg/m^3 and power coefficient may be taken as 0.4. Calculate the extractable power from the wind.

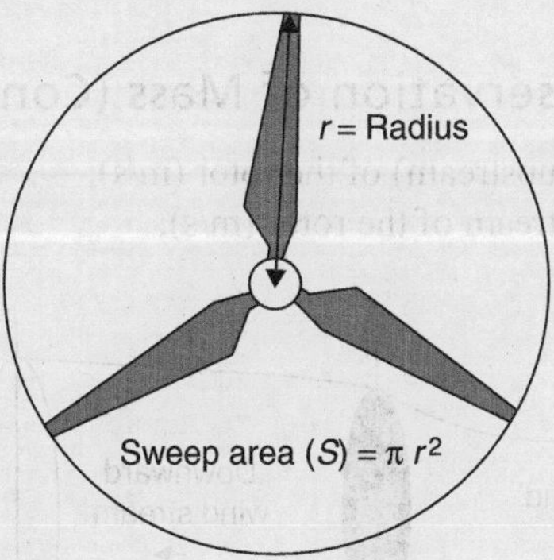

Figure P6.1 Three-bladed wind rotor

Solution Given data are as follows:

Blade length, $L = 52$ m; wind speed, $v = 12$ m/s; air density, $\rho = 1.23$ kg/m^3; power coefficient, $C_p = 0.4$.

Substituting the value for blade length as the radius of the swept area in Equation (6.10) we have:

Radius of blade = L = radius of blade = 52 m

A = swept area = $\pi r^2 = \pi(52)^2 = 8{,}495\ m^2$

Using the equation (6.9), $P_{available} = C_p\,(1/2\rho A v^3)$

$= 0.4 \times ½ \times 1.23 \times 8{,}495 \times (12)^3$

$= 3.6$ MW

6.9 SIMPLE WIND TURBINE THEORY

The simplest model is based on a momentum theory developed over a century ago to predict the performance of ship propellers. The adaptation of this early theory to wind turbines was undertaken by Betz (1927), and the turbine being represented by a uniform actuator disc that generates a discontinuity of pressure in the stream tube of air flowing through it. A simple representative showing the overall control volume is given in Figure 6.17.

6.9.1 Assumptions

1. The rotor does not possess a hub; this is an ideal rotor, with an infinite number of blades, which have no drag. Any resulting drag would only lower this idealized value.
2. The flow in and out of the rotor is axial. This is a control volume analysis, and to construct a solution, the control volume must contain all flow going in and out; however, failure to account for that flow would violate the conservation equations.
3. This is incompressible flow. The density remains constant, and there is no heat transfer from the rotor to the flow or vice versa.
4. The rotor is also mass less. Therefore, angular momentum imparted to either the rotor or the air flow behind the rotor is not considered, that is, wake effect was not taken into account.

6.9.2 Application of Conservation of Mass (Continuity Equation)

Let V_1 = the speed of wind in front (upstream) of the rotor (m/s);

V_2 = the speed of wind in downstream of the rotor (m/s);

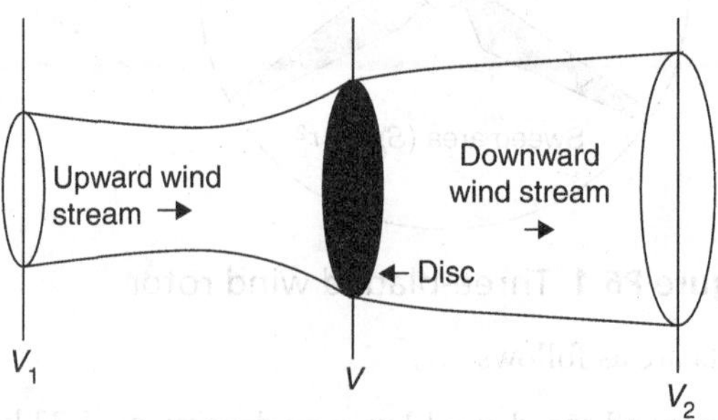

Figure 6.17 Schematic of fluid flow through a disk-shaped actuator

V = the speed at the fluid power device (average speed of the rotor disc) (m/s);
ρ = the fluid density (kg/m^3);
S = swept area of turbine (m^2).

The Betz equation is analogous to the Carnot cycle efficiency in thermodynamics suggesting that a heat engine cannot extract all the energy from a given source of energy and must reject part of its heat input back to the environment. While the Carnot cycle efficiency can be expressed in terms of the Kelvin isothermal heat input temperature T_1 and the Kelvin isothermal heat rejection temperature T_2:

$$\eta_{\text{CARNOT}} = (T_1 - T_2)/T_1 = (1 - T_2/T_1) \tag{6.11}$$

The Betz equation deals with the wind speed upstream of the turbine V_1 and the downstream wind speed V_2. The limited efficiency of a heat engine is caused by heat rejection to the environment. The limited efficiency of a wind turbine is caused by braking of the wind from its upstream speed V_1 to its downstream speed V_2, while allowing a continuation of the flow regime. The additional losses in efficiency for a practical wind turbine are caused by the viscous and pressure drag on the rotor blades, the swirl imparted to the air flow by the rotor, and the power losses in the transmission and electrical system.

In addition, uniformity is assumed over the whole area swept by the rotor, and the speed of the air beyond the rotor is considered to be axial. The ideal wind rotor is taken at rest and is placed in a moving fluid atmosphere. Considering the ideal model shown in Figures 6.17 and 6.18, the cross-sectional area swept by the turbine blade is designated as S, with the air cross-section upwind from the rotor designated as S_1 and downwind as S_2.

The wind speed passing through the turbine rotor is considered uniform as $V = (V_1 + V_2)/2$, with its value as V_1 upwind and as V_2 downwind at a distance from the rotor. The extraction of mechanical energy by the rotor occurs by reducing the kinetic energy of the airstream from upwind to downwind, or simply applying a braking action on the wind. Consequently, the cross sectional area of the airstream increases from upstream of the turbine to the downstream location; further, if the airstream is considered as a case of incompressible flow, the conservation of mass or continuity equation can be written as follows:

$$= \rho S_1 V_1 = \rho S V = \rho S_2 V_2 = \text{constant} \tag{6.12}$$

This expresses the fact that the mass flow rate is a constant along the wind stream. Continuing with the derivation, Euler's Theorem gives the force exerted by the wind on the rotor as

$$F = ma = m\mathrm{d}V/\mathrm{d}t = \dot{m}\,\rho V = \rho SV(V_1 - V_2) \tag{6.13}$$

The incremental energy or the incremental work done in the wind stream is given by

$$\mathrm{d}E = F\mathrm{d}x$$

From which, the power content of the wind stream is given as

$$P = \mathrm{d}E/\mathrm{d}t = F\mathrm{d}x/\mathrm{d}t = FV \tag{6.14}$$

Substituting for the force F from Equation (6.13), the extractable power from the wind is

$$P = \rho SV^2(V_1 - V_2) \tag{6.15}$$

The power as the rate of change in kinetic energy from upstream to downstream is given by

$$P \simeq \Delta E/\Delta t \simeq (1/2\, mV_1^2 - 1/2\, mV_2^2)/\Delta t = 1/2\ \dot{m}\ (V_1^2 - V_2^2) \tag{6.16}$$

Using the continuity equation (Equation 6.12), it can be written as

$$P = (1/2)\,\rho SV\,(V_1^2 - V_2^2) \tag{6.17}$$

Equating the two expressions for the power P in Equations (6.17) and (6.15) yields

$$P = 1/2\rho SV(V_1^2 - V_2^2) = \rho SV^2(V_1 - V_2) \tag{6.18}$$

The last expression (6.17) implies that

$$(1/2)\,(V_1^2 - V_2^2) = (1/2)(V_1 - V_2)(V_1 + V_2) = V\,(V_1 - V_2) \tag{6.19}$$

where $V = (V_1 + V_2)/2$; V, S, and ρ are not equal to zero. Or

$$V = (V_1 + V_2)/2, \text{ since } (V_1 - V_2) \neq 0 \text{ OR } V_1 \neq V_2 \tag{6.20}$$

This in turn suggests that the wind velocity at the rotor may be taken as the average of the upstream and downstream wind velocities. It also implies that the turbine must act as a brake, reducing the wind speed from V_1 to V_2, but not totally reducing it to $V = 0$, at which point the equation is no longer valid. To extract energy from the wind stream, its flow must be maintained and not totally stopped.

The last result allows writing new expressions for force F and power P in terms of the upstream and downstream velocities by substituting for the value of V as

$$F = \rho S\,V\,(V_1 - V_2) = (½)\,\rho S\,(V_1^2 - V_2^2) \tag{6.21}$$

$$P = \rho\,SV^2\,(V_1 - V_2) = (1/4)\,\rho\,S(V_1 + V_2)^2\,(V_1 - V_2) \tag{6.22}$$

$$= (1/4)\,\rho\,S(V_1^2 - V_2^2)\,(V_1 + V_2) \tag{6.23}$$

Let us introduce the 'downstream velocity factor,' or 'interference factor' b as the ratio of downstream speed V_2 to upstream speed V_1 as

$$b = V_2/V_1 \tag{6.24}$$

From Equation (6.21), force F can be expressed as

$$F = (1/2)\,\rho SV_1^2\,(1 - b^2) \tag{6.25}$$

Extractable power P in terms of the interference factor b can be expressed as

$$P = (1/4)\rho\,S(V_1^2 - V_2^2)\,(V_1 + V_2) = (1/4)\,\rho\,SV_1^3\,(1 - b^2)\,(1 + b) \tag{6.26}$$

The most important observation pertaining to wind power production is that the extractable power from the wind is proportional to the cube of the upstream wind speed (V_1^3) and is a function of the interference factor b.

The 'power flux' or rate of energy flow per unit area, sometimes referred to as 'power density' is defined as

$$P_d = P/S = [(1/2)\rho\,SV^3]/S = (1/2)\rho\,V^3, \quad [\text{J/m}^2\text{s}]\ [\text{Wm}^2] \tag{6.27}$$

The kinetic power content of the undisturbed upstream wind stream with $V = V_1$ and over a cross sectional area S becomes

$$W = (1/2)\rho\,SV_1^3, \quad [\text{J/s or W}] \tag{6.28}$$

The performance coefficient or efficiency is the dimensionless ratio of extractable power P to kinetic power W available in the undisturbed stream:

$$C_p = P/W \tag{6.29}$$

The performance coefficient is a dimensionless measure of the efficiency of a wind turbine in extracting the energy content of a wind stream. Substituting the expressions for P from Equation (6.26) and for W from Equation (6.28) gives:

$$C_p = P/W = [(1/4)\,\rho S V_1^3\,(1 - b^2)\,(1 + b)]/\,[(1/2)\,\rho S V_1^3]$$

$$= (1/2)\,(1 - b^2)\,(1 + b) \tag{6.30}$$

Equation (6.30) is graphically shown in Figure 6.18 and it represents the performance coefficient C_p as a function of the interference factor $b = V_2/V_1$.
From which, the following observations are made:

1. When $b = 1$, $V_1 = V_2$ and the wind stream is undisturbed, leading to a performance coefficient of zero.
2. When $b = 0$, $V_1 = 0$, the turbine stops all the air flow and the performance coefficient is equal to 0.5.
3. It can be noticed from the graph that the performance coefficient reaches a maximum around $b = 1/3$.

6.9.3 Condition for Maximum Performance Coefficient

A condition for maximum performance coefficient (C_p) can be obtained by differentiating Equation (6.30) with respect to the interference factor b. Using rule of differentiation [$d(uv)/dx = u dv/dx + v du/dx$] and setting the derivative equal to zero yields

$$d(C_p)/dt = (1/2) d/db\,[(1 - b^2)\,(1 + b)] = (1/2)\,[(1 - b^2) - 2b\,(1 + b)]$$

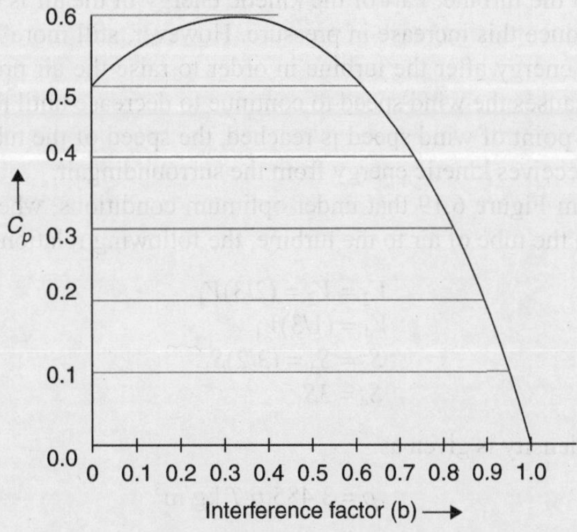

Figure 6.18 Performance coefficient C_p as a function of the interference factor

$$= (1/2)\,(1 - b^2 - 2b - 2b^2) = (1/2)\,(1 - 3b^2 - 2b)$$
$$= (1/2)\,(1 - 3b)\,(1 + b) = 0 \tag{6.31}$$

Equation (6.31) has two solutions. The first solution is $(1 + b) = 0$ or $b = V_2/V_1 = -1$ or $V_2 = -V_1$; this is the trivial solution.

The second solution is $(1 - 3b) = 0$;

$$\text{or } b = V_2/V_1 = 1/3$$
$$\text{or } V_2 = 1/3\ V_1 \tag{6.32}$$

Equation (6.32) is the practical physical solution. Further, this equation shows that for optimal operation, downstream velocity V_2 should be equal to one-third of upstream velocity V_1. Using Equation (6.32), the maximum or optimal value of the performance coefficient C_{popt} becomes:

$$C_{popt} = (1/2)\,(1 - b^2)\,(1 + b) = (1/2)\,(1 - (1/3)^2)(1 + 1/3) = 16/27 = 0.59259 = 59.26\%$$

This is referred to as the Betz criterion or the Betz limit. It was first formulated in 1919 and applies to all wind turbine designs. It is the theoretical power fraction that can be extracted from an ideal wind stream. Modern wind machines operate at a slightly low practical non-ideal performance coefficient. Albert Betz was a German physicist who calculated that by turning a rotor, no wind turbine could convert more than 59.3% of the kinetic energy of the wind into mechanical energy.

6.9.4 Wind Speed and Pressure Variations in Ideal Wind Turbine

Consider a tube of moving air (see Fig. 6.19) with initial or undisturbed diameter d_1, speed V_1, and pressure p_1 as it approaches the turbine. The speed of the air decreases as the turbine is approached, causing the tube of air to enlarge to turbine diameter d_2.

The air pressure will rise to a maximum just in front of the turbine and will drop below atmospheric pressure behind the turbine. Part of the kinetic energy in the air is converted to potential energy in order to produce this increase in pressure. However, still more kinetic energy will be converted to potential energy after the turbine in order to raise the air pressure back to atmospheric pressure. This causes the wind speed to continue to decrease until the pressure is in equilibrium. Once the low point of wind speed is reached, the speed of the tube of air will increase back to $V_4 = V_1$, as it receives kinetic energy from the surroundingair.

As it is evident from Figure 6.19 that under optimum conditions, when maximum power is being transferred from the tube of air to the turbine, the following relationships hold:

$$\begin{aligned} V_2 &= V_3 = (2/3)V_1 \\ V_4 &= (1/3)V_1 \\ S_2 &= S_3 = (3/2)S_1 \\ S_4 &= 3S_1 \end{aligned} \tag{6.33}$$

An expression for air density is given as

$$\rho = 3.485\ p/T\ \text{kg/m}^3 \tag{6.34}$$

In this equation, p is the pressure in kPa and T is the temperature in K.

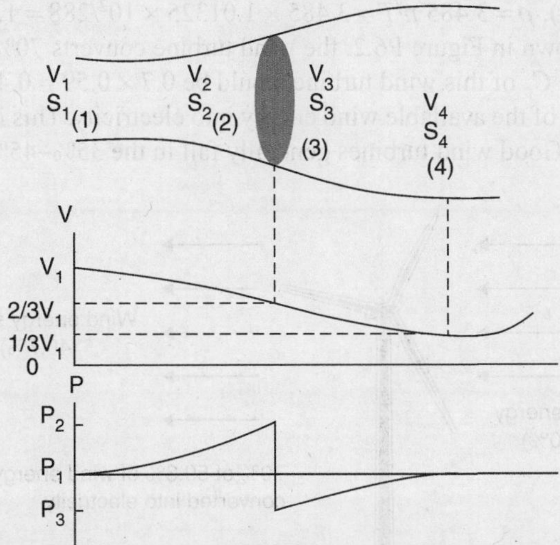

Figure 6.19 Wind speed (*V*) and pressure (*p*) variations in ideal wind turbines

The power in the wind is then

$$P = (1/2)\,\rho SV^3 = (1/2)3.485 \times p/T \times S \times V^3 = (1.742 \times pS \times V^3)/T \qquad (6.35)$$

where S = swept area in m^2 and V = wind speed in m/s. For air under standard conditions, 101.3 kPa and 273 K, this reduces to

$$P = 0.647SV^3 \text{ W} \qquad (6.36)$$

The mechanical power extracted is the difference between the input and output power in the wind:

$$P_{\text{mideal}} = P_1 - P_4 = (1/2)\,\rho\,[S_1 V_1^3 - S_4 V_4^3)$$

Since $V_4 = (1/3)\,V_1$, $P_{\text{mideal}} = (1/2)\,\rho\,[(8/9)\,S_1\,V_1^3$

This states that factor (8/9) of the power in the original tube of air is extracted by an ideal turbine. However, this tube is smaller than the turbine; this can lead to confusing results. The normal method of expressing this extracted power is in terms of the undisturbed wind speed V_1 and the turbine area S_2. This method yields

$$P_{\text{mideal}} = (1/2)\,\rho\,[(8/9)\,\{2/3S_2\}\,V_1^3] = (1/2)\,\rho\,[(16/27)\,S_2V_1^3] \text{ W}$$

Example 6.2

Wind speed at a particular location is 10 m/s at 1 atm and 15°C. Calculate density of air.

Solution 1 atm = 1.01325×10^5 Pa = 1.01325×10^2 kPa

Temperature in K = 15 + 273 = 288 K

Using Equation (6.34), $\rho = 3.485\, p/T = 3.485 \times 1.01325 \times 10^2/288 = 1.226$ kg/m^3.

In the diagram shown in Figure P6.2, the wind turbine converts 70% of the Betz Limit into electricity. Therefore, C_p of this wind turbine would be $0.7 \times 0.59 = 0.41$. Therefore, this wind turbine converts 41% of the available wind energy into electricity. This is actually a pretty good coefficient of power. Good wind turbines generally fall in the 35%–45% range.

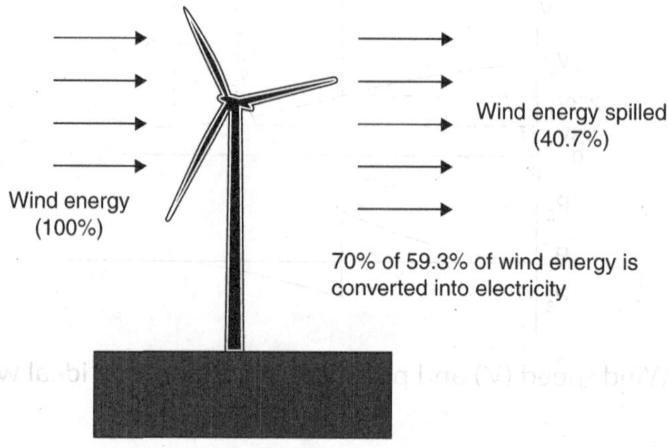

Figure P6.2 Wind energy conversion

6.10 CHARACTERISTICS OF WINDMILL ROTORS (ROTOR DESIGN)

In order to understand the effects of differences in rotor design, it is useful to describe how the blades of a rotor react to the wind, and to define some of the standard design parameters. Several technical parameters are discussed in the following sections.

6.10.1 Pitch

The blades of a rotor are curved so that they deflect the wind, as illustrated in Figure 6.20. The lift force created causes the rotor to rotate. In order to generate the maximum amount of lift, the blades must be set at an appropriate angle (β) to the wind called the pitch.

Since the tips of the blades travel faster than points near the axis, the angle of the wind 'seen' by the blade changes with the radius, as shown in Figure 6.20. The rotor is most efficient if this angle 'seen' by the blade is as large as possible without being so large than the rotor stalls. To make the angle large all the way along the blade, it must be twisted. For the same reason, a rotor designed to rotate fast, such as a two- or three-bladed wind turbine and has its blades set at a small pitch.

6.10.2 Tip-speed Ratio (TSR)

It is defined as the ratio of the speed of the extremities (see Fig. 6.21) of a windmill rotor to the speed of the free wind. It is a variable expressing the ratio between the peripheral blade speed and the wind speed denoted by TSR and computed as

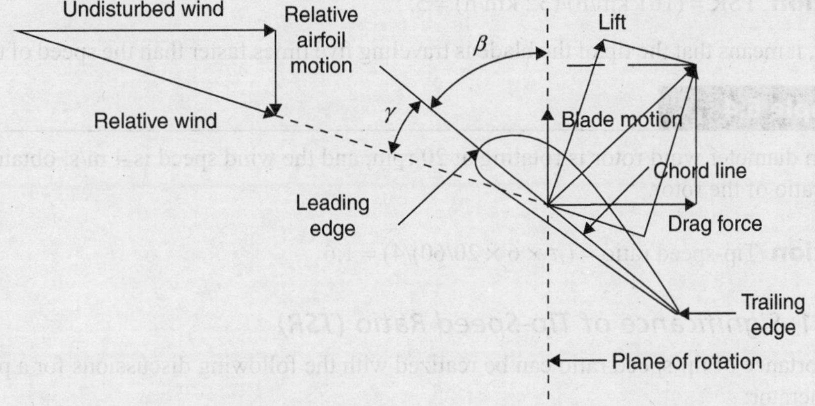

Figure 6.20 Definition of pitch angle β and angle of attack γ

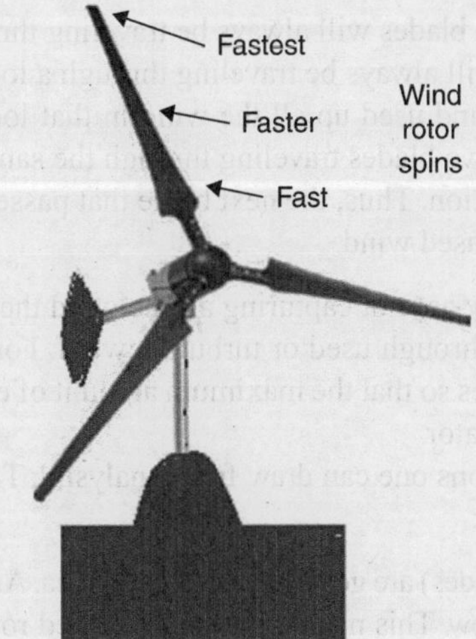

Figure 6.21 Wind speed spins at extremities

$$\text{TSR} = \text{blade tip speed/wind speed} = v_{RB}/V = (\pi DN/60)/V$$
$$= 0.052 \times \text{rotor diameter in m} \times \text{rotational speed in rpm/wind speed in m/s}$$

where L = rotor blade length (m); v_{RB} = rotor speed (m/s) = $R2\pi N/60$; V = wind speed (m/s); R = radius of wind rotor = rotor blade length (L) = $D/2$; D = wind rotor diameter (m); and N = wind rotor speed (rpm).

Speed movement along the rotor blade is shown in Figure 6.21.

Example 6.3

If the tip of a wind rotor blade is traveling at 100 miles/h (161 km/h or 45 m/s) and the wind speed is 20 miles/h (32 km/h or 9 m/s), obtain the tip-speed ratio.

Solution TSR = (161 km/h)/(32 km/h) = 5.

Simply, it means that the tip of the blade is traveling five times faster than the speed of the wind.

Example 6.4

If a 6 m diameter wind rotor is rotating at 20 rpm, and the wind speed is 4 m/s, obtain the tip-speed ratio of the rotor.

Solution Tip-speed ratio = $(\pi \times 6 \times 20/60)/4) = 1.6$

6.10.2.1 Significance of Tip-Speed Ratio (TSR)

The importance of tip-speed ratio can be realized with the following discussions for a particular wind generator:

1. If the blade set spins too slowly, then most of the wind will pass by the rotor without being captured by the blades.
2. If the blades spin too fast, then the blades will always be traveling through used or turbulent wind. This is because the blades will always be traveling through a location that the blade in front of it just travelled through (and used up all the wind in that location). It is important that enough time lapses between two blades traveling through the same location so that new or unused wind can enter this location. Thus, the next blade that passes through this location will be able to harness fresh or unused wind.

In short, if the blades are too slow, they are not capturing all the wind they could and if they are too fast, then the blades are spinning through used or turbulent wind. For this reason, TSRs are employed when designing wind turbines so that the maximum amount of energy can be extracted from the wind using a particular generator.

There are many important conclusions one can draw from analysing TSR. Following is a few of the most basic and important points:

1. Rotors with many blades (i.e., 11 blades) are generally not a good idea. An 11-bladed rotor would have an optimal TSR that is very low. This means that an 11-bladed rotor would operate most efficiently at extremely low rpms. Because nearly all generators (permanent magnet alternators) are not optimized for extremely low rpms, there is no advantage or reason to use a rotor with many blades. It should also be remembered that rotors with more number of blades are capturing used or turbulent wind at high TSR and are, thus, extremely inefficient if used as a high-rpm blade set. This is a very important point because many people intuitively think that more blades equal a faster and more efficient blade set. However, the laws of physics say that this is not true.
2. If one already has a generator or a motor and it requires high rpm to reach charging voltage, then the best choice is a two or three-bladed rotor. These rotors operate more efficiently at high rpm. Further, it is necessary keep the blades as short as pragmatically possible because shorter blades obviously spin faster than the longer blades.

6.10.2.2 Effect of the Number of Rotor Blades on the Tip-speed Tatio (TSR)

The optimal TSR depends on the number of rotor blades (n) of the wind turbine. The smaller the number of rotor blades, the faster the wind turbine must rotate to extract the maximum power from the wind. It has empirically been observed that for an n-bladed rotor, a figure Z is approximately equal to 50% of the rotor radius. Thus,

$$Z \approx 0.5r, \text{ and hence, } (\text{TSR})_{\text{OPT}} \approx (2\pi/n)/(r/Z) \approx 4\pi/n \qquad (6.37)$$

Therefore, for $n = 2$, the optimal TSR is calculated to be 6.28.
For $n = 3$, the optimal TSR is reduced to be 6.28;
For $i = 4$, the optimal TSR further reduced to be 6.28.

Highly efficient aerofoil rotor blade design can increase these optimum values by as much as 25%–30% increasing the speed at which the rotor turns, and therefore, generating more power. A well-designed typical three-bladed rotor would have a tip-speed ratio of around 6 to 7.

The optimum tip-speed ratio depends on the number of blades in the wind turbine rotor. The fewer the number of blades, the faster the wind turbine rotor needs to turn to extract maximum power from the wind. Drag devices always have tip-speed ratios less than one and hence turn slowly, whereas lift devices can have high tip-speed ratios (up to 13:1) and hence turn quickly relative to the wind.

The following are the disadvantages of a high TSR:

1. Blade tips operating at 80 m/s or greater are subject to leading edge erosion from dust and sand particles and would require special leading edge treatments like helicopter blades to mitigate such damage;
2. Noise, both audible and inaudible, is generated;
3. Vibration, especially in two- or one-bladed rotors;
4. Reduced rotor efficiency due to drag and tip losses;
5. Higher speed rotors require much larger braking systems to prevent the rotor from reaching a runaway condition that can cause disintegration of the turbine rotor blades.

Example 6.5

An rpm of 450 at the tip of a 1 m blade using digital tachometer has been measured. How far does the tip of the blade travel in 1 h? If the wind is blowing at 20 miles/h, obtain the tip-speed ratio.

Solution Distance travelled by the blade in one revolution = $2\pi r = 2\pi 1$ m = 6.28 m.

Number of revolutions made by the blade in 1 h
= (450 r/min)(60 min/h) = 27,000 revolutions/h

Total distance travelled by the blade in 1 h
= (27,000 rotations) (6.28 m/rotation) = 169,560 m
= 169,560 m(1 mile)/(1,609 m) = 105 miles

Therefore, TSR = (blade tip speed)/(wind speed) = (105 mph)/ (20 mph) = 5.3

6.10.3 Solidity

Solidity is usually defined as the percentage of the circumference of the rotor that contains material rather than air. It is, in effect, the fraction of the swept area of the rotor, which is filled with metal.

The general equation of solidity is given by

% Solidity = [number of blades × blade width × blade length (or radius)/swept area]100
% Solidity = 31.8 × number of blades × blade width/rotor diameter

Example 6.6

If a 6 m diameter rotor has 24 blades, each 0.35 m wide, calculate its solidity.

Solution Solidity = $24 \times 0.35 \times 100/\pi \times 6 = 45\%$

The greater the solidity of a rotor, the slower it needs to turn to intercept the wind. A two- or three-bladed wind turbine has a very low solidity, and therefore, it needs to rotate quickly to intercept the wind. Otherwise, a lot of wind energy would be lost through the large gaps between the blades.

High-solidity machines carry a lot of material and have coarse blade angles. They generate much higher starting torque than low-solidity machines but are inherently less efficient than low-solidity machines, as shown in Figure 6.22. The extra materials also cost more money. However, low-solidity machines need to be made with more precision that leads to little difference in costs.

6.10.4 Coefficient of Performance

It is the proportion of the power in the wind that the rotor can extract (it is also called power coefficient or efficiency; symbol C_p) and its variation as a function of tip-speed ratio is commonly used to characterize different types of rotor. It is physically impossible to extract all the energy from the wind, without bringing the air behind the rotor to a standstill. Consequently, there is a maximum value of C_p of 59.3% (known as the Betz limit), although in practice, real wind rotors have maximum C_p values in the range of 25%–45%.

The performance coefficient of a rotor is the fraction of wind energy passing through the rotor disc, which is converted into shaft power. This is a measure of the efficiency of the rotor and it varies with the tip-speed ratio. Typical performance coefficient versus tip-speed ratio curves for rotors of varying solidity is shown in Figure 6.22. Each type of rotor has a unique performance coefficient versus tip-speed ratio curve.

6.10.5 Torque

Torque is the turning force produced by the rotor. It depends on the solidity and tip-speed ratio of the rotor. Most wind turbines extract power from the wind in mechanical form and transmit it to the load by rotating shafts. These shafts must be properly designed to transmit this power. When power is being transmitted through a shaft, a torque T will be present. This torque is given by

$$T = P/\omega \text{ (N)} \tag{6.38}$$

ω = angular velocity in rad/sec.

$$= (\text{TSR} \times V)/r \tag{6.39}$$

The Torque equation of wind rotor is from eq(6.38) and (6.39)

$$T = (1/2)\, \rho . r . C_p . S . V^2/\text{TSR} \tag{6.40}$$

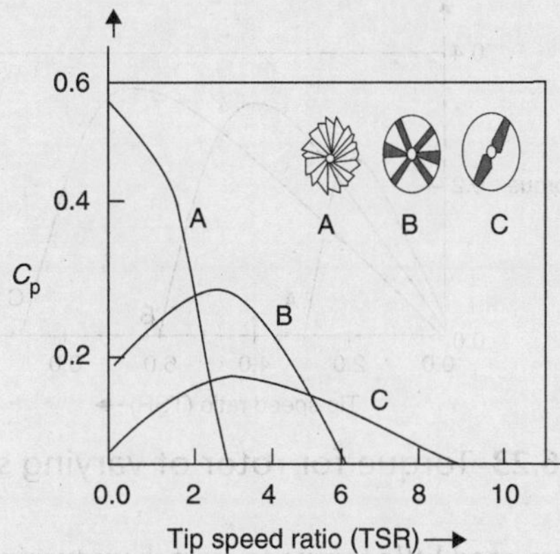

Figure 6.22 C_p versus TSR for rotor of varying solidity

where T = torque output (N); ρ = air density (kg/m^2); C_p = power coefficient; S = swept area of wind rotor (m^2); V = wind velocity (m/s);

$$\text{TSR} = v_{RB}/V = (\pi DN/60)/V \tag{6.40}$$

From Equation (6.22), the power that can be obtained in terms of upstream and downstream wind velocities is

$$P = (1/4)\,\rho S\,(V_1^2 - V_2^2)\,(V_1 + V_2) \tag{6.41}$$

The maximum power that can be extracted from the wind stream (under the condition $V_2 = (1/3)$ V_1) is

$$P_{MAX} = 1/4 \rho S\,(V_1^2 - [(1/3)\,V_1]^2\,)\,(V_1 + 1/3 \times V_1)$$
$$= 1/2 \times 16/27 \times \rho S V_1^3 \tag{6.42}$$

Since, swept area $S = \pi D^2/4$, $\qquad P_{MAX} = 2/27 \times \rho \pi D^2 V_1^3 \qquad (6.43)$

Maximum torque

$$T_{MAX} = P_{MAX}/v_{RB}.\ \text{Substituting from Equation (6.40)}$$
$$T_{MAX} = (2/27)\,\rho \pi D^2 \times V_1^3/(\pi DN/60) = (2/27)\,\rho D V_1^3 \tag{6.44}$$

Rearranging Equation (6.44), we get

$$T_{MAX} = (2/27)\,\rho D \times V_1^3/\text{TSR} \tag{6.45}$$

High solidity rotors with low tip-speed ratios (like multibladed wind pump rotors) produce much more torque than low-solidity high-speed machines (like wind turbines). Figure 6.23 illustrates this.

The important features to note are that the high speed machine has a slightly high maximum performance coefficient, but a low starting torque. Conversely, the high solidity rotor produces a high starting torque but has a slightly low maximum performance coefficient.

The choice of rotor depends on the load characteristics. A positive displacement pump, likc the piston pumps used in boreholes, demands a high starting torque than running torque, and therefore, a high-solidity rotor is almost essential unless some method of unloading the rotor to help it to start is included.

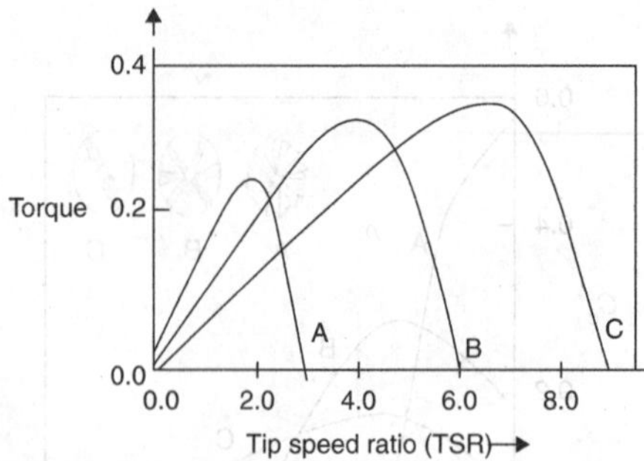

Figure 6.23 Torque for rotor of varying solidity

However, electricity generators need little torque to start them turning and they need to be driven at high speed, so generally a high-speed, low solidity rotor is used for this type of load. Positive displacement pumps that are invariably used with wind pumps need a fairly high torque to start, but will then continue to run with a low torque. The rotor of wind pump will always operate at a speed such that the torque produced exactly matches the torque required by the pump. For this reason, the torque characteristics of wind pump are important. In order to produce a high starting torque, a high solidity rotor is needed. This is why the wind pumps are almost always designed with high solidity multibladed rotors.

For a reciprocating positive displacement pump, approximately three as much torque is needed to start it than to keep it running. This means that even if a wind pump will operate at low wind speeds, it will need a gust of high wind speed to actually start it.

Example 6.7

The undisturbed wind speed at a place is 13.415 m/s^2. The wind speed at the turbine rotor and at exit is 6% and 30%, respectively, of the undisturbed wind. The rotor diameter is 9 m and air density is 1.293 kg/m^3. Calculate

1. Power available in the undisturbed wind
2. Power in the wind at exit
3. Power developed by the turbine rotor
4. Coefficient of performance

Solution 1. Power available in the undisturbed wind

Using Equation (6.1), $P_{un} = P = .½. (\rho SV^3) = (1/2) \times 1.293 \times [\pi \times (9)^2/4] \times (13.415)^3$ $= 41.128 \times 2{,}414.2 = 99.292$ kW

2. Power in the wind at exit

$V = 60\%$ of $13.415 = 8.049$ and $V_2 = 30\%$ of $13.415 = 4.0245$. Therefore,

$$P_E = (1/2) \times 1.293 \times [\pi \times (9)^2/4] \times 8.049 \times (4.0245)^2$$
$$= (1/2) \times 1.293 \times 63.617 \times 8.049 \times (4.0245)^2 = 5.381 \text{ kW}$$

3. Power developed by the turbine rotor, where $V_1 = 12.074$ m/s² (calculated)

   From Equation (6.23), $P_d = (1/4)\,\rho S\,(V_1^2 - V_2^2)\,(V_1 + V_2) = (1/2)\,\rho SV\,(V_1^2 - V_2^2)$

   $= 41.128 \times 8.049 \times (12.074^2 - 4.0245^2) = 42.897$ kW

4. Coefficient of performance

   $$C_p = P_d/P_{un} = 42.897/99.292 = 0.432$$

Example 6.8

Following data are given for a propeller-type HAWT:
Speed of wind = 10 m/s; air density = 1.226 kg/m³; rotor diameter = 120 m
Rotor speed = 40 rpm; coefficient of performance = 40%.
Calculate

1. Total power density in wind system,
2. Total power available in the wind, kW
3. Maximum extractable power
4. Maximum Torque and axial thrust

Solution

1. Using Equation (6.35), total power in the wind system $P = (1/2)\,\rho SV^3$

   Power density $(p_d) = P/S = (1/2)\,r\mathbf{V}^3 = 1/2 \times 1.226 \times (10)^3 = 613$ W/m²

2. Total power available in the wind = power density (p_d) × swept area of the wind turbine (S).

   Swept area of the wind turbine $(S) = \pi \times D^2/4 = \pi \times (120)^2/4 = 11{,}304$ m²

   Therefore, $P = p_d \times S = 613 \times 11{,}304 = 6{,}930$ kW.

3. Given that wind power coefficient $(C_p) = 40\%$,

   Maximum extractable power = $C_p \times P = 0.4 \times 6{,}930 \times 10^3 = 2{,}771$ kW.

4. Using Equation (6.21), the axial force $F = \rho SV\,(V_1 - V_2) = (½)\,\rho S(V_1^2 - V_2^2)$.

   The maximum axial force will occur at maximum efficiency when $V_2 = (1/3)\,V_1$

   Thus, $F_{max} = (\pi/9) \times \rho D^2 \times (\,V_1\,)^2 = (\pi/9) \times 1.226 \times (120)^2 \times (10)^2 = 619.95$ kN

   From Equation (6.44), $T_{MAX} = (2/27) \times \rho \times \pi \times D^2 \times V_1^3/(\pi DN/60) = (2/27) \times \rho \times D \times V_1^3$

   $= (2/27) \times 1.226 \times 120 \times (10)^3 = 10.897$ kN

6.11 TYPES OF GENERATORS USED WITH WIND TURBINES

The generator is the device inside the turbine that actually generates the electricity. It takes the kinetic energy generated by the wind rotor and translates it into electricity. Inside the generator, coils of copper wire called the armature are rotated in a magnetic field to produce electricity.

Wind generators are used for battery charging, AC utility, water or space heating, and direct motor drive for water pumping. A typical system would have the wind generator on a tall tower (tilt-up, self-supporting, pipe, or lattice), battery bank for storage, inverter to change DC to 120 V AC. Depending on the type and size of inverter and wind system, the inverter can power all the loads, part of the loads, or sell the surplus back to the utility company. Outputs will vary with wind speed, turbulence, and blade diameter and generator size. An average wind speed of 8 mph or more is recommended for a battery charging system and 12 mph or more for a utility intertie system. Most wind generators will begin producing power at 7–10 mph and will reach full output at 25–30 mph. Air density will influence output. Hot temperatures or high altitude lowers air density, which lowers the power generated and thus, it reduces expected output about 3% per 1,000 feet above sea level. Wind generators are mounted at least 30 feet above the nearby trees and buildings within a 300 feet radius. Tilt-up towers are precise because no climbing is needed. A wind generator, PV modules, and hydroelectric unit can all be used to charge the same battery. The wind often blows on cloudy days when solar electric modules have low or no output.

Generators can be designed to produce either AC or DC, and they are available in a large range of output power ratings. If the turbine is designed to produce DC, it will have an additional component inside the housing called a rectifier that will convert AC to DC. Another approach for getting DC is to pass AC to a separate device called a rectifier that can convert the AC to DC.

Grid-tied and off-the-grid systems have different requirements when it comes to wind turbines. In a grid-tied system, it is essential that AC output matches that of the electrical grid itself. Most home and office appliances operate on 120 V (or 240 V), 60/50 Hz AC. Heavy wind turbines (as those in a wind farm) generate AC, because it is easy to connect them to the grid. If wind turbine is used as part of an off-the-grid standalone system, it is required to store the power it generates in batteries when not using it directly. Most of the home turbines generate DC (through a DC generator or an AC generator and a rectifier. This is because, they are often used to charge batteries. The usual way to get AC out of a home DC wind turbine is using a power inverter, just as a photovoltaic system. There are many different kinds of generators that could be used in a wind turbine.

6.11.1 Induction Generator

An induction generator is a type of electrical generator that is mechanically and electrically similar to an induction motor. Induction generators produce electrical power when their shaft is rotated faster than the synchronous frequency of the equivalent induction motor. Induction generators are often used in wind turbines and some micro-hydro installations. Induction generators are mechanically and electrically simpler than other generator types. They are also more rugged, requiring no brushes or commutator.

Induction generators are not self-exciting, meaning they require an external supply to produce a rotating magnetic flux; the power required for this is called reactive current. The external supply can be supplied from the electrical grid or from the generator itself, once it starts producing power or we can use a capacitor bank to supply it. The rotating magnetic flux from the stator induces currents in the rotor, which also produces a magnetic field. If the rotor turns slower than the rate of the rotating flux, the machine acts like an induction motor. If the rotor is turned faster, it acts like a generator, producing power at the synchronous frequency. In the United States, it would be 60 Hz, and in India, it is 50 Hz.

The common downside of using an induction generator in a wind turbine is gearing. Typically, it is required that an induction motor runs above 1,500 RPM to meet the generation requirements. Hence, a gearing is almost always needed.

Wind turbine is one of the oldest technologies used in wind turbine generators. It consists of an induction generator connected to the rotor blades via a gearbox. This type of turbine is very rugged and very simple in its construction. The induction generator used in most of the turbines is usually type A or type B, operating in a low slip range between 0% and 1%.

Many turbines use dual-speed induction generators where two sets of windings are used within the same stator frame. The first set is designed to operate in a low rotational speed (corresponds to low wind speed operation), and the second set is designed to operate in a high rotational speed (corresponds to high wind speed operation). Since the start-up current is high, many wind turbines employed a phase controlled soft-start to limit start-up currents. This soft start consists of back-to-back thyristors in series with each phase of the induction generator.

The natural characteristic of an induction generator is that it draws reactive power from the utility supply. Thus, this type of turbine requires reactive power compensation implemented in the form of switched capacitors in parallel with each phase of the winding. Operation without switched capacitors can lead to excessive reactive power drawn from the utility, thus creating a significant voltage drop across the transmission line, and results in low voltage at the terminals of the induction generators.

The size of the capacitors switched in and out is automatically adjusted according to the operating point of the induction generator. At high wind speed, the generated power increases and the operating slip of the induction generator is high, and as a result, the reactive power required is also large. It is customary to keep the operation of the induction generator at close to the unity power factor.

6.11.2 Permanent Magnet Alternators

Most home wind generators are permanent magnet alternators. They produce three-phase, wild, high voltage AC. Here, the term wild means that the voltage changes with the wind speed. Without controlling this voltage, it cannot be used.

To control this current, it is converted to DC with the use of a bridge rectifier. Once converted, the power is connected to a diversion charge controller. The primary job of the diversion controller is to maintain a constant voltage range that can be used for battery charging or trying into the grid.

Permanent magnets alternators (PMA) have one set of electromagnets and one set of permanent magnets. Typically, the permanent magnets will be mounted on the rotor with the electromagnets on the stator. Permanent magnet motor and generator technology have advanced greatly in the past few years with the creation of rare earth magnets (neodymium, samarium-cobalt, and alnico). Generally, the coils will be wired in a standard three-phase star or delta.

Permanent magnet alternators are very efficient, in the range of 60%–95%, typically around 70% though. As a generator, they do not require a controller as a typical three-phase motor would need. It is easy to rectify the power from them and charge a battery bank or use with a grid tie.

6.11.3 Synchronous Generators

Like the DC generator, the operation of the synchronous generator is also based on Faraday's law of electromagnetic induction and works in a similar way to the automotive alternator. The difference this time is that the synchronous generator produces a three-phase AC voltage output from its stator windings, unlike the DC generator that produces a single DC or DC output. Single-phase synchronous generators are also available for low-power domestic wind turbine synchronous generator systems.

Basically, the synchronous generator is a synchronous electro-mechanical machine used as a generator and consists of a magnetic field on the rotor that rotates and a stationary stator containing multiple windings that supply the generated power. The rotors' magnetic field system (excitation) is created by using either permanent magnet mounted directly onto the rotor or energized electromagnetically by an external DC flowing in the rotor field windings. This DC field current is transmitted to the synchronous machine's rotor via slip rings and carbon or graphite brushes. Unlike the previous DC generator, synchronous generators do not require complex commutation, and thus allowing for a simple construction.

6.11.4 DC Generators

DC generators are commonly used for home-built wind turbines. On a DC generator, the electromagnets spin on the rotor with the power coming out of what is known as a commutator. This does cause a rectifying effecting outputting lumpy DC, but this is not an efficient way to 'rectify' the power from the windings, it is used because it is the only way to get the power out of the rotor. A good motor can reach a good efficiency, but they should be typically at most 70%.

There are many great advantages for using DC generators. One of the biggest reasons is because typically not requiring any gearing and still gets a battery charging voltage in light wind.

6.11.5 Applications of Wind Turbines

The four leading applications of wind power are as follows:

1. Autonomous electric power supplies, where the wind generator operates in conjunction with electrolytic batteries and perhaps also a diesel generator and inverter.
2. Parallel operation with a much large power network, where the system frequency and/or pitch variation is used to control the turbine speed at nominally constant level. The network is, in this case, essential the reserve capacity for periods of low or zero wind speed.
3. Operation with a large power network through an inverter. Here, WIG may run at variable speed, delivering variable frequency; further, a regulator and inverter are used to condition the power for delivery to the system. The network is again available to the user as reserve capacity under low wind conditions.
4. Water heating using resistive immigration elements. In this case, the WIG may run at variable speeds, delivering variable voltage and frequency.

SUMMARY

- Windmills have been used for many centuries for pumping water and milling grain. If the mechanical energy is used directly by machinery, such as for a pumping water or grinding stones, the machine is usually called a windmill.
- The terms *wind power* and *wind energy* describe the process by which the kinetic energy available in wind is converted **into electricity**.
- Wind energy is a freely available renewable source of energy having no air pollutants or greenhouse gases.

- Wind farm refers to as grouping of a large number of wind turbines as a single-wind power plant to generate bulk electrical power.
- The power available in wind is directly proportional to the cube of the wind velocity.
- The power available in wind is directly proportional to the air density. Since air density in summer is lower than that in winter, a wind generator will produce lesser power in summer than that in winter at the same wind speed.
- The efficiency of a wind turbine is measured as the ratio between the energy extracted from the wind to perform useful work (e.g., electricity) and the total kinetic energy of the wind without the presence of a wind turbine.
- A German physicist Albert Betz concluded in 1919 that no wind turbine can convert more than 16/27 (59.3%) of the kinetic energy of the wind into mechanical energy turning a rotor. This is known as the Betz limit or Betz' law.
- Tip-speed ratio is defined as the ratio of the speed of the extremities (peripheral blade speed) of a windmill rotor to the speed of the free wind.
- Solidity is usually defined as the percentage of the circumference of the rotor that contains material rather than air; this means the fraction of the swept area of the rotor that is filled with metal.
- The coefficient of performance is the proportion of the power in the wind that the rotor can extract (it is also called power coefficient or efficiency, C_p) and its variation as a function of tip-speed ratio is commonly used to characterize different types of rotor.

REVIEW QUESTIONS

1. Explain the momentum theory in wind power generation. Give the classification of rotor used for wind generation.
2. What prohibits large scale utilization of wind power for electricity generation?
3. What is the basic principle of wind energy conversion?
4. Derive the expression for power developed due to wind.
5. Prove that in the case of HAWT, maximum power can be obtained when exit velocity = 1/3 wind velocity and $P = 8/27\ PAV^3$
6. Describe the main consideration in selecting a site for wind generators.
7. How are EC system classified?
8. Discuss the advantages and disadvantages of WEC systems.
9. Describe horizontal axis-type aerogenerators.
10. Discuss the advantages and disadvantages of horizontal and vertical axis windmill.
11. What methods are used to overcome the fluctuating power generation of windmill?
12. What are advantages of vertical-axis windmill over horizontal type? Describe a rotor for relatively low velocity wind.

13. Write short notes on
    (a) Application of wind energy (b) Savonius rotor
    (c) Darrieus rotor (d) Wind energy storage.
14. State the essential features of a probable site for a wind farm.
15. Describe the construction of a three-bladed horizontal shaft wind turbine generator unit. Explain the terms yaw control, pitch control, and tethering control.
16. Compare the Darrieus rotor wind turbine with a three-bladed horizontal shaft wind turbine with regard to advantages and disadvantages, practical size, etc.
17. Describe a Darrieus rotor wind turbine generator unit.
18. Based on simple turbine theory, derive expression for power development by wind turbine rotor. Further, show that an ideal wind turbine cannot extract more than about 60% of the energy in the free wind present at the turbine rotor location.
19. The undisturbed wind speed at a location is 13.415 m/s. The wind speed at turbine rotor and at the exit is 60% and 30%, respectively, of the undisturbed wind speed. The rotor diameter is 9 m and wind density is 1.293 kg/m^2.Calculate
    (a) Power developed by the turbine rotor and
    (b) Coefficient of performance
20. Derive a suitable relationship for the power developed in a wind power plant in terms of the wind velocity, sweep area of the windmill, and its conversion efficiency.
21. A wind power generating unit uses a four-bladed, Dutch-type windmill designed to work at a maximum velocity of 90 km/h and with a maximum sweep diameter of 10 m. If the conversion efficiency of the windmill is 30%, determine the maximum power that can be obtained from the generating unit. Assume the average value of air density under average ambient condition to be 1.293 kg/m^2.
22. Define the power coefficient $C_{p,}$ of a wind turbine. Show that for a horizontal-axis wind machine, ideally C_p will be a maximum when $V_2/V_1 = 1/3$, where V_2 = air velocity downwind of the blades and V = air velocity for upstream of the wind machine.
23. Give a detailed analysis for power delivery by a wind energy converter having aerodynamically constructed rotor blades. On the basis of the analysis, discuss the operation of a Darrieus rotor.
24. Draw a block diagram indicating the basic components of a wind electrical generator scheme. Describe the operation of wind electrical generation schemes involving (a) induction generator, (b) synchronous generator, and (c) double output induction generator.
25. The Sandia 17 m turbine has a diameter of 16.7 m, an area of 187 m^2 and is spinning at 42 rpm. The ambient temperature is 15°C and the pressure is 83 kPa. For each of the wind speeds 5, 7, 12, and 16 m/s, find
    (a) tip-speed ratio.
    (b) coefficient of performance.
    (c) predicted mechanical power output.
26. A large turbine is rated at 2,500 kW at standard conditions (0° and 101.3 kPa). What would be its rated power at the same rated wind speed, if the temperature was 20°C and the turbine was located at 1,500 m above the sea level?

Chapter 7

Geothermal Energy

Geothermal energy is the clean and sustainable heat resource from the Earth. It ranges from the shallow ground to hot water and hot rock found a few kilometres beneath the earth's surface and even deeper down to the extremely high temperatures of molten rock called magma.

The shallow ground of the earth's surface maintains a nearly constant temperature between 10°C and 16°C almost everywhere. The geothermal heat pumps can tap this resource to heat and cool buildings. Well bores can be drilled into underground geothermal reservoirs of hot water for hot water application and for the generation of electricity using binary cycle. Hot water near the surface of earth can be used directly for heating buildings, growing plants in greenhouses, drying crops, heating water at fish farms, and several industrial processes like pasteurization of milk.

Hot dry rock geothermal resources occur at depths of 4–7 km everywhere beneath the earth's surface and at lesser depths in certain areas. The utilization of these resources involves injecting cold water down one well, circulating it through hot fractured rock, and drawing off the heated water from another well.

High cost of tapping heat from the hot dry rocks and magma from deep depth and existing technology do not yet allow commercial applications of geothermal energy.

Geothermal energy is the earliest branch of power generation and these energy resources utilize the earth's deep heat to provide a significant contribution to the power budget of every country. In general, earth's deep heat may and must be utilized for the following applications:

1. Electrical power generation.
2. The extraction of rare elements such as iodine, bromine, boron, lithium, and caesium.

KEY CONCEPTS

- Geothermal energy and classification
- Geothermal resource utilization for direct heat applications and for electrical power
- Conversion technology for electrical power generation
- Prospects of geothermal fields in india
- Electrical and mechanical features of geothermal power plants and their operations
- Environmental and other associated problems

3. Utilization in complex air conditioning, medical purposes, mineral water bottling, etc.
4. Thermal waters used in many branches of economy-heating premises, hot water supply, technological premises, industrial applications, etc.
5. The thermal waters widely used in agriculture, mainly in vegetable growing, cattle breeding, drying seeds, etc.

Utilization of greater proportion of geothermal resource base depends on achieving one or more of the following techniques:

1. Breakthrough in drilling technology that would permit low-cost drilling of bores of depths greater than 3 km.
2. Development of techniques for artificial stimulation that would increase the productivity of the geothermal reservoirs.
3. Technical advances that would allow electrical generation from low-temperature and low-pressure reservoirs.
4. Expansion of the use of low-grade geothermal resources for space heating and air conditioning, product processing, agriculture, and desalination schemes.
5. Maintenance of the geothermal energy system that is environmentally acceptable to human being.

7.1 GEOTHERMAL SYSTEMS

The interior of the earth is shown in Figure 7.1. Under the Earth's crust, there is a layer of hot and molten rock called magma. Heat is continually produced there, mostly from the decay of natural radioactive materials such as uranium and potassium. The amount of heat within 10,000 m of the earth's surface contains 50,000 times more energy than that of all oil and natural gas resources in the world.

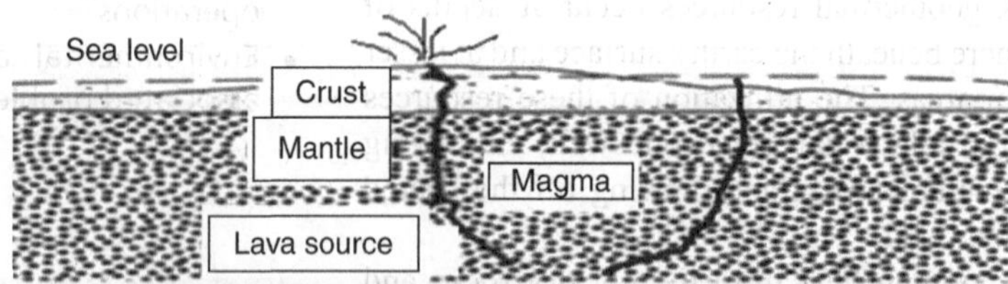

Figure 7.1 Interior of the earth

The temperature gradient between the magma and the earth's surface causes magma's heat to flow slowly towards the surface where it is lost to atmosphere. When magma is cooled, earth's crust is formed, which has reached an average thickness of only 32 km. Due to the highly irregular topography of the earth's surface, it is obvious that the crust has been subjected to considerable strains. As a result, mountain ranges are formed. Earthquakes quickly shift the earth's surface over large areas and some areas are in slow but perceptible motion relative to others. If magma breaks through the surface at the weak spots in the crust, the result is a volcano. In some other areas, deep faults in the crust permit surface water to come in contact with the hot solidifying magma. Additionally, the molten magma itself contains water which it releases as it solidifies. The heated water, or steam, rises to the surface and is emerged as hot springs.

The geothermal resource base is defined as total heat greater than 15°C in the earth's crust; but only a small portion of this storage of heat base can properly be considered as a resource. The magnitude of these resources depends on the evaluation of many physical, technological, economical, environmental, and governmental factors. Physical factors that control the distribution of heat at depth can be evaluated approximately; however, more complicated are the assumptions of technology, economics, and governmental policy. Differences among these assumptions, in large part, are responsible for the varying magnitudes of different geothermal resources.

Since nature allows only a tiny fraction of geothermal energy to leak through the surface, this energy has to be tapped. At present, geothermal exploitation is confined to thermal areas (geysers, hot springs, etc.) where the downward temperature gradient is exceptionally high when compared with the average of about 1°C per 31 m of depth.

In non-thermal areas, there are still immense reserves of heat below the ground level; however, at such depths, these cannot be tapped economically by the existing means. However, it is not inconceivable that, in the course of ways, may be found to penetrate some distance into magma (molten rock that forms the earth's crust) possibly with the help of underground nuclear explosions at a commercially acceptable cost.

Geothermal developments are unique in the sense that all activities related to resource production are localized to the immediate vicinity of the power plant or other utilization facility. Support operations such as mining, fuel processing, transportation, and other handling facilities do not exist. Thus, the environmental effects are solely site-dependent in origin.

7.2 CLASSIFICATIONS

From the improved scientific study of different geothermal systems developed so far, these systems have been grouped under the following two categories:

1. *Vapour-dominated (dry steam) geothermal systems*: This system are uncommon and poorly understood when compared to liquid-dominated system. Vapour-dominated system requires relatively potent heat supplies and low initial permeability. After an early hot water stage, a system becomes vapour dominated; that is, when net discharge starts to exceed recharge, steam boils from a declining water table and some steam escapes to the atmosphere. However, most of the steam condenses below the surface, where its surface of vaporization can be conducted upwards. The main vapour-dominated reservoir actually is a very deep water table, where the transmission is in the upward direction. Most liquid condensate flows down to the water table, but some may be swept out with steam in channels of principal upflow. Liquid water favours small pores and channels because of the high surface tension relative to that of steam. Steam is largely excluded from smaller spaces, but greatly dominates the larger channels and discharge from the wells.
2. *Liquid-dominated (hot waters) geothermal systems*: Many parts of the world consist of only hot water geothermal field at comparatively modest temperatures and low enthalpies. Despite the unsuitability of hot water for use in power production, there have been direct

heating applications far exceeding in total heat equivalent of existing geothermal electricity-generating facilities.

At Larderello (Italy) and Geyser plants (North America), wells deeper than 100 m and near centres of activity were found to yield lightly superheated steam. However, some wells on the borders of active systems produced hot water and steam in non-commercial quantities and pressures.

At Wairakei (New Zealand) and in Mexico (State of Baja California), liquid-dominated geothermal field has been discovered. New Zealand first demonstrated the generation of geothermal power. At Wairakei field subsurface, hot waters at temperatures up to 260°C are erupted through wells to the surface; some water flashes to steam as temperature and pressure decrease to the operating pressures, commonly from 3 to 6 kg/m^2. This steam, generally 10%–20% of the total mass, is separated from the mineral water and directed through turbines to generate electricity. Same method is also being adopted in Mexico fields.

Whatever may be the type of geothermal system, the geothermal steam contains 1%–3% of various gases, consisting of carbon dioxide principally, but with large amounts of hydrogen sulphide and less amounts of hydrogen, methane, nitrogen, etc. These gases may pose severe corrosion problems. However, these have been considered in some cases as a gaseous by-product of electric generation; further, the feasibility of commercial production of carbon dioxide, sulphur, ammonium sulphate, and ammonium carbonate in certain geothermal fields have already been demonstrated.

7.3 GEOTHERMAL RESOURCE UTILIZATION

In the following sections, we discuss about the extensive use of geothermal energy, one can imagine.

7.3.1 Direct Use of Low Grade Geothermal Energy

1. *Aquaculture and horticulture*: Geothermal renewable energy is used in aquaculture and horticulture in order to raise plants and marine life that require a tropical environment. The steam and heat are all supplied by geothermal energy. Many farmers use geothermal power to heat their greenhouses. In Tuscany, Italy, farmers have used water heated by geothermal energy for hundreds of years to grow vegetables in the winter. Hungary is also a major user of geothermal power. Eighty percentage of the energy demand from vegetable growers is met by using geothermal energy. It is also used in fishing farms. The warm water spurs the growth of animals ranging from alligators, shellfish, tropical fish, and amphibians to catfish and trout. Fish growers from countries like Oregon, Idaho, China, Japan, and even Iceland use geothermal power.
2. *Industry and agriculture*: Industries are another consumers of geothermal energy. Their uses vary from drying fruits, vegetables, and wood, dying wool to extracting gold and silver from ore. It is also used to heat sidewalks and roads to prevent freezing in the winter. Thus, geothermal power generation is playing a major role in industry and agriculture. Timber is dried using heat acquired from geothermal energy, and paper mills use it for all stages of processing. There are many potential uses of geothermal energy in the industry.
3. *Food processing*: The earth naturally contains an endless supply of heat and steam, which can be utilized to sterilize equipment and rooms. This would put an end to the use of chemicals

for this purpose. There are many potential uses of geothermal energy in food processing, but as yet, this renewable energy source has yet to be utilized to a large degree in this sector.

4. *Providing heat for residential use*: The most common use of geothermal energy is for heating residential districts and businesses. The first uses of geothermal fluid for heating a district in United States dates back to 1893. However, the French dominance, by almost 500 years as per the records, indicates that they were using geothermal energy in the 15th century. In the last few years, this renewable energy has caught the interest of an increasing number of house owners. Geothermal power generation provides more than just heat in summer; but a complete temperature control system that enables you to cool your home in winter as well. This significantly reduces heating and cooling bills, and keeps the home at a comfortable temperature year round.

Direct geothermal heating systems contain pumps and compressors, which may consume energy from a polluting source. This parasitic load is normally a fraction of the heat output, so it is always less polluting than electric heating. However, if the electricity is produced by burning fossil fuels, then the net emissions of geothermal heating may be comparable to directly burning the fuel for heat. For example, a geothermal heat pump powered by electricity from a combined-cycle natural gas plant would produce about as much pollution as a natural gas condensing furnace of the same size Therefore, the environmental value of direct geothermal heating applications is highly dependent on the emissions intensity of the neighbouring electric grid. Low temperature means temperatures of 149°C or less. Low-temperature geothermal resources are typically used in direct-use applications, such as district heating, greenhouses, fisheries, mineral recovery, and industrial process heating. Approximately 70 countries made direct use of 270 petajoules (PJ) of geothermal heating in the beginning of this century. More than half went for space heating, and another third for heated pools. The remainder supported industrial and agricultural applications. Global installed capacity was 28 GW, but capacity factors tend to be low (30% on average) since heat is mostly needed in winter. The abovementioned figures are dominated by 88 PJ of space heating extracted by an estimated 1.3 million geothermal heat pumps with a total capacity of 15 GW. Heat pumps for home heating are the fastest growing means of exploiting geothermal energy, with a global annual growth rate of 30% in energy production. Direct heating appliances of geothermal energy are more efficient than electricity generation, as the former requires low temperature heat resources. Heat may come from co-generation via a geothermal electrical plant or from smaller wells or heat exchangers buried in shallow ground. As a result, geothermal heating is economic at many more sites than geothermal electricity generation.

Where natural hot springs are available, the heated water can be piped directly into radiators. If the ground is hot and dry, heat exchangers can collect the heat. However, even in areas where the ground is colder than room temperature, heat can still be extracted with a geothermal heat pump more cost-effectively and cleanly than by conventional furnaces. These devices draw on much shallower and colder resources than traditional geothermal techniques, and they frequently combine a variety of functions, including air conditioning, seasonal energy storage, solar energy collection, and electric heating. Geothermal heat pumps can be used for space heating essentially anywhere.

Geothermal heat supports many applications. District heating applications use networks of piped hot water to heat many buildings across entire communities. In the USSR, the energy that is represented by geothermal hot water is used for heating of buildings, soil warming, green houses, and medical baths, and it is amounted to 15 million tons coal equivalent in 1970. In Hungarian basin, district heating, animal husbandry, and industrial processes benefit from geothermal heat.

In Iceland, the principal uses are for space heating. In Reykjavík, Iceland, spent water from the district heating system is piped below the pavement and sidewalks to melt the snow. Geothermal desalination has been demonstrated. At Rotorua (New Zealand), lithium bromide absorption units enable air conditioning to be applied to a 100-room hotel.

7.3.2 Electricity Generation

It provides not just heat and steam, but electricity itself. Geothermal power generation is completely clean, and releases no harmful gas emissions whatsoever. Geothermal fluid is a good electricity generator as well. Conversion technology for electricity generation is as follows:

1. *Flashed Steam Plants*: The water 'flash' boils and the steam is used to turn turbines.
2. *Dry steam plants*: These plants rely on the natural steam that comes from the underground reservoirs to generate electricity.
3. *Binary power plants*: These plants use the water to heat a 'secondary liquid' that vaporizes and turns the turbines. The vaporized liquid is then condensed and reused.
4. *Hybrid power plants*: In these plants, binary and flash techniques are utilized simultaneously.

Low-temperature resources can generate electricity using binary cycle electricity-generating technology. Utilization of the earth's geothermal energy resources for production of electric power has been commercially developed in North America at the Geysers, north of San Francisco. A 12-MW unit commercial operation was started in 1960. Geothermal electric plants were traditionally built exclusively on the edges of tectonic plates where high temperature geothermal resources are available near the surface. The development of binary cycle power plants and improvements in drilling and extraction technology enable enhanced geothermal systems over a much greater geographical range. Demonstration projects are operational in Landau-Pfalz, Germany, and Soultz-sous-Forêst, France, while an earlier effort in Basel, Switzerland was shut down after it triggered earthquakes. Other demonstration projects are under construction in Australia, the United Kingdom, and the United States of America. Presently, it has an installed capacity of 11,000 MW approximately. Although the total installed capacity of geothermal power plant throughout the world presently is only around 11,000 MW, DE White of Geological Survey of America has estimated that world's geothermal field will be capable of producing more than 60,000 MW of electricity.

At Larderello, boric acid is an important gaseous by-product of electricity generation while this field also demonstrated the feasibility of commercially producing carbon dioxide, sulphur, ammonium sulphate, and ammonium carbonate.

A dual-purpose application at Panzhetsho (USSR) generates electricity from steam and employs separated hot water for heating purpose in the local settlement. The success achieved by these activities has stimulated further exploitation in the USA, the USSR, Iceland, Japan, EI Salvador, and the Philippines, while reconnaissance programs have been inaugurated in Chile, EI Salvador, India, Kenya, and Ethiopia. The present scenario of installed geothermal electricity plant is given in Table 7.1.

Table 7.1 Installed Geothermal Electricity Plant

Country	Capacity (MW) 2007	Capacity (MW) 2010	Percentage of National Production
USA	2,687	3,086	0.3
Philippines	1,969.7	1,904	27
Indonesia	992	1,197	3.7
Mexico	953	958	3
Italy	810.5	843	1.5
New Zealand	471.6	628	10
Iceland	421.2	575	30
Japan	535.2	536	0.1
Iran	250	250	
El Salvador	204.2	204	25
Costa Rica	162.5	166	14
Nicaragua	87.4	88	
Russia	79	82	
Turkey	38	82	
Papua-New Guinea	56	56	
Guatemala	53	52	
Portugal	23	29	
China	27.8	24	
France	14.7	16	
Ethiopia	7.3	7.3	
Germany	8.4	6.6	
Austria	1.1	1.4	
Australia	0.2	1.1	
Thailand	0.3	0.3	
TOTAL	9,981.9	10,959.7	

7.3.3 Multi-purpose Total Energy Utilization of Geothermal Resources

It could result in

1. Steam separated in a conventional manner and used to supply cheap water
2. Carbon dioxide extracted from the gases are used for refrigeration and food processing.
3. Hydrogen sulphide refined to obtain sulphur.
4. Hot waters form the wells could supply a desalination plant (one such plant is in operation at EI Tatio in northern Chile); the fresh water would not only meet the needs of the local population, but has a high market value for irrigation.
5. The hot water would also provide air conditioning and refrigeration. Finally, the effluent from the desalination could be applied to mineral extraction and the resulting minerals processed in taking advantage of the availability of cheap base electricity.

7.4 RESOURCE EXPLORATION

Geothermal exploration involves outlining broad regions where the heat flow is significantly greater than 1.5×10^{-6} cal cm^{-2} s^{-1}. Most of these regions with high heat flow are in zones of early volcanic and tectonic activity, and most of them are characterized by hot springs.

Techniques for identifying potentially economic concentrations of geothermal energy within broad regions of high heat flow are not well developed. Important consideration included distribution of hot springs, evaluation of volcano and tectonic setting, and chemical analysis of hot springs fluids. In particular, the silica content and the ratio of sodium, potassium, and calcium provide information about the minimum subsurface temperature to be expected. The methods that play major role in geothermal exploration techniques are geological, geochemical, electrical, seismic, gravitational, magnetic, and thermal methods.

Some important geological research problems are concerned with the determination of age, size, and magnetic type of igneous occurrences related to convective hydrothermal and permeable rock systems and the relationship of convective hydrothermal systems to broad region of elevated heat flow. Geological study of the surface is done by the collection and analysis of samples. With the refinement of aerial photography, geological study has greatly been simplified. Aerial photography is the most useful method to detect the geothermal sources by means of remote sensing and thermal scanning. Such photographs are invaluable in the geothermal surveys. The use of radar is of great help for obtaining good quality pictures through cloud covers. Infrared scanners can distinguish negligible temperature differences. Thus, aerial surveys with infrared scanners have great future in the detection of geothermal resources.

Several geophysical techniques have proved useful in the final delineation of geothermal targets. Of these techniques, perhaps most unambiguous is the direct measurement of temperature gradient at the depth 25–100 m. However, it can be misleading primarily owing to the effect of seasonal changes in the temperature and to the shallow movement of groundwater.

Several techniques that measure the electrical conductivity at depth have had great success in geothermal exploration. The conductivity at depth varies directly with temperature, porosity, salinity of terrestrial fluid, and content of clays and zoolatrous. All these factors tend to be high within good, and consequently, the electrical conductivity in these geothermal reservoirs is relatively high. Electrical conductivity at depth can be measured by electrical (galvanized) or electromagnetic (induction) methods. Among the electrical techniques, only direct methods are reliable owing to skin effects of an AC current. Electrical resistivity techniques should be directed towards the understanding of the variation of porosity, water salinity, and temperature in actual geothermal reservoirs, improving electric field techniques and procedures for extracting true resistivity values from field data, and developing complementary exploration techniques that will improve the interpretation of resistivity data. Other electrical exploration research should include further work on the resistivity, self-potential, electromagnetic, telluric, and magnetotelluric techniques.

Seismic methods are proving useful in locating fractured and permeable zones in geothermal areas. Micro-earthquakes at relatively shallow focal depths are concentrated along the fracture or fault zones in many geothermal areas. In addition, some geothermal areas appear to have a high level of seismic ground noise, analysis of aerial distribution of this noise may outline prospective zones of geothermal production. Seismic method should be studied to characterize energy absorption attenuation and frequency shift in known high temperature system. Seismic noise studies should be undertaken to evaluate temporal and social noise vibrations, characteristics of recognized noise sources, noise spectrum, and direction of noise propagation and apparent velocity.

Several other geophysical techniques have proved useful in special circumstances. Additional research is needed to determine the source of gravity and negative anomalies associated with known geothermal areas and to decide whether these anomalies can be used as indicators of the internal temperature of surface.

Latest technique of rotary percussion drilling is promising for geothermal resource exploration. However, present drilling technology needs improvement in techniques such as isolating testing intervals in unconsolidated sands and in fractured reservoirs, inexpensive core recovery with reserve fluid in place, logging at temperatures above 180°C, and transmission of information from the borehole face to the surface. The development of inexpensive, low density, low viscosity, non-thermally sensitive, high thermal conductivity, and high surface tension of the fluid would lower drilling costs and leave the borehole face in a more nearly undisturbed state.

Hydrological studies are required to understand fully the effect of groundwater movement in a local geothermal gradient. Geothermal research is needed in chemical, physical, and thermodynamical properties of aqueous solution, as temperature ranging from 100°C to 400°C, and in the determination of chemical composition of in situ geothermal fluid as a function of in situ temperature, and in isotopic relationship between the water and the various dissolved constituents, particularly, gases

7.4.1 Prospects of Geothermal Fields in India

India also has vast potential for geothermal energy scattered all over the country. There are more than 300 hot springs scattered all over the country. These thermal springs are mostly in Bihar, Bombay, Ratnagiri, Himachal Pradesh, and Ladakh. Some important hot springs are located at Puga, Manikaran, Badrinath, Surajkund, Ratnagiri, Gundala, Kalwa, etc. These thermal springs were not given proper attention till the middle of 20th century. However, in the wake of increasing energy demand in recent years, geothermal energy resources have also attracted attention, and thus, government of India has set up a hot springs committee in 1966, consisting of engineers, geologists, and scientists from related fields, to examine the possibility of tapping this new source of energy. After an initial investigation, this committee recommended that exploration work should be undertaken to start with Puga hot springs in Laddakh and Manikaran hot springs in Kullu region of Himachal Pradesh. Geothermal fields at Puga Valley (about 25 MW), Manikaran, Rajgir, etc., have been estimated. However, the latest scientific investigation indicated that Indian geothermal field may possibly be a liquid-dominated field (liquid like hot water) and hence a careful study is essential to determinate whether such low temperature and low pressure resource are economical for electrical generation using binary cycle or it may be used in its basic form.

In the light of abovementioned recommendations, government of India asked the USA for help, and thus, a joint project has been launched in 1973. With the investigation carried out so far, there are ample prospects of electricity generation at Puga valley, Manikaran, Rajgir, etc. All the Indian geothermal fields are expected to be exploited in near future and will be fully utilized for the advancement of the country.

The latest scientific study indicates that Indian geothermal field is streams of water under pressure from depths varying from 11 to 32 m have been encountered in the majority of bore holes, at Puga Valley. The temperature of water and wet mixture of water and steam coming out of bore holes at Puga valley ranging from 50°C to 110°C have been reported. The overall natural heat flow from the Puga valley is estimated approximately to be 6,000 kcal/s (about 25 MW).

The temperature of Manikaran thermal springs (located in the Parbati river valley of Kullu district, Himachal Pradesh) ranges from 69°C to 93°C. A considerable amount of steam and gaseous fumes with sulphurous smell come out along with hot waters. In these hot waters, sodium, HCO_8, and chlorine are the dominating ions. Sulphur, mercury, and boron are also reported to be present in the hot water.

The thermal springs of Rajgir group (in Patna and Gaya districts) are perhaps one of the purest class of thermal waters reported anywhere in the world. The temperature of Rajgir group of thermal springs ranges from 35.5°C to 42.5°C. In Surajkund (Hazaribagh), thermal spring's temperature of 87°C and the high flow at 173 Kilolitre/hour was recorded from the springs at Silakund.

Further, the Oil and Natural Gas Commission has detected hot water and steam at depth of 1,500 to 2,000 m during oil exploration in Cambay region of Gujarat.

7.5 GEOTHERMAL-BASED ELECTRIC POWER GENERATION

Geothermal-based electric power generation technology represents the entire process of turning hydrothermal resources into electricity. Of the four available to developers, one of the fastest growing is the binary cycle, which includes a Rankine cycle engine.

7.5.1 Dry Steam-based Geothermal Power Plants

Figure 7.2 shows a dry steam plant. The basic cycle for steam plants remains similar to the structure that was first operated in 1904 in Larderello, Italy. Even so, incremental technology improvements continue to advance these systems.

Dry steam plants have been operating for over 100 years—longer than any other geothermal conversion technology, though these reservoirs are rare. In a dry steam plant like those at The Geysers in California, steam produced directly from the geothermal reservoir runs the turbines

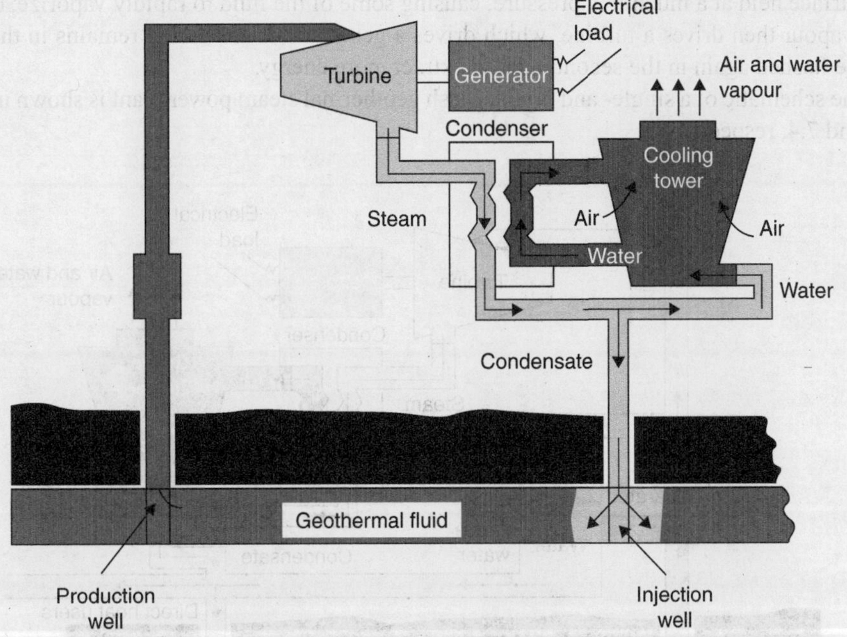

Figure 7.2 Dry steam geothermal electric power plant

that power the generator. Dry steam systems are relatively simple, requiring only steam and condensate injection piping and minimal steam cleaning devices. A dry steam system requires a rock catcher to remove large solids, a centrifugal separator to remove condensate and small solid particulates, condensate drains along the pipeline, and a final scrubber to remove small particulates and dissolved solids. Today, steam plants make up a little less than 40% of U.S. geothermal-electricity production, all located at The Geysers in California.

7.5.2 Flash Geothermal Power Plants

The term flash steam refers to the process where high-pressure hot water is flashed (vaporized) into steam inside a flash tank by lowering the pressure. This steam is then used to drive around turbines. Flash steam is today's most common power plant type. The first geothermal power plant that used flash steam technology was the Wairakei Power Station in New Zealand, which was built already in 1958.

At high pressure below earth's surface, the water exists as compressed liquid. Pipeline is installed to tap into the resource. When the compressed liquid water reaches the surface at atmospheric pressure then, a portion of it immediately flashes to steam. The steam portion is redirected into a steam turbine, where power is produced. The exhaust is then piped to a condenser where it is returned to liquid. This hot liquid water can then be used for further heating applications prior to the reinjection into the rock.

Flash steam plants are the most common type of geothermal power generation plants in operation today. Fluid at temperatures greater than 182°C is pumped under high pressure into a tank at the surface held at a much low pressure, causing some of the fluid to rapidly vaporize, or 'flash.' The vapour then drives a turbine, which drives a generator. If any liquid remains in the tank, it can be flashed again in the second tank to extract more energy.

The schematic of a single- and double-flash geothermal steam power plant is shown in Figures 7.3 and 7.4, respectively.

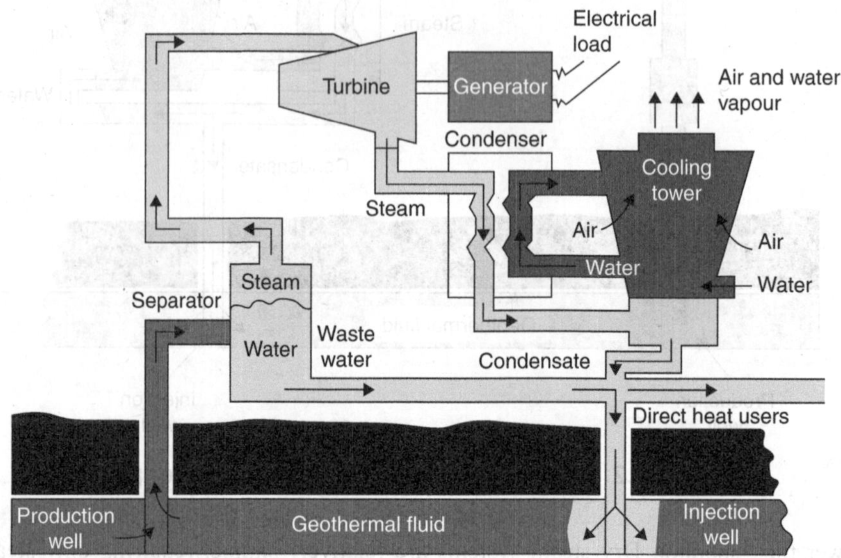

Figure 7.3 Single-flash geothermal steam-electric power plant

7.5.2.1 Advantages

1. Very low emissions
2. Safe and reliable
3. Immune to varying weather conditions
4. Cost effective over life of plant
5. Sustainable
6. Small footprint
7. No fuel cost

7.5.2.2 Disadvantages

1. High initial cost
2. Increased risk of seismic activity
3. Location sensitive
4. Risk of overexploiting resources

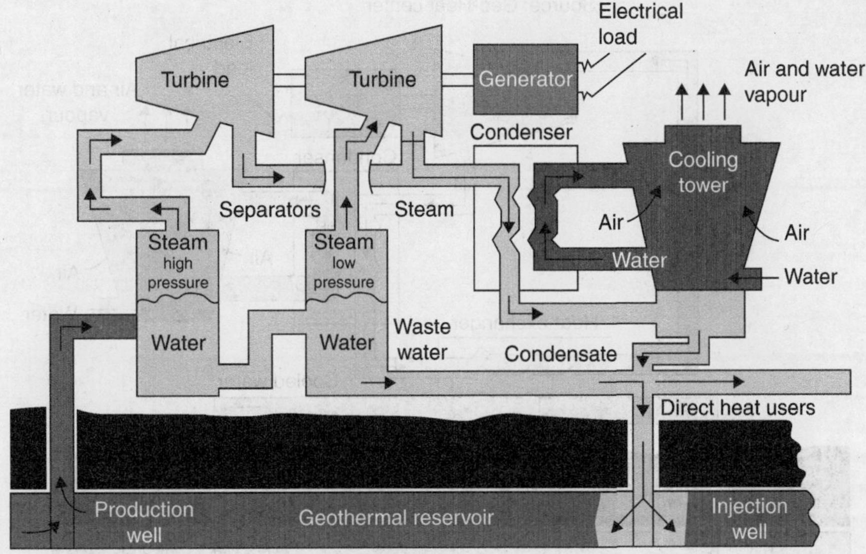

Figure 7.4 Double-flash geothermal steam-electric power plant

At a flash facility, hot liquid water from deep in the earth is under pressure, and thus kept from boiling. As this hot water moves from deeper in the earth to shallower levels, it quickly loses pressure, boils, and 'flashes' to steam. The steam is separated from the liquid in a surface vessel (steam separator) and is used to turn the turbine, and the turbine powers a generator. The most common type of power plant to date is a flash power plant, where a mixture of liquid water and steam is produced from the wells. About 45% of geothermal electricity production in USA comes from flash technology. Flash power plants typically require resource temperatures in the range of 177°C–260°C.

7.5.3 Binary Cycle-based Geothermal Plants

In the binary process, the geothermal fluid, which can be either hot water, steam, or a mixture of the two, heats another liquid such as isopentane or isobutane (known as the 'working fluid'), that boils at a lower temperature than water. The two liquids are kept completely separate through the use of a heat exchanger that is used to transfer heat energy from the geothermal water to the working fluid. When heated, the working fluid vaporizes into gas and (like steam) the force of the expanding gas turns the turbines that power the generators. Technology developments during the 1980s have advanced lower temperature geothermal electricity production. These plants, known as 'binary' geothermal plants, today make use of resource temperatures as low as 74°C (assuming certain parameters are in place) and as high as 177°C. Approximately 15% of all geothermal power plants utilize binary conversion technology. It is shown schematically in Figure 7.5.

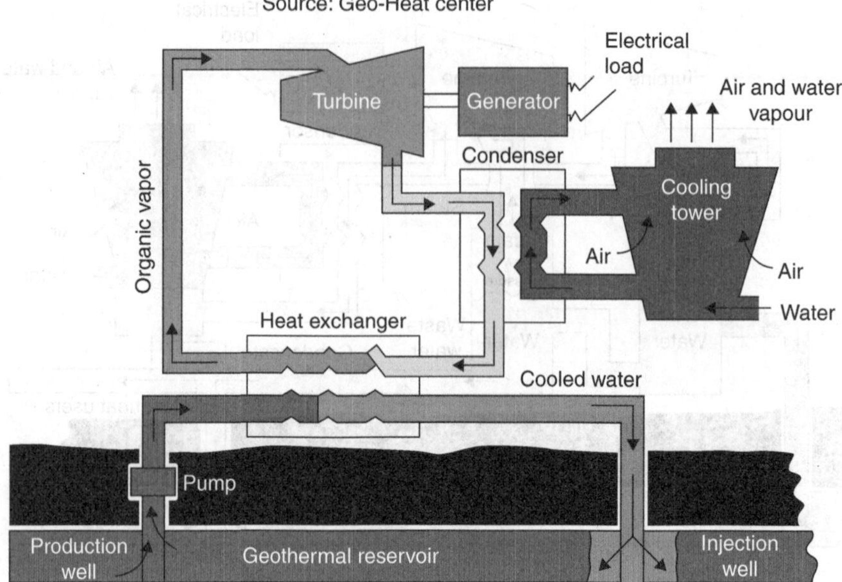

Figure 7.5 Binary cycle-based geothermal electric power plant

7.5.4 Electrical and Mechanical Features

Electrical and mechanical features of a geothermal power plant are essentially the same as of a fossil plant. However, gaseous by-product present in the steam are of corrosive nature and are particularly troublesome. Thus, the materials used in the geothermal plant should have high resistance to corrosion and possess sufficient electrical and mechanical properties. Relays, motor control equipment, excitation gears, and switchgear should be located in clean room and kept under positive pressure. Another precaution is to eliminate the copper commutator of a motor-driven exciter by using a static-type excitation system.

For the successful operation of the Geyser plants in USA, the following requirements for material were determined and they are, more or less, essential for any geothermal plants:

1. There are no special requirements in the steam path through the pipeline and turbine. The pipe is ordinary carbon steel and turbine is almost entirely of standard material for low pressure, turbine manufacture.
2. When the steam is condensed and is exposed to oxygen, as in the direct contact condenser and afterwards, corrosion problems can be severe. All the condensates are lined with an austenitic stainless steel. Condensate piping above grade is aluminium or stainless steel. Pumps have stainless steel impellers and bowls or volutes. Valves are aluminium or stainless steel. Below the grade, piping is cement asbestos with plastic lining or glass-reinforced plastic.
3. Copper is severely attacked by hydrogen sulphide in water as well as that in the atmosphere in the vicinity of the plants. In electrical equipment and in various instruments and controls, this

presents problems in obtaining equipment in which copper is protected or in which other materials can be used. Copper alloys (bronze and brass) tarnish badly and must be used with caution.

4. The condensate attacks concrete and concrete structures that would be exposed to condensate are coated with bituminous compounds, epoxy materials, or neoprene sheets where conditions are severe.
5. Galvanized coating performs fairly well when compared to the performance of cadmium coating.

7.5.5 Operation of Geothermal Plants

1. Geothermal plants are simpler to operate than fuel-fired steam plants because there are no boilers with all the necessary auxiliaries and controls.
2. Geothermal plants can be operated without round the clock attention. All the operation and some minor maintenance can be performed by roving operators. Thus, for example, the geysers unit is being maintained.
3. Since the unit may be operated unattended, it should be necessary to have an annunciator to generate an alarm in the transmitting system so that alarm may be transmitted to an attendant station where appropriate action could be taken based on the information received.
4. Unit should be inspected at an interval less than that normally used for fossil units. The stop valves of the turbines must be tested three or four times a day to assure that there is no binding from impurities in the steam. They must also be lubricated frequently.
5. Turbine tube oil and generator heat exchangers should be single-pass or U-tube design in the cooling water side, otherwise there is apt to be plugging in the reversing water box from colloidal sulphur and rock dust in cooling water, which will be condensate from the cooling towers.
6. The steam jet ejector creates high noise levels that requires the use of ear protectors when working near them.
7. Because of the atmospheric difference between day and night, better turbine vacuum is obtained at night, when cooling tower is more effective. If the rated generator load were carried during the day, the generator could be overloaded at night unless the unit were provided with load-limited devices. These devices sense the generator load and maintain it at a level. Geyser plants are equipped with such devices because the plant operates without attendants.

7.6 ASSOCIATED PROBLEMS

A major problem of geothermal power is the estimation of the power life of the reservoir to make a reasonably accurate decision on the size of station to be built. The financial life of such a station should be sufficiently long in which the borrowed money methods have been developed for predicting the reservoir with unique features, and the prediction of life of reservoir is only determined on the basis of historical development; thus, the assessment of field life remains a subject of current interest.

The second problem is associated with the separation of steam from the steam-water mixtures at the well heads and transmission of steam only through a long pipeline to the power house.

In spite of large and extensive commercial development at Larderello and Geysers, the origin and nature of the geothermal systems that yield dry steam and why they differ from the abundant hot water systems are yet unknown.

Another important problem is the selection of materials that are suitable for geothermal systems and plants. Materials should have large resistance to corrosion for the gaseous products and properties to fulfil the electromechanical and other requirements.

As the automatic-start control is rather expensive, manual-start control is used. Thus, an operating condition serious enough to trip the unit requires investigation at the plant before the unit is restarted.

7.7 ENVIRONMENTAL EFFECTS

Considerable attention has been made that can have relatively small effect on the environment. Although environmentalists have reasons to believe that geothermal energy may prove to be the cleanest source of convertible power readily available, certain undesirable effects can extend for several kilometres from the geothermal field itself; thus, they introduce environmental problems into the surrounding regions. These damages to environment that are inherent in geothermal development need careful study and solutions must be found to control them at reasonable costs. The potential effects on the environment of most immediate concern are gaseous and particulate emission, land pollution, subsidence potential, and seismic consideration, biological, and social effects. Research should be directed towards accurate determination and evaluation of their characteristics and magnitude.

7.7.1 Gaseous and Particulate Emission

Fluids drawn from the deep earth carry a mixture of gases, notably carbon dioxide (CO_2), hydrogen sulphide (H_2S), methane (CH_4), and ammonia (NH_3). These pollutants contribute to global warming, acid rain, and noxious smells if released. Existing geothermal electric plants emit an average of 122 kg of CO_2 per megawatt-hour (MWh) of electricity and a small fraction of the emission intensity of conventional fossil fuel plants. Plants that experience high levels of acids and volatile chemicals are usually equipped with emission-control systems to reduce the exhaust emissions.

In addition to the dissolved gases, hot water from geothermal sources may hold in solution trace amounts of toxic chemicals such as mercury, arsenic, boron, and antimony. These chemicals precipitate as the water cools and can cause environmental damage, if released. The modern practice of injecting cooled geothermal fluids into the earth to stimulate production has an additional benefit of reducing this environmental risk.

Some of the noxious materials creating air pollution from the existing geothermal steam power plants are hydrogen sulphides, mercuric compounds, radioactive materials such as lead 210 and radon 222. Research must be applied to identify the chemical composition of the non-condensable gas derived from geothermal reservoirs and to quantity permissible exposure limit. Based on these findings, methods for containment and safe disposal of these materials must be developed. Concomitant with these research efforts, there is a need to develop the analytical techniques and instrumentation to monitor and control the discharge of these materials.

Similar research must be directed towards the emission of particulate matter. Particulate emission from an operating geothermal power plant may not readily appear, but it can occur.

The basic needs of research are thus to identify, quantity, and regulate the disposal of gaseous and particulate matter from geothermal resources.

7.7.2 Land Pollution

In relation to the problem of land surface pollution, research is to be directed towards preventing the degradation of usable soil and towards the control of on-site surface disposition of pollutant that may be transported subsequently from the site of production to the surrounding environment. Types of the geothermal field complicate these problems. With regard to vapour-dominated system, research is needed to identify and quantify all pollutants (such as Hg, As, Se, and Pb^{210}) in the vapour phase of a geothermal source and at each site proposed for development.

In connection with the water-dominated geothermal system, the effects of accidental run off of geothermal fluids from the production site to surrounding land areas need to be assessed with an identification of surface-deposited material that may harm plant, soil sterility, or be subjects to biological magnification and entrance into food chains. Blowout contingency plans to minimize land pollution by chemical deposition and to control possible erosion are needed at each developed site.

7.7.3 Subsidence Effect

It has been investigated that subsidence occurs in some areas when a fluid is removed from the ground, while in other areas, the removal of equal quantities of fluid produced no measurable subsidence. Subsidence effects are better understood from studies of booth petroleum and groundwater reservoir; however, very little has been gathered from the production of geothermal fluids. The tools and techniques for these studies are presently available.

7.7.4 Seismic Hazards

Mostly, the geothermal resource areas are closely associated with the regions of high geologic activity, which is manifested most commonly as earthquakes. Studies have shown that if fluid pressures are changed in regions of tectonics stresses, fluid pressures are also changed in regions of earthquake activity. Present research is being directed towards resolving many questions regarding seismic activities and some of the information can be applied to geothermal field. However, seismic monitoring stations should be established near productive geothermal areas to determine if patterns emerge; these patterns appear to be related to the removal or injection of fluids from geothermal reservoirs.

7.7.5 Water Pollution

A possible risk associated with the development and utilization of geothermal resources is the contamination of surface and groundwater by geothermal fluids. Although earlier tools and techniques developed can be applied, but specific research is required to identify those chemical constituents, which may have a detrimental effect. Sample collection, analysis, and procedures should be developed where they are presently lacking or are too expensive for widespread field application with particular reference to injection of geothermal fluids. Chemical and isotopic studies should be undertaken to determine if geothermal fluids will return to the reservoir from which they are produced or if they migrate into other reservoirs.

7.7.6 Biological Effects

Numerous unknown effects exist regarding the impact of geothermal operation upon the bionature to prospective resource area as well as areas presently under exploration or development. Given the delicate balance of natural environment, damage to many species of plant and animal life can take place through changes of chemical balance in soil and water, the use of toxic substances in industrial application, the destruction of such specialized habitants as thermal pools alpine meadow, the interruption of migratory patterns, long term alteration in humidity, and the introduction of human presence, and activity into formerly unaffected regions. These and other factors need critical study in representatively selected geothermal resource areas to determine necessary procedures for the adequate protection of plants and animal life in regions of development.

7.7.7 Social Effects

Serious social effects arising from geothermal resource development, which need research, involve problems of noise and land use. Sociological, economical, and planning studies are greatly needed to determine public policy for the equitable resolution of conflicts of land use arising in these cases.

SUMMARY

- Geothermal energy is the thermal energy contained in the rock and fluid (that fills the fractures and pores within the rock) and below the earth's crust called magma.
- In a binary cycle power plant, the heat from geothermal water is used to vaporize a working fluid in separate adjacent pipes. The vapour (like steam) powers the turbine generator. In the heat exchanger, heat is transferred from the geothermal water to a second liquid. The geothermal water is never exposed to the air and is injected into the periphery of the reservoir.
- Electrical and mechanical features of a geothermal power plant are essentially the same as of a fossil fuel power plant.
- Gaseous by-product present in the geothermal fluid are of corrosive nature and are particularly troublesome.
- Flash steam geothermal is the most common power plant type.
- Carbon dioxide extracted from the gases are used for refrigeration and food processing. Hydrogen sulphide is refined to obtain sulphur.

REVIEW QUESTIONS

1. What are the different sources of geothermal energy?
2. Define a geothermal source. What are the types of geothermal fluids?
3. Categorize the different geothermal sources.

4. What do you mean by dry, wet, and hot water geothermal systems? Discuss the field of applications of these systems.
5. What is geothermal gradient?
6. What is hot, dry rock geothermal source? How is it used? Explain the concept of hot, dry rock geothermal power plant.
7. What is a geopressure deposit?
8. Describe a binary cycle geothermal power plant.
9. What is the limitation of a flashed steam system? What are the advantages of double flash system?
10. Describe a binary cycle system for liquid-dominated system.
11. What are the main applications of geothermal energy?
12. What are the possible sources of geothermal pollution? How are these avoided?
13. What is geothermal power? State and explain essential types of geothermal system in commercial use around the world.
14. What are the difficulties in the large-scale utilization of geothermal energy? What development could increase the role of geothermal energy in future?
15. Discuss different systems used for generating the power using geothermal energy, in brief.
16. Explain the difference between the geothermal power plant and the thermal power plant.
17. Briefly discuss environmental effects of geothermal energy.

Chapter 8

Solid Wastes and Agricultural Refuse

Waste has always been the negative side of the economy. In production and consumption, it is wasteful matter always available in energy activity system that is rejected as useless, harmful things that damage the environment. Waste is also known as the garbage waste, rubbish, refuse, etc. The social task of waste management has been to either minimize or to completely get rid of it. Traditional way of waste removal is to carry away them through sewers and dustbins, dispatched in the air through burning, dumped in disused quarries or the oceans, or fly-tipped in gutters or behind hedges.

The availability of free places for dumping waste in near future and associated environmental problems have created the need to find some new ways for efficient waste utilization and environmental protection. For example, landfill sites are a significant cause of global warming and a source of groundwater pollution, because of their methane emissions. However, incinerators also produce hazards. Their emissions of acid gases, mercury, dioxins, and furans have very harmful effects.

KEY CONCEPTS

- Waste is wealth
- Incinerators, pyrolysis, anaerobic digestion, and recycling methods of wastes disposal
- Waste recovery management systems
- Sources and types of wastes
- Advantages and disadvantages of wastes recycling
- Tips on reducing waste and conserving resources (the three R's—reduce, reuse, and recycle)
- Plastic recycling and plastic resin identification codes (RID)

8.1 WASTE IS WEALTH

All the wastes and waste management concepts are, therefore, now changing. Globally, the focus is to modify all resources from waste to wealth or from trash to cash; both are as good as having the better of two words. Three basic drivers of change are turning waste and waste management into a dynamic, fast changing, and international economic sector. This transformation presents new choices and opportunities and provides lessons and pointers for industrial, social, and environmental policy in the new post-industrial landscape. The following are some of the driving forces of change:

1. Growing concern about the hazards of waste disposal
2. Broad environmental concerns, especially global warming and resource depletion
3. Economic opportunities created by new waste regulations and technological innovation
4. Since the fuel shortage today is widely recognized not only as a political problem but also as agricultural problems as they are also consuming the great proportions of fuels and hence they have started thinking that fuels can be produced as an agricultural crop.

The following are the three main concepts have been introduced for converting waste materials to usable fuel and energy with main concern of minimizing the environmental damage.

1. *Heat energy generation*: Waste is used as supplemental boiler fuel and heat energy is obtained by the direct combustion of the waste to heat energy.
2. *Bioenergy generation*: It is a modern method of hazards control of waste disposal and for the recovery of fuels and energy (such as methane). Biological methane generation have commercially potential for energy resource recovery.
3. *Eco-modification through recycling*: Improved and efficient design of process and products reduces the health hazards and increases the resource productivity. Recycling conserve material, fuel, and energy by lengthening their life span.
4. *Fuel and energy generation from forest and agricultural and municipal wastes*: In order to cope up with fuel and energy shortages, all such wastes are considered as raw resources for converting them into improved fuels and energy.

8.1.1 Incinerators

Incinerator is precisely a furnace where waste is burnt to produce energy. Burning waste in incinerators only reduces the volume of solid waste, but it does not dispose the toxic substances contained in the waste and creates the largest source of dioxins.

The burning of waste produces heat that boils the water. Thus, the steam obtained is used to convert heat energy into electrical energy by thermo electromechanical converters. As already stated, the flue gases coming out of simple incinerator contain toxic gases (hazardous gases such as furans and dioxins).

Modern incinerators are equipped with pollution improvement systems to remove health hazardous gases. Incinerator combustion temperature of about 1000°C is maintained for complete combustion of wastes to reduce chlorine-enriched organic substances. Flue gases are sent through scrubbers for the removal of dangerous chemicals. A high chimney having cooling systems is installed as it removes the hazardous gases. Cost and efficiency are considered as the main parameters for selecting incinerator as a method of waste disposal. However, they can be thought of as a sustainable energy production system.

Incineration with recovery of energy is considered the best method of waste management and dominates over plain incineration and landfill. Incineration converts solid waste into ash, flue gas, and heat. Still with the incineration, some quantity of about 10% waste is produced that of original wastes.

8.1.1.1 Process of Incineration

Electricity generation is the most important useful energy obtained from incinerators. Incinerators have the common mode of operation even there are many variations in the incineration process.

1. *Auditing of wastes*: Understanding the quality, quantity, and composition of wastes is the important step in selecting the appropriate disposal options.
2. *Proper incinerator selection*: It is essential to assess the suitability of existing incinerators capability to disposal of newly added wastes and improving and replacing by new (modern) incinerator. Building and equipment design is appropriate. The recommended configuration is a dual-chamber controlled air incinerator.
3. *Proper operation of incinerator*: Operating the incinerator correctly ensures optimum combustion conditions. It includes separation and sorting of wastes, weighing and mixing it for specific calorific value, and closing and opening of incinerator doors for complete combustion of wastes. Personnel safety is very important.
4. *Removal of dangerous chemicals and toxic gases*: Combustions of wastes at high temperature reduce chlorine-enriched organic substances. The installation of high chimney with cooling system and scrubbers removes chlorine compounds and dangerous toxic gases contained in flue gases.
5. *Safe handling and disposal of incinerator residues*: Ash (or residues) obtained from bottom of chamber of the incinerator containing materials that health hazardous and environmental polluting materials and disposed of as per standard norms (in batch production). Incinerated waste is finally disposed in landfills.

8.1.1.2 Advantages of Incineration

1. Incineration is a practical method of disposal that minimizes high transport costs of wastes to landfills.
2. It largely reduces the carbon footprint during the transportation of wastes.
3. Large reduction in the amount of waste volume after incineration requires less space for landfills, thus saving money.
4. Gases and leachates produce in landfills by waste are completely eliminated.
5. Incinerated wastes are totally free of any environmental risk.
6. It is used for hazardous and clinical (containing pathogenic bacteria) waste treatments.
7. It is used for generation of electricity.
8. It can be operated under any weather conditions.
9. Incineration is cheaper in the long term and has long life span.

8.1.1.3 Disadvantages of Incineration

1. Incineration can generate hazardous gases such as furans and dioxins (an unwanted by-product of chlorine), which causes cancer, affects the functioning of hormones, and damages the immune system.

2. It can generate CO_2 and odour sometimes.
3. Expensive to build and operate and requires high energy.

8.1.2 Pyrolysis

Pyrolysis provides an alternative to methods of municipal waste disposal (such as anaerobic digestion, landfill storage, and more specifically incineration). In this technology, organic waste is burnt at relatively low temperatures to produce char (like charcoal), oils, and combustible gases. The oils can be used as a chemical feedstock and as fuel. Feedstock includes mixed waste, plastics, tires, and sewage sludge. Essentially, it involves chemically mining (a form of treatment that chemically decomposes organic waste materials by heat in the absence of oxygen under pressure and at operating temperatures above 430°C) the waste to produce elements that can be used for energy generation or chemical inputs.

In practice, it is not possible to achieve a completely oxygen-free atmosphere. Because some oxygen is present in any pyrolysis system, a small amount of oxidation occurs. If volatile or semi-volatile materials are present in the waste, thermal desorption will also occur. Organic materials are transformed into gases, small quantities of liquid, and a solid residue containing carbon and ash. The off-gases may also be treated in a secondary thermal oxidation unit (secondary combustion chamber), flared, and partially condensed. Particulate removal (such as fabric filters or wet scrubbers) is also required. Several types of pyrolysis units are available, including the rotary kiln, rotary hearth furnace, and fluidized bed furnace. These units are similar to incinerators except that they operate at low temperatures and with less air supply.

Pyrolysis transforms hazardous organic materials into gaseous components, small quantities of liquid, and a solid residue (coke) containing fixed carbon and ash. Pyrolysis of organic materials produces combustible gases, including carbon monoxide, hydrogen and methane, and other hydrocarbons. If the off-gases are cooled, liquids condense producing an oil or tar residue and contaminated water.

Pyrolysis liquids can be used directly (e.g., as boiler fuel and in some stationary engines) or refined for high quality uses such as motor fuels, chemicals, adhesives, and other products.

Direct pyrolysis liquids may be toxic or corrosive.

8.1.2.1 Advantages of Pyrolysis

1. Volume of the waste is significantly reduced.
2. Produces useful products for multiple applications. Solid, liquid, and gaseous fuel can be produced from the waste.
3. Storable or transportable fuel or chemical feedstock is obtained.
4. It is safer and more environment friendly than incineration, land filling, and many other gasification processes.
5. Environmental problem is reduced. Produces few air emissions due to limited use of oxygen.
6. Contamination of air emissions is easy to control because syngas is cleaned after production to get rid of any contaminants.
7. Desirable process as energy is obtained from renewable sources such as municipal solid waste or sewage sludge.

8. Replaces coal and natural gas as viable fuel sources, causing a reduction in climate change,
9. Can be easily implemented in CHP systems. More efficient than incineration.
10. It is sustainable process by reusing thrown away products that can be recycled again and again.
11. Pyrolysis plants are flexible and easy to operate because they are modular. They are made up of small units that can be added to or taken away when the mass or volume of organic matter changes.

8.1.2.2 Disadvantages of Pyrolysis

1. Generates possible toxic residues such as inert mineral ash, inorganic compounds, and unreformed carbon.
2. Potential to produce a number of possible toxic air emission such as acid gases, dioxins and furans, nitrogen oxides, sulphur dioxide, and particulates.
3. Pyrolysis plants require a certain amount of materials to work effectively.

8.1.2.3 Pyrolysis Versus Incineration

They are fundamentally different as given in Table 8.1:

Table 8.1 Differences between Pyrolysis and Incineration

Pyrolysis	Incineration
It is the thermal degradation in the absence of oxygen.	It requires oxygen in the form of air.
Pyrolysis is carried out at lower temperatures, 450°C–500°C and there is no combustion.	Incineration is carried out at higher temperatures (about 850°C or higher) and combustion takes place.
It produces the liquid fuel of useful high quality although the syngas is quite rich in carbon monoxide and hydrogen as well.	Incineration produces energy as heat (low level) that can be used to create steam for generating electrical power, industrial process, or district heating.
Pyrolysis is a controlled chemical process (temperature, pressure, batch or continuous system, catalyst, etc.) in order to produce valuable secondary raw materials (solid, liquid, or gas) or energy.	Incineration is to reduce the quantity of waste to be landfilled.
The pyrolysis system for treatment of MSW and other wastes demonstrates excellent practical performance in controlling the emission of harmful substances like dioxins.	Like coal combustion, the incineration of MSW produces carbon dioxide, as well as nitrogen and sulphur oxides and a range of other gas phase organic and inorganic air emissions.

8.1.3 Anaerobic Digestion

Anaerobic digestion is a series of biological processes in which microorganisms break down biodegradable material in the absence of oxygen. Anaerobic digester is an airtight chamber in which organic waste is decomposed and transformed into biogas by a biological process called anaerobic digestion.

One of the end products is biogas, which is combusted to generate electricity and heat, or can be processed into renewable natural gas and transportation fuels.

A range of anaerobic digestion technologies are converting livestock manure, municipal wastewater solids, food waste, high strength industrial wastewater and residuals, fats, oils and grease (FOG), and various other organic waste streams into biogas for 24 × 7.

Separated digested solids can be composted, utilized for dairy bedding, directly applied to cropland, or converted into other products. Nutrients in the liquid stream are used in agriculture as fertilizer.

8.1.4 Recycling

Waste control and reduction depends on complex flows and simple or specialist treatment. It is organized around material streams and creates a circular flow of separate materials as an alternative to the linear flow of mass waste. Its central concept is the 'closed loop'. As a result, the innovations of eco-modified recycling are in collection systems rather than high tech plants. The cost of collection and sorting has been one of the main barriers for increasing recycling among households and small traders.

It involves the collection of used and discarded materials in order to process these materials and make them into new products. It reduces the amount of waste that is thrown into the community dustbins, thereby making the environment clean and the air fresh to breathe. Recycling is an economic development as well as an environmental tool. Reuse, recycling, and waste reduction offer direct development opportunities for communities. When collected with skill and care, and upgraded with quality in mind, discarded materials are local resources that can contribute to local revenue, job creation, business expansion, and the local economic base.

Recycling-based economic development has been the heart of Waste to Wealth program of national economy. Recycling and reusing reduce the pressure on primary resources. In some sectors, such as machinery, cars, and household appliances, there has been a long-term practice of scrap recycling; however, substantial amounts are still landfilled along with precious metals and other materials in electronic goods. Alongside its potential for the environment, economy, and local regeneration, recycling also offers many social benefits.

8.1.5 Bioenergy Conversion

Bioenergy conversion seems to be the most promising energy conversion techniques, specifically for India in near future probably because of the following points:

1. Absence of or the difficulties related to the installation of centralized power supply systems.
2. Increasing energy demand for energy even in remote rural areas or isolated parts of the country.
3. Basic need for large amounts of protein for food and feeding purpose.
4. Inexpensive methods available for collecting and storing of energy.

Energy schemes utilizing plant (biomass) as source of liquid fuel (such as ethanol or methanol) are therefore worth attempting in addition to electrical power generation. The production of usable energy through algal and similar crops includes the following three important conversion steps.

1. Photosynthesis production of organic matters.
2. Collection and processing of plant materials.
3. Fermentation of the organic matters, leading to liquid and gaseous fuels and storage.

Sugar crops, trees, grains, and grasses are various aquatic fuel sources and have relative potentials on each other utilized in any biomass production schemes. Sugar crops and algal crops seem to be the most promising crops of importance suitable for bioenergy conversion in India.

8.2 KEY ISSUES

The following are the key issues that must be investigated before the economic viability of a refuse-derived fuel (RDF) scheme:

1. Collection of waste from doorsteps, commercial places, community dump, and final disposal sites.
2. The volume and nature of refuse to be processed.
3. The type of efficient RFD process required and market for fuel products.
4. The required potential users and the revenue obtainable.
5. The economy of the alternative method of disposal of the refuse.
6. The utilization of solar thermal energy for increasing the temperatures of digesters.

8.3 WASTE RECOVERY MANAGEMENT SCHEME

A simple waste, refuse resource recovery scheme can be understood from Figure 8.1, which represents the various important scheme components as energy use and solid waste generation, transportation, storage, energy recovery, treatment, and final disposal of the waste.

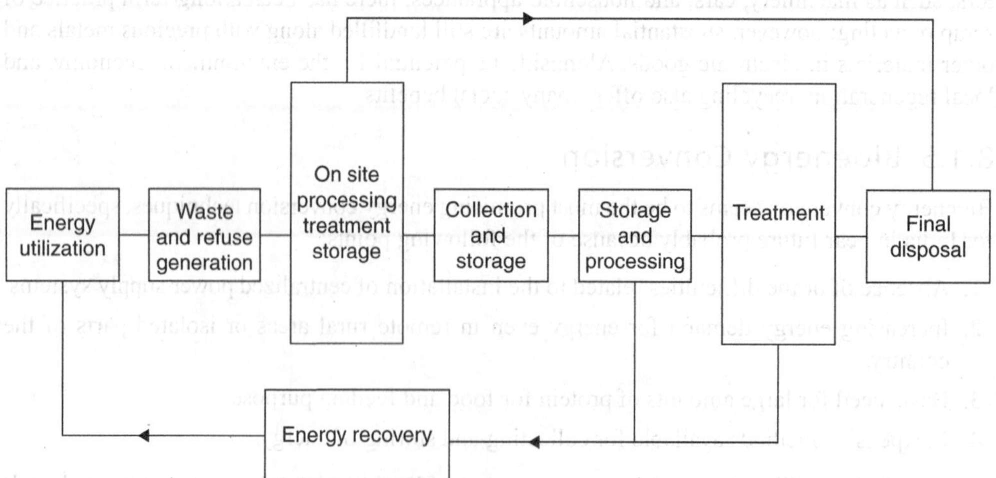

Figure 8.1 Schematic representation of waste refuse energy management

The major part of waste obtained after the energy utilization are non-organic that have diversified nature and characteristics, and thus, their identification and separation from the main waste stream by improved techniques are an essential parameter of any energy recovery scheme. On-site processing of waste for the reduction of in-home compactors and industrial shredders through improved technology should be employed, which may be environmentally acceptable. Collection and transportation components of the waste energy conversion scheme are the most expensive components owing to many varying social, technical, and other reasons. A careful cost analysis and implementation of this vital component will minimize the running cost of the scheme. The storage of waste for resource recovery and final disposal after suitable treatment is another component of scheme and selection of storage station and other associated problems invite careful attention. Normally, two types of energy recovery systems are used:

1. Separation of metals, paper, and glass from the remaining waste through the process such as size reduction, screening, vibrating sorting, and electronic scanning; however, a truly homogeneous, inexpensive separation system will provide competitive input to waste energy utilization.
2. Conversion of the remaining waste product to usable form of energy and energy conversion may include the following:
   (a) Generation of methane gas (biogas conversion) or other fuels (biological conversion)
   (b) Generation of electricity either from (a) or through thermo-mechanical process
   (c) Composting of fertilizers

8.3.1 Treatment

Here, the treatment means that those process designed to reduce waste to innocuous forms without or after energy recovery. The most familiar techniques are the burning of waste at high temperatures in the presence of oxygen (known as incineration) and the breaking down of the complex compounds using heat in the absence of oxygen (known as pyrolysis). However, treatment techniques should be selected so as to be accepted socially, environmentally, and economically. The cheapest method for final disposal of waste before or after energy recovery is a systematic burial in ground.

8.4 ADVANTAGES AND DISADVANTAGES OF WASTE RECYCLING

Significant advantages and disadvantage of waste recycling are discussed in this section.

8.4.1 Advantages of Waste Recycling

Recycling is a process of using old or waste products into new products; this is an important step towards energy conservation (to reduce energy usage and reduce the consumption of fresh raw materials) and reduction in pollution (to reduce air, water, land pollution, and greenhouse emissions).

1. *Reduced damage to environment*: This is the foremost advantage of recycling and this promotes environmental protection in a balanced manner. For instance, let us consider the case

of cutting down trees for paper production; here, individuals can create balance by recycling old used papers and new paper products made from trees. In such a way, deforestation and felling is reduced. Natural resources are conserved this way.

2. *Reduced consumption of energy*: Large amount of energy is consumed when raw materials are processed during manufacturing. Therefore, recycling helps reduce energy consumption making production process beneficial and cost effective. It leads to reduced utilization of raw materials. It ensures additional energy availability and saving money. It reduces the creation of waste at source.
3. *Reduced environmental impact and pollution*: At present, industrial waste is major source of pollution. Recycling industrial products such as plastics and cans help a lot in cutting down levels of pollution for the reason that these materials are being reused instead of being thrown away irresponsibly. It saves on requirement of open landfill spaces, the surroundings clean and healthy. It also reduces environmental impact of traditional methods of waste treatment and disposal.
4. *Mitigate global warming*: Recycling aids in alleviating or lessening global warming and its harsh effects. Today, massive waste is being burned producing large amount of greenhouse gas emissions. Therefore, recycling is an effective way of ensuring that the process of burning is reduced and waste are regenerated and converted to useful and eco-friendly products without creating harmful impact to the environment.
5. *Promotes sustainable utilization of resources*: Recycling promotes sustainable and wise use of resources. This activity helps ensure that there is no discriminate use of materials and resources saving them for possible use in the future.

8.4.2 Disadvantages of Waste Recycling

1. *High cost of recycling*: The establishment of separate facilities in order to process products and make them reusable is cost effective. This might somehow trigger pollution in terms of transporting the materials and cleaning activities.
2. *Durability and small life span of recycled items*: The durability and efficiency of recycled products does not guarantee 100%. Recycled products are sometimes taken from cheap and overused materials; therefore, there is no assurance that it can last for long.
3. *Unsafe and unhygienic process*: Recycling sites and processes are often unhygienic and unsafe and this might pose dangers to your health.

8.4.3 Status of Municipal Solid Wastes Management in India

Municipalities in India spend hardly between 10% and 50% of their budget on solid waste management (SWM), but most of this is consumed in the salaries of sanitation workers and transport of waste, while a minute proportion is spent on its scientific disposal. The abysmal state of affairs with regard to the collection and transport of waste is all too well known. However, the implications of the negligence in waste treatment and disposal, such as untreated and unprocessed garbage left in open dumpsites, and its grave consequences for public health and the environment are not fully understood. They are the main cause of river water, land, and air pollutions.

8.4.4 Tips on Reducing Waste and Conserving Resources (the Three R's—Reduce, Reuse, and Recycle)

The three R's (reduce, reuse, and recycle) help approaching system acceptability index to unity, and thus, to cut down on the amount of dissipated energy (waste). They conserve natural resources, landfill space, and energy.

The three R's save land and money as waste to dispose of waste in landfills. Identifying a new landfill has become difficult and more expensive due to environmental regulations and public opposition.

8.4.4.1 Reduce

Reduce means using fewer resources in the first place. This is the most effective of the three R's and the place to begin. The best way to manage waste is to not produce it. This can be done by shopping carefully and being aware of a few guidelines:

1. Use of disposable goods (such as paper and plastic plates, cups, napkins, razors, and lighters.) must be avoided. Throwaways of disposable goods contribute to the waste disposal problem and cost more because they must be replaced again and again.
2. Well-built durable goods or that carry good warranties should be purchased. They will last longer, save money in the long run, and save landfill space.
3. Electronic mail, online railway reservation, banking services, etc., minimize the wastage of paper.
4. Use of cloth napkins, jute bags, and dish cloth towels should be encouraged in place of paper and plastic made products.
5. Avoid overly packaged goods. The packaging is a total throw-away materials.
6. Things made with toxic materials must be avoided.
7. Use of water and electricity must be minimized. Highly energy efficient electrical and electronic goods should be purchased.

8.4.4.2 Reuse

It makes economic and environmental sense to reuse products. Sometimes, it takes creativity. Reusing keeps new resources from being used for a while longer, and old resources from entering the waste stream.

Before recycling or disposing of anything, it must be considered that whether it has life left in it or not?

1. Reusing an item means that it continues to be a valuable, useful, productive item, and replaces new items that would utilize more water, energy, timber, petroleum, and other limited natural resources in their manufacture.
2. Lumber, tools, windows, doors, light fixtures, paint, plumbing supplies and fixtures, architectural pieces, fencing, hardware, and many other items needed for constructing or refurbishing a building can be used from available waste building materials.
3. Desks, tables, chairs, filing cabinets, credenzas, shelving units, stacking trays, tape dispensers, notebook binders and other equipment and supplies can be reused in offices, schools, hospitals, non-profit organizations, and others.

4. Reuse products for the same purpose. Save paper and plastic bags, and repair broken appliances, furniture, and toys.
5. A product that can be used repeatedly instead of a version that is only used once and thrown away should be purchased.

8.4.4.3 Recycle

The third R in the hierarchy is for recycle, which in terms of waste is the reprocessing of disposed materials into new and useful products. Items that are commonly recycled include glass, plastic, paper, and metal. When recycled, some of these materials are used to create more of the same original product, while other materials are used to create entirely different products after recycling.

Recycling is a series of steps that takes a used material and processes, remanufactures, and sells it as a new product. Begin recycling at home and at work.

1. A recycled plastic soda bottle is chipped, melted, and made into fibre, which becomes a jacket or sleeping bag stuffing.
2. Recycling household waste is a relatively simple and inexpensive way to contribute to a better global environment.

8.5 SOURCES AND TYPES OF WASTES

The following are some of the wastes:

1. *Residential wastes*: These are single family or multifamily dwellings. They constitute kitchen wastes, paper and cardboards, clothes and leather materials, plastics and rubber materials, glass, wood and metal crockery and furniture, electrical and electronics appliances and gadgets, etc.
2. *Municipal services wastes*: They include general wastes collected from street sweeping, park, recreational places, sludge; landscaping, and tree trimming.
3. *Industrial and commercial wastes*: They are housekeeping and food wastes, packaging and demolition material wastes, scraps, hazardous wastes, wood, cardboard paper, plastics, etc.
4. *Building construction and demolition*: They constitute various types of wastes such as wood, concrete, steel, and dust.
5. *Agriculture*: It consists of dairy and agriculture farm crop wastes, hazardous pesticides, etc.

8.6 RECYCLING OF PLASTICS

Plastics play an important role in almost every aspect of our lives. Plastics are durable; their toughness and inertness are what make them so useful. Unfortunately, they are so durable that they break down very slowly in a landfill. Plastics are used to manufacture everyday products such as beverage containers, toys, and furniture. The widespread use of plastics demands proper end of plastic life management. The largest amount of plastics is found in containers and packaging (e.g., soft drink bottles, lids, shampoo bottles), but they also are found in durable (e.g., appliances, furniture) and nondurable goods (e.g., diapers, trash bags, cups and utensils, medical devices). The recycling rate for different types of plastic varies greatly. Plastics are a versatile material that can be a valuable asset to recycling program.

Plastics can be divided into two major categories:

1. *Thermosets*: A thermoset solidifies or 'sets' irreversibly when heated. They are useful for their durability and strength and are, therefore, used primarily in automobiles and construction applications. Other uses are adhesives, inks, and coatings.
2. *Thermoplastics*: A thermoplastic softens when exposed to heat and returns to original condition at room temperature. Thermoplastics can easily be shaped and moulded into products such as milk jugs, floor coverings, credit cards, and carpet fibres.

According to most estimates, 80% of post-consumer plastic waste is sent to landfill, 8% is incinerated, and only 7% is recycled.

Since the production of plastics uses 8% of the world's oil production, it is in the best interests to recycle plastics. In addition to reducing the amount of plastics waste requiring disposal, recycling plastic will reduce the consumption of non-renewable fossil fuels, energy, the amount of solid waste going to landfill, and the amount of carbon emissions.

8.6.1 Recycling of Plastics

Recycling plastic material is one of the important environmental agenda defined in the three R's. Instead of simply reusing the material as it is, successful chemical reusing is a more effective way for reducing the use of natural resources and environmental damage incurred thereof. An effective process and a pertinent effective plant that successfully converts plastic wastes into wax-free hydrocarbon such as naphtha and diesel oil is an important plastic recycling system.

Plastics from Municipal Solid Wastes (MSW) are usually collected from curbside recycling bins or drop-off sites. Then, they go to a material recovery facility, where the materials are sorted by plastic type, baled, and sent to a reclaiming facility. At the facility, any trash or dirt is sorted out, then the plastics are washed and ground into small flakes. A flotation tank may be used to further separate contaminants based on their different densities. Flakes are then dried, melted, filtered, and formed into pellets. The pellets are shipped to product manufacturing plants, where they are made into new plastic products.

8.6.2 Plastic Resin Identification Code

Plastics come in a variety of colours and chemical formulations, all with different recycling needs. Turn the product over and look for the recycling symbol, a triangle with a number from 1 to 7 inside, as shown in Figure 8.2. The plastic resin identification code is a set of symbols placed on plastics to identify the polymer type. It was developed by the Society of the Plastic Industry (SPI) in 1988 and is used internationally. It was transferred to ASTM International in 2010. The primary purpose of the codes is to allow efficient separation of different polymer types for recycling. Separation must be efficient because the plastics must be recycled separately. Even one item of the wrong type of resin can ruin a mix.

There are seven different types of plastic resins that are commonly used to package household products. The identification codes listed in Figure 8.2 can be found on the bottom of most plastic packaging. It (RIC system) offered a way to identify the resin content of bottles and containers commonly found in the residential waste stream. Plastic household containers are usually marked with a number that indicates the type of plastic. Consumers can then use this information to determine whether or not certain plastic types are collected for recycling in their area. Contrary to common belief, just because a plastic product has the resin number in a triangle, which looks very similar to the recycling symbol, it does not mean it is collected for recycling.

RIC Code	Type of Resin Content	Image Symbol	Properties	Uses and availability	Recycle End Use
1	PETE or PET	1 PETE 01 PET	It is inexpensive, light weight, low risk of leaching breakdown. PET plastic composes 18% of the world's polymer production	Soda and water containers, bottled beverages. Polyester fibre and Dacron and thin films (Mylar)	Easy to recycle and used for textile, polyester carpet, quilts and jackets, bottle market
2	HDPE	2 HDPE 02 PE-HD	HDPE is called natural since that is it is natural colour.	Milk, detergent, and oil bottles, toys, and plastic bags	Easy to recycle. Used for plastic pipes, lumber, flower pots, trash cans, non-food application bottles.
3	PVC or V	3 V 03 PVC	The most damaging of all plastics. It releases carcinogenic dioxins into the environment when manufactured or incinerated. Health hazardous	Food wrap, vegetable oil bottles, blister packages. Polyvinyl chloride pipe, fencing, shower curtains, lawn chairs, non-food bottles, and children's toys	Difficult to recycle. End use is panelling, flooring, speed bumps, decks, and roadway gutters.
4	LDPE	4 LDPE 04 PE-LD	It is chemically similar to HDPE but it is less dense and more flexible.	Many plastic bags. Shrink wrap, garment bags, and polyethylene film	Recyclable for plastic trash bags, grocery sacks, plastic tubing, agricultural film, plastic lumber for decks and fencing.
5	PP	5 PP 05 PP	Highly resistant to most acids and alkalis. Lightweight, high tensile strength	Refrigerated containers, some bags, most bottle tops, some carpets, some food wrap. Bumpers, car interior trim, industrial fibres, and carry-out beverage cups	Recyclable. Used into brooms, auto battery cases, bins, pallets, signal lights, ice scrapers, and bicycle racks.

Figure 8.2 Resin identification code

6	PS	6 PS 06 PS	High impact resistance, food contact acceptable, flame retardant,good processability, general purpose, high heat resistance), good flow, expandable, fast moulding. Inexpensive	Throwaway utensils, meat packing, protective packing. Desk accessories, cafeteria trays, plastic utensils, toys, video cassettes and cases, clamshell containers, packaging peanuts, and insulation board, expanded polystyrene products, and Styrofoam	Can be recycled into egg cartons, vents, foam packing, and insulation.
7	Other	7 Other 07 O	Usually layered or mixed plastic.	Acrylic, nylon, polycarbonate, and poly lactic acid (a bio plastic), and multilayer combinations of different plastics. Bottles, plastic lumber applications, headlight lenses, and safety shields or glasses, iPod cases, and bullet-proof materials	No recycling potential—must be landfilled. Only a few can be recycled into plastic lumber and other custom-made products.
PET: polyethylene terephthalate, HDPE: high-density polyethylene, LDPE: low-density polyethylene, PVC or V: polyvinyl chloride, PP: polypropylene, PS: polystyrene, Other: mixed plastics					

Figure 8.2 Continued

8.6.3 Benefits of Plastic Recycling

The following are the benefits of plastic recycling:

1. *Energy and natural resource conservation*: The first and foremost benefit of recycling of plastics is the conservation of petroleum products (oil and natural gas) and water used for making new plastic products. A large amount of petroleum and natural gas can be saved by recycling plastics. Oil and natural gas are two of the main components used for the production of the raw materials used to make plastic. These natural resources are not only in limited supply, but also in high demand for other important uses such as powering automobiles and producing electricity. Because plastic materials are already in such abundance, recycling these existing materials means that fewer natural resources are used for production of new plastic materials.
2. *Environmental protection*: The benefits of recycling of plastics include reduction in emission of greenhouse gases in its manufacturing by burning petroleum and natural gases. As the use of petroleum products are reduced in manufacturing of new plastics, the greenhouse gas emission will also reduce. Factories that produce plastic products from raw materials also produce a great deal of harmful greenhouse gases. By using recycled plastic materials, production time is greatly reduced, which means that less greenhouse emissions are making their way into the atmosphere.

Plastics being a major contributor to the worldwide waste can cause serious environmental concerns because of their non-degradable nature that keeps them intact for a very long time.

One of the important benefits of recycling of plastics is that it saves life of animals, birds, and aquatic creatures from fatal due to ingestion of plastics. Pollution of air, soil, and water is greatly reduced by recycling of plastics.

3. *Reducing the dumping (landfill) spaces*: Plastics are not a biodegradable material, which means that it can sit in a landfill for hundreds or thousands of years. Recycling plastic products also keeps them out of landfills and allows the plastics to be reused in manufacturing new products. Best of all, plastics can be indefinitely recycled, which means they could potentially be eradicated from landfills altogether. Landfill space is largely reduced by recycling of plastic.
4. *Energy conservation*: By using recyclable plastic materials, factories can produce new plastic products using approximately 2/3 less energy than with raw material production.

8.6.4 Thermal Depolymerization

While interest in combusting and gasifying plastics appears to be growing, there is another route to making practical use of all the waste plastics modern society produces. A catalytic pyrolysis system has been developed to convert waste plastics into liquid hydrocarbons, coke and gas, which can then be used as boiler fuel for power generation. Power generation of approximately 5 kW is possible at 1 *l* of mixed oil.

Thermal depolymerization (TDP) is a depolymerization process using hydrous pyrolysis for the reduction of complex organic materials (usually waste products of various sorts, such as biomass and plastics) into light crude oil. It mimics the natural geological processes that are involved in the production of fossil fuels. Under pressure and heat, long chain polymers of hydrogen, oxygen, and carbon decompose into short-chain petroleum hydrocarbons with a maximum length of around 18 carbons.

Pyrolysis is a process of thermal degradation of a material in the absence of oxygen. Plastic is fed into a cylindrical chamber. The pyrolytic gases are condensed in a specially designed condenser system to yield a hydrocarbon distillate comprising straight and branched chain aliphatic, cyclic aliphatic, and aromatic hydrocarbons, and liquid is separated using fractional distillation to produce the liquid fuel products. The plastic is pyrolysed at 370°C–420°C.

The following are the essential steps involved in the pyrolysis of plastics:

1. Evenly heating the plastic to a narrow temperature range without excessive temperature variations.
2. Purging oxygen from pyrolysis chamber.
3. Managing the carbonaceous char by-product before it acts as a thermal insulator and lowers the heat transfer to the plastics.
4. Careful condensation and fractionation of the pyrolysis vapours to produce distillate of good quality and consistency.

8.6.4.1 Catalytic Pyrolysis of Waste Plastics

In this method, a suitable catalyst is used to carry out the cracking reaction. The presence of catalyst lowers the reaction temperature and time. The process results in much narrower product distribution of carbon atom number and peak at lighter hydrocarbons that occurs at lower temperatures. The cost should be further reduced to make the process more attractive from an economic perspective. Reuse of catalysts and the use of effective catalysts in lesser quantities can optimize this option. This process can be developed into a cost-effective commercial polymer recycling process for solving the acute environmental problem of disposal of plastic waste. It also offers the higher cracking ability of plastics and the lower concentration of solid residue in the product.

8.6.5 Plastic Waste to Oil Production in India

If all goes well, in India, Surat will be the first in the state of Gujarat to convert its plastic waste into crude oil and pellets, which could be further used as fuel substitute to power industrial units, vehicles, power plants, boilers, and generators.

Process claimed to be relatively simple. Forced air, heated by a gas burner, is used to indirectly heat the feedstock inside the process vessel. The air is continually recycled in a loop to minimize heat loss.

The process vessel is isolated from oxygen and is exposed to a negative pressure (vacuum) environment. The energy transferred to the plastic feedstock from the burner is used to depolymerize, or 'crack' the plastic into synthetic crude oil.

Oil is chromatographically removed from the waste plastic and aggregated from several vessels for on-site micro-refinement or sent to existing commercial refinement facilities.

Waste products are recycled for energy usage (gases), treated, and reused or disposed (liquids), or made available for commercial use (solids).

8.6.6 Waste Plastic to Electricity Generation

Non-biodegradable plastics break down in a waste combustor to create an alternative source of fuel to generate electricity. Self-sustainability is the key to the double-tank combustor design. Plastic waste is first processed in an upper tank through pyrolysis, which converts solid plastic into gas. Next, the gas flows to a lower tank, where it is burned with oxidants to generate heat and steam. The heat sustains the combustor while the steam can be used to generate electric energy. Self-sustainability is the key to the double-tank combustor design. Plastic waste is first processed in an upper tank through pyrolysis, which converts solid plastic into gas. Next, the gas flows to a lower tank, where it is burned with oxidants to generate heat and steam. The heat sustains the combustor while the steam can be used to generate electric energy.

8.6.7 Advantages of Recycling of Plastics

1. It is good for the environment.
2. It is good for practicing green living.
3. It saves the energy, resource, and some money.

8.6.8 Disadvantages of Recycling of Plastics

1. Plastics are bad for the environment because it takes hundreds of years to biodegrade.
2. It takes money, time, and energy to recycle plastics itself. Plastics are made from petroleum and thus have high significant heating value than coal. The recycled plastics can recover energy and can produce enough electricity to meet power requirements.

SUMMARY

- Waste is any material that is unwanted, discarded, or useless. These wastes causes dangerous environmental damage and other problems related with their dumping and disposal throughout the entire world.
- Efforts are, therefore, directed to make all these wastes as wealth to the extent possible, and the methods are designed to protect the environment
- Incineration is a waste treatment technology that involves burning commercial, residential, and hazardous waste. It converts discarded materials, including paper, plastics, metals, and food scraps into bottom ash, fly ash, combustion gases, air pollutants, wastewater, wastewater treatment sludge, and heat.
- Incineration is a practical method of disposal that saves a lot of money on transport of waste to landfills. It also reduces the carbon footprint that such transport leaves behind.
- Incinerator can be understood more precisely as a furnace where waste is burnt.
- Pyrolysis is a controlled chemical process (temperature, pressure, batch or continuous system, catalyst, etc.) in order to produce valuable secondary raw materials (solid, liquid, or gas) or energy.
- Pyrolysis is the process of thermal degradation of wastes in the absence of oxygen.
- Anaerobic digestion is a series of biological processes in which microorganisms break down biodegradable material in the absence of oxygen.
- The three R's (reduce, reuse, and recycle) help approaching the system acceptability index to unity, and thus, they help to cut down on the amount of dissipated energy (waste).

REVIEW QUESTIONS

1. What are solid wastes? Explain their effects on environment. Also discuss main sources of solid wastes.
2. What are the major classifications of solid wastes
3. State and explain the methods of municipal solid wastes management.
4. Define wastes source reduction and its necessity and benefits. Also explain practical methods of source reduction.
5. Define and explain recycling of wastes and its benefits.
6. Define the following terms: Incineration, Pyrolysis, and Gasification. What is the difference among these processes?
7. What is the advantage and disadvantage of the following solid treatment technique?
   (i) Incineration
   (ii) Composting
   (iii) Land filling
   (iv) RDF (Refused Derived Fuel)
8. Explain the role of incineration, Pyrolysis and composting in solid waste management.

Chapter 9

Biomass Energy

Solar energy by means of photosynthesis stores energy in trees and plants that can be converted into liquid fuels suitable for internal combustion engines. Similarly, ethanol could be produced from cellulose on large scale. There is no doubt that rising energy costs will lead to more concentrated research of such biological system; such that, energy gains made via plant photosynthesis using intensive systems are subsequently more than that lost in the conversion of biomass energy content into storable high energy fuels (i.e., ethanol or methane). The growth of sugarcane and its fermentation to ethanol may be considered to be the most favourable for the marginal net energy production process, which is suitable for Indian climatic conditions. Biomass is used for heating, electric power generation, and combined heat and power. Several methods are used for the conversion of biomass into useful energy, such as electricity generation by direct burning of biomass, synthesis gas production by gasification, and methane gas production by anaerobic digestion. However, following issues may be thoroughly investigated for implementing biomass production scheme:

1. The types of vegetation best suited for an intensive energy plantation and bio generation selection criteria.
2. The type and availability of land for growing energy crops and plant material productions.
3. Harvesting for conceptual plantation.
4. Techno-economic comparison of firing crops directly for electric power generation with conversion to clean fuel gas (methane or low BTU gas) either at the farm site or at the selected market.

KEY CONCEPTS

- Biomass production
- Composition and properties of producer gas
- Theory and process of gasification
- Gasifier types and comparison
- Use of a biomass gasifier
- Biomass feed (fuel)

9.1 BIOMASS PRODUCTION

Sun is the primary source of all kinds of available raw energy resources including biomass. The sunlight energy is transferred to biosphere by the photosynthesis process that occurs in plants, algae, and some types of bacteria.

Plant matter created by the process of photosynthesis is called biomass. Photosynthesis is a natural radiation. In its simplest form, the final reaction of this process can be represented as follows:

$$6H_2O + 6CO_2 + \text{solar light energy} \rightarrow C_6H_{12}O_6 + 6CO_2$$

It is seen that in the process, water and carbon dioxide are converted into organic material.

The term biomass refers to those organic matters that are stored in plant and trees in the form of carbohydrate (sugar). It is then transferred through food chains in humans, animals, and other living creatures and their wastes.

The term biomass includes all plant life: trees, agricultural plants, bush, grass and algae, and their residues after processing. Biomass may be obtained from forests woods, agricultural lands, arid lands, and even waste lands. It may be obtained in a planned or unplanned manner. The term is also generally understood to include animal and human waste.

Biomass has the advantage of controllability and availability when compared to many other renewable energy options. There are a variety of ways of obtaining energy from biomass. These may be broadly classified as direct methods and indirect methods.

9.1.1 Direct Methods

Raw materials that can be used to produce biomass energy are available throughout the world in the following forms:

1. Forest wood and wastes
2. Agricultural crops and residues
3. Residential food wastes
4. Industrial wastes
5. Human and animal wastes
6. Energy crops

Properly managed forests will always have more trees, and agricultural and energy crops management will always have crops; further, the residual biological matter are taken from those crops.

Raw biomass has a low energy density based on their physical forms and moisture contents and their direct use are burning them to produce heat for cooking. The twin problems of traditional biomass use for cooking and heating are the energy inefficiency and excessive pollution.

Inefficient way of direct cooking applications, inconvenient and inefficient methods of raw biomass transportation and storage and high environmental pollution problems made them unsuitable for efficient and effective use. This necessitated some kind of pre-processing and conversion technology for enhancing the usefulness of biomass.

9.1.2 Indirect Methods

Biomass can also be used indirectly by converting it either into electricity and heat or into a convenient usable fuel in solid, liquid, or gaseous form. The efficient conversion processes are as follows:

1. *Thermo-electrical conversion:* The direct combustion of biomass material in the boiler produces steam that is used either to drive a turbine coupled with an electrical generator to produce electricity or to provide heat for residential and industrial system. However, the boiler equipment are very expensive and energy recovery is low. Fortunately, improved pollution controls and combustion engineering have advanced to the point that any emissions from burning biomass in industrial facilities are generally less when compared to the emissions produced when using fossil fuels (coal, natural gas, and oil).
2. *Biomass conversion to fuel:* Under present conditions, economic factors seem to provide the strongest argument of considering biomass conversion to fuel such as fermentation and gasification. In many situations, where the price of petroleum fuels is high or where supplies are unreliable, the biomass gasification can provide an economically viable system, provided the suitable biomass feedstock is easily available. Biomass conversion processes can be classified under two main types:
   (a) Thermo-chemical conversion includes processes such as destructive distillation, pyrolysis, and gasification.
   (b) Biological conversion includes processes such as fermentation and anaerobic digestion.

Gasification produces a synthesis gas with usable energy content by heating the biomass with less oxygen than needed for complete combustion. Pyrolysis yields bio-oil by rapidly heating the biomass in the absence of oxygen. Anaerobic digestion produces a renewable natural gas (methane gas) when organic matter is decomposed by bacteria in the absence of oxygen.

As a result, it is often advantageous to convert this waste into more readily usable fuel form like producer gas. Hence, it is the attractiveness of gasification.

The efficiency of a direct combustion or biomass gasification system is influenced by a number of factors such as including biomass moisture content, combustion air distribution and amounts (excess air), operating temperature and pressure, and flue gas (exhaust) temperature.

9.2 ENERGY PLANTATION

An interesting approach for the large-scale planned use of wood is the 'energy plantation' approach. In this scheme, selected species of trees are planted and harvested over regular intervals of time in a phased manner so that wood is continuously available for cooking or allied purposes.

Energy plantations include, amongst others, pine, cottonwood, hybrid poplar, sweetgum, and eucalyptus. Much of the emphasis has been on hardwood plantations due to their ability to coppice, continued genetic improvement programs as well as the opportunity to combine fast growth and wood. Some important trees grown in India for this purpose are eucalyptus, babool, and casuarinas.

A rich experience of commercial energy plantations management system in varied climatic conditions has emerged during the past 4–5 decades. Improvements in soil preparation, planting, cultivation methods, species matching, biogenetics and pest, disease and fire control have led to enhanced yields.

It has been suggested that electrical power be produced by the energy plantation approach, the wood grown in this manner being used as a fuel for the boilers of a conventional power plant. The technology of biomass-based electric power plants is well established in the USA and Europe and there are over 500 such plants use wood, wood waste, and various types of agricultural waste. When a photosynthetic conversion efficiency of around 1% is assumed, it is estimated that a 1,000 MW power plant may require an area of about 1,000 km^2 for the energy plantation. Although this is a large area, it should not be difficult to provide in most countries since the land required need not displace agricultural land. However, care has been taken so that there is no danger of monoculture weakening the ecological system.

9.3 BIOMASS GASIFICATION

Biomass gasification is a process of partial combustion in which solid biomass usually in the form of pieces of wood or agricultural residue is converted into a combustible gas mixture.

Gasification, which is incomplete combustion of carbonaceous fuels, can be represented with the following sub-stoichiometric equation.

Biomass + air → carbon monoxide (CO) + carbon dioxide (CO_2) + methane (CH_4) + hydrogen (H_2) + nitrogen (N_2) + water vapour.

Gasification produces a synthesis gas with usable energy content is produced by gasification in which biomass is heated with less oxygen than that needed for complete combustion.

As a result, a gaseous mixture of carbon monoxide (CO), carbon dioxide (CO_2), methane (CH_4), hydrogen (H_2), and nitrogen (N_2) called producer gas is obtained.

Producer gas can be used

1. to run internal combustion engines (both compression and spark ignition)
2. as substitute for furnace oil in direct heat applications and
3. to produce, in an economically viable way, methanol

Methanol is an extremely attractive chemical that is useful both as fuel for heat engines as well as chemical feedstock for industries. Since any biomass material can undergo gasification, this process is much more attractive than ethanol production or biogas where only selected biomass materials can produce the fuel.

Gasification processes involved with biomass are as follows:

1. *Drying of fuels*: It is the process of drying biomass before it is fed into gasifier.
2. *Pyrolysis*: It is a process of breaking down biomass into charcoal by applying heat to biomass in the absence of oxygen.
3. *Combustion*: All the heat required for different processes of gasification are made available from combustions.
4. *Cracking*: In this process, breaking down of large complex molecules (such as tar) takes place when heated into lighter gases.
5. *Reduction*: Oxygen atoms are removed in this process from the combustion products (hydrocarbon) molecules and returning them to combustible form again.

9.3.1 Low Temperature Gasification

When gasification of biomass is carried out at 750°C to 1,100°C, it is referred to as low temperature gasification. The gas produced has relatively high level of hydrocarbons. It is used directly to either burn for steam production and generation of electricity or cleaned and used in internal combustion engine or combined heat power (CHP).

The producer gas is a mixture of carbon monoxide (CO), carbon dioxide (CO_2), hydrogen (H_2), methane (CH_4), and nitrogen from air. The gas mixture composition depends on gasifiers.

9.3.2 High Temperature Gasification

It is carried out in temperature range of 1,200°C–1,600°C and gas product is referred to as synthesis gas (Syngas). It contains high proportion of CO and H_2 and is convertible to high quality synthetic diesel biofuel compatible for use in diesel engines

9.3.3 Composition and Properties of Producer Gas

The producer gas is affected by various processes as abovementioned, and hence one can expect variations in the gas produced from various biomass sources. The composition of producer gas is highly dependent upon the inputs to the gasifier and gasifier design.

Table 9.1 lists the composition of gas produced from various sources. Nitrogen affects the maximum dilution of gas and almost 50%–60% of gas is composed of noncombustible nitrogen. The use of oxygen instead of air will be beneficial for gasification with due regards to the costs of oxygen. Nevertheless, production of a high energy quality methanol may justify the cost of oxygen.

On an average, 1 kg of biomass produces about 2.5 m^3 of producer gas at S.T.P and consumes about 1.5 m^3 of air for combustion. For complete combustion of wood, about 4.5 m^3 of air is required.

Table 9.1 Composition of Producer Gas from Various Biomass Feed

Bio Mass Feed	Gasifier	% Volume					Calorific value MJ/m^3
		CO	H_2	CH_4	CO_2	N_2	
Charcoal	Downdraft	28–31	5–10	1–2	1–2	55-60	4.60–5.65
Wood with 12–20% moisture content	Downdraft	17–22	16–20	2–3	10–15	55–50	5.00–5.86
Wheat straw pellets	Downdraft	14–17	17–19	–	11–14	–	4.50
Coconut husks	Downdraft	16–20	17–19.5	–	10–15	–	5.80
Coconut shells	Downdraft	19–24	10–15	–	11–15	–	7.20
Pressed Sugarcane	Downdraft	15–18	15–18	–	12–14	–	5.30
Charcoal	Updraft	30	19.7	–	3.6	46	5.98
Corn cobs	Downdraft	18.6	16.5	6.4	–	–	6.29
Rice hulls pelleted	Downdraft	16.1	9.6	0.95	–	–	3.25
Cotton stalks cubed	Downdraft	15.7	11.7	3.4	–	–	4.32

The average energy conversion efficiency of gasifiers is defined as

$$\eta_{gs} = \frac{\text{Calorific value of } \frac{\text{gas}}{\text{kg}} \text{ of fuel}}{\text{Average calorific value of 1 kg of fuel}} \tag{9.2}$$

Example 9.1

One kilogram of wood produces 2.5 m^3 of gas with average calorific value of 5.4 MJ/m^3. Average calorific value of wood (dry) is 19.8 MJ/kg. Calculate the average conversion efficiency of wood gasifier.

Solution From Eq. (9.2), $\eta_{gs} = (2.5 \times 5.4 \times 10^6)/(19.8 \times 10^6 \times 1) = 68.18\%$.

The average energy conversion efficiency of wood gasifiers is about 60%–70%.

9.3.4 Temperature of Gas

The average gas temperature produced by gasifier is about 300°C–400°C and it may even attain a higher temperature of approximately 500°C, if partial combustion of gas is taking place. Partial combustion of biomass can be eliminated by increasing air flow rate higher than the design value.

9.4 THEORY OF GASIFICATION

Gasification may be considered as a special case of pyrolysis where destructive decomposition of biomass (wood wastes) by heat is converted into charcoal, oils, tars, and combustible gas. It is referred to as the partial combustion of solid fuel (biomass) and takes place at temperatures of about 1,000°C. The reactor used for gasification is called a gasifier.

The complete combustion of biomass produces biomass gasses that generally contain nitrogen, water vapour, carbon dioxide, and surplus of oxygen. However, in gasification (with incomplete combustion), as shown in Figure 9.1, product gas contains gases such as carbon mono oxide (CO), hydrogen (H_2), and traces of methane and non-useful products such as tar and dust. The production of these gases is obtained by the reaction of water vapour and carbon dioxide through a glowing layer of charcoal. Thus, the key to gasifier design is to create conditions such that

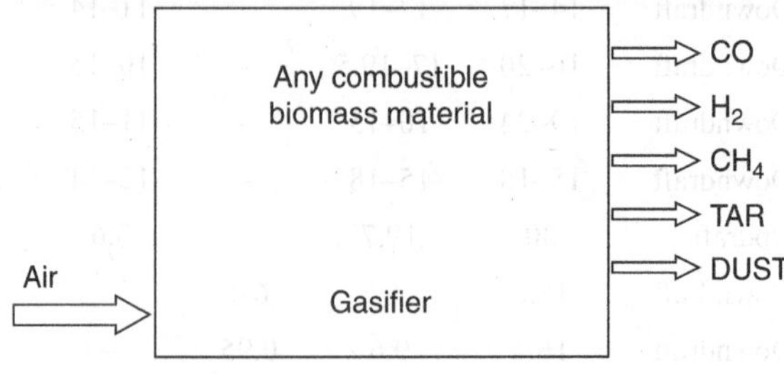

Figure 9.1 Products of gasifiers

1. Biomass is reduced to charcoal
2. Charcoal is converted at suitable temperature to produce CO and H_2

Typically, the volumetric composition of biomass-based producer gas is as follows:

CO : 20%–22%,
H_2 : 15%–18%,
CH_4 : 2%–4%,
CO_2 : 9%–11% and
N_2 : 50%–54%.

9.5 GASIFIER AND THEIR CLASSIFICATIONS

Biomass gasifier may be considered as a chemical reactor in which biomass goes through several complex physical and chemical processes and producer or syngas is produced and recovered.

There are two distinct types of gasifier:

1. *Fixed bed gasifier*: In this gasifier, biomass fuels move either countercurrent or concurrent to the flow of gasification medium (steam, air, or oxygen) as the fuel is converted to fuel gas. They are relatively simple to operate and have reduced erosion.

   Since there is an interaction of air or oxygen and biomass in the gasifier, they are classified according to the way air or oxygen is introduced in it. There are three types of gasifier as shown in Figure 9.2.

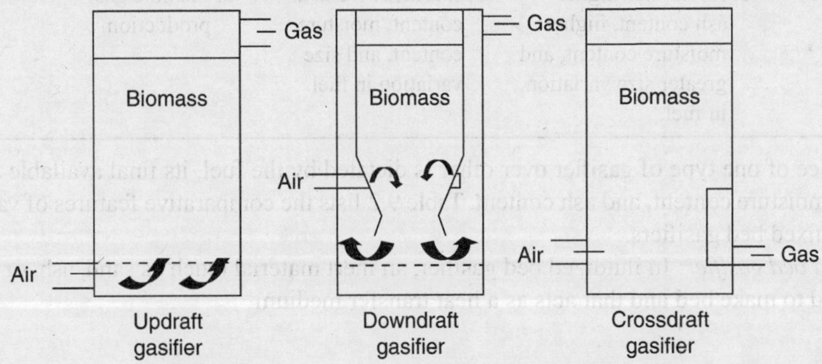

Figure 9.2 Types of fixed bed gasifiers

(a) *Downdraft gasifiers*: In the downdraft gasifier, the air is passed from the layers in the downdraft direction. Single throat gasifiers are mainly used for stationary applications, whereas double throat gasifier is used for varying loads as well as automotive purposes.
(b) *Updraft gasifiers*: Updraft gasifier has air passing through the biomass from bottom and the combustible gases come out from the top of the gasifier.
(c) *Cross draft gasifiers*: It is a very simple gasifier and is highly suitable for small outputs. With slight variation, almost all the gasifiers fall in the abovementioned categories.

Table 9.2 Advantages and Disadvantages of Fixed Bed Gasifiers

	Updraft Gasifier	Downdraft Gasifier	Cross-draft Gasifier
Comparative features	1. It works on coal, briquettes, and other fuels (fuel flexibility).	1. It works on woody biomass and charcoal (fuel specific).	1. Type of fuel usage restricted to only low ash fuels such as wood, charcoal, and coke.
	2. Comparatively low quality gas having tar and particulate matter.	2. High quality gas.	2. Good quality gas.
	3. Suitable for thermal applications.	3. Suitable for power (IC engines) and thermal applications.	3. Suitable for heat and power applications.
	4. Gas is drawn out of the gasifier from the top of the fuel bed, while the gasification reactions take place near the bottom. The air comes in at the bottom and produced syngas leaves from the top of the gasifier.	4. Air is introduced into downward flowing packed bed or solid fuels and gas is drawn off at the bottom. Hence, fuel and gas move in the same direction.	4. Air enters from one side of the gasifier, and fuel is released from the opposite side.
	5. It tolerates higher ash content, higher moisture content, and greater size variation in fuel.	5. It is sensitive to ash content, moisture content, and size variation in fuel.	5. Flexible gas production.

The choice of one type of gasifier over other is dictated by the fuel, its final available form, its size, moisture content, and ash content. Table 9.2 lists the comparative features of various types of fixed bed gasifiers.

2. *Fluidized bed gasifier*: In fluidized bed gasifier, an inert material (such as sand, ash, or char) is utilized to make bed and that acts as a heat transfer medium

9.6 CHEMISTRY OF REACTION PROCESS IN GASIFICATION

Four distinct processes take place in a gasifier when fuel makes its way to gasification:

1. *Drying zone of fuel*: In this zone, the moisture content of biomass is removed to obtain the dry biomass. Some organic acids also come out during the drying process. These acids give rise to corrosion of gasifiers.
2. *Pyrolysis zone*: In this zone, the tar and other volatiles are driven off. The products depend upon temperature, pressure, residence time, and heat losses. However, following general remarks can be made about them.

(a) Up to the temperature of 200°C, only water is driven off.
(b) Between 200°C and 280°C carbon dioxide, acetic acid, and water are given off.
(c) The real pyrolysis, which takes place between 280°C and 500°C, produces large quantities of tar and gases containing carbon dioxide. Besides light tars, some methyl alcohol is also formed.
(d) Between 500°C and 700°C, the gas production is small and contains hydrogen.

3. *Combustion(oxidation) zone*: In this zone, carbon from the fuel combust and forms carbon dioxide with the oxygen in the air by the reaction:

$$C + O_2 \rightarrow CO_2 + \text{Heat}$$

Because of the heat emitted during the reaction, the temperature rises until a balance between heat supply and heat loss occurs.

4. *Reduction zone*: The hot gas passes through the reduction zone after the combustion zone. As there is no free oxygen in this zone that causes inflammable carbon dioxide gas to react with the carbon in the fuel and forms flammable carbon monoxide gas. This reaction is endothermic (demands heat) and occurs at temperature exceeding about 1,000°C. Carbon monoxide is the most important flammable elements in the produced gas obtained from the reduction reaction as

$$C + CO_2 + \text{heat} \rightarrow 2CO \tag{9.3}$$

Another important endothermic reaction in the reduction zone is the water–gas shift reaction. It is the reaction of water vapour and carbon to give carbon monoxide and hydrogen

$$C + H_2O + \text{Heat} \rightarrow CO + H_2 \tag{9.4}$$

Both gasses are flammable, and the heating value of the gas is increased. If there is still surplus of water in the reduction zone, then carbon monoxide may react with water vapour and form carbon dioxide and hydrogen. This reaction is exothermic (emits heat) and decreases the heating value of the produced gas. The reaction is

$$CO + H_2O - \text{Heat} \rightarrow CO_2 + H_2 \tag{9.5}$$

Equations (9.3) and (9.4) are main reduction reactions and being endothermic have the capability of reducing gas temperature. Consequently, the temperatures in the reduction zone are normally 800°C–1,000°C. The lower the reduction zone temperature (~700°C–800°C), lower is the calorific value of gas.

The gas also contains measurable amounts of particulate material and tar. The heating value of the gas ranges from 4,000 to 5,000 kJ/m^3, which is a relatively low value when compared to the heating value of other gaseous fuels like natural gas.

The conversion efficiency of a gasifier is defined as the ratio of the heat content in the producer gas to the heat content in the biomass supplied and is usually around 75%.

Although there is a considerable overlap of the processes, each can be assumed to occupy a separate zone where fundamentally different chemical and thermal reactions take place.

Figure 9.3 shows schematically an updraft gasifier with different zones and their respective temperatures. Figure 9.4 for downdraft and Figure 9.5 for cross-draft also show these regions.

9.7 UPDRAFT GASIFIERS

The oldest and simplest type of gasifier is the counter current or updraft gasifier shown schematically in Figure 9.3. The air intake is at the bottom and gas leaves at the top (the counter current flow). The reactive agent is injected at the bottom of the reactor and ascends to the top, while the fuel is introduced at the top and descends to the bottom. The combustion reactions occur near the grate at the bottom that are followed by reduction reactions somewhat higher up in the gasifier. In the upper part of the gasifier, heating and pyrolysis of the feedstock occur as a result of heat transfer by forced convection and radiation from the lower zones. Gases, tar, and other volatile compounds are dispersed at the top of the reactor, while ash is removed at the bottom. The syngas typically contains high levels of tar, which must be removed or further converted to syngas for use in applications other than direct heating.

Updraft gasifiers are widely used to gasify biomass resources and generally use steam as the reactive agent, but slagging can be severe if high ash fuels are used. They are unsuitable for use with fluffy, low-density fuels.

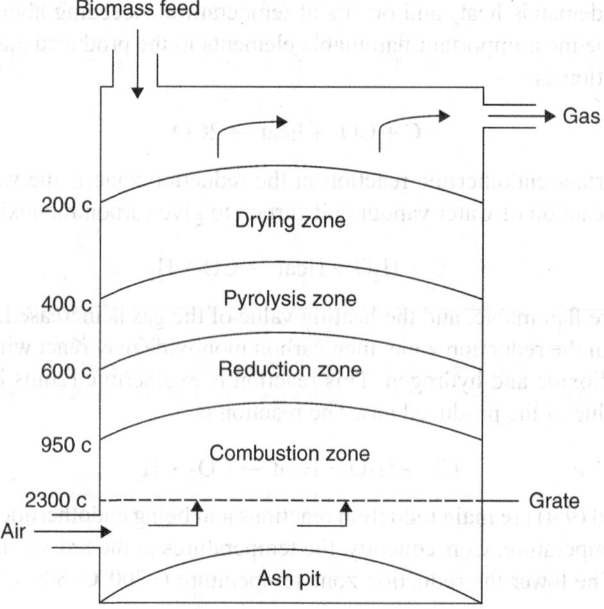

Figure 9.3 Updraft gasifier

These gasifiers are best suited for applications where moderate amounts of dust in the fuel gas are acceptable and a high flame temperature is required. Typical applications where the updraft gasifiers have been successfully used are as follows:

1. Packaged boilers
2. Thermal fluid heaters
3. Aluminium melting/annealing furnaces
4. All kinds of fryer roaster

9.8 DOWNDRAFT GASIFIER

In this gasifiers, the primary gasification air is introduced at or above the oxidation zone in the gasifier and the producer gas is removed at the bottom of the apparatus, so that fuel and gas move in the same direction, as schematically shown in Figure 9.4. The biomass feed (such as wood waste) and its gasification air both flow in the same downward direction through the gasifiers' fuel bed.

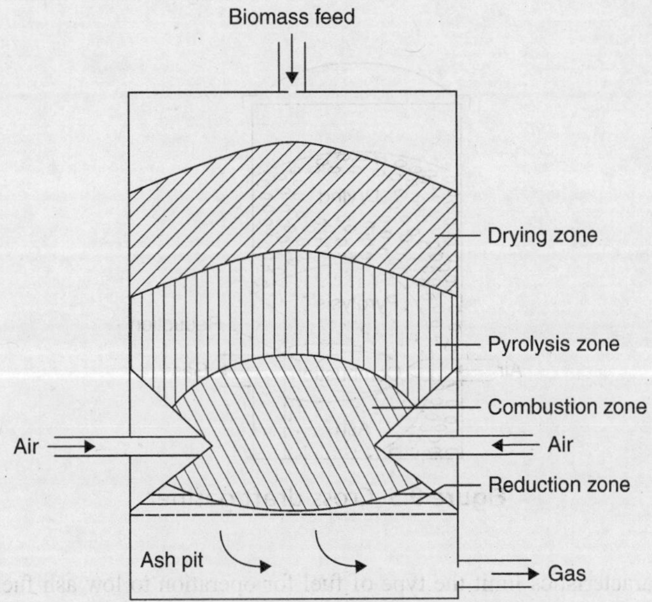

Figure 9.4 Downdraft gasifier

The biomass feed is admitted at the top similar to the updraft gasifier. As the feed progresses down through the gasifier, it dries and its volatiles are pyrolysed. The char is directed into a reduced-diameter cylindrical throat section at the bottom of the gasifier. Gasification air is injected into the throat through openings in the throat wall. Due to the high temperatures existing at the throat section, tars and oils could be cracked, which tend to form in producer gas, particularly when the biomass is wetter than about 20% moisture content (wet basis). The producer gas leaves at the bottom of the gasifier. The start-up time of about 5–10 min is necessary to ignite and bring plant to working temperature with good gas quality is shorter than updraft gas producer.

Downdraft gasifiers are widely used in the following applications:

1. Continuous baking ovens (bread, biscuits, and paint)
2. Batch type baking oven (rotary oven for bread)
3. Dryers and curing (tea, coffee, mosquito coil, and paper drying)
4. Boilers
5. Thermal fluid heaters

6. Annealing furnaces
7. Direct fired rotary kilns
8. Internal combustion engines

9.9 CROSS-DRAFT GASIFIER

Figure 9.5 is a schematic representation of cross-draft gasifier. Unlike downdraft and updraft gasifiers, the ash bin, fire, and reduction zone in cross-draft gasifiers are separated.

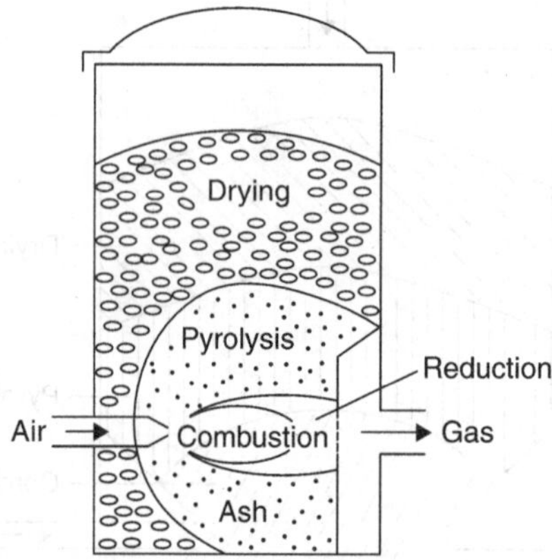

Figure 9.5 Cross-draft gasifier

These design characteristics limit the type of fuel for operation to low ash fuels such as wood, charcoal, and coke. The relatively high temperature in cross-draft gas producer has an obvious effect on gas composition such as high carbon monoxide, and low hydrogen and methane content when dry fuel like charcoal is used. Cross-draft gasifier operates well on dry air blast and dry fuel.

Typically, the gasifier is a vertical cylindrical vessel of varying cross section. The biomass is fed in at the top at regular intervals of time and is converted through a series of processes into producer gas and ash, as it moves down slowly through various zones of the gasifier.

9.10 FLUIDIZED BED GASIFICATION

Fluidized bed gasification has been successfully used to convert prepared wastes (i.e., wood wastes, bark, agricultural wastes, and RDF (Refused Derived Fuel) into a clean fuel gas that can be used to fire various types of industrial equipment. Past applications have included gasification of wastes to provide gas for dryers previously fired on natural gas.

The fluidized bed gasifier is illustrated schematically in Figure 9.6. This gasifier is an improved version of fixed bed gasifiers. The bed made of an inert material (such as sand, ash, or char) initially and it is heated and the fuel is introduced when the temperature has reached the appropriate level. The bed material transfers heat to the fuel and blows the reactive agent through a distributor

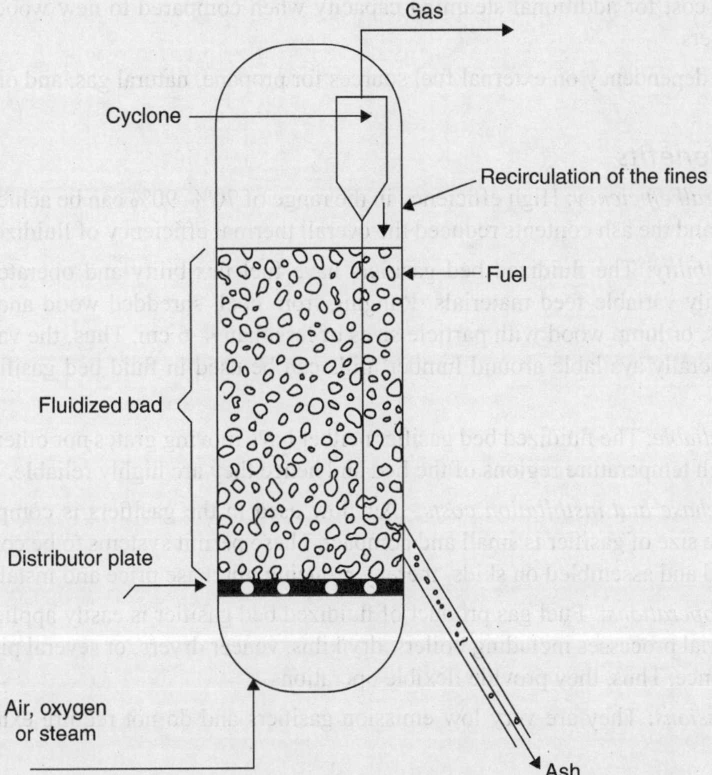

Figure 9.6 Fluidized bed gasifier

plate at a controlled rate. Fluidized bed gasifiers have no distinct reaction zones (as in the case of fixed bed gasifiers) and drying, pyrolysis and gasification occur simultaneously. The fuel particles are introduced at the bottom of the reactor, very quickly mixed with the bed material and almost instantaneously heated up to the bed temperature. As a result of this treatment, the fuel is pyrolysed very fast, resulting in a component mix with a relatively large amount of gaseous materials. Further gasification and tar-conversion reactions occur in the gas phase.

For biomass feeds which have high ash content and the ash has low melting point, fluidized bed combustion seems to gasify them.

9.10.1 Advantages and Benefits

The fluidized bed gasification process offers several substantial benefits when compared to simple burning processes and other forms of gasification.

9.10.1.1 Advantages

1. Reduced cost of boiler or dryer or kiln operation by using wood and/or bark wastes rather than gas or oil.

2. Reduced cost for additional steaming capacity when compared to new wood and or bark-fired boilers.
3. Reduced dependency on external fuel sources for propane, natural gas, and oil.

9.10.1.2 Benefits

1. *High overall efficiency*: High efficiency in the range of 70%–90% can be achieved. Moisture contents and the ash contents reduced the overall thermal efficiency of fluidized bed gasifier
2. *Fuel flexibility*: The fluidized bed gasifiers have fuel flexibility and operate satisfactorily with highly variable feed materials. Ranging from coal, shredded wood and bark to saw-dust fines, or lump wood with particle sizes of less than 4–6 cm. Thus, the various types of fuels generally available around lumber mills can be used in fluid bed gasifiers with good results.
3. *Highly reliable*: The fluidized bed gasifier neither have moving grates nor other moving parts in the high temperature regions of the bed and hence they are highly reliable.
4. *Low purchase and installation costs*: Air flow used in the gasifiers is comparatively low, and hence size of gasifier is small and compact. These permit systems to be completely shop fabricated and assembled on skids, thereby reducing purchase price and installed costs.
5. *Flexible operations*: Fuel gas product of fluidized bed gasifier is easily applied to a variety of industrial processes including boilers, dry kilns, veneer dryers, or several pieces of equipment at once. Thus, they provide flexible operations.
6. *Low emissions*: They are very low emission gasifiers and do not require exhaust clean up devices.

9.11 USE OF BIOMASS GASIFIER

The output of a biomass gasifier can be used for a variety of direct thermal applications such as cooking, drying, heating water, and generating steam. It can also be used as a fuel for internal combustion engines to obtain mechanical shaft power or electrical power.

If used as a fuel for internal combustion engines, it has to be cleaned first for complete removal of particulate material and tar. A cleaning system consisting of cyclone, a scrubber, and a filter is used for the purpose. If the engine is a spark-ignition engine, it can operate with producer gas alone. The gas is sucked from the gasifier and cleaner unit by the engine suction along with a proportionate amount of air. It is then compression-ignition engine, as it operates in the 'dual-fuel' mode. Here, the engine sucks in a mixture that is compressed and a small amount of diesel sprayed in. Combustion initiates with the diesel droplets and then spreads to the mixture of the gaseous fuel and air. The phrase 'dual-fuel' implies that both diesel and producer gas are simultaneously used.

India is one of the leading countries in the world in the field of biomass gasification. Biomass gasifier systems are available in a wide range of capacities and standard facilities for testing and evaluating gasifiers have been set up. For thermal applications, systems with outputs ranging from 60,000 to 5×10^6 kJ/h are available, while for electrical power generation, systems with outputs ranging from 3 to 500 kW are also available.

The largest biomass gasification system produces 5×10^6 kJ/h (1,450 kW) output in the thermal mode and 500 kW in the electric power generation mode. It uses biomass in the form of wood blocks (25 to 100 mm long and up to 70 mm in diameter) at the rate of 500 kg/h and produces 1,250 m^3/h of gas. The internal combustion engine is a compression ignition engine operating in the dual-fuel mode. It uses only 25% of the diesel normally required by the engine if operating with diesel alone.

9.11.1 Liquid Fuels

When compared to gaseous fuels such as producer gas or biogas, liquid fuels are somewhat harder to obtain from biomass sources. One of the methods is the production of methanol from wood or straw. The process involves the gasification of plant matter followed by chemical synthesis. Another method is the conversion of certain food grains and crops such as sugarcane, maize, cassava and tapioca by fermentation into ethanol. When blended with petrol, ethanol is good alternate fuel for automotive engines. This fact has received considerable attention as a means of overcoming the oil crisis. However, if one examines the requirement of land for growing the agricultural products concerned, it is obvious that the method can be of substantial benefit only to a country having a large surplus of land. For this reason, Brazil has adopted this method on a large scale and produces significant amounts of ethanol for use as an alternate fuel. However, the position in India is quite different since the availability of land is limited.

In plants, algae and certain types of bacteria, the photosynthetic process results in the release of molecular oxygen and the removal of carbon dioxide from the atmosphere that is used to synthesize carbohydrates (oxygenic photosynthesis). Other types of bacteria use light energy to create organic compounds but do not produce oxygen (anoxygenic photosynthesis). Photosynthesis provides the energy and reduced carbon required for the survival of virtually all life on our planet, as well as the molecular oxygen necessary for the survival of oxygen consuming organism. In addition, the fossil fuels, currently being burned to provide energy for human activity, were produced by ancient photosynthetic organisms.

9.12 GASIFIER BIOMASS FEED CHARACTERISTICS

Most of gasifier manufacturers claim that a gasifier is available and can gasify any biomass feed. However, there is no such thing as a universal gasifier. A gasifier is in real sense very much biomass feed specific and it is tailored accordingly.

Following biomass feed characteristics or parameters dictate the quality and classification of gasifiers:

1. Energy content of the fuel
2. Bulk density
3. Moisture content
4. Dust content
5. Tar content
6. Ash and slogging characteristic

9.12.1 Energy Content and Bulk Density of Fuel

The higher the energy content and bulk density of fuel, the similar is the gasifier volume; as for one biomass fuel charge, power can be obtained for longer time duration.

9.12.2 Moisture Content

Moisture content is very trivial components of biomass fuels and it is determined by the type of fuel, its origin, and treatment. It is desirable to use fuel with low moisture content to minimize heat loss due to its evaporation.

Besides impairing the gasifier heat budget, high moisture content also puts load on cooling and filtering equipment by increasing the pressure drop across these units because of condensing liquid. Thus, in order to reduce the moisture content of fuel, some pre-treatment of fuel is required. Generally, desirable moisture content for fuel should be less than 20%.

9.12.3 Dust Content

All gasifier fuels produce undesirable dust that can clog the internal combustion engine and hence it has to be removed. The gasifier design should be such that it should not produce dust beyond specified limits.

The higher the dust produced, more is the load put on filters necessitating their frequent flushing and increased maintenance.

9.12.4 Tar Content

Tar is one of the most unpleasant constituents of the gas as it tends to deposit in the carburettor and intake valves causing sticking and troublesome operations. It is a product of highly irreversible process taking place in the pyrolysis zone. There are approximately 200 chemical constituents that have been identified in tar so far.

Very little research work has been done in the area of removing or burning tar in the gasifier so that relatively tar free gas comes out. Thus, the major effort has been devoted to cleaning this tar by filters and coolers.

9.12.5 Ash and Slagging Characteristics

The mineral content in the fuel that remains in oxidized form after complete combustion is usually called ash. The ash content of a fuel and the ash composition has a major impact on trouble-free operation of gasifier. Ash basically interferes with the gasification process in two ways:

1. It fuses together to form slag and this clinker stops or inhibits the downward flow of biomass feed.
2. Even if it does not fuse together, it shelters the points in fuel where ignition is initiated, and thus lowers the fuel's reaction response.

Ash and tar removal are the two most important processes in gasification system for its smooth running. However, slagging can be overcome by two types of operation of gasifier:

1. Low temperature operation that keeps the temperature well below the flow temperature of the ash.

2. High temperature operation that keeps the temperature above the melting point of ash.

The first method is usually accomplished by steam or water injection, while the latter method requires provisions for tapping the molten slag out of the oxidation zone. Each method has its advantages and disadvantages and depends on specific fuel and gasifier design.

Keeping in mind the abovementioned characteristics of fuel, only two fuels have been thoroughly tested and proven to be reliable. They are charcoal and wood.

As charcoal is tar free and has relatively low ash content property, it was the preferred fuel during World War II and still remains so. However, there is a major disadvantage of charcoal in terms of energy. Charcoal is mostly produced from wood and in the conversion of wood to charcoal, about 50% of original energy is lost.

9.12.6 Biomass Feed (Fuel)

The major biomass sources presently used are as follows:

1. Sugarcane and corn, wheat, sugar beet, sweet sorghum, and cassava to produce bioethanol.
2. Rapeseed, sunflower seeds, soybean, canola, peanuts, jatropha, coconut, and palm oil for biodiesel production.
3. Wide range of cellulosic materials (such as grassy crops, woody plants, by-products from the forestry and agricultural sector including wood residues, stems, and stalks and municipal wastes constitute the so-called second generation of feedstock).
4. Wastes and residues constitute a large source of biomass. These include solid and liquid municipal wastes, manure, lumber and pulp mill wastes, and forest and agricultural residues.

Low water content (dried) biomass feedstock is burnt to generate heat and electricity. Wood wastes in the paper and pulp industries and bagasse from the sugarcane industry are used in ethanol fermentation.

A variety of raw materials that include agricultural wastes, municipal solid wastes, market garbage, and waste, water from food and fermentation industries (all organic materials containing carbohydrates, lipids, and proteins) are feedstock for anaerobic digester for methane production.

9.13 APPLICATIONS OF BIOMASS GASIFIERS

The main applications of biomass gasifier products are as follows:

1. *Motive power*: Gasifier products are used to provide shaft power to industrial and agricultural equipment and machinery such as
   (a) Diesel engine operation on dual or 100% modes.
   (b) Water pumps
   (c) Tractors, harvesters, etc.
   (d) Running of high efficiency Stirling engines.
2. *Direct heat applications*: Gasifiers heat has direct heat applications such as
   (a) Drying of agricultural crop and food products such as large cardamom, ginger, rubber, and tea at low temperature range of about 85°C–125°C.
   (b) Baking of tiles and potteries in the moderate temperature range of about 800°C–900°C.

(c) For melting metals and alloys in non-ferrous in the temperature range of 700°C–1,000°C.
(d) As boiler fuels provide steam or hot water for process industries such as silk reeling, dyeing, turmeric boiling, cooking, jiggery making, etc.

3. *Electrical power generation*: Electric power generation from few kilowatts to hundreds of kilowatts either for local consumption or for grid power is being installed based on gasifier products. Small-scale electricity generation systems also provide an attractive alternative to electric supply company.
4. *Chemical production*: Production of chemicals such as methanol and formic acid from producer gas.

9.14 COOLING AND CLEANING OF GAS

For efficient and effective use of gas for numerous applications, it should be cleaned of tar and dust, free from moisture content and cooled. Therefore, cooling and cleaning of the gas is one of the most important processes in the whole gasification system. The failure or the success of producer gas units depends completely on their ability to provide a clean and cool gas to the engines or for burners.

The temperature of gas coming out of generator is normally between 300°C and 500°C. The energy density of gas can be increased to a large extent by cooling it. Most coolers are gas to air heat exchangers where the cooling is done by free convection of air on the outside surface of heat exchanger. Some heat exchangers provide partial scrubbing of gas for the removal of moisture and tar contents. Thus, ideally, the gas going to an internal combustion engine should be cooled to nearly ambient temperature and shall be free from tar and moisture contents.

Normally, there are three types of filters used for cleaning of gas, as shown in Figure 9.7, which is schematically a downdraft gasification system with cleaning and cooling train.

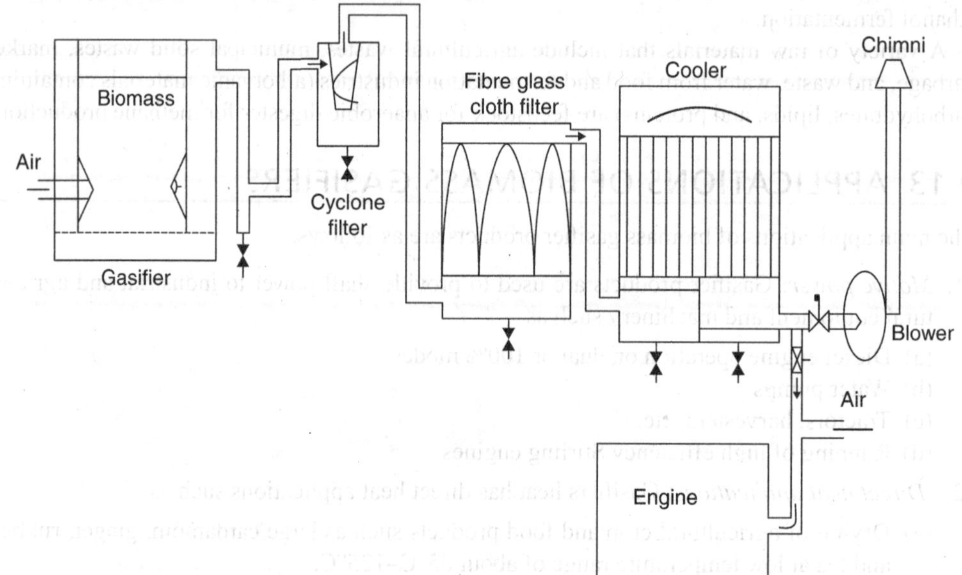

Figure 9.7 Schematic diagram of producer gas plant

They are classified as dry, moist, and wet.

1. *Cyclone filters*: They are designed according to the rate of gas production and its dust content. They are useful for particle size of 5 μm and greater. Since 60%–65% of the producer gas contains particles above 60 μm in size, the cyclone filter is an excellent cleaning device.
2. *Wet scrubber*: Even after cyclone filtering, the gas still contains fine dust, particles, and tar. It is further cleaned by passing through a wet scrubber where gas is washed by water in countercurrent mode. The scrubber also acts like a cooler, from where the gas goes to cloth or cork filter for final cleaning.
3. *Cloth filters*: It is a fine filter. Any condensation of water on it stops the gas flow because of an increase in pressure drop across it. Thus, in quite a number of gasification systems, the hot gases are passed through the cloth filter, and then only do they go to the cooler. Since the gases are still above the dew point, no condensation takes place in filter.

SUMMARY

- Biomass gasification offers the most attractive alternative energy system for agricultural purpose
- Most preferred fuels for gasification have been charcoal and wood. However biomass residues are the most appropriate fuels for on-farm systems and offer the greatest challenge to researchers and gasification system manufacturers.
- Very limited experience has been gained in gasification of biomass residues.
- Most extensively used and researched systems have been based on downdraft gasification. However it appears that for fuels with high ash content fluidized bed combustion may offer a solution. At present no reliable and economically feasible systems exist.
- Biggest challenge in gasification systems lies in developing reliable and economically cheap cooling and cleaning trains.
- Maximum usage of producer gas has been in driving internal combustion engine, both for agricultural as well as for automotive uses. However, direct heat applications like grain drying is very attractive for agricultural systems.
- A spark ignition engine running on producer gas on an average produces 0.55–0.75 kWh of energy from 1 kg of biomass.
- Compression ignition (diesel) engines cannot run completely on producer gas. Thus to produce 1 kWh of energy they consume 1 kg of biomass and 0.07 l of diesel. Consequently, they effect 80%–85% diesel saving.
- Future applications like methanol production using producer gas in fuel cell are under investigation.

REVIEW QUESTIONS

1. What is the meaning of biomass? Further, discuss its multipurpose utilization
2. How biomass conversion takes place?
3. State the names of raw biomass from waste and cultivated crops.
4. State the significant aquatic biomass resources.

5. State the important forest biomass resources
6. State the important rural biomass.
7. State the names of various agricultural wastes used in Fluidized Bed Combustion Boiler.
8. Discuss the applications of aquatic energy sources.
9. What are the basic steps involved in Waste Recycling?
10. Describe the entire recycling route from the collection of waste to final supply of recycle material.
11. Write the applications of biomass gasifiers.
12. Explain issues in biomass energy productions.
13. Classify and explain methods of obtaining energy from biomass.
14. Define biomass gasification. Further, explain the average energy conversion efficiency of gasifiers.
15. What is producer gas? Further, discuss its properties.
16. List different types of gasifiers and compare them.
17. State and explain processes of biomass gasification. Further, define average energy conversion efficiency of gasifiers.
18. Draw the schematic representation of either of the following gasifiers. Further, explain their working and application areas.
    (a) Updraft gasifier
    (b) Downdraft gasifier
    (c) Cross-draft gasifier
    (d) Fluidized bed gasifier
19. State and explain parameters that dictate the quality and classification of gasifiers.
20. Describe the advantages and benefits of fluidized bed gasifier?

Chapter 10

Biogas Energy

Biogas is a mixture of different gases, such as methane, carbon dioxide, hydrogen, etc., produced by the biological breakdown of organic matter in the absence of oxygen. It is a renewable energy source, and in many cases, it exerts a very small carbon footprint. Biogas can be produced by either anaerobic digestion with anaerobic bacteria, which digest material inside a closed system, or fermentation of biodegradable materials. It is primarily methane, carbon dioxide, small amounts of hydrogen sulphide, moisture, and siloxanes.

Anaerobic digestion is a process that breaks down organic matter into simpler chemical components in the absence of oxygen. This process has proved to be very effective to treat organic wastes for minimizing environmental pollution. The common organic wastes are listed as follows:

1. Sewage sludge
2. Organic farm wastes
3. Municipal solid wastes
4. Organic industrial and commercial wastes
5. Forests and agricultural wastes

The digestion process itself takes place in digester, which is classified in terms of temperature, water content of feedstock and the number of stages (single or multi-stage). The by-products of anaerobic digestion, namely biogas and digestate, can be used to create a source of income.

KEY CONCEPTS

- Biogas and its composition
- Anaerobic digestion and process stages of anaerobic digestion
- Construction and working of biogas plants
- Types of biogas plants
- Factors affecting the selection of a particular model of a biogas plant
- Biogas plant feeds and their characteristics

10.1 INTRODUCTION

In an anaerobic digestion process, organic matters are broken down into simpler chemical components in the absence of oxygen, and this process has proved to be very useful for organic wastes treatment in order to minimize environmental pollution. In this process, breaking down of organic matters takes place in anaerobic digesters that are classified based on feedstock, water content, temperature, and number of stages.

The common organic wastes feed for anaerobic digesters include human and animal excreta, forests and agricultural wastes, sewage sludge, organic farm wastes, municipal solid wastes, organic industrial and commercial wastes.

Biogas and manure are the by-products of anaerobic digestion that provides energy and organic fertilizers for improving comforts and income. Anaerobic digestion is also used for waste management in industrial and domestic sectors.

10.2 BIOGAS AND ITS COMPOSITION

Biogas is a clean, non-polluting, and low-cost fuel. It contains about 50%–70% methane, which is inflammable. A methane gas molecule has one atom of carbon and four atoms of hydrogen (CH_4) and is the main constituent of popularly known biogas. A colourless, odourless, inflammable gas also been referred to as sewerage gas, clear gas, marsh gas, refuse-derived fuel (RDF), sludge gas, gobar gas (cow dung gas), and bio energy. It produces about 9,000 kcal of heat energy per cubic metres of gas burnt and specifically used for cooking, heating, and lighting.

The composition of biogas is shown in Table 10.1, which mainly composed of 50% to 70% methane (CH_4), 30% to 40% carbon dioxide (CO_2), and traces of other gases.

Biogas is lighter than air by about 20% and has an ignition temperature in the range of 650°C to 750°C burns with clear blue flame similar to that of liquefied petroleum gas (LPG) and burns with 60% efficiency in a conventional biogas stove. Its calorific value is 20 MJ/m^3.

Its equivalence with other energy and fuels are as follows:

A 1,000 cubic feet of processed biogas is equivalent to about

1. 600 cubic feet of natural gas
2. 4.6 gallons of diesel oil
3. 5.2 gallons of gasoline
4. 6.4 gallons of butane

Table 10.1 Composition of Biogas

S. No.	Substances	Symbol	%
1	Methane	CH_4	50–70
2	Carbon dioxide	CO_2	30–40
3	Hydrogen	H_2	5–10
4	Nitrogen	N_2	1–2
5	Water vapour	H_2O	0.2–0.3
6	Hydrogen sulphide	H_2S	Minute traces

It is also estimated that for a simple family size of five persons and four cows and buffaloes, animal dung (leaving the use of human excreta for social problems) will produce about 175 cubic feet of biogas per day which will be sufficient for family requirements of cooking and lighting. In addition, rural housewives using the biofuel are spared the irritating smoke coming out of traditional cooking on raw biomass material and reduced labour required for cleaning the cooking equipments and utensils. The digested material, which comes out of the plant, is enriched manure.

10.3 ANAEROBIC DIGESTION

It is a biological process that produces a gas (commonly known as biogas) in the absence of oxygen and has major components of methane (CH_4) and carbon dioxide (CO_2).

Anaerobic digestion of methane gas production is a series of processes in which microorganism break down biodegradable material in the absence of oxygen which completes through following steps:

1. In the first step, the organic matter (e.g. plants residues, human and animal wastes and residues) is decomposed (hydrolysis) to break down the organic material into usable-sized molecules such as sugar.
2. Conversion of decomposed matter into organic acids is the second step.
3. Finally, organic acids are converted to biogas (methane gas).

10.3.1 Process Stages of Anaerobic Digestion

The biological and chemical stages of anaerobic digestion are shown in Figure 10.1. These are divided into the following four main stages:

1. Hydrolysis
2. Acedogenesis
3. Acetogenesis
4. Methanogenesis

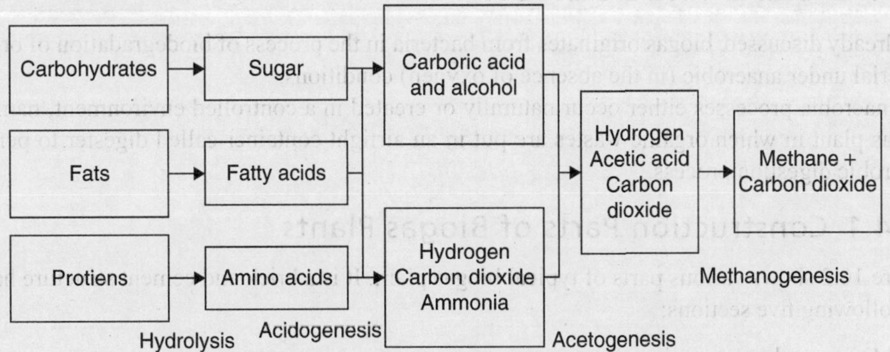

Figure 10.1 Process of anaerobic digestion

The four main stages are explained as follows.

10.3.1.1 Hydrolysis

The process of breaking large biomass organic chains into their smaller constituent parts such as sugar, fatty acids, and amino acids and dissolving the smaller molecules into solution is called hydrolysis. This process assists bacteria in anaerobic digesters to access the energy potential of the material. Hydrolysis of these high-molecular-weight polymeric components of biomass completes the first step in anaerobic digestion.

Hydrogen and acetate products of first stage are directly used by methanogens. Other molecules with a chain length larger than that of acetate (e.g. volatile fatty acids) must first be catabolized into compounds and then used by methanogens.

10.3.1.2 Acidogenesis

Acidogenesis is the biological process in which the remaining components are broken down by acidogenetic (fermentative) bacteria. It creates voltaic fatty acids together with ammonia, carbon dioxide, and hydrogen sulphide, and other by-products.

10.3.1.3 Acetogenesis

In this stage of anaerobic digestion, simple molecules created through the acidogenesis phase are further digested to produce more acetic acid, carbon dioxide, and hydrogen.

10.3.1.4 Methanogenesis

Finally, the process of biogas production is completed by methanogenesis. In this stage of anaerobic digestion, the methanogens use intermediate products of the preceding stages and convert them into methane, carbon dioxide, and water which makes the majority of the biogas emitted from the system. Methanogenesis is sensitive to both high and low pH values.

A simplified generic chemical equation for the overall processes outlined earlier is as follows:

$$C_6H_{12}O_6 \rightarrow 3CO_2 + 3CH_4 \tag{10.1}$$

The remaining indigestible material cannot be used by microbes and any dead bacterial remains constitute the digestate.

10.4 BIOGAS PRODUCTION

As already discussed, biogas originates from bacteria in the process of biodegradation of organic material under anaerobic (in the absence of oxygen) conditions.

Anaerobic processes either occur naturally or created in a controlled environment, namely a biogas plant in which organic wastes are put in an airtight container called digester to perform anaerobic digestion process.

10.4.1 Construction Parts of Biogas Plants

Figure 10.2 shows various parts of typical biogas plant. It is a brick and cement structure having the following five sections:

1. Mixing tank
2. Digester tank

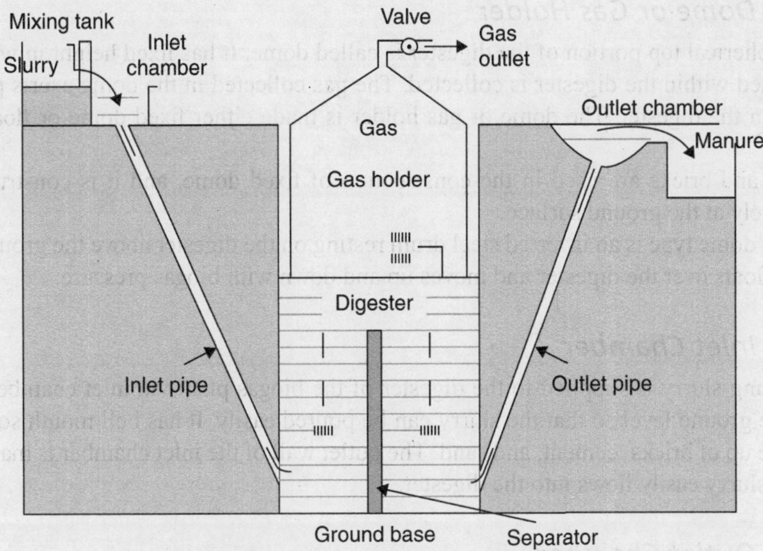

Figure 10.2 A typical biogas plants

3. Dome or gas holder
4. Inlet chamber
5. Outlet chamber

10.4.1.1 Mixing Tank

It is the first part of biogas plants located above the ground level in which the water and cow dung are mixed together in equal proportions (the ratio of 1:1) to form the slurry that is fed into the inlet chamber.

10.4.1.2 Digester Tank

It is a deep underground well-like structure and is divided into two chambers by a partition wall in between. It is the most important part of the cow dung biogas plants where all the important chemical processes or fermentation of cow dung and production of biogas takes place.

The digester is also called as fermentation tank. It is cylindrical in shape and made up of bricks, sand, and cement built underground over the solid foundation. Two openings are provided on the opposite sides and at the specified height of digester for inflow of fresh cow dung slurry and outflow of used slurry as manure.

The two long cement pipes are used as follows:

1. Inlet pipe opening into the inlet chamber for inputting the slurry in digester tank.
2. Outlet pipe opening into the overflow tank (outlet chamber) for the removal of spent slurry from the digester tank.

A separator is also placed in the middle of digester tank to improve effective fermentations of feedstock.

10.4.1.3 Dome or Gas Holder

The hemispherical top portion of the digester is called dome. It has fixed height in which all the gas generated within the digester is collected. The gas collected in the dome exerts pressure on the slurry in the digester. The dome or gas holder is made either fixed dome or floating dome type.

Cement and bricks are used in the construction of fixed dome, and it is constructed using approximately at the ground surface.

Floating dome type is an inverted steel drum resting on the digester above the ground surface. The drum floats over the digester and moves up and down with biogas pressure.

10.4.1.4 Inlet Chamber

The cow dung slurry is supplied to the digester of the biogas plant via inlet chamber, which is made at the ground level so that the slurry can be poured easily. It has bell mouth sort of shape and is made up of bricks, cement, and sand. The outlet wall of the inlet chamber is made inclined so that the slurry easily flows into the digester.

10.4.1.5 Outlet Chamber

The digested slurry from the biogas plants is removed through the outlet chamber. The opening of the outlet chamber is also at the ground level. The slurry from the outlet chamber flows to the pit made especially for this purpose.

10.4.1.6 Gas Outlet Pipe and Valve

The gas holder has an outlet at the top which could be connected to gas stoves for cooking or gas-lighting equipments or any other purpose. Flow of the gas from the dome via gas pipe can be controlled by valve. The gas taken from the pipe can be transferred to the point of use.

10.4.1.7 Foundation

The foundation forms the base of the digester where the most important processes of biogas plant occur. It is made up of cement, concrete, and bricks strong enough so that it should be able to provide stable foundation for the digester walls and be able to sustain the full load of slurry filled in it. The foundation should be waterproof so that there is no percolation and leakage of water.

10.4.2 Working of Biogas Plant

The working principle of biogas plant can be explained in Figure 10.2. The various steps of working principle of biogas plants are as follows:

1. Cattle dung and water are mixed together thoroughly in equal proportion (in the ratio of 1:1) to form the slurry in the mixing tank. Then, this slurry is poured into the digester via inlet chamber up to the cylindrical portion level of the digester.
2. The fermentation of slurry starts in the digester tank, and after completion of different anaerobic digestion processes, biogas is formed.

3. The gas continuously produced in digester tank is accumulated at the top of the digester in the dome or gas holder.

   Normally, the outlet gas valve remains closed, and hence, the accumulated biogas in the dome exerts pressure on the slurry which starts moving in the inlet and outlet chamber due to which the level of slurry drops in digester and increases in the outlet chamber. This process continues till the slurry reaches to highest possible level in the inlet and outlet chamber because of increased gas pressure.
4. If the gas valve is still kept closed the biogas will further get accumulated in the dome and develop high pressure enough in the gas to start escaping through the inlet and outlet chambers to the atmosphere. The biogas creates bubbles in the slurry in inlet and outlet chambers during its escape, and froth is also formed.
5. An increase in the volume of slurry in the inlet and outlet chambers helps to calculate the amount of biogas generated within the digester.
6. Gas pipe valve can be opened partly or fully to provide biogas for different applications. Under this situation, slurry level in the digester increases while the level in inlet and outlet chambers reduces.
7. When the gas is being taken out from the gas outlet at the top of the dome, the slurry from the outlet chamber is removed and equivalent amount of fresh slurry is inducted into the digester to continue the process of fermentation and the formation of the biogas. Therefore, more is the biogas required, more continuous will be the fresh slurry of cow dung and water required. The size of the digester tank also decides the amount of the gas that can be generated by the biogas plant.

10.4.3 Types of Biogas Plants

Fixed dome and floating dome construction are the two types of biogas plants. Based on these types, several biogas plant models are developed.

10.4.3.1 Fixed Dome Type

Schematic of a fixed dome biogas plant is given in Figure 10.3. It consists of following parts.

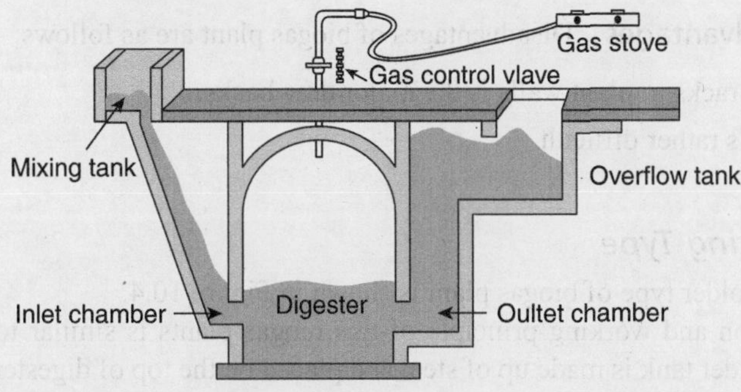

Figure 10.3 Fixed dome type biogas plant

1. *Mixing tank*: In mixing tank, the water and cattle dung are mixed together thoroughly in the ratio of 1:1 to form the slurry.
2. *Inlet chamber*: The mixing tank opens underground into a sloping inlet chamber.
3. *Digester*: Digester is a huge tank with a dome type ceiling. The ceiling of the digester has an outlet with a valve for the supply of biogas. The inlet chamber opens from below into the digester tank. The digester opens from below into an outlet chamber which is opened from the top into a small overflow tank.

10.4.3.1.1 Working Principle The various forms of organic biodegradable biomass are collected and mixed with equal amount of water properly in the mixing tank to form slurry. The slurry is fed into the digester tank through inlet chamber and pipe, and the digester is partially filled by about half of its height. The feeding of slurry is then discontinued for about 60 days when anaerobic bacteria present in the slurry decomposes or ferments the biomass in the presence of water. Biogas is then formed and starts accumulating in the upper dome area of the biogas plants, and the pressure is exerted on the spent slurry to force it flow into the outlet chamber. Finally, the spent slurry overflows into the overflow tank from where it is manually removed and used as manure for agricultural crops and plants.

Gas control valve at the top of dome is opened partially or fully to supply required gas for particular applications. A functioning plant is fed continuously with the prepared slurry to obtain a continuous supply of biogas.

10.4.3.1.2 Advantages Advantages of fixed dome-type biogas plant are as follows:

1. The costs of a fixed dome biogas plant are relatively low as compared to floating dome type.
2. It is simple in construction as no movable dome exists.
3. It is made up of concrete, bricks, and cements and long life of the plant (20 years or more) can be expected.
4. Underground and almost ground surface dome construction saves space and protect from physical damage to the plant.
5. The anaerobic digestion processes in the digester are little influenced by temperature fluctuation in day and night.

10.4.3.1.3 Disadvantages Disadvantages of biogas plant are as follows:

1. Porosity and cracks in plant walls is the major drawbacks.
2. Maintenance is rather difficult.

10.4.3.2 Floating Type

The floating gas holder type of biogas plant is shown in Figure 10.4.

The construction and working principle of this biogas plants is similar to fixed dome type except that gas holder tank is made up of steel and placed on the top of digester circular tank and is movable up and down also shown in Figure 10.2.

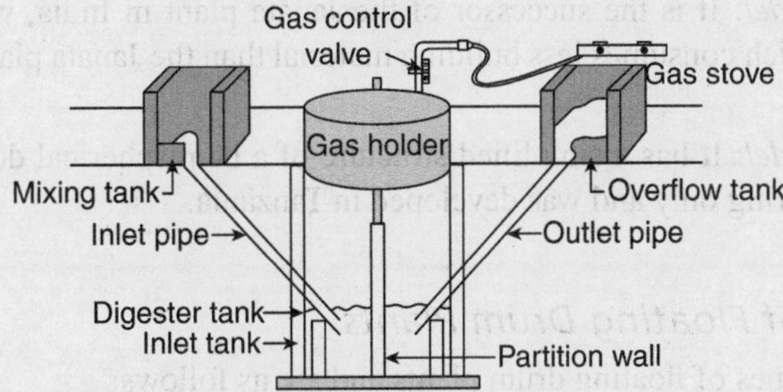

Figure 10.4 Floating dome-type biogas plant

10.4.3.2.1 Advantages Floating dome-type biogas plant has the following advantages:

1. Very efficient
2. Simple maintenance scheduling possible

10.4.3.2.2 Disadvantages Floating dome-type biogas plant has the following disadvantages:

1. Expensive
2. Steel drum may rust
3. Requires regular maintenance

10.4.4 Different Models of Biogas Plants

Several types of biogas plants are available in sizes and the capacities ranging from about 2 to 180 m^3 gas output per day. There are hardly three million biogas plants of small capacities are installed in India because of social acceptability and other problems.

The design model as shown in Figures 10.2 and 10.4 is called KVIC (Khadi and Village Industries Commission) model. A unit of this type with a gas capacity of 2 m^3/day costs approximately ₹15000. A number of other designs ranging in cost from ₹10000 to 25000 for the same capacity have also been developed. Large-sized community biogas plant is encouraged to assure better and economical utilization of animal and human wastes Control of several operating parameters, such as temperature and alkalinity of the slurry, sludge liquidity and build-up of scum on the surface of the slurry, gives good performance. Such control is obviously easier to achieve in community-sized plants.

10.4.4.1 Types of Fixed Dome Biogas Plants

Different types of fixed dome biogas plants are as follows:

1. *Chinese fixed dome type*: It has arch-type fixed dome as shown in Figure 10.3. The digester consists of a cylinder with round bottom and top. Several millions of such biogas plants have been constructed in China.
2. *Janata model*: In response to the Chinese fixed dome plant, Janata model was the first fixed-dome design constructed in India. It is not constructed anymore. The mode of construction leads to cracks and gas leakage in the gas holder, and hence, this model has no social acceptability.

3. *Deenbandhu model*: It is the successor of the Janata plant in India, with improved crack-proof design, which consumes less building material than the Janata plant with a hemisphere digester.
4. *CAMARTEC model*: It has a simplified structure of a hemispherical dome shell based on a rigid foundation ring only and was developed in Tanzania.

10.4.4.2 Types of Floating Drum Plants

There are different types of floating drum plants and are as follows:

1. KVIC model is the oldest and most widespread floating drum biogas plant from India.
2. Pragati model is developed with a hemisphere digester.
3. Ganesh model is constructed with angular steel and plastic foil.
4. Arati biogas model has low-cost floating drum plants made of plastic water containers or fiberglass drums.
5. BORDA model combines the static advantages of hemispherical digester with the process stability of the floating drum and the longer lifespan of a water jacket plant.

10.5 BENEFITS OF BIOGAS

A biogas energy system has whole range of benefits for the users, the society, and the environment. It includes the following:

1. *Production of energy (heat, light, and electricity)*: The calorific value of biogas is about 6 kWh/m^3, which is equivalent to about half a litre of diesel oil. The net calorific value depends on the efficiency of the burners or appliances. It replaces the conventional and traditional cooking and heating fuels and therefore permits the conservations of energy and fuels.

   The small- and medium-sized units (up to 6 m^3) are generally used for providing gas for cooking and lighting purposes. Large units (or communal units) produce this gas in large quantities and can be used to power engines and generators for mechanical work or power generation.
2. *Transformation of organic wastes into high-quality organic fertilizer*: The biogas plant is considered as a perfect fertilizer-making machine. There is no better way to digest or compost manure and other organic material than in a biogas plant. Output from the digester (digested manure) is actually a high-quality organic fertilizer. It has been analysed that the fertilizer, which comes from a biogas plant, contains three times more nitrogen than the best compost made through open air digestion.

   This nitrogen is already present in the manure. The nitrogen is preserved when waste is digested in an enclosed biogas plant, whereas the same nitrogen evaporates away as ammonia during open air composting. The biogas plant does not make extra nitrogen, it does not create nitrogen, and it merely preserves the nitrogen that is already there.
3. *Health benefits of biogas and the improvement of hygienic conditions (reduction of pathogens, worm eggs, and flies)*: Significant health benefits are achieved by the use of pure biogas. It has been found that non-biogas users have more respiratory diseases than those who use biogas plants.

Respiratory illness, eye infection, asthma, and lung problems have largely decreased in the family having installing a biogas plant for heating, cooking, and other work. The improvement in hygienic cooking on biogas also has economic benefits.

The principal disease spreading organisms, such as typhoid, paratyphoid, cholera and dysentery bacteria, hookworm, bilharzias tapeworm, and roundworm, are killed in biogas plants. Cooking on biogas can have effects on nutritional patterns too. It increases cooked food digestibility.

4. *Reduction of workload, mainly for women, in firewood collection and cooking*: Time and human labour energy is greatly reduced in searching, collecting, and carrying the firewood home from long distance places and cleaning of cooking equipments and utensils.

   Biogas plants also improve health conditions in the homes. Home remains free from smokes and dust, and more hygienic conditions are maintained, and the space required for keeping firewood materials is also minimized.

5. *Environmental advantages through protection of forests, soil, water, and air*: A biogas plant directly saves the use of forest wood and forest residues and helps in deforestation.

   The widespread production and utilization of biogas is expected to make a substantial contribution to soil protection and amelioration. First, biogas could increasingly replace firewood as a source of energy. Second, biogas systems yield more and better fertilizer. As a result, more fodder becomes available for domestic animals. This, in turn, can lessen the danger of soil erosion attributable to overgrazing.

6. *Global environmental benefits of biogas technology*: Biogas is a renewable source of energy which has important climatic effects. As the demand for fossil fuel required for heating and cooking is reduced by the use of bio gas, emissions of carbon dioxide are also largely reduced. Also, capturing-uncontrolled methane emission significantly reduces the global warming.

10.6 FACTORS AFFECTING THE SELECTION OF A PARTICULAR MODEL OF A BIOGAS PLANT

Various factors affecting the selection of a particular model of a biogas plant are explained as follows:

1. *Cost*: The principal and maintenance costs of biogas plans should be as low as possible (in terms of the production cost per unit volume of biogas) both to the user and to the society
2. *Simplicity in design*: The design should be simple not only for construction purposes but also for operation and maintenance. This is an important consideration especially in countries where the rate of literacy is low and the availability of skilled human resource is scarce.
3. *Durability*: Longer lifespan of biogas plants is essential in situations where people are yet to be motivated for the adoption of this technology, and the necessary skill and materials are not readily available, and it is necessary to construct plants that are more durable, although this may require a higher initial investment.
4. *Suitability for use with available raw inputs*: The design should be compatible with the type of inputs that would be used. If plant materials such as rice straw, maize straw, or similar agricultural wastes are to be used, then the batch feeding design or discontinuous system should be used instead of a design for continuous or semi-continuous feeding.

5. *Inputs and outputs use frequency*: Frequency of utilization of biogas and feedstock inputting in biogas plants, influence the selection of a particular design, and the size of various components of biogas plants.

10.7 BIOGAS PLANT FEEDS AND THEIR CHARACTERISTICS

Any biodegradable organic material can be used as inputs for processing inside the biodigester. However, for economic and technical reasons, some materials are more preferred as inputs than others. If the inputs are costly or have to be purchased, then the economic benefits of outputs such as gas and slurry will become low.

Economic value of biogas and its slurry and reduced environmental cost of biodegradable wastes disposal in landfill are the two benefits of biogas energy.

One of the main attractions of biogas technology is its ability to generate biogas out of organic wastes that are in abundance and freely available. Cattle dung is most commonly used as an input mainly because of its availability. The potential gas production from some animal dung is given in Table 10.2.

In addition to the animal and human wastes, plant materials are also used to produce biogas and biomanure. Since different organic materials have different biochemical characteristics, their potential for gas production also varies.

Basic requirements for gas production or for normal growth of methanogens are achieved by mixing two or more of different organic materials and feeding to the biogas plants can be used together provided that some are met. Some characteristics of these raw organic inputs materials having significant impact on the level of gas production are described below.

10.7.1 Carbon/Nitrogen (C/N) Ratio

The relationship between amount of carbon and nitrogen present in organic materials is expressed in terms of the carbon/nitrogen (C/N) ratio. A C/N ratio ranging from 20 to 30 is considered optimum for anaerobic digestion. For organic materials with very high C/N ratio, the nitrogen will be consumed rapidly by methanogens for meeting their protein requirements and left over carbon content of the material will not have any reaction process. This will reduce the biogas production.

For very low C/N, nitrogen will be liberated and accumulated in the form of ammonia (NH_4) which will increase the pH value of the content in the digester. A pH values higher than 8.5 will start showing toxic effect on methanogens population.

C/N ratio of a few commonly used materials are presented in Table 10.3.

Table 10.2 Gas Production Potential of Various Types of Dung

Types of Dung	Gas Production Per Kg Dung (m^3)
Cattle (cows and buffaloes)	0.023–0.040
Pig	0.040–0.059
Poultry (Chickens)	0.065–0.116
Human	0.020–0.028

Table 10.3 C/N Ratio of a Few Commonly Used Materials

Raw Materials	C/N ratio
Duck dung	8
Human excreta	8
Chicken dung	10
Goat dung	12
Pig dung	18
Sheep dung	19
Cow dung/buffalo dung	24
Water hyacinth	25
Elephant dung	43
Straw (maize)	60
Straw (rice)	70
Straw (wheat)	90
Saw dust	above 200

As evident from Table 10.3 animal waste, particularly cattle dung has an average C/N ratio of about 24. The plant materials, such as straw and sawdust, contain a higher percentage of carbon/nitrogen ratio whereas the human excreta have a C/N ratio as low as 8.

In order to bring the average ratio of the composite input to a desirable level materials with high C/N ratio could be mixed with those of low C/N ratio. In China, it is customary to load rice straw at the bottom of the digester upon which latrine waste is discharged as a means to balance C/N ratio.

10.7.2 Advantages

It includes the following:

1. Clean fuel of high calorific value and has a convenient ignition temperature.
2. No residue, smoke, and dust produced.
3. Non-polluting. Significant health benefits are achieved by the use of clean biogas.
4. Economical benefits of biogas and high-quality manure.
5. Provides nutrient rich (N and P) manure for plants.

10.7.3 Limitations

The limitations are as follows:

1. Initial cost of installation of the plant is high.
2. Inadequacy of organic raw materials and its continuity of supply.
3. Social acceptability.
4. Maintenance and repair of bio gas plants.

10.7.4 Uses

1. It is used as a domestic fuel.
2. It is used as a fuel for motive power.
3. It is used for electricity generation.

SUMMARY

- Biogas is a biofuel and usually means a mixture of methane, carbon dioxide, and hydrogen produced by using bacteria to break down organic materials.
- In India, Pakistan, and many other countries, the biogas produced from the anaerobic digestion of manure in small-scale digestion facilities is called gobar gas.
- Biogas can be produced from raw materials such as agricultural waste, manure, municipal waste, plant material, sewage, green waste, and food waste.
- Waste material is digested in the absence of oxygen (which is called anaerobic conditions) by bacteria. It works best at around 35°C~40°C.
- An anaerobic digester that treats farm wastes, energy crops, and dung and human excreta is commonly called a biogas plant.
- Biogas can be used as a relatively low-cost fuel for the generation of energy and heating purposes (such as cooking, lighting, running engine, and generating electrical power).
- Biogas, which is essentially gobar gas in India, did not face much of problem of social acceptability, and in practice, there is no inhibition for using this gas for cooking. Unfortunately, the same cannot be said about methane produced from human excreta. Even connecting the toilet to cattle dung plants is resisted by almost all sections of people.

REVIEW QUESTIONS

1. What is biogas? Discuss its composition and property.
2. Explain the availability of biogas plant feedstock. Discuss the effect of C/N ratio on biogas production.
3. Explain the meaning of anaerobic digestion and discuss processes involved during anaerobic digestion.
4. Describe the construction and working of a biogas plant, its material aspects, and utilization of plant products.
5. What is meant by anaerobic digestion? Explain the factors which affect biodigestion.
6. The data given below is for a family biogas digester suitable for the output of five cows, the retention time is 20 days, and temperature is 30°C dry matter consumed per day is 2 kg and biogas yield is 0.24 m^3 per kg. The efficiency of the burner is 60% methane proportion is 0.8. Heat of combustion of methane = 28 MJ/m^3, density of dry material in fluid is 50 kg/m^3.

Calculate the following:

(a) the volume of biogas digester

(b) power available from digester.

7. Explain the constructional details of KVIC digester.
8. Explain advantages and uses of biogas.
9. Explain advantages and limitations of biogas plants.

Chapter 11

Tidal Energy

Tides are periodic rises and falls of large bodies of water. Gravity is one major force that creates tides. In 1687, Sir Isaac Newton explained that ocean tides result from the gravitational attraction of the sun and moon on the oceans of the earth. Spring tides are especially strong tides that occur when the earth, the sun, and the moon are in a line. The gravitational forces of the moon and the sun both contribute to the tides. Spring tides occur during the full moon and the new moon. Neap tides are especially weak tides. They occur when the gravitational forces of the moon and the sun are perpendicular to one another with respect to the earth. Neap tides occur during quarter moons. Tidal energy is a form of hydropower that converts the energy of the tides into electricity or other useful forms of power. The tide is created by the gravitational effect of the sun and the moon on the earth causing cyclical movement of the seas. Therefore, tidal energy is an entirely predictable form of renewable energy. Until recently, the common plant for tidal power facilities involved erecting a tidal dam, or barrage, with a sluice across a narrow bay or estuary. As the tide flows in or out, creating uneven water levels on either side of the barrage, the sluice is opened and water flows through low-head hydro turbines to generate electricity. For a tidal barrage to be feasible, the difference between high and low tides must be at least 5 m.

KEY CONCEPTS

- Tidal energy resource
- Tidal energy availability world wide and in india
- Energy availability in tides and calculation of tidal power
- Tidal power basin
- Turbines for tidal power
- Problems faced in exploiting tidal energy

11.1 GENERAL

Energy naturally present in ocean water bodies or in their movement can be used for the generation of electricity. This is achieved broadly in the following ways:

1. *Tidal energy*: During the rising period of tides, water is stored in a water reservoir constructed behind dams on

shore. The potential energy of stored water body is used to generate electrical energy similar to that in a conventional hydropower plant. For the tidal energy method to work effectively, the tidal difference (difference in the height of the high and low tides) should be at least 4m. We discuss tidal energy in this chapter.

2. *Wave energy*: Using the kinetic (dynamic) energy of the ocean, waves is utilized to rotate an underwater power turbine and generate electricity thereon as an underwater wind farm. This will be discussed in Chapter 12.
3. *Ocean thermal energy*: Chapter 13 focusses on the temperature difference between warm ocean surface water and deep sea cold water is used to generate electricity. This is similar to geothermal power generation where heat trapped in the earth surface is converted into electrical energy.

11.2 TIDAL ENERGY RESOURCE

Tides are the waves caused due to the gravitational pull of the moon and also the sun (although its pull is very low). The rise of seawater is called high tide and fall in seawater is called low tide and this process of rising and receding of water waves happen twice a day and cause enormous movement of water.

Thus, enormous rising and falling movement of water is called tidal energy, which is a large source of energy and can be harnessed in many coastal areas of the world. Tidal dams are built near shores for this purpose in which water flows during high tide and water flows out of dam during low tides. Thus, the head created results in turning the turbine coupled to electrical generator.

Tidal energy has been developed on a commercial scale among the various forms of energy contained in the oceans. When the moon, the earth, and the sun are positioned close to a straight line, the highest tides called spring tides occur. When the earth, moon, and sun are at right angles to each other (moon quadrature), the lowest tides called neap tides occur.

The water mass moved by the moon's gravitational pull when moon is very close to ocean and results in dramatic rises of the water level (tide cycle). The tide starts receding as the moon continues its travel further over the land, away from the ocean, reducing its gravitational influence on the ocean waters (ebb cycle).

11.3 TIDAL ENERGY AVAILABILITY

Gravitational forces between the moon, the sun, and the earth cause the rhythmic rise and fall of ocean waters throughout the world. Those result in tide waves. The moon exerts more than twice as great a force on the tides as the sun due to its much closer position to the earth. As a result, the tide closely follows the moon during its rotation around the earth, creating diurnal tide and ebb cycles at any particular ocean surface. The amplitude or height of the tide wave is very small in the open ocean where it measures several centimetres in the centre of the wave distributed over hundreds of kilometres. However, the tide can increase dramatically when it reaches continental shelves, bringing huge masses of water into narrow bays, and river estuaries along a coastline. For instance, the tides in the Bay of Fundy in Canada are the greatest in the world, with amplitude between 16 and 17 m near shore. High tides close to these figures can be observed at many other sites worldwide, such as the Bristol Channel in England, the Kimberly coast of Australia, and the Okhotsk Sea of Russia. Table 11.1 gives ranges of amplitude for some locations with large tides.

Table 11.1 Highest Tides (Tide Ranges) of the Global Ocean

Country	Site	Highest Tide Range (m)
Canada	Bay of Fundy	16.2
EnglandZ	Severn Estuary	14.5
France	Port of Granville	14.7
France	La Rance	13.5
Argentina	Puerto Rio Gallegos	13.3
Russia	Bay of Mezen (White Sea)	10.0
Russia	Penzhinskayaguba	13.4
India	The Gulf of Cambay, Gujarat	11
India	Gulf of Kutch, Gujarat	8
India	The Ganges Delta in the Sundarban, West Bengal	5

Source: NOAA Federal

Tidal energy projects are extremely site specific. The quality of the topography of the basin also needs to facilitate civil construction of the power plant. It is a clean mechanism and does not involve the use of fossil fuels. However, environmental concerns exist mainly to do with high silt formation at the shore (due to preventing tides from reaching the shore and washing away silt) and disruption to marine life near the tidal basin. Wave energy projects have lesser ecological impact than tidal wave energy projects.

In terms of reliability, tidal energy projects are believed to be more predictable than those harnessing solar or wind energy, since occurrences of tides are fully predictable. Table 11.2 provides glimpses of few potential sites for tidal power generation.

Table 11.2 A Few Potential Sites for Tidal Power Generation

Country	Site	Average Tide Height (m)	Basin Area (m^2)	Estimated Power Potential (MW)
Argentina	San-Jose	6.0	780	7,000
Australia	Secure	8.4	130	570
Australia	Walcoti	8.4	260	1,750
Korea	Carolina Bay	4.7	90	480
Russia	Mezen	5.66	2,640	15,000
Russia	Tugur	5.38	1,080	6,790
UK	Severn	8.3	490	6,000
UK	Mersey	8.4	60	700
USA	Cook Inlet	4.35	3,100	18,000
USA	Passamaquoddy	5.55	300	400

Source: NOAA Federal

11.4 TIDAL POWER GENERATION IN INDIA

Long coastline with the estuaries and gulfs in India has a strong tidal range and height to move turbines for electrical power generation. Important site location and estimated power potential of a few Indian tidal energy plant is given in Table 11.3.

Table 11.3 Indian Tidal Energy Plant

Site Location	Tide Heights (m)	Estimated Power Potential (MW)
The Gulf of Cambay, Gujarat	11 (6.7 av)	7,000
Gulf of Kutch, Gujarat	8 (5.23 av)	12,000
The Ganges Delta in the Sundarban, West Bengal	5 (2.97)	8,000

Many organizations and government agencies are busy in the construction of tidal power plants on all those location and harnessing tidal energy at full capacity. There is an ample prospect for tidal power development in India. It has been investigated that Gulf of Cambay may prove the biggest tidal energy reservoir for India. Extensive exploration on the western coast in Gulf of Kutch (at Mandva), Gulf of Combay (at Hazira), Maharashtra (at Janjira and Dharmata) and also in Hoogali, Chhatarpur, and Puri on Eastern coast may be worth attempting.

Nevertheless, the possibility of developing tidal power scheme in India may be examined in the following all aspects:

1. Economic aspects of tidal power schemes when compared to the conventional schemes.
2. Problems associated with the construction and operation of plant.
3. Problems related to the hydraulic balance of the system in order to minimize the fluctuation in the power output.
4. Environmental effects of the schemes.

11.5 LEADING COUNTRY IN TIDAL POWER PLANT INSTALLATION

Worldwide installed capacity of few countries are approximately shown in Table 11.4.

Table 11.4 Installed Tidal Power Capacities of Few Countries

Country	Site Location	Installed Capacity (MW)
France	La Rance	24 bulb-type turbines, each of 10 MW rating. Total = 240 MW
UK	The Severn Barrage	A total of 214 turbines each of 40 MW rating. Total = 8,560 MW
Russia	Kislaya Guba	0.4 MW
Canada	Annapolis	18 MW
China		3.9 MW

11.6 ENERGY AVAILABILITY IN TIDES

Potential energy and kinetic energy are the two energy components of energy of the tide waves. The potential energy is the work done in lifting the mass of water above the ocean surface.

This energy can be calculated as

$$E = \rho g A \int z\,dz = 0.5\,\rho g A h^2 \quad (11.1)$$

where E = the energy; g = acceleration of gravity; ρ = the seawater density (which equals its mass per unit volume); A = sea area under consideration; z = vertical coordinate of the ocean surface; and h = the tide amplitude.

Taking an average of the product of $(\rho g) = 10.15$ kN/m^3 for seawater, energy for a tidal cycle per m^2 of ocean surface can be approximated as:

$$E = 1.4\,h^2 \quad \text{(Watt-hour)} \quad (11.2)$$

or

$$= 5.04\,h^2 \quad \text{(kJ)} \quad (11.3)$$

Since extracting all the available stream power could be environmentally damaging, it is necessary to use a factor that expresses the usable power percentage with apparently no damaging consequences. It is called α (significant extraction factor) that may vary from 0.2 to 0.6.

The kinetic energy (KE) of the water mass (m) is its capacity to do work by virtue of its velocity (V). It is defined by

$$\text{KE} = 0.5mV^2 \quad (11.4)$$

The total energy of tide waves equals the sum of its potential and kinetic energy components.

Estimation and understanding of the potential energy availability of the tides are key for designing conventional tidal power plants using water dams for creating artificial upstream water heads. Such power plants exploit the potential energy of vertical rise and fall of the water.

The kinetic energy of the tide has to be known for designing other types of tidal power plants (like Soating), which harness energy from tidal currents or horizontal water. They do not involve the installation of water dams.

11.6.1 Calculation of Tidal Power

Potential tidal power can be reckoned based on a mathematical calculation. Let us assume that the surface area of the reservoir as stable between the full stored water level and the emptied floor, the energy produced by the ebbing water could be expressed as

$$d(w) = \rho g h d(v) = \rho g A h d(h) \quad (11.5)$$

Here, d(w) = energy unit; ρ = density of seawater (about [1.02–1.04] $\times 10^3$ kg/m^3); g = acceleration of gravity (9.8 m/s^2); A = surface area of the reservoir (m^2) assumed as a constant from high tide to low tide; h = instant water level height (m); v = volume of reservoirs (m^3).

Therefore, its power could be written as

$$P = \int d(w) / \int dt = \rho\, g A H^2 / 2T \quad (11.6)$$

Here, P = potential power (W); T = tidal period (s); and H = tidal ranges (m).

Let us assume $\rho = 1.04 \times 10^3$ kg/m^3 and T = 6 h.

This formula can be simplified as $P = 0.226AH^2$ (W) (11.7)

11.6.2 Tidal Stream Generator

It is often referred to as a tidal energy converter (TEC). It is a machine that extracts kinetic energy from moving masses of water in particular tides. A TEC device extracts energy from a tidal flow much in the same way that a windmill extracts energy from the wind. The following equation can be used to calculate the power output of either devices,

$$P = \frac{1}{2} \rho A \eta_T V^3 \tag{11.8}$$

where P = power (W); ρ = density of seawater (kg/m^3); A = capture area (m^2); η_T = combined efficiency from water to electric wire; and V = flow speed (m/s).

The abovementioned power calculation is valid, provided the capture area of the tidal energy conversion device(s) is small in comparison to the cross-sectional area of the channel. The significant difference between wind and tidal power calculations is the typical range of flow speeds and the density of the fluid.

11.7 TIDAL POWER BASIN

The basin system is the most practical method of harnessing tidal energy. It is created by enclosing a portion of sea behind erected dams. The dam includes a sluice that is opened to allow the tide to flow into the basin during tide rise periods and the sluice is then closed. When the sea level drops, traditional hydropower technologies (water is allowed to run through hydro turbines) are used to generate electricity from the elevated water in the basin. From Equation 11.7, we can observe that the tidal power varies as the square of the head and since the head varies with the tidal range, the power available at different sites shows very wide variation. In order to overcome this wide variation in availability of tidal power, various tidal basin systems have, therefore, been developed. They are discussed in the following sections.

11.7.1 Single-basin System

This is the simplest way of power generation and the simplest scheme for developing tidal power is the single-basin arrangement as shown in Figure 11.1. Single water reservoir is closed off by constructing dam or barrage. Sluice (gate), large enough to admit the water during tide so that the loss of head is small, is provided in the dam.

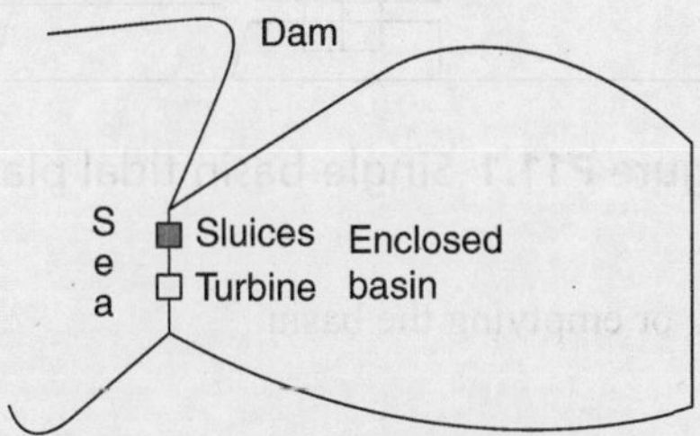

Figure 11.1 Single-basin system

The single-basin system has two configurations, namely:

1. *One-way single-basin system*: The basin is filled by seawater passing through the sluice gate during the high tide period. When the water level in the basin is higher than the sea level at low tide period, then power is generated by emptying the basin water through turbine generators. This type of systems can allow power generation only for about 5 h and is followed by the refilling of the basin. Power is generated till the level of falling tides coincides with the level of the next rising tide.
2. *Two-way single basin*: This system allows power generation from the water moving from the sea to the basin, and then, at low tide, moving back to the sea. This process requires bigger and more expensive turbine.

Single-basin system has the drawbacks of intermittent power supply and harnessing of only about 50% of available tidal energy.

Example 11.1

For a typical tidal power plant shown in Figure P11.1, the basin area is 25×10^6 m^2. The tide has a range of 10 m. However, turbine stops working when the head on it falls below 2 m. Assume that density of seawater is 1,025 kg/m^2, acceleration due to gravity is 9.81 m/s^2, combined efficiency of turbine and generator is 75%, and period of energy generation is 6 h and 12.5 min. Calculate:

1. Work done in filling or emptying the basin.
2. Average power
3. The energy generated in one filling process.

Solution

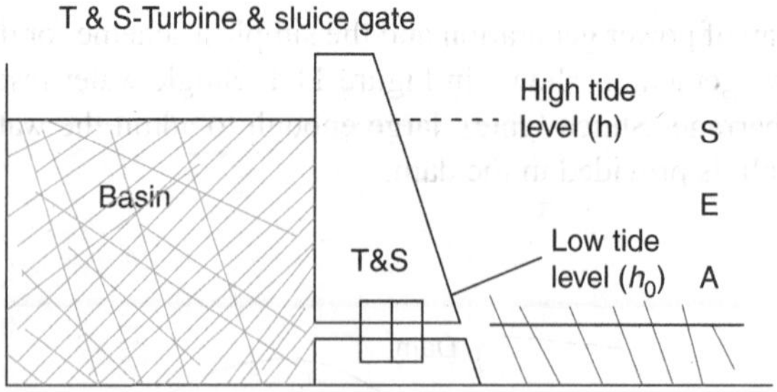

Figure P11.1 Single-basin tidal plant

1. Total work done in filling or emptying the basin

$$W = \int_{h_0} \rho g A h \, d(h) = \tfrac{1}{2}\rho g A(h^2 - h_0^2)$$

h

$$= (1/2) \times [1{,}025 \times 9.81 \times 25 \times 10^6 \times (10^2 - 2^2)]$$
$$= 17.6 \times 10^{12}$$

2. Average power

$$P_{av} = W/t = 17.6 \times 10^{12}/22{,}350 = 787.32 \times 10^6 \text{ W}$$

3. Energy generated

$$E = 0.75 \times 787.32 \times 106 \times 3{,}600/100 = 2.123 \times 10^9 \text{ kWh}$$

11.7.2 Two-basin Systems

An improvement over the single-basin system is the two-basin system. In this system, a constant and continuous output is maintained by suitable adjustment of the turbine valves to suit the head under which these turbines are operating.

A two-basin system regulates power output of an individual tide, but it cannot take care of the great difference in outputs between spring and neap tides. Therefore, this system provides a partial solution to the problem of getting a steady output of power from a tidal scheme.

This disadvantage can be overcome by the joint operation of tidal power and pumped storage plant. During the period, when the tidal power plant is producing more energy than required, the pumped storage plant utilizes the surplus power for pumping water to the upper reservoir. When the output of the tidal power plant is low, the pumped storage plant generates electric power and feeds it to the system. This arrangement, even though technically feasible, is much more expensive, as it calls for high installed capacity for meeting a particular load.

This basic principle of joint operation of tidal power with steam plant is also possible when it is connected to a grid. In this case, whenever tidal power is available, the output of the steam

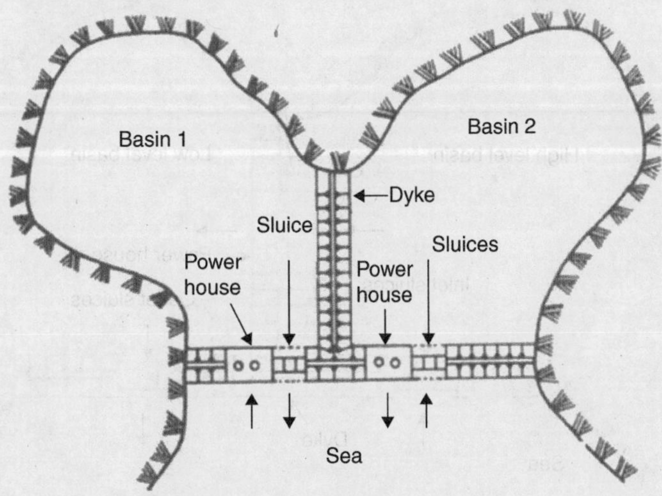

Figure 11.2 Two-basin system

plant will be reduced by that extent that leads to saving in fuel and reduced wear and tear of steam plant. This operation requires the capacity of steam power plant to be equal to that of tidal power plant and makes the overall cost of power obtained from such a combined scheme very high. In the system shown in Figure 11.2, the two basins close to each other, operate alternatively. One basin generates power when the tide is rising (basin getting filled up) and the other basin generates power while the tide is falling (basin getting emptied). The two basins may have a common power house or may have separate power house for each basin. In both the cases, the power can be generated continuously. The system could be thought of as a combination of two single-basin systems, in which one is generating power during tiding cycle, and the other is generating power during emptying.

11.7.3 Co-operating Two-basin Systems

This scheme consists of two basins at different elevation connected through the turbine. The sluices in the high- and low-level basin communicate with seawater directly, as shown in Figure 11.3. The high-level basin sluices are called the inlet sluices and the low level asoutlet sluices.

The basic operation of the scheme is as follows:

1. The rising tide fills the high-level basin through the sluiceways.
2. When the falling seawater level is equal to the water level in the high-level basin, the sluiceways are closed to prevent the outflowing high-level basin water back to the sea.
3. The water from high-level basin is then allowed to flow through the turbine generators to the low-level basin.
4. When the falling seawater level becomes lower than the rising water level in the low-level basin, the sluiceways are opened to allow water to flow into the sea from the low-level basin. This process continues until the water level in the low-level basin equals to the rising sea level. Then, the sluiceways are closed to prevent the filling of low-level basin from the seawater.

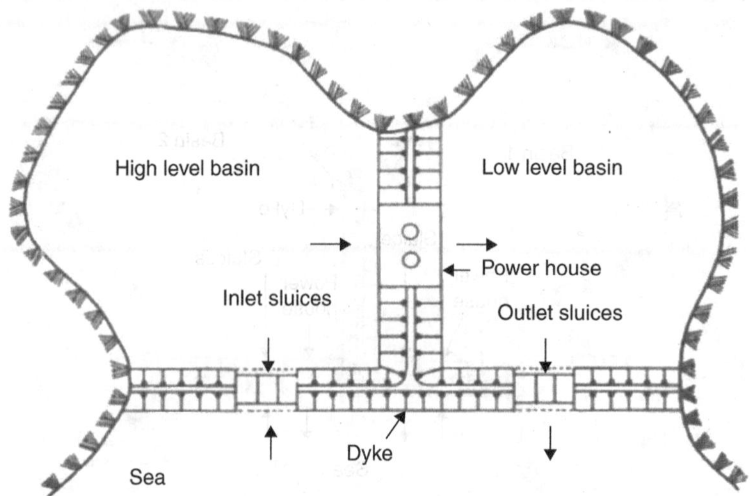

Figure 11.3 Co-operating two-basin systems

5. When the seawater again rises during the next rising tide equals to low level of high-level basin, sluices of high-level basin is again open for filling of water in high-level basin. Thus, the cycle is repeated.

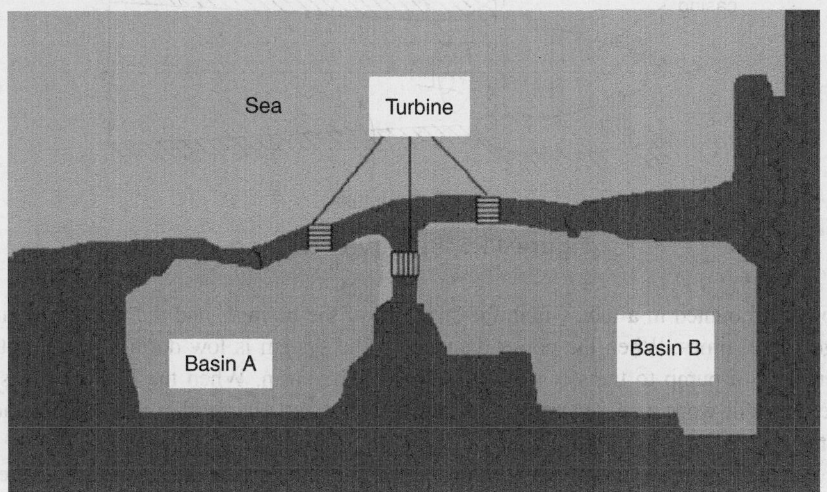

Figure 11.4 Co-ordinating two basin systems

Figure 11.4 gives another schematic diagram of co-coordinating two-basin tidal power stations. With two basins, one is filled at high tide and the other is emptied at low tide. Turbines are placed between the basins and between the basin and the sea. This two basin systems allow continuous power generation. However, they are very expensive to construct due to the cost of the extra length.

11.8 TURBINES FOR TIDAL POWER

Tidal power plants operate using a rapidly varying head of water, and therefore, their turbines must have high efficiency at varying head.

These are as follows:

1. The Kaplan type of water turbine operates quite favourably under these conditions.
2. The propeller type of turbine is also suitable because the angle of the blades can be altered to obtain maximum efficiency while water is falling.
3. A compact reversible horizontal turbine (bulb-type turbine) has been developed by French Engineer and it acts with equal efficiency both as a pump and as a turbine.

11.8.1 Bulb-type Turbine

The bulb-type turbine shown in Figure 11.5 consists of a steel shell completely enclosing the generator that is coupled to the turbine runner.

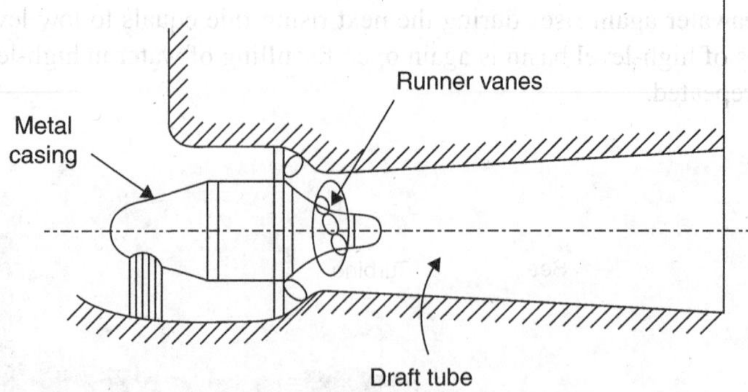

Figure 11.5 Bulb-type turbine

The turbine is mounted in a tube within the structure of the barrage, and the whole machine being submerged at all times. When the power demand on the system is low during the rising tides, the unit operates as a pump to transfer water from sea to the basin. When the load on this system is high, the unit will work as a generator, and deliver the stored energy that is a valuable additional input to the system.

Bulb turbines incorporated the generator–motor unit in the flow passage of the water. These turbines are used at the La Rance power station in France. The main drawback is that water flows around the turbine, making maintenance difficult.

11.8.2 Commercial Status of Tidal Stream Devices (as on 2009)

Table 11.5 Status of Tidal Stream Devices

Company	Class	Technology	Country	Year	Stage
Aqua Marine Power	Tidal	Horizontal Axis Turbine	UK	2007	Prototype
Verdant Power	Tidal	Horizontal Axis Turbine	US	2000	Commercial
Marine Current Turbines	Tidal	Horizontal Axis Turbine	UK	2000	Commercial
SMD Hydrovision	Tidal	Horizontal Axis Turbine	UK	2003	Prototype
Open-Hydro	Tidal	Open Centre Turbine	Ireland	2006	Pre-commercial
Hammerfest Strom	Tidal	Horizontal Axis Turbine	Norway	2007	Pilot

11.9 ADVANTAGES AND DISADVANTAGES OF TIDAL POWER

The following are the advantages of tidal power:

1. About two-third of earth's surface is covered by water, there is scope to generate tidal energy on large scale.

2. Techniques to predict the rise and fall of tides as they follow cyclic fashion and prediction of energy availability is well established.
3. The energy density of tidal energy is relatively higher than other renewable energy sources.
4. Tidal energy is a clean source of energy and does not require much land or other resources as in harnessing energy from other sources.
5. It is an inexhaustible source of energy.
6. It is an environment friendly energy and does not produce greenhouse effects.
7. Efficiency of tidal power generation is far greater when compared to coal, solar, or wind energy. Its efficiency is around 80%.
8. Despite the fact that capital investment of construction of tidal power is high, running and maintenance costs are relatively low.
9. The life of tidal energy power plant is very long.

The following are the disadvantages of tidal power:

1. Capital investment for construction of tidal power plant is high.
2. Only a very few ideal locations for construction of plant are available and they too are localized to coastal regions.
3. Unpredictable intensity of sea waves can cause damage to power generating units.
4. Aquatic life is influenced adversely and can disrupt the migration of fish.
5. The energy generated is not much as high and low tides occur only twice a day and continuous energy production is not possible.
6. The actual generation is for a short period of time. The tides only happen twice a day so electricity can be produced only for that time, approximately for 12 h and 25 min.
7. This technology is still not cost effective and more technological advancements are required to make it commercially viable

11.10 PROBLEMS FACED IN EXPLOITING TIDAL ENERGY

1. Usually the places where tidal energy is produced are far away from the places where it is consumed. This transmission is expensive and difficult.
2. *Intermittent supply*: Cost and environmental problems, particularly barrage systems are less attractive than some other forms of renewable energy.
3. *Cost*: The disadvantages of using tidal and wave energy must be considered before jumping to conclusion that this renewable, clean resource is the answer to all our problems. The main detriment is the cost of those plants.
4. *Altering the ecosystem at the bay*: Damages such as reduced flushing, winter icing, and erosion can change the vegetation of the area and disrupt the balance. Similar to other ocean energies, tidal energy has several prerequisites that make it only available in a small number of regions.

For a tidal power plant to produce electricity effectively (about 85% efficiency), it requires a basin or a gulf that has a mean tidal amplitude (the differences between spring and neap tide) of 7 m or more. It is also desirable to have semi-diurnal tides where there are two high and low tides everyday.

A barrage across an estuary is very expensive to build and affects a very wide area—the environment is changed for many miles upstream and downstream. Many birds rely on the tide uncovering the mud flats so that they can feed. There are few suitable sites for tidal barrages.

1. Only provides power for around 10 h each day, when the tide is actually moving in or out.
2. Present designs do not produce a lot of electricity, and barrages across river estuaries can change the flow of water, and consequently, the habitat for birds and other wildlife.
3. Expensive to construct.
4. Power is often generated when there is little demand for electricity.
5. Limited construction locations.
6. Barrages may block outlets to open water. Although locks can be in stalled, this is often a slow and expensive process.
7. Barrages affect fish migration and other wildlife; many fish like salmon swim up to the barrages and are killed by the spinning turbines.
8. Fish ladders may be used to allow passage for the fish, but these are never 100% effective.
9. Barrages may also destroy the habitat of the wildlife living near it.
10. Barrages may affect the tidal level—the change in tidal level may affect navigation, recreation, cause flooding of the shoreline, and affect local marine life.
11. Tidal plants are expensive to build.
12. They can only be built on ocean coastlines; this mean that for communities that are far away from the sea, it is useless.

SUMMARY

- Tidal energy is clean, renewable, and sustainable form of energy resource. It has no impact on climate because it does not produce any greenhouse gases.
- The tidal basin system is a portion of the sea enclosed behind a dam or dams and water is allowed to run through turbines, as the tide subsides.
- The tidal basin system is either single-basin or two-basin or co-operated two-basin system.
- La Rance Barrage is the world's first tidal power station. The facility is located on the estuary of the Rance River, in Brittany, France. It is opened on the 26th November 1966.
- Large investment and long time is required to build barrage. These are the main causes of slowing down the tidal energy development.
- The biggest and main use of tidal energy is in the generation of electricity.
- Tidal barrages and power plants have several environmental drawbacks including changes to marine and shoreline ecosystems, most notably fish populations.

REVIEW QUESTIONS

1. What are the special problems in the construction of barrage for tidal scheme?
2. A typical tidal power station has 24 generators each of 10 mW operating at maximum head of 3.5 m. It generates for 6 h twice a day. The density of seawater is 1,025 kg/m^3, and the efficiency of turbine and generator each is 93%. Assuming that power decreases linearly and that the reservoir has rectangular cross-sectional area, calculate the following:
   (a) Basin capacity in m^3
   (b) Annual energy production
3. Explain the 'single-basin' and 'two-basin' systems of tidal power harnessing. Further, discuss their advantages and limitations
4. A tidal power plant of single-basin type has a pool area 80 × 104 m^2. The tidal has a range of 8 m. However, the turbine stops operating when head on it falls below 2 m. Calculate the energy generated in one filling process in kW-h, if the turbine–generator efficiency is 90%.
5. Derive an expression for total energy generated by a modulated single-pool tidal power plant with a tidal range of R meters. The water level in the pool rises during filling process as 0.06 $R\,(t - t_1)$' where t_1 is the instant when power generation starts and t is time in hours. The basin has an area of A km^2 and water density is 1,025 kg/m^3. Power generation stops at t_2, when head falls below generation threshold. Assume sinusoidal tidal pattern.
6. Explain the difference between tidal range and head of water in tidal scheme.
7. Derive an expression for tidal energy per tidal cycle for a simple single-pool single effect tidal scheme.
8. State the types of tidal energy conversion schemes.
9. Explain the principle of operation of a simple single-effect tidal power plant and give a graph of sequential operating modes.
10. Explain the use of additional pumping feature in a single-pool single effect tidal scheme.
11. Discuss the relative merits and limitations of tidal power.
12. What are the difficulties in tidal power developments?
13. Write a short note on Sundarban tidal power scheme.

Chapter 12

Sea Wave Energy

KEY CONCEPTS

- Motion in the sea waves and parameters of wave propagation
- Power associated with sea waves and wave formula
- Wave energy availability in world and in india
- Devices for harnessing wave energy
- Key issues of wave power development
- Environmental considerations

The energy in ocean waves mainly comes in an irregular and oscillating form at all times of the day and night. Solar energy causes winds to blow over vast ocean areas, which in turn cause waves to form, gather, and travel huge distances to the shoreline of continents. The wave height, period, and direction are primarily dependent on the wind properties (speed, direction, and duration) and also the geometry of the sea (fetch length and depth). There is surprisingly little loss of energy in deep-water ocean waves, so as they travel to distant shores they continue to collect more and more wind energy. However, as waves approach relatively shallow water, their energy is greatly dissipated due to ground effects and this causes the dynamic, chaotic, and highly variable environment known as wave breaking close to shore.

Kinetic energy, the energy of motion, in waves is tremendous. An average 4-foot, 10-s wave striking a coast puts out more than 35,000 horsepower per mile of coast. Waves get their energy from the wind. Wind comes from solar energy. Waves gather, store, and transmit this energy thousands of miles with little loss. As long as the sun shines, wave energy will never be depleted. It varies in intensity, but it is available all the times.

12.1 GENERAL

Waves get their energy from the solar energy through the wind. Wave energy will never be depleted as long as the sun shines. Energy intensity may, however, have variation but it is available 24 h a day in the entire year. They are caused by the wind blowing over the surface of the ocean with enough consistency and force in many areas of the world to provide continuous waves along the shore line. It contains tremendous energy potential and wave power devices extract energy from either the surface motion of ocean waves or from pressure fluctuations below the surface. The movement of the ocean water and the changing water wave heights and speed of the swells are the main sources of wave energy. Kinetic energy in the wave motion is tremendous that can be extracted by the wave power devices from either the surface motion of ocean waves or from pressure fluctuations below the ocean surface.

12.2 MOTION IN THE SEA WAVES

When the wind blows across smooth water surface, air particles from the wind grab the water molecules they touch. Stretching of the water surface by the force or friction between the air and the water creates capillary waves (small wave ripples). Surface tension acts on these ripples to restore the smooth surface, and thereby, waves are formed.

The combination of forces due to the gravity, sea surface tension, and wind intensity are the main factors of origin of sea waves as shown in Figure 12.1, which illustrates the formation of sea waves by a storm. Wave size is determined by wind speed and fetches (defined as the distance over which the wind excites the waves) and by the depth and topography of these abed (which can focus or disperse the energy of the waves). Sea waves have a regular shape at far distance from the fetch and this phenomenon is called swell. Wave formation makes the water surface further rough and the wind continuously grips the roughened water surface, and thus, waves are intensified.

A wave is a forward motion of energy and not the water in deep sea. In true sense, the seawater does not move forward with a wave. Waves are characterized by the following parameters, as shown in Figure 12.2.

1. *Crest*: The peak point (the maximum height) on the wave is called the crest.
2. *Trough*: The valley point (the lowest point) on the wave is called the trough.
3. *Wave height* (H): Wave height is a vertical distance between the wave crest and the next trough (m).
4. *Amplitude* (a): It is defined as $H/2$ (m).
5. *Wave length* (λ): It is the horizontal distance either between the two successive crests or troughs of the ocean waves (m).

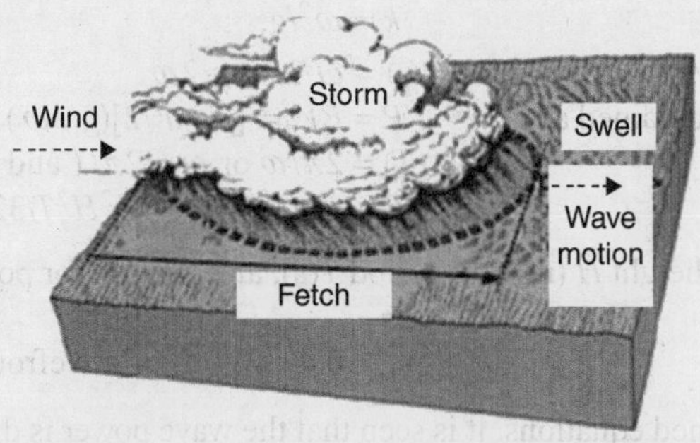

Figure 12.1 Sea wave formation by storm

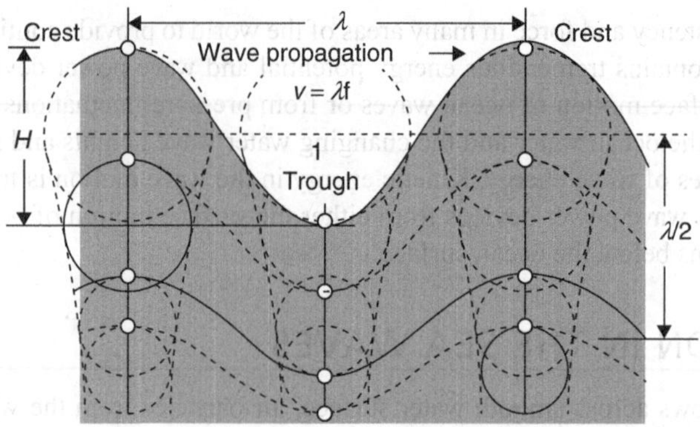

Figure 12.2 Sea wave propagation

6. *Wave propagation velocity* (v): The motion of seawater in a direction (m/s).
7. *Wave period* (T): It measures the size of the wave in time(s). It is the time required for two successive crests or two successive troughs to pass a point in space.
8. *Frequency* (f): The number of peaks (or troughs) that pass a fixed point per second is defined as the frequency of wave and is given by $f = 1/T$ (cycle/s).

12.3 POWER ASSOCIATED WITH SEA WAVES

It has been concluded by researchers through linear wave motion theory that the kinetic and potential energy (E) of a wave per meter of crest and unit of surface can be approximated as

$$E = \rho g a^2/2 \tag{12.1}$$

where ρ = density of water; g = gravitational acceleration; and a = amplitude of the wave (approximately equals to half its wave height H).

The power that a meter of crest holds can be obtained by multiplying the amount of energy transported by the group velocity.

In deep water, dispersion relation (k) is given as

$$k = \omega^2/g, \tag{12.2}$$

Further, group velocity $(V_g) = \omega/2k = g/2\omega$ (12.3)

The total power (P) is obtained as $P = EV_g = [\rho g\, a^2/2](g/2\omega) = \rho g^2 a^2/4\omega$ (12.4)

Further, wave period $(T) = 2\pi/\omega$ or $\omega = 2\pi/T$ and $a = H/2$

Therefore, $P = \rho g^2 a^2/4\omega = \rho g^2 H^2 T/32\pi$ (12.5)

For irregular waves of height H (m) and period T(s), an equation for power per unit of wavefront can be derived as

$$P_{\text{irregular}} = 0.4 \text{ (kW/m) of wavefront} \tag{12.6}$$

From the abovementioned equations, it is seen that the wave power is directly proportional to the square of wave height.

Example 12.1

A 2–m sea wave has a 6 s period and occurs at the surface of 100-m deep water. Assume seawater density equals to 1,025 kg/m^3. Calculate the energy and power densities of the wave.

Solution

Wavelength of the sea wave, $\lambda = v\,T$ (m) (P12.1.1)
where v is wave propagation velocity (m/s); and T = time period of the wave (s).

The wave time period (T) and wave propagation velocity (v) depend upon the wavelength and depth of seawater. The relationship between the wavelength and the time period can be approximated as

$$\lambda = 1.56\,T^2 \quad \text{(P12.1.2)}$$

Therefore, $\lambda = 1.56 \times 6^2 = 56.16$ m

Therefore, velocity of wave propagation, $v = \lambda/T = 56.16/6 = 9.36$ m/s and height of the wave is the height from the crest to trough, $H = 2$ m

Wave frequency, $f = 1/T = 1/6$

Energy density = $(1/8)\,(\rho g H^2) = 1{,}025 \times 4 \times 9.81/8 = 5{,}027.625$ J/m^2

Power density= E/T = E.f = 5,027.625/6 = 837.9375 W/m^2

12.3.1 Another Wave Power Formula

In deep water, where the water depth is larger than half the wavelength, the wave energy flux power is

$$P = (1/2)\,[\rho g^2 H^2 T/32\pi] = \rho g^2 H^2 T/64\pi \approx 0.5\ (\text{kW/m}^3\text{s})\, H^2 T \quad (12.7)$$

where P = wave power per unit of wave-crest length, (k/m); H = significant wave height (m); T = wave period (s); ρ = density of water; and g = acceleration due to gravity.

Example 12.2

An ocean swell, a few kilometres away from the coastline and in deep seawater, has wave height of 3 m and wave period of 8 s. Obtain power of the wave energy flux per unit of wave crest length.

Solution

From Eq. (12.7), $P \approx 0.5 \times 3^2 \times 8 \approx 36$ kW/m

12.4 WAVE ENERGY AVAILABILITY

The density of water is about 800 times higher than air, and therefore, the energy density of ocean waves are significantly several times more than air. The amount of energy available in ocean waves is tremendously high, and hence, it is considered as a renewable, zero emission source of power. Estimates of the global ocean wave energy are more than 2 TW (which means 17,500 TWh/year) according to the World Energy Council.

It has been reported that the total available US wave energy resource is 23 GW, which is more than twice as much as Japan, and nearly five times as much as Great Britain. The West Coast of US is the most promising area with wave energy densities in the range of 25–40 kW/m.

The ocean wave along the western coast of Europe is characterized by particularly high energy. It has over half the wave energy potential of Europe and has power up to extent of 75 kW/m off the coastal area of Ireland and Scotland.

Wave power is distinct from the diurnal flux of tidal power and the steady gyre of ocean currents. This huge amount of renewable and environmentally acceptable wave energy, if extracted and utilized, has competitiveness with fossil and nuclear fuels. Generally, extreme latitudes and west coasts of continents are the best wave location. A view of global wave atlas (based on satellite data) and another world wave map are shown in Figure 12.3.

Wave energy is converted into electricity by placing wave energy converter on the surface of the ocean. The electrical energy generated is the most often used in desalination plants, power supply to electrical consumers, and energizing water pumps.

They are mostly using the first generation oscillating water columns (OWS) converters. Other technologies such as the Japanese Pendulor and the Tapchan can also be fit in this category.

These ocean wave energy technologies rely on the up-and-down motion of waves to generate electricity.

Several installations have been built in Scotland, Portugal, Norway, the USA, China, Japan, Australia, and India. The next generation of devices comprises new, modular floating devices, but these require further research and/or demonstration.

A few installation of wave power converts are as follows:

1. The first wave-power patent was for a 1799 proposal by a Parisian named Monsieur Girard and his son got patented the first wave power converter in 1979 to use direct mechanical action to drive pumps, saws, mills, or other heavy machinery.
2. During the first decades of the 19th century, a device was put in operation in Algeria that captured wave oscillation and transformed it into usable form by using a system of cams and gears.
3. A 10-kW complaint flap pilot plant was installed in the Baltic sea in 1917 and later on dismantle.

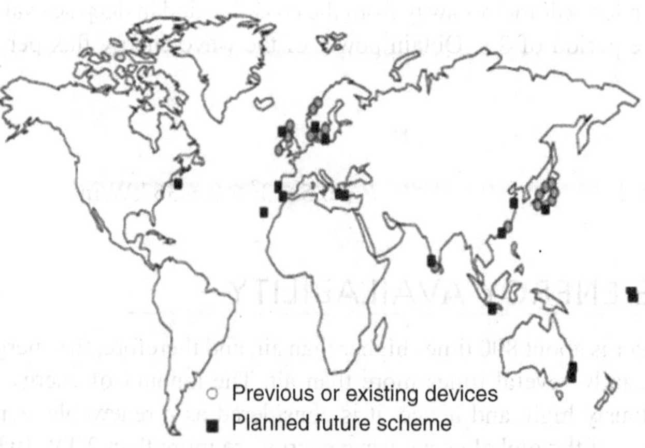

Figure 12.3 Location of wave power plant

4. Pelamis became the world's first offshore wave machine to generate electricity and fed into the grid, when it was first connected to the UK grid in 2004.
5. Salter Duck wave converter was developed around 1980 in UK.
6. A 120 kW (Oscillating wave column) prototype (The Mighty Whale) with 3 OWCs in a row has been operating since 1998 (1.5 km off Nansei Town, Japan) at 40 m depth
7. A 2 MW (AWS) system off the coast of Portugal.
8. The prototype (Wave Dragon) is deployed in Nissum Bredning, an inlet in the northern part of Denmark.
9. A 40 m long prototype (McCabe Wave Pump) was deployed in 1996 off the coast of Kilbaha, County Clare, Ireland
10. A typical 30 MW (Pelamis) installation would occupy a square kilometre of ocean and provides sufficient electricity for 20,000 homes.
11. A 750 kW project (Pelamis) off Islay, Scotland.
12. A 2 MW (Pelamis) project off the coast of Vancouver Island, Canada.
13. A 5 MW (perhaps the world's first commercial wave energy plant) developed by Wave Gen is located in Isle of Islay, Scotland.

12.4.1 Wave Energy Availability in India

The coastal area of Maharashtra has an annual wave potential ranging between 4 kW/m and 8 kW/m wavefront, which is quite high as 12–20 kW/m during the monsoon.

The wave energy potential of the most feasible sites in Maharashtra is given in Table 12.1 for offshore location.

Table 12.1 Offshore Location of Wave Power in Maharashtra

Site	Average Wave Power (kW/m) (Annual)	Average Wave Power (kW/m) (June–August)
Malvan Rock	6.91	16.73
Kura Inset	5.79	13.74
Redi	6.35	16.57
Vengurla Rock	8.01	20.61
Square Rock	6.79	16.64

Coastal average power location site in Maharashtra is given in Table 12.2.

Table 12.2 Coastal Average Power Location Site in Maharashtra

Site	Average Wave Power kW/m (Annual)	Average Wave Power kW/m (June–August)
Vijaydurg	5.86	13.58
Girye	5.90	14.21

(Continued)

Table 12.2 Continued

Site	Average Wave Power kW/m (Annual)	Average Wave Power kW/m (June–August)
Ambolgarh	5.74	13.48
PawaPoint	5.36	13.10
Kunkeshwar	5.64	13.35
Wagapur	5.70	13.10

The Vengurla and Malvan rocks and Redi are on the top among the offshore locations. In the coastal location, however, Pawa and Ratnagiri top the list followed by Girye and Miyet point.

Vizhinjam fishing harbour, Kerala, is the site of a unique demonstration plant that converts sea wave energy to electricity and is given to the local grid. This plant has oscillating water column (OWC) converter in 1990.

12.5 DEVICES FOR HARNESSING WAVE ENERGY

There are three basic technologies for converting wave energy to electricity. They are as follows:

1. *Terminator devices*: It is a wave energy device oriented perpendicular to the direction of the wave and has one stationary and one moving part. The moving part moves up and down like a car piston in response to ocean waves and pressurizes air or oil to drive a turbine. An oscillating water column (OWC) converter is an example of terminator device. These devices generally have power ratings of 500 kW to 2 MW, depending on the wave parameters and the device dimensions.
2. *Attenuator devices*: These devices are oriented parallel to the direction of the waves and are long multi-segment floating structures. It has a series of long cylindrical floating devices connected to each other with hinges and anchored to the seabed. They ride the waves like a ship, extracting energy by using restraints at the bow of the device and along its length. The segments are connected to hydraulic pumps or other converters to generate power as the waves move across. Pelamis wave energy converter is one of the known examples of attenuator devices.
3. *Point absorber*: It is a floating structure with parts moving relative to each other owing to wave action but it has no orientation in any defined way towards the waves instead absorbs the wave energy coming from any direction. It utilizes the rise and fall of the wave height at a single point for energy conversion. The pressurized water creates up and down bobbin-type motion and drives a built-in turbine generator system to generate electricity. AquaBuOY WEC is an example of point absorber devices.
4. *Overtopping devices*: These devices have reservoirs like a dam that are filled by incoming waves, causing a slight build-up of water pressure. Gravity causes released water from reservoir to flow back into the ocean through turbine coupled to an electrical generator. Salter Duck WEC is the example of overtopping devices.

12.5.1 Float or Buoy Devices

This system is shown in Figure 12.4. A series of anchored buoys rise and fall with the wave that creates mechanical energy to drive electrical generator for generation of electricity, which is transmitted to ocean shore by underground cables.

Figure 12.4 Float or buoy device

12.5.2 Oscillating Water Column Devices

An oscillating water column device (OWC device) is shown in Figure 12.5. It is a form of terminator in which water enters through a subsurface opening into a chamber, trapping air above. The wave action causes the captured water column to move up and down like a piston, forcing the air though an opening connected to a turbine to generate power.

It is a shoreline-based oscillating water column (OWC) build in UK. Further, it is installed at Islay. It is a concrete structure partially submerged in seawater and encloses a column of air on top of a column of water. The water columns in partially submerged chamber rise and fall, when sea waves impinge on the device. This wave action alternatively compresses and depressurizes the air column, which is allowed to flow to and from the atmosphere via a turbine. The energy can then be extracted from the system and used to generate electricity.

Wells' turbines as shown in Figure 12.6 are used to extract energy from the reversing air flow. It has the property of rotating in the same direction regardless of the direction to the airflow.

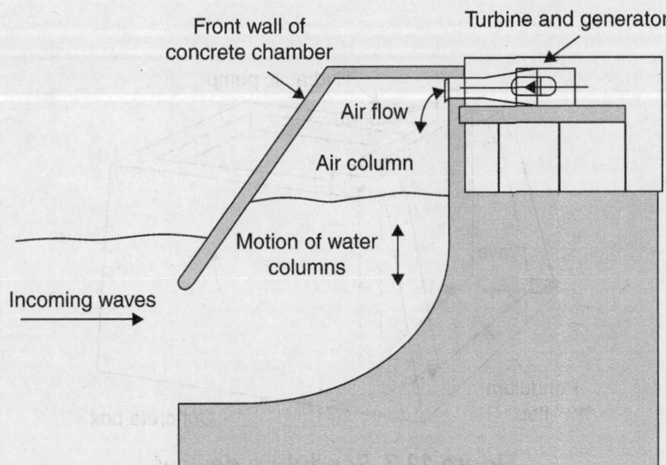

Figure 12.5 Schematic of an oscillating water column device

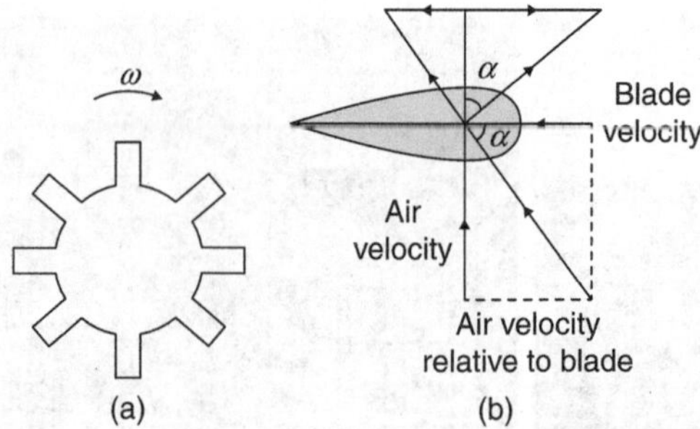

Figure 12.6 Well's turbine (a) plan view of blades and (b) velocity and force triangles in frame of reference of blades

12.5.3 Pendulum System

The pendulum system is a shoreline device that consists of a parallelepiped concrete box, which is open to the sea at one end, as shown in Figure 12.7.

A pendulum flap is hinged over this opening, which swings back and forth by the actions of the waves. The back and forth motion of pendulum is then used to power a hydraulic pump and an electric generator.

12.5.4 TAPCHAN (Tapered Channel)

The schematic arrangement of TAPCHAN device (a Norwegian system) is shown in Figure 12.8. It has a tapered channel connected to a reservoir constructed above the sea level at a height of 3–5 m. They are relatively low power output devices and suitable for deep-water shore line and low tidal range. It is a very simple device. Waves collect into a channel, which tapers into a large reservoir. As the wave width decreases, the wave amplitude increases according to the principles of conservation of energy and this enables the waves to travel up a ramp and pour into the reservoir as shown in Figure 12.8.

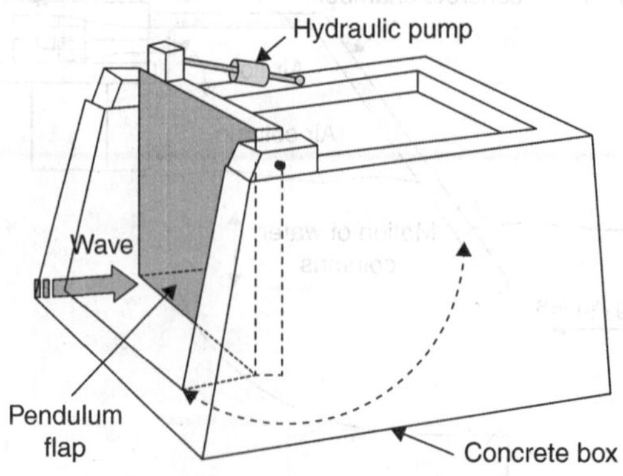

Figure 12.7 Pendulum devices

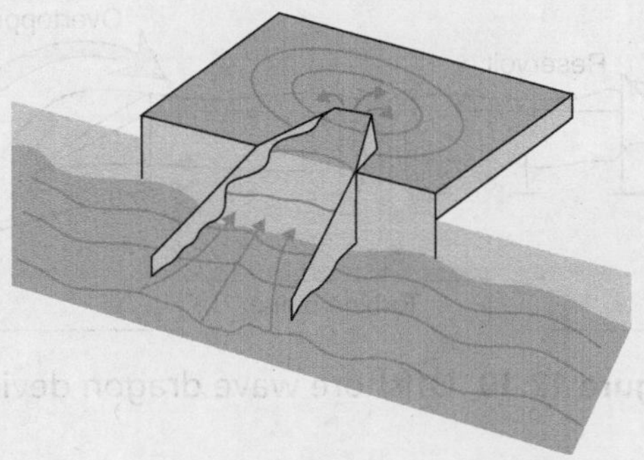

Figure 12.8 TAPCHAN

The potential energy of water stored in the reservoir is extracted by releasing the reservoir water back to the sea through a low head Kaplan turbine coupled to an electrical generator.

12.5.5 Salter's Duck System

Salter Duck WEC is the example of overtopping devices. It was invented in Scotland in 1970 to extract mechanical energy from the ocean waves. The schematic cross section of Salter Duck is given in Figure 12.9.

It is an egg-shaped device that moves with the motion of the waves. The shape of leading edge of the duck is in such a way that the approaching sea wave pressure is exerted on the duck. It forces the duck to rotate about a central axis and the tip of the cam bobs up and down in the water. As the Salter Duck moves (or bobs or rocks) up and down on the sea waves, pendulum connected to electrical generator swings forward and backward to generate electricity. Two sets of cables are attached to the device, one to a pendulum inside the device and the other to a fixed arm outside the device. The cables attached to the internal pendulum contain hydraulics that pumps as the device moves back and forth with the waves. This movement of the pressurized oil pumped into hydraulic machine that drives electric generators.

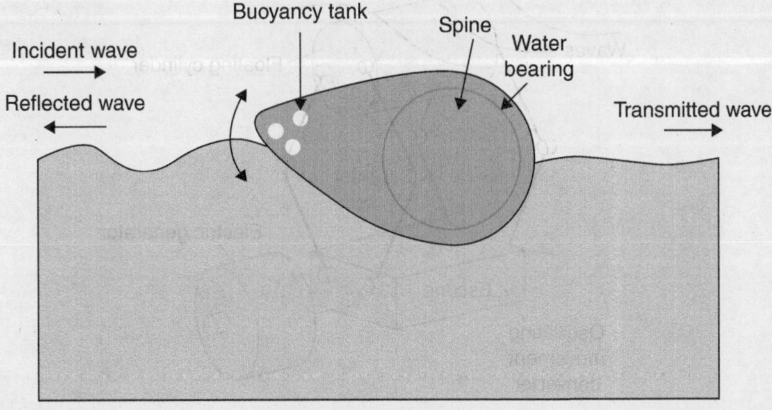

Figure 12.9 Salter duck

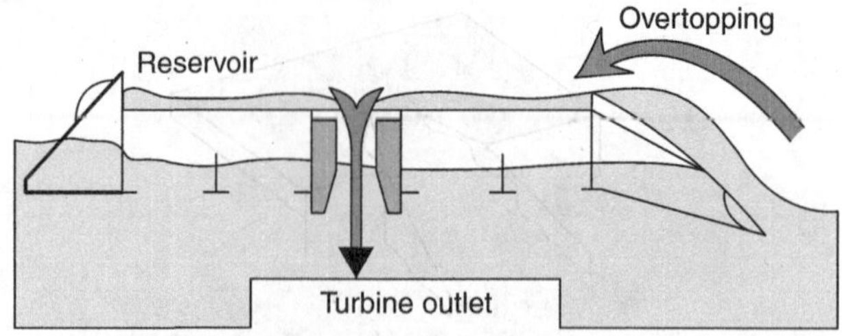

Figure 12.10 Offshore wave dragon devices

12.5.6 Offshore Wave Dragon System

The wave dragon is an overtopping device that elevates ocean waves to a reservoir above sea level as shown in Figure 12.10. Water is let out through a number of turbines, and in this way, it is transformed into electricity. The basic idea of this system consists of two large 'arms' that focus waves up a ramp into a reservoir. The water returns to the ocean by the force of gravity via a low head hydro turbine that drives an electric generator.

12.5.7 Bristol Cylinder

The Bristol cylinder operates under the sea level, as shown in Figure 12.11. It consists of a floating cylinder that collected the wave's movement. The cylinder is mechanically connected to the energy unit by flexible joints and rods. The rods are moving slowly with cylinder and the reciprocating motion is transferred to the axels in converter unit.

When transferring converter movements with mechanical arms and rotation to the generator, the efficiency should be kept as high as possible.

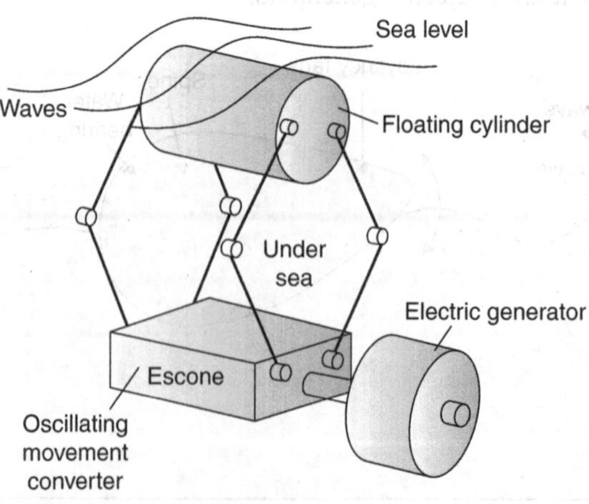

Figure 12.11 Bristol cylinder devices

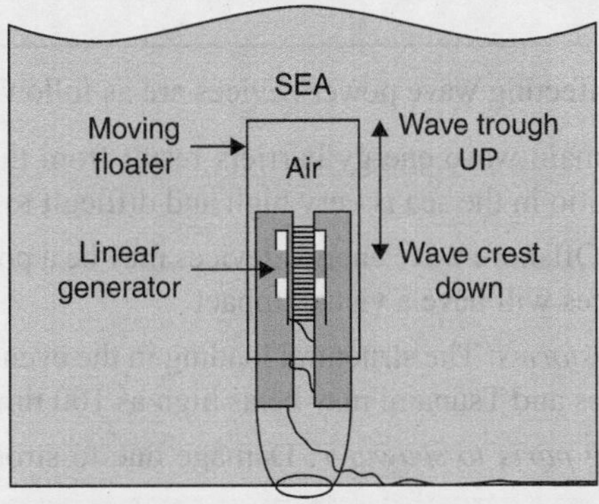

Figure 12.12 Archimedes waves swing device

12.5.8 Archimedes Wave Swing Devices

The Archimedes wave swing device (shown in Fig. 12.12) is an underwater buoy of which the upper part (floater) moves up and down in the wave, while the lower part stays in position. The floater (air-filled chamber) is pushed down under a wave crest (top) and moves up under a wave trough (valley). The interior of the system is pressurized with air and serves as an air spring. The mechanical power is converted into electrical power by means of a power take-off system (PTO). The PTO consists of a linear electrical generator and a nitrogen filled damping cylinder.

It has the advantage of being a 'point' absorber that absorbs power from waves travelling in all directions, and extracts about 50% of the incident wave power in addition to the advantage of being able to survive despite rough sea conditions on the surface.

12.6 ADVANTAGES AND DISADVANTAGES OF WAVE POWER

12.6.1 Advantages

1. Sea waves have high energy densities and provide a consistent stream of electricity generation capacity.
2. Wave energy is clean source of renewable energy with limited negative environmental impacts.
3. It has no greenhouse gas emissions or water pollutants.
4. Operating cost is low and operating efficiency is optimal.
5. Damage to ocean shoreline is reduced.

12.6.2 Disadvantages

1. High construction costs.
2. Marine life is disrupted and displaced.
3. Damage to the devices from strong storms and corrosion create problems.
4. Wave energy devices could have an effect on marine and recreation environment.

12.7 KEY ISSUES

In general, the key issues affecting wave power devices are as follows:

1. *Energy barriers*: The main wave energy barriers result from the energy carrier itself. The peak-to-average load ratio in the sea is very high and difficult to predict.
2. *Navigational hazards*: Offshore wave energy devices may be a potential navigation hazard to ships. Near shore devices will have a visual impact
3. *Survivability in violent storms*: The structural loading in the event of extreme weather conditions, such as hurricanes and Tsunami may be as high as 100 times the average loading.
4. *Vulnerability of moving parts to seawater*: Damage due to strong storms and corrosion is expected.
5. *High capital and operational cost*: High construction costs induce high power generation costs, thus making the technology uncompetitive.

The incidence of wave power at deep ocean sites is three to eight times the wave power at adjacent coastal sites, but the cost of electricity transmission from deep ocean sites is often prohibitively high.

SUMMARY

The sea has long been seen as a source of energy, and the idea of harvesting energy from sea waves was first patented in 1799, in Paris, by Girard and his son.

- Blowing wind and pressure fluctuations below the surface are the main reasons for causing waves. However, consistency of waves differs from one area of ocean to another. Some regions of oceans receive waves with enough uniformity and force. Ocean wave energy can be captured directly from surface waves.
- It is an irregular and oscillating low-frequency energy source that can be converted to grid frequency and can then be added to the electric utility grid.
- Energy in waves comes from the movement of the ocean and the changing heights and speed of the swells.
- Wave power is the transport of energy by ocean sea waves, and the capture of that energy to do useful work; for example, electricity generation, water desalination, or pumping of water into reservoirs.
- Machinery able to exploit wave power is generally known as a wave energy converter (WEC).
- Large amounts of money and research are required for wave power to catch up with wind and solar energy.

REVIEW QUESTIONS

1. Discuss the principle and working of sea wave energy conversion system.
2. Discuss the performance and limitations of sea wave energy conversion plants.

3. State the expression for energy and power in ocean waves.
4. Discuss limitations of ocean wave energy.
5. Describe principle of oscillating air column ocean wave machine.
6. Describe principle of oscillating hydraulic piston accumulator wave energy convertor.
7. Describe Dam-atoll concept of wave machine.
8. Compare ocean waves and ocean tides with reference to the period energy density and energy conversion plants.
9. State main criteria for deciding the location of ocean wave plants.

Chapter 13

Ocean Thermal Energy Conversion

Ocean thermal energy conversion (OTEC) is a method to produce electricity by using the temperature differences between warm ocean surface and cool deep ocean water to run a heat engine. If temperature difference is greater, then more energy will be produced. About 70% of the earth's surface is covered by oceans, which are continuously heated by the sun. Extracting the solar energy stored in an ocean is carried out by exploiting the temperature difference between warm surface water and cold deep sea water.

Low-grade heat from renewable energy sources is considered to be a good candidate to generate electricity. Among those sources, OTEC and solar energy are typically utilized in converting low-grade heat into power generation and other applications.

OTEC systems use the ocean's natural thermal gradient to drive a power-producing cycle. As long as the temperature difference between warm surface water and cold deep sea water is greater than about 20°C, an OTEC system can produce a significant amount of power. Suitable locations for OTEC systems in the world have been identified. It was found that natural ocean thermal gradients necessary for OTEC operation generally exist between latitudes 200°N and 200°S. OTEC can, therefore, be sited anywhere across about 60 million square kilometres of tropical oceans anywhere there is deep cold water lying under warm surface water. This generally means between the Tropic of Cancer and the Tropic of Capricorn. Surface water in these regions, warmed by the sun, generally stays at 25°C or above. Ocean water more than 1,000 metres below the surface is generally at about 5°C.

KEY CONCEPTS

- Principles of ocean thermal energy conversion (OTEC) plants
- Closed-cycle, open-cycle, and hybrid-cycle OTEC
- Basic rankine cycle and its working
- Selection of working fluids
- Carnot cycle
- Application of OTEC in addition to production of electricity
- Advantages and disadvantages of OTEC

It would not be profitable to use an OTEC power plant in the Baltic Sea, because the average temperature is about 8°C–10°C.

13.1 INTRODUCTION

Low-temperature heat obtained from renewable energy resources, such as solar thermal, geothermal, ocean thermal, etc. is presently converted into electricity and utilized for direct heating applications. About 70% of earth's surface is covered by ocean which is continuously heated by solar heat. Solar heat is stored as uneven distribution of heat between warm surface water and cold deep ocean water (called gradient) from where it is harnessed as ocean thermal energy.

OTEC sites that are located between the Tropic of Cancer and Tropic of Capricorn (23.5°N and 23.5°S of equator) found to be best locations. Ocean water with temperature gradient of 5°C and more is known as ocean thermal energy.

However, significant amount of electric power can be generated in the location where a temperature difference of 20°C and above exists between warm surface water and cold deep water.

In many regions, ocean surface water is generally maintained at 25°C or above and more than 1,000 metres below the surface is generally at about 5°C. Since average temperature in Baltic Sea is about 10°C, setting up of OTEC electrical power plant is not profitable.

Therefore, OTEC is an energy technology that converts solar radiation to electric power through heat of ocean water. These systems use ocean's natural thermal gradient. As long as the temperature difference between the warm surface water and the cold deep water below 600 metres by about 20°C, an OTEC system can produce a significant amount of power. Thus, oceans are vast renewable resources with the potential to produce thousands of kW of electric power.

The cold deep sea water used in the OTEC system is also rich in nutrients, and it can be used to cultivate plant and marine organism near the shore or on land.

13.2 PRINCIPLE OF OCEAN THERMAL ENERGY CONVERSION

The basic principle of ocean thermal energy conversion (OTEC) is explained as follows:

The warm water from the ocean surface is collected and pumped through the heat exchanger to heat and vapourize a working fluid, and it develops pressure in a secondary cycle. Then, the vapourized working fluid expands through a heat engine (similar to a turbine) coupled to an electric generator that generates electrical power. Working fluid vapour coming out of heat engine is condensed back into liquid by a condenser. Cold deep ocean water is pumped through condenser where the vapour is cooled and returns to liquid state. The liquid (working fluid) is pumped again through heat exchanger and cycle repeats. It is known as closed-cycle OTEC.

If ocean surface water is high, enough propane or similar material is used as working fluid; otherwise, for low-temperature surface water, fluid such as ammonia with low boiling point is used.

In an open-cycle OETC, warm ocean surface water is pumped into a low-pressure boiler to boil and produce steam. Then, the steam is used in steam turbine to drive an electrical generator for producing electrical power. The cold deep sea water is used in condenser to condense steam.

Some fractions of electrical power generated by OTEC plants are used for operating and controlling equipments involved in power plants, and high electrical power is used for feeding to several other energy consumers.

13.3 OCEAN THERMAL ENERGY CONVERSION PLANTS

There are two different kinds of OTEC power plants, namely land-based power plant and floating power plant.

13.3.1 Land-based Power Plant

The land-based power plant will consist of a building as shown in Figure 13.1.

It is constructed on shore and accommodates all parts of OTEC plants. It requires laying down long pipes from plant site on shore to two extreme points of necessary temperature gradient. One pipe is used to collect warm ocean surface water through screened enclosure near the shore. Another long pipe lay down on the slope deep into the ocean to collect cold water.

A third pipe is used as outlet to discharge used water again in ocean via marine culture ponds deep down the ocean. Cost of pipe installation and maintenance is very expensive, and land-based plant is also very expensive. Since large electricity is used to pump water through long pipes, the net electricity reduces considerably.

Land-based OTEC plant has the advantage of savings on electrical transmission line and connectivity to electrical power grid.

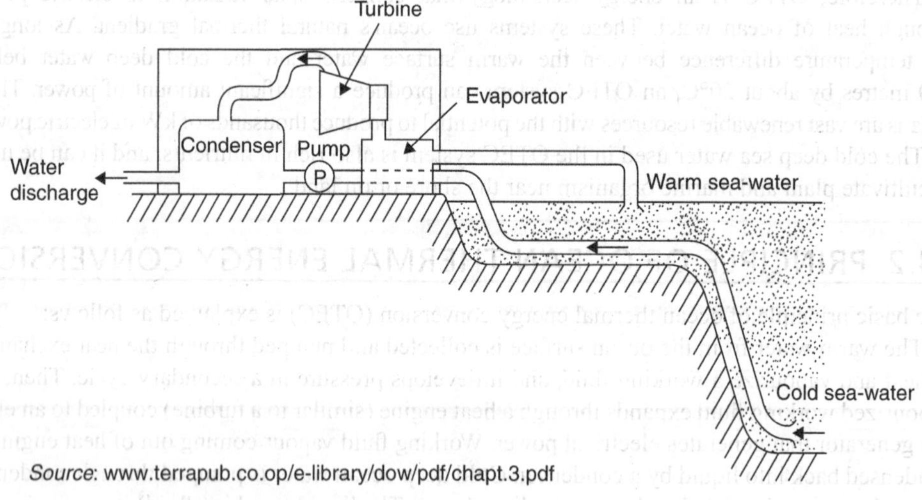

Source: www.terrapub.co.op/e-library/dow/pdf/chapt.3.pdf

Figure 13.1 Land-based OTEC power plant

13.3.2 Floating Power Plant

Floating power plant is built on a ship platform exactly where required temperature gradient sufficient for OTEC plant is available. The working principle of ocean thermal energy conversion (OTEC) is same as that of land-based power plant. Undoubtedly, the cost savings exist on piping system, but long transmission line is required to transmit electrical power from plant to sea shore.

Owing to high installation cost of long underwater power cables and its inefficiency and many other associated problems, floating OTEC plants are considered for the production of fuels, such as hydrogen, on the platform itself by the electrolysis of water.

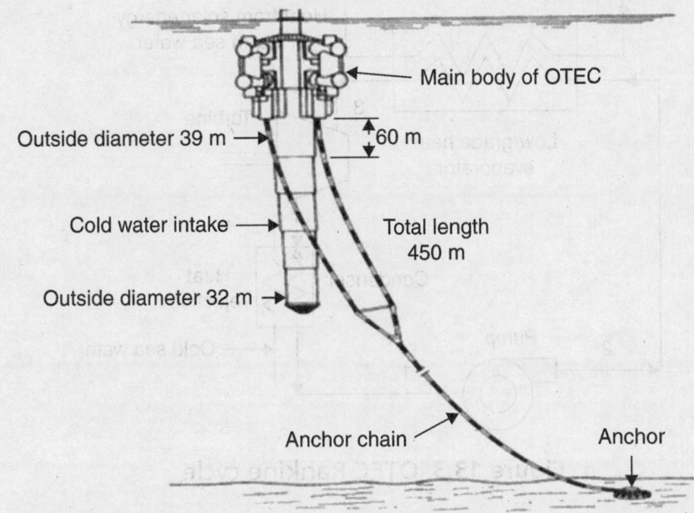

Source: http://www.britannica.com/EBchecked/topic/424415/ocean-thermal-energy-conversion-OTEC

Figure 13.2a Submersible OTEC plant designed by lockheed

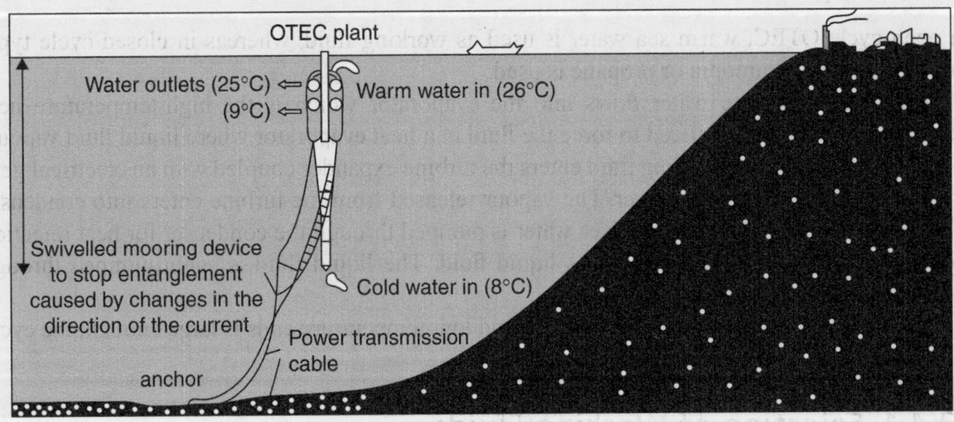

Source: http://energy.se/goran/cng/alten/proj/97/ot/ot.html

Figure 13.2b Floating OTEC power plant

Cold water pipe is the largest single item in the land-based plant design, as the slopes are seldom larger than 15° or more. If 1,000-metres-long vertical pipe with 10 to 15 m diameter used in floating plant, the length of land-based plant considering slope will be about three times.

13.4 BASIC RANKINE CYCLE AND ITS WORKING

The basic Rankine cycle shown in Figure 13.3 consists of the following:

1. An evaporator
2. A turbine expander

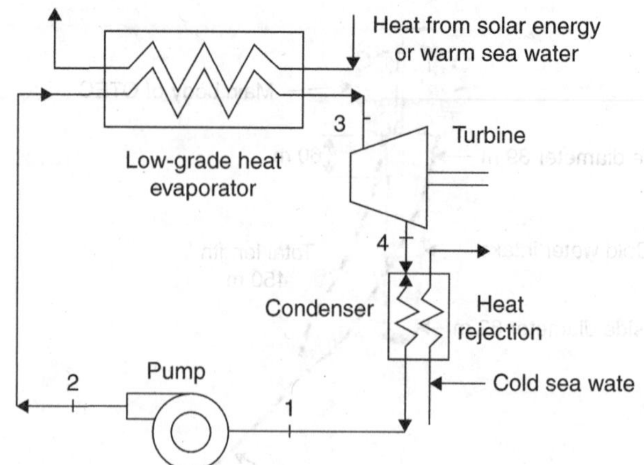

Figure 13.3 OTEC Rankine cycle

3. A condenser
4. A pump
5. A working fluid

In open-cycle OTEC, warm sea water is used as working fluid, whereas in closed-cycle type, low-boiling point ammonia or propane is used.

Warm ocean surface water flows into the evaporator which is the high-temperature heat source. A fluid pump is utilized to force the fluid in a heat evaporator where liquid fluid vapourizes. Then, the vapour of boiling fluid enters the turbine expander coupled with an electrical generator to generate electrical power. The vapour released from the turbine enters into condenser where it condenses. The cold deep sea water is pumped through the condenser for heat rejection from vapour fluid and condenses it as liquid fluid. The liquid fluid is again pumped through evaporator and cycle repeats.

As temperature difference between high- and low-temperature ends is large enough, the cycle will continue to operate and generate power.

13.4.1 Selection of Working Fluids

The steam Rankine cycle and organic Rankine cycle are the two main types used in OTEC systems, and the choice of working fluids plays an important role in design and performance of OTEC. Water is the only working fluid for steam Rankine cycle, but a large number of working fluid is available for organic Rankine cycle. The working fluid has the following properties:

1. *Chemical stability and compatibility*: Certain organic fluids are more prone to decompose when subjected to high pressure and temperature which results in material corrosion of different parts of plants, explosion etc. Thus, working fluid should be chemically stable and compatible with materials and structures of OTEC plants.
2. *Heat transfer coefficient*: Low-thermal resistance of working fluids improves heat transfer.

3. *Flash point*: A working fluid with a high flash point should be used in order to reduce flammability.
4. *Specific heat*: A working fluid with a low specific heat should be used to reduce load on the condenser.
5. *Latent heat*: A working fluid with a high latent heat should be used in order to raise the efficiency of heat recovery.
6. *Safety*: Working fluid should be non-corrosive, non-toxic, and non-inflammable having maximum allowable concentration and explosion limit for safe and efficient operation of OTEC plants.
7. *Environmental acceptability*: Low-toxicity working fluid minimizes water contamination. The environmental risk of OTEC plant is low.
8. *Cost and availability*: The ease of availability and low cost of working fluid is also important.

13.5 CLOSED CYCLE, OPEN CYCLE, AND HYBRID CYCLE

There are three types of OTEC cycle designs, namely open cycle, closed cycle, and hybrid cycle.

1. In an open cycle, warm sea water is pumped into a flash evaporator as working fluid where it boils at low pressure and converts into steam. This steam expands through low-pressure turbine which drives an electrical generator and generates electricity. The steam released from turbine condensed in a condenser by deep sea cold water as non-saline water. When non-condensable gases are separated and exhausted, the non-saline water is either pumped in marine culture ponds for freshwater applications or finally discharged in sea surface water.
2. In closed cycle, organic fluid flows in a separate closed-cycle loop called organic Rankine cycle. Warm sea surface water pumped through another pipe vapourizes working fluid in heat exchangers to drive turbine generator, The fluid vapour condenses into liquid form by deep sea water pumped in condenser by a separate pumping system, The process of pumping liquid fluid in an evaporator cycle is repeated.
3. A hybrid cycle is a combination of both closed and open cycle.

13.5.1 Open-cycle OTEC

An open-cycle OTEC uses the warm ocean surface water as working fluid. It is a non-toxic and environment friendly fluid. The major components of this system are shown in Figure 13.4. It consists of evaporator, low-pressure turbine coupled with electrical generator, condenser, marine culture ponds, non-condensable gas exhaust, and pumps. Evaporator used in an open-cycle system is a flash evaporator in which warm sea water instantly boils or flash in the chamber that has reduced pressure than atmosphere or vacuum. It results in reduced vapourization pressure of warm sea water. A large turbine is required to accommodate large volumetric flow rates of low-pressure steam, which is needed to generate electrical power, and is used with other plant components in a similar manner. During vapourization process in an evaporator, oxygen, nitrogen, and carbon dioxide dissolved in sea water are separated and are non-condensable. They are exhausted by non-condensable gas exhaust system. Condenser is used to condense vapour or

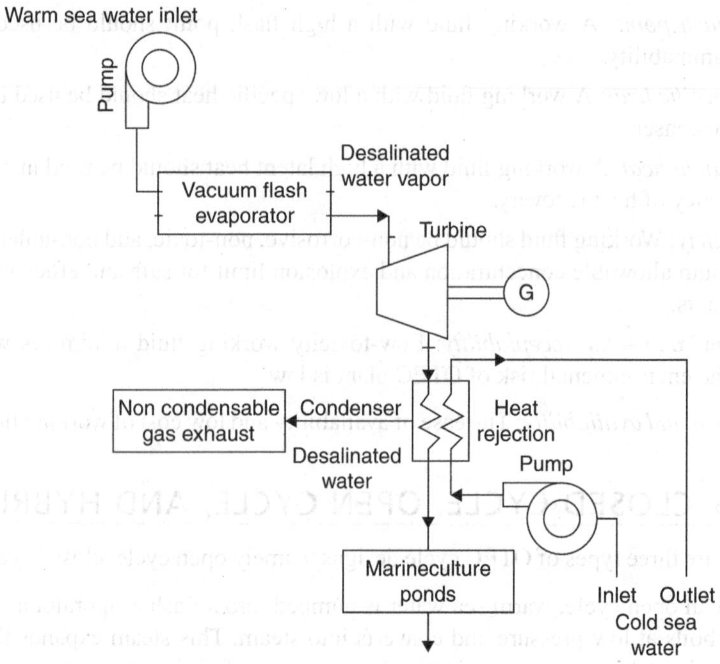

Figure 13.4 Open-cycle OTEC

steam released from steam turbine is condensed by cold deep sea water and returned back to sea. If a surface condenser is used, condensed steam (desalinated water) remains separated from cold sea water and is pumped into marine culture ponds. To avoid leakage of air in atmosphere and to prevent abnormal operation of plants, perfect sealing of all components and piping systems is essential. The working principles of open-cycle OTEC plants are explained as follows with the help of Figure 13.4.

1. The warm ocean surface water is pumped into flash evaporator where it is partially flashed into steam at a very low pressure. The remaining warm sea water is discharged into the sea.
2. The low-pressure vapour (steam) expands in turbine to drive a coupled electrical generator to produce electricity. A portion of electricity generated is consumed in plants to run pumps and for other work, and the remaining large amount of electricity is stored as net electrical power.
3. The steam with many gases (such as oxygen, nitrogen, and carbon dioxide) released from the turbine separated from sea water in an evaporator is pumped into condenser. The steam is cooled in a condenser by cold deep sea water.
4. The condensed non-saline water is discharged either directly in deep sea cold water or through the marine culture pond.
5. The non-condensable gases are compressed to pressure and exhausted simultaneously.
6. The warm ocean surface water is continuously pumped into evaporator and cycle repeats.

13.5.2 Closed-cycle OTEC

The schematic of closed-cycle OTEC is shown in Figure 13.5. It has different arrangement when compared to open-cycle OTEC. Organic fluid with low boiling point is used as working fluid. Ammonia liquid is the most widely used working fluid. Working fluid flows in a closed loop and perfectly sealed piping system. Working fluid circulates around the loop continuously. Warm ocean surface water flows through completely separate piping system and discharges in upper surface of ocean. Warm surface sea water and working fluid piping are placed very closely to each other in a heat exchanger to transfer warm sea water heat into working fluid. The cold deep sea water piping system is in contact with working fluid piping system in a condenser where working fluid condenses to its liquid state. Other components of both open- and closed-cycle OTECs are similar. Working principles of closed-cycle OTEC are as follows:

1. Working fluid is pumped through heat exchangers in a closed loop cycle which is perfectly leakage proof.
2. Warm sea surface water is pumped through separate pipe in heat exchanger in close contact with fluid closed loop cycle
3. Warm sea water transfer its heat energy to working fluid in heat exchanger and working fluid vapourizes.
4. The fluid vapour makes the turbine to rotate and drive an electrical generator to produce electricity.
5. Fluid vapour leaving the turbine is cooled and condensed as liquid fluid and is pumped again to repeat cycle.
6. Cold deep sea water is pumped through a separate pipe in condenser for providing efficient cooling of working fluid.

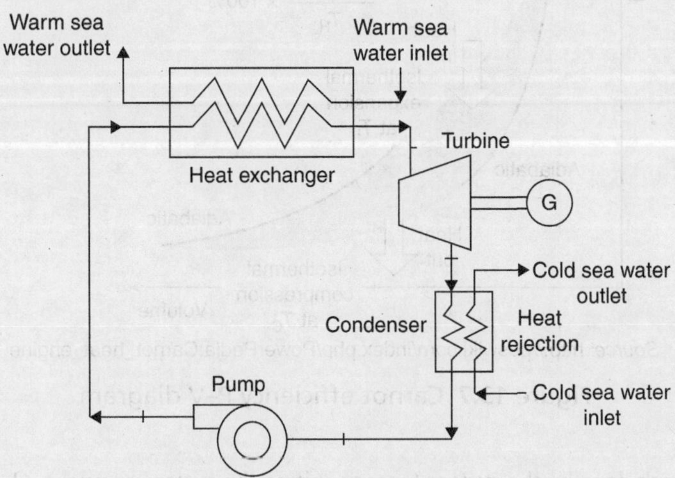

Figure 13.5 Closed-cycle OTEC plant

13.5.3 OTEC Hybrid Cycle

As shown in Figure 13.6, a hybrid cycle combines the features of both closed-cycle and open-cycle systems. Warm sea water is pumped into a vacuum chamber where it is used to flash and produces steam. Working fluid in another closed cycle loop is evaporated and vapourized by steam in vacuum chamber. The fluid vapour rotates the turbine and drive an electric generator to produce electricity.

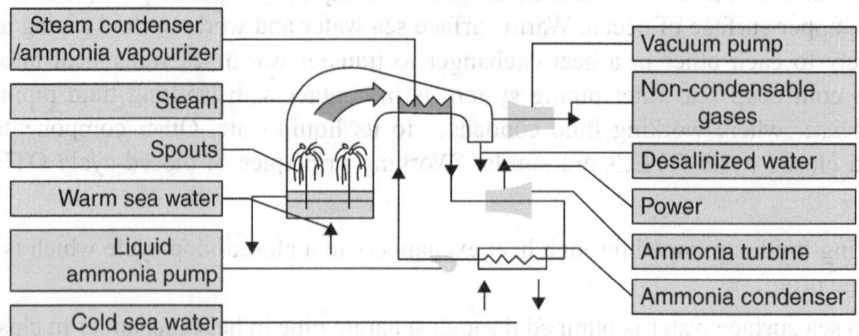

Figure 13.6 OTEC hybrid cycle

13.6 CARNOT CYCLE

The Carnot cycle is the most efficient thermodynamical cycle by exploiting the warm sea surface water and cold deep sea water.

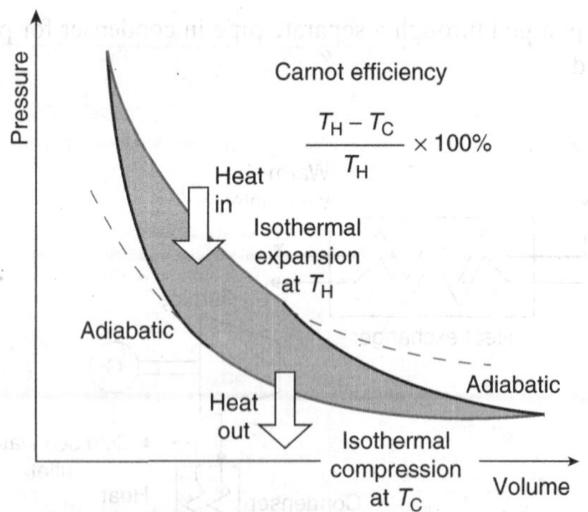

Source: http://peswiki.com/index.php/PowerPedia:Carnot_heat_engine

Figure 13.7 Carnot efficiency P–V diagram

Let W be the work done by the system (energy exiting the system as work), Q_H be the heat put into the system (heat energy entering the system),

T_C be the absolute temperature of the sea surface and
T_H be the absolute temperature of the deep sea water hot reservoir.
Carnot efficiency (η) is given by the following equation:

$$\eta = W/Q_H = 1 - T_C/T_H \qquad (13.1)$$

13.7 APPLICATION OF OTEC IN ADDITION TO PRODUCE ELECTRICITY

OTEC schematic diagram and applications are shown in Figure 13.8. Ocean thermal converting plants provide several products for use by mankind. These are explained as follows:

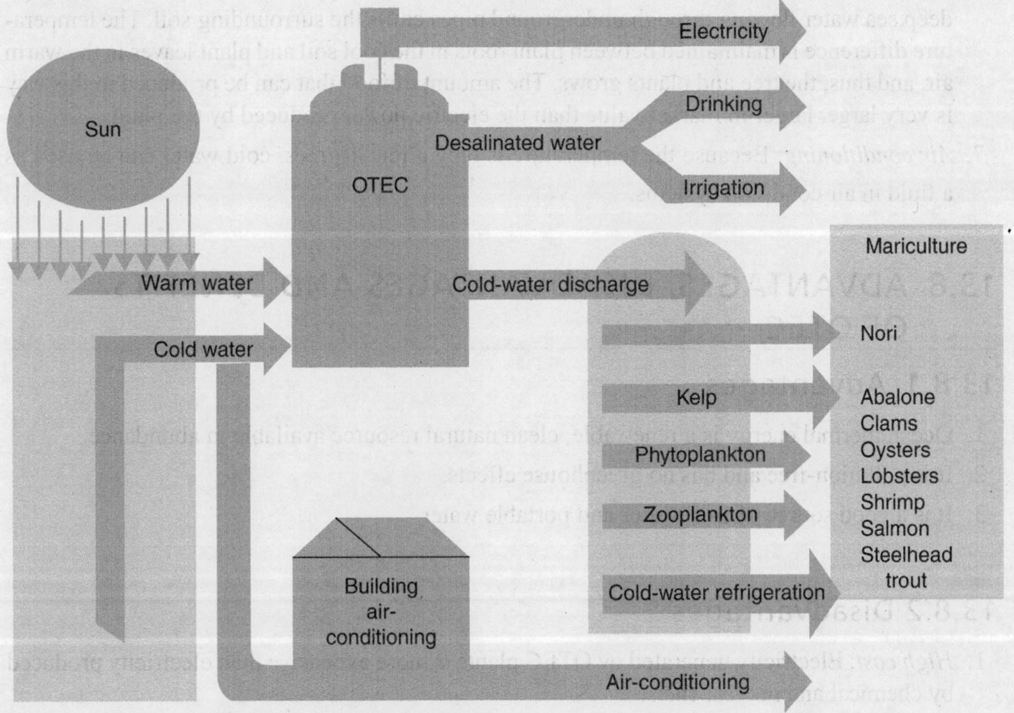

Source: http://www.homelandsecuritynewswire.com/

Figure 13.8 OTEC plant and applications

1. *Electricity*: Electrical energy is the primary product of OTEC plants. Laying down long transmission and distribution cables up to the sea shore for domestic and industrial applications is not practical from economic view point. OTEC plants are, therefore, considered for other products and applications.
2. *Hydrogen production*: Electricity produced from OTEC plants is used for separating water in hydrogen and oxygen by the method of electrolysis of water. Hydrogen is considered

as the second best usable form of energy after electricity. Use of deep sea cold water and OTEC electricity for hydrogen production signifies the important applications of OTEC plants.

3. *Ammonia and methanol production*: OTEC electricity can be used to obtain by-products, such as ammonia and methanol, that can be transported either by tankers or through pipe lines to on shore applications
4. *Desalinated water*: Desalinated water is produced in an open-cycle and hybrid-type OTEC plants through surface condenser. It is freshwater and widely used as water resource for drinking, agriculture, and industry.
5. *Aquaculture*: Nutrient-rich cold deep sea water provides sufficient environment for fish farming which may create a profitable business activities.
6. *Chilled soil agriculture*: Chilled soil agriculture is another application of OTEC plants. Cold deep sea water flowing through underground pipes chills the surrounding soil. The temperature difference is maintained between plant roots in the cool soil and plant leaves in the warm air, and thus, the tree and plants grows. The amount of food that can be produced in this way is very large, larger in market value than the electric power produced by the plant.
7. *Air conditioning*: Because the temperature is only a few degrees, cold water can be used as a fluid in air condition systems.

13.8 ADVANTAGES, DISADVANTAGES AND BENEFITS OF OTEC

13.8.1 Advantages

1. Ocean thermal energy is a renewable, clean natural resource available in abundance.
2. It is pollution-free and has no greenhouse effects.
3. It is a good source of freshwater and portable water.

13.8.2 Disadvantages

1. *High cost*: Electricity generated by OTEC plants is more expensive than electricity produced by chemical and nuclear fuels.
2. *Complexity*: OTEC plants must be located where a difference of about 20°C occurs year round. Ocean depths must be available fairly close to shore-based facilities for economic operation. Floating plant ships could provide more flexibility.
3. *Acceptability*: For the large-scale production of electricity and other products, OTEC plants are poorly acceptable due to their high costs.
4. *Ecosystem damage*: It is obvious by setting OTEC plants.
5. *Lower efficiency*: A higher temperature difference between ocean surface warm water and cold deep ocean water is required for highly efficient operation of plant.

13.8.3 Benefits as a Measure of the Value of OTEC

Economic and other benefits are the value of OTEC plants. These include the following:

1. It is a clean, renewable natural resource available in plenty.
2. It has no environmental problems and greenhouse effects.
3. It is a source of base load electricity and fuels such as hydrogen, methanol, and ammonia.
4. It provides freshwater for drinking, agriculture, and industry.
5. It encourages chilled agriculture and aquaculture.
6. Self-sufficiency, no environmental effects, and improved sanitation and nutrition are the added benefits for island.

SUMMARY

- Basic thinking behind ocean thermal energy conversion (OTEC) was first suggested in 1881 by a French physicist Jacques d'Arsonval which involves extracting useful energy from the solar heat stored in the oceans.
- Ocean thermal energy conversion is a potential source of renewable energy that creates no emissions.
- It uses the ocean's temperature difference between warm surface water and cold deep sea water to generate both electricity and potable water. (See more at: http://www.otecorporation.com/ocean-thermal-energy.)
- OTEC is an energy technology that converts solar radiation to electric power. These systems use the ocean's natural thermal gradient. As long as the temperature difference between warm surface water and cold deep sea water below 600 meters by about 20°C, an OTEC system can produce a significant amount of power.
- The cold sea water used in the OTEC process is also rich in nutrients, and it can be used to cultivate both marine organisms and plant life near the shore or on land.
- Sea water air conditioning (SWAC) is a clean method of air conditioning buildings, using cold deep sea water in place of polluting standard refrigerants.
- The main advantages of OTEC are that the method is fuel-free, has a low environmental impact, can supply pure water for both drinking and agriculture, can supply refrigeration and cooling, and can provide a coastal community with reliable energy. The disadvantages include high capital cost, potential for hostile ocean environment during construction and use, and an overall lack of familiarity with OTEC technology.
- There have been several analyses of the feasibility of full-scale implementation of OTEC. While some of these investigations are contradictory to each other, research with actual mini OTEC plants is proving that OTEC systems will one day become a feasible, efficient, and renewable source of energy.
- It is very expensive.

REVIEW QUESTIONS

1. Write a short note on 'ocean thermal energy'.
2. Explain the difference between fixed dome-type and floating dome-type biogas plant.
3. What is the basic principle of OTEC?
4. What are the main types of OTEC power plants? Describe their working principle in brief.
5. Describe the 'closed-cycle' OTEC systems. Write its advantages over 'open-cycle systems'.
6. Explain Carnot efficiency for an OTEC plant with the help of a thermodynamic cycle on a T–S plane.
7. What is the limitation of open-cycle OTEC systems?
8. State the merits and demerits of OTEC plants.
9. State the expression for energy and power in ocean waves.
10. Explain how the ocean temperature differences can be used to generate electrical power.
11. Discuss in detail OTEC systems based on (i) open cycle (ii) closed cycle.

Chapter 14

Fuel Cell

A fuel cell is a device that converts the chemical energy from a fuel into electricity through a chemical reaction with oxygen or another oxidizing agent. A fuel cell consists of two electrodes, namely an anode and a cathode. Anode is positively charged and cathode is negatively charged in an electrolytic cell. The reactions that produce electricity take place at the electrodes. Every fuel has an electrolyte, which carries electrically charged particles from one electrode to the other, and a catalyst, which speeds the reactions at the electrodes. In fuel cells, hydrogen is used as the most common fuel, but hydrocarbons, such as natural gas and alcohols like methanol, are sometimes used.

Fuel cells are different from batteries in that they require a constant source of fuel and oxygen to run, but they can produce electricity continuously as long as these inputs are supplied. In other words, it can be stated that both batteries and fuel cells are electrochemical devices. Similarly, both have a positively charged anode, a negatively charged cathode, and an ion-conducting material called an electrolyte. Fuel cells are classified by the nature of electrolyte material. Electrochemical devices generate electricity without combustion of the fuel and oxidizer, as opposed to what occurs with traditional methods of electricity generation.

Generally, fuel cell construction consists of a fuel electrode (anode) and an oxidant electrode (cathode) separated by an ion-conducting membrane. Oxygen passes over one electrode and hydrogen over the other, generating electricity, water, and heat. Fuel cells chemically combine the molecules of a fuel and oxidizer without burning or having to dispense with the inefficiencies and pollution of traditional combustion. A single fuel cell generates a tiny amount of

KEY CONCEPTS

- Schematic of fuel cell
- Basic characteristics of fuel cell
- Hydrogen–oxygen fuel cell and performance
- Schematic and working of different types of fuel cell
- Sources of overvoltage in a fuel cell
- Fuels for fuel cells
- Series–parallel connection of fuel cell

direct current (DC) electricity. In practice, many fuel cells are usually assembled into a stack. Principles are the same for cell or stack.

14.1 INTRODUCTION

A fuel cell is an electrochemical device that generates electricity from an electrochemical reaction from fuel such as hydrogen and oxygen or another oxidizing agent. Other fuels used are hydrocarbons such as natural gas and alcohols. They are similar to battery but require constant input of fuel and oxygen for continuous electricity generation.

Fuel cells are available in very small size (single fuel cell) with electricity-generating capacity of a few watts up to a big size (fuel cell stack) producing megawatts. All types of fuel cells have similar component parts in their design as they utilize two electrodes separated by a solid or liquid electrolyte that carries electrically charged particles between them. A catalyst is often used to speed up the reactions at electrodes. In addition, they are classified based on electrolyte material used. As they are static equipment and do not involve any intermediary steps of energy conversion, they are highly efficient and environmentally acceptable direct energy conversion method.

The by-products of fuel cell apart from electricity generation are water, heat, and small amount of nitrogen oxide and other emission depending upon fuel source used.

14.2 SCHEMATIC OF FUEL CELL

The schematic diagram of fuel cell is given in Figure 14.1. It consists of the following parts:

1. Anode
2. Cathode
3. Electrolyte

14.2.1 Anode

It is a positive electrode and facilitates electrochemical oxidation of fuel.

14.2.2 Cathode

It is a negative electrode and promotes electrochemical reduction of oxidant.

14.2.3 Electrolyte

It is a solution of liquid, gases with salts and carries electrically charged particles between them. Any substance having free ions to make the substance electrically conductive is called electrolyte.

When a salt, such as normal salt, is placed into a solvent, such as water, electrolyte solutions are formed. The salt dissolves into its ion components by dissociation or dissolving by the following dissociation reaction,

$$NaCl_{(s)} \rightarrow Na^{+}_{(aq)} + Cl^{-}_{(aq)} \tag{14.1}$$

Dissociation is a process of separating ionic compounds into smaller ions, and dissolving is defined as a process of breaking up of ionic crystal into ions in water.

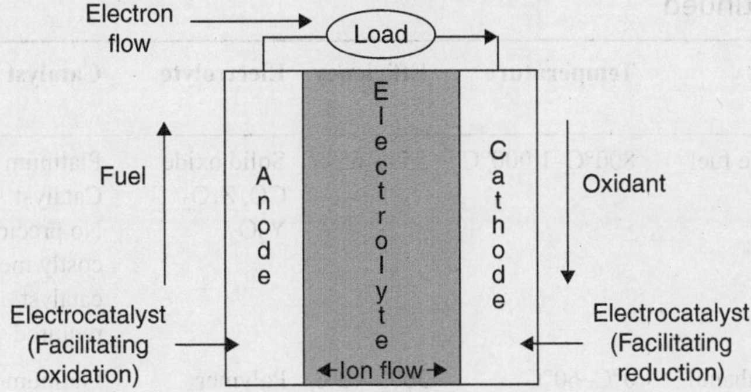

Figure 14.1 Schematic of fuel cell

Many substances, such as carbon dioxide gas, dissolve into water, react with it, and produce ions. Apart from various methods involved in the formation of electrolytes, the electrolyte liquid conducts electricity when electrodes are connected to an external load.

Electrolytes with high concentration of ion are called concentrated, whereas electrolytes with low ion concentration are called dilute.

Electrolytes also act as a barrier between fuel and oxidant, and they never mix with each other and no combustion occurs. The fuel cell efficiency is, therefore, not limited by Carnot efficiency and has very high efficiency of about 40%–60% or (extended to 80% or so if waste heat is efficiently removed). Electron generated by oxidation reaction at the anode moves through external load circuit (flow of electricity generated) to the cathode and completes the reduction reaction.

Polyelectrolytes and molten salt electrolytes have also been developed.

Ions generated during oxidation or reduction are transported from one electrode to the other through ionically conductive but electronically insulating electrolyte.

Electrolyte used determines the operating temperature and dictates the classification of fuel cell.

Table 14.1 lists the various types of fuel cells along with electrolyte used, operating temperature, and efficiency.

Table 14.1 Types of Fuel Cells

S. No.	Fuel Cells	Temperature	Efficiency	Electrolyte	Catalyst	Charge Carriers
1.	Alkaline fuel cell	60°C–90°C	50%–60%	35%–50% KOH	Nickel	Hydroxyl ions(OH^-)
2.	Polymer electrolyte fuel cell	50°C–80°C	50%–60%	Polymer membrane	Platinum	H^-
3.	Phosphoric acid fuel cell	160°C–220°C	40%–50%	Phosphoric acid	Platinum	H^-
4.	Molten carbonate fuel cell	620°C–660°C	60%–65%	Molten carbonate	Nickel	CO_2^-

(Continued)

Table 14.1 Continued

S. No.	Fuel Cells	Temperature	Efficiency	Electrolyte	Catalyst	Charge Carriers
5.	Solid oxide fuel cell	800°C–1,000°C	55%–65%	Solid oxide CO, ZrO_2/ Y_2O_3	Platinum (Pt) Catalyst No precious costly metal catalyst required	O^-
6.	Direct methanol fuel cell	0°C–60°C	30%–40%	Polymer membrane	Platinum–Ruthenium Pt–Ru	H^-

14.3 BASIC CHARACTERISTICS

The following basic characteristics together with the aforementioned general characteristics differentiate between fuel cell types:

1. Charge carrier
2. Performance degradation by contamination
3. Fuels
4. Factors affecting the fuel cell performance
5. Fuel reforming

These characteristics are explained as follows:

14.3.1 Charge Carrier

It is the ion that passes through electrolyte from the anode to cathode or vice versa. A single proton hydrogen ion (H^+) is the charge carrier for several types of fuel cell, such as PEMFC, DMFC, and PAFC but differ for many types of fuel cell, such as OH^- for AFC, CO_3^- for MCFC, and O^- for SOFC.

14.3.2 Performance Degradation by Contamination

Fuel cell contamination means the effect of impurities, and it is an important issue in fuel cell operation and applications. Severe performance degradation is caused in fuel cell because of different molecules behaviour exhibited. Sulphur-containing compounds, such as hydrogen sulphides and carbonyl sulphides are the major containment of performance degradation for all types of fuel cells. All fossil fuels are the natural sources of sulphur compounds. The presence of several other gases, such as carbon monoxide, methane, etc., adversely affects the fuel cell performances.

14.3.3 Fuels

The primary fuels used in fuel cells are hydrogen. Hydrocarbons, such as methanol, methane, and carbon monoxide, when used as a fuel have additional effects on the performance of fuel cells.

Since solid oxide fuel cell operating temperature is high, carbon monoxide (CO) is dangerous and poisonous for fuel cells, such as PEMFC, operating at low temperature. In general, electrolyte and catalyst of fuel cell determine the acceptability of gases as fuel with due regards to fuel cell performance.

14.3.4 Factors Affecting the Fuel Cell Performance

Fuel cell performance is largely affected by the following:

1. Choice and composition of electrolytes
2. Operating temperatures.
3. Geometry of fuel cell, area, and shape of electrodes and their coatings
4. Catalyst used
5. Gas pressure

14.3.5 Fuel Reforming

Fuel reforming refers to transformation of fossil fuels to hydrogen. Steam reforming is one of the methods in which steam is mixed with fossil fuels at about 750°C. The reforming reaction for natural gas composed primarily of methane (CH_4) is given as follows:

$$CH_4 + 2H_2O \rightarrow CO_2 + 4H_2 \tag{14.2}$$

Carbon monoxide (CO) is the fuel used in high-temperature fuel cell, such as MCFC and SOFC, and the water–gas shift reaction is given as follows:

$$CO + H_2O \rightarrow CO_2 + H_2 \tag{14.3}$$

The ultimate fuel is hydrogen itself.

The fuel reforming is performed immediately before it is inputted in fuel cell either on large or on intermediate or on small scales. Hydrogen is the most widely used fuels for low-temperature fuel cells, but there is neither ready availability of bulk hydrogen source and storage system nor infrastructure is available for hydrogen transportation and delivery.

The fuel reforming is also conducted at both intermediate (Gasoline station) and small scale (fuel cell-powered vehicle).

Reforming of fossil fuels produces a mixture of hydrogen and other molecular components such as nitrogen and carbon dioxide. The purity of this hydrogen will depend on developments in techniques to cost-effectively separation of H_2 from other gases. This hydrogen would probably be delivered to customers as a high-pressure gas.

In the longer term, most, if not all, of the hydrogen used to power fuel cells could be generated from renewable resources such as wind or solar energy. The electricity generated at a wind farm could be used to split water into hydrogen and oxygen. This electrolysis process would produce pure hydrogen and oxygen. If hydrogen energy technology for bulk production, pipeline transportation, and effective utilization is commercialized, hydrogen will be the energy of future.

14.4 FUEL CELL FUNCTIONALITY

For a typical fuel cell (e.g., H_2–O_2 cell) as shown in Figure 14.3, an oxidizer (oxygen of air) and a fuel (hydrogen) combine in a simple electrochemical reaction to form water and generate electricity which flows in an electrical load connected across the two electrodes. Oxidizer and fuel is continuously supplied to cathode and anode, respectively, of fuel cell. It is a quiet, efficient, and reliable source of power as there are no moving parts. An ion-conducting material is used as electrolyte that separates the two electrodes. Anode reaction takes place at the anode, and electrons and protons (hydrogen ions) are separated. The protons pass through the electrolytes, and electrons pass through external load circuit and reaches cathode. At the cathode, electrons and protons recombine to form water by cathode reaction. Electron flow in external circuit is DC that supplies power to electrical loads. These two reactions are given as follows:

At anode

$$2H_2 \rightarrow 4H^+ + 4e^- \tag{14.4}$$

At cathode

$$O_2 + 4H^+ + 4e^- \rightarrow 2H_2O \tag{14.5}$$

Overall cell reaction

$$2H_2 + O_2 \rightarrow 2H_2O \tag{14.6}$$

14.4.1 Electrical Output

The DC electrical output parameters of fuel cells are DC voltage, current, and power. They are explained as follows:

14.4.1.1 Output Voltage

A single fuel cell gives very low voltage of about 0.6 V to 0.9 V. During fuel cell operation to supply electrical load, the output voltage further reduces to a lower value because of the internal impedance drops caused in fuel cell with increase in load currents.

A large numbers of individual fuel cells are combined in a stack to obtained desire voltage.

14.4.1.2 Output Current

A single fuel cell also gives low output current which is directly proportional to the areas of electrodes.

The current carrying capability of fuel cell is increased either by increasing the size of electrodes or by making the electrodes surface porous or selecting the electrodes materials of very fine particle size.

14.4.1.3 Output Power

The output power of a typical single fuel cell is very low and is about 1 W/cm^2.

DC power of fuel cell is calculated by the following formula:

$$\text{Power (watts)} = \text{Voltage (volts)} \times \text{Current (amps)}$$

By using multiple stack of fuel cells, power output can be obtained in megawatt range.

14.5 FUEL CELLS VERSUS TRADITIONAL ELECTRICITY GENERATION

Fuel cells work on the principle of direct energy conversion in single stage to obtain electrical energy from chemical energy as follows:

$$\text{Chemical energy} \rightarrow \text{Electrical energy}$$

It does not involve the burning of fuel, but electrical energy is directly obtained from chemical energy by electrochemical process.

Traditional or conventional method of electrical energy conversion is an indirect method of energy conversion and involves intermediary conversion to energy forms and finally to electrical energy as follows:

$$\text{Chemical energy} \rightarrow \text{Heat} \rightarrow \text{Mechanical energy} \rightarrow \text{Electrical energy}$$

Fossil fuels and nuclear fuels are burnt to produce heat and high-pressure steam. The high-pressure steam expands in a turbine which starts rotating and ultimately drives the coupled electrical generator to convert mechanical energy into electrical energy.

14.6 PERFORMANCES OF FUEL CELLS VERSUS OTHERS

The single-stage direct energy conversion of chemical energy to electrical energy in fuel cell is a highly efficient process than three-stage conversion of chemical energy into electrical energy in traditional methods that involve combustion or burnings.

The energy release by chemical reaction in fuel cell is a function of change in Gibbs free energy. The Carnot law and efficiency of heat engine is not applicable to fuel cell. The maximum theoretical efficiency of fuel cells is much higher than conventional indirect methods of conversion

Fuel cells have less environmental damage than conventional methods of electrical generation.

14.7 FUEL CELL CONSTRUCTION (HISTORICAL DEVELOPMENT)

The conceptualization of a fuel cell was demonstrated by Humphry Davy in 1801. This was followed by pioneering work of Christian Friedrich Schönbein in 1838 and discovered the operating principle of fuel cells. In 1839, William Grove, a chemist, physicist, and lawyer, developed a crude device known as gas voltaic battery. He conducted a series of experiments on his gas voltaic battery and proved that electric current could be produced from an electrochemical reaction between hydrogen and oxygen over a platinum catalyst. The term fuel cell was first used in 1889 by Charles Langer and Ludwig Mond, who researched fuel cells using coal gas as a fuel.

Fuel cells are widely used as main electrical energy source or backup power for commercial, industrial, and residential buildings and in remote or inaccessible areas. They also find wide application in automobiles, buses, forklifts, airplanes, boats, motorcycles, and submarines.

14.7.1 William Grove's Fuel Cell

In 1839, the operating principle of fuel cell was discovered by a German scientist Christian Friedrich Schonbein. His principle was improved into a practical shape by Sir William Robert Grove of Walsh by developing the first fuel cell prototype, namely gas voltaic cell, in February 1839. The schematic diagram of his fuel cell stack, as sketch by himself, is shown in Figure 14.2, which is very similar to the present phosphoric acid fuel cell.

The above original fuel cell design was improved by W. Thomas Grubb in 1955 by using sulphonated polystyrene ion-exchange membrane as the electrolyte.

In 1991, the first hydrogen–oxygen fuel cell automobile was developed by Roger Billings.

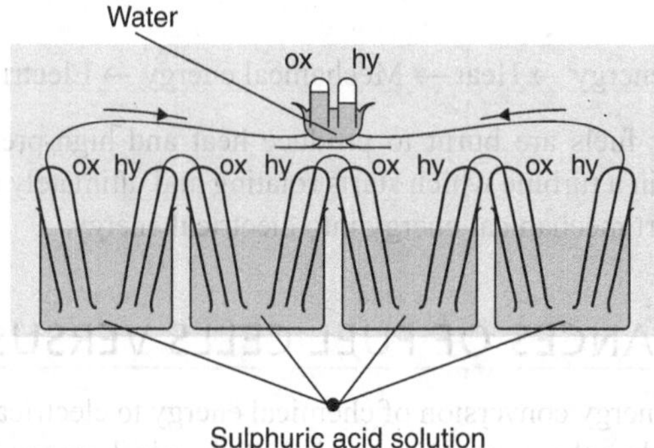

Figure 14.2 Sketch of William Grove's fuel cell

14.7.2 Hydrogen–Oxygen Fuel Cell

This type of fuel cells was developed in 1960. As the electrolyte used for this device is aqueous alkaline solution such as potassium hydroxide, the procedure for electricity consumption is rather expensive. They were used in Apollo space programme for producing drinking water in addition to electricity. The operating efficiency of this fuel cell is higher of the order of 60%. Although they are inexpensive, their power output is low and the catalyst is degraded by the carbon dioxide present in the atmosphere. The basic construction and operating principle of the fuel cell is illustrated in Figure 14.3.

Hydrogen is continuously routed to anode electrode where it splits into proton ions (H^+) and electrons (e^-) on the catalyst layer by the anode reaction given by the Eq. (14.4). The electrons flow out through the external load circuit to generate electricity and reaches cathode electrode. The proton ions separated at anode by anode reaction move through potassium hydroxide (KOH) electrolyte to the cathode electrode, where oxidizer (oxygen or air) is continuously supplied.

At the cathode, oxygen, electrons, and protons recombine to form water by cathode reaction given by the Eqs (14.5) and (14.6).

Therefore, a fuel cell combines hydrogen and oxygen to produce electricity. Water is the only by-product of this cell.

The theoretical open circuit voltage of hydrogen–oxygen fuel cell is 1.23 V at 25°C and considering internal voltage drop of fuel cell it reduces to about 1 V. It further reduces to value in between 0.5 to 0.8 under loaded condition.

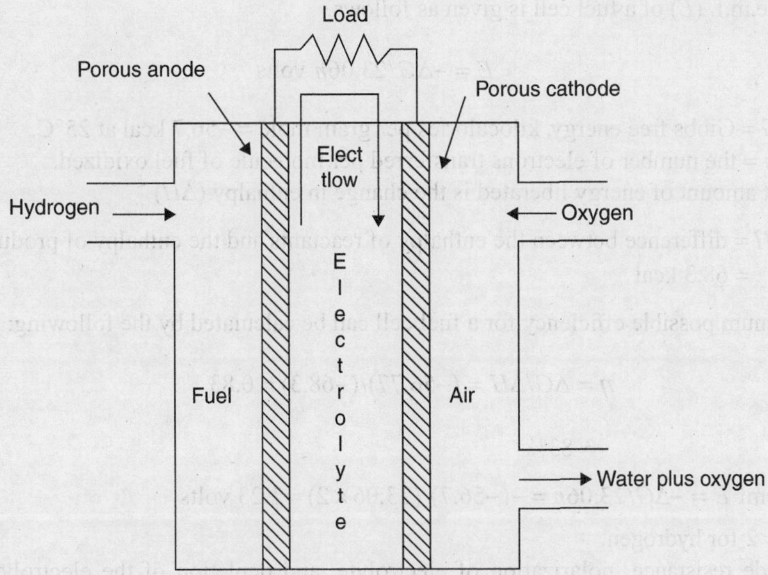

Figure 14.3 Hydrogen–oxygen fuel cell

14.7.2.1 EMF of Fuel Cell

In hydrogen–oxygen fuel cell, hydrogen is the fuel and oxygen is the oxidant. Electrodes are porous and are connected to the load. At anode, the hydrogen fuel split into positive hydrogen ions and electrons as per the following chemical reaction:

$$H_2 \rightarrow 2H^+ + 2e^- \quad (14.7)$$

After flowing through the load circuit, electrons and hydrogen ions combine with oxygen at the cathode as per the following reaction:

$$2H^+ + 2e^- + 1/2O_2 \rightarrow H_2O \quad (14.8)$$

The overall cell reaction is given as follows:

$$2H_2 + O_2 \rightarrow H_2O \quad (14.9)$$

The oxidation product is water, which is removed from the cell.

The input/output model of a fuel cell is shown in Figure 14.4. The energy change theoretically available from the fuel cell is ΔG, which is referred to as Gibbs free energy charge.

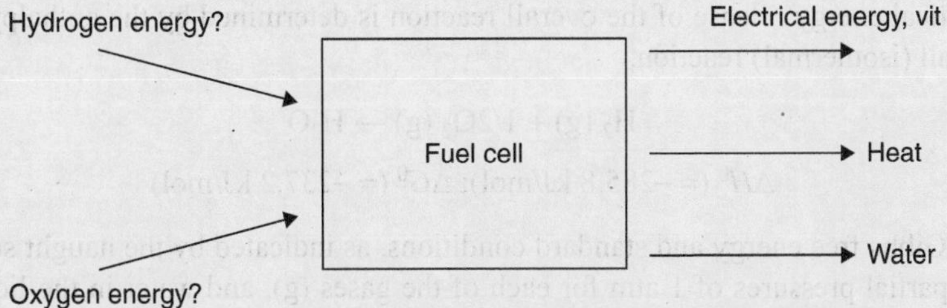

Figure 14.4 Input/output model of a fuel cell

The ideal e.m.f. (E) of a fuel cell is given as follows.

$$E = -\Delta G/23.06n \text{ volts} \quad (14.10)$$

Where ΔG = Gibbs free energy, kilocalories per gram mole = –56.7 kcal at 25°C.
n = the number of electrons transferred per molecule of fuel oxidized.
The net amount of energy liberated is the change in enthalpy (ΔH)

ΔH = difference between the enthalpy of reactants and the enthalpy of products.
= 68.3 kcal

The maximum possible efficiency for a fuel cell can be calculated by the following:

$$\eta = \Delta G/\Delta H = (-56.77)/(-68.3) = 0.83$$

$$= 83\%$$

And cell emf $E = -\Delta G/23.06n = -(-56.7)/(23.06 \times 2) = 1.23$ volts

Where $n = 2$ for hydrogen.

Electrode resistance, polarization of electrolyte, and depletion of the electrolyte limit the actual voltage produced.

14.7.2.2 Moles–g mole and kg mole

Mole is defined as a measure of the 'amount' of a substance that takes into account its molar mass.

The molar mass of hydrogen is 2.01588 ± 0.00014 g/mol (say 2) and of water is 18.01528 ± 0.00044 g/mol (say 18)

A mole of any substance always has the same number of entities (e.g. molecules)

$$= 6.022 \times 10^{23}$$

It is called Avogadro's number. This is represented by the letter N.

A 'mole of electrons' = 6.022×10^{23} electrons.

The charge on one electron (e) = 1.602×10^{-19} C

The quantity (N, e) is called Faraday constant and is designated by the letter F.

$$F = N \times e = [6.022 \times 10^{23} \times 1.602 \times 10^{-19}] = 96{,}485 \text{ Coulomb}$$

The theoretical energy release of the overall reaction is determined by the enthalpy change ΔH in the overall (isothermal) reaction,

$$H_2\,(g) + 1/2O_2\,(g) \rightarrow H_2O \quad (14.11)$$

$$\Delta H^0\ (= -285.8 \text{ kJ/mol});\ \Delta G^0\ (= -237.2 \text{ kJ/mol})$$

ΔG^0 is the Gibbs free energy and standard conditions, as indicated by the naught superscript, at T = 25°C, partial pressures of 1 atm for each of the gases (g), and water in the liquid state (l).

This last distinction is important. For high-temperature fuel cell, water remains in vapour form and the lower heating value is used as follows:

$$\Delta H = -241.8 \text{ kJ/mol}$$

and $$\Delta G = -228.6 \text{ kJ/mol.}$$

The higher heating value is used for the fuel cells operating below the boiling point of water.

$$\Delta G = \Delta H - T \times \Delta S$$

The standard change in entropy is –0.163 kJ/mol/K. An energy balance on a fuel cell shows that

$$d/dt(dQ + dW_{\text{elect}}) = d/dt(dH + dKE + dPE) \tag{14.12}$$

For steady-state operation and neglecting changes in kinetic energy (*dKE*) and potential energy (*dPE*), Eq. (14.12) is modified as follows.

$$d(Q) = T \times d(S) \tag{14.13}$$

Therefore, $W_{\text{elec,}}$ the electrical power output, is

$$W_{\text{elect}} = \Delta H - T \cdot \Delta S = \Delta G \tag{14.14}$$

The fuel cell efficiency is defined as the ratio of maximum work output to the enthalpy input and is given as follows.

$$\eta_{\text{FC}} = \Delta G / \Delta H \tag{14.15}$$

Using the standard free energy and enthalpy given previously

$$\Delta G = -237.2 \text{ kJ/mole and } \Delta H = -285.8 \text{ kJ/mole}$$

It shows the maximum thermodynamic efficiency under standard conditions = [–237.2 kJ/mole]/[–285.8 kJ/mole] = 83%.

This reversible potential changes with changing pressure as the Nernst equation.

14.7.2.3 Derivation of Ideal Fuel Cell Voltage

$$\Delta G = \Delta H - T \cdot S \tag{14.16}$$

$$= 285{,}800 \text{ J} - (298 \text{ K})(-163.2 \text{ J/K})$$

$$= 237{,}200 \text{ J}$$

$$\Delta E = -\Delta G / n \cdot F \tag{14.17}$$

$$= -237{,}200 \text{ J}/(2 \times 96{,}487 \text{ J/V})$$

$$= -1.23 \text{ V}$$

For the hydrogen/air fuel cell at 1 atmosphere pressure and 25°C (298 K), the cell voltage is 1.23 V.

As temperature rises from room temperature to that of an operating fuel cell (80°C or 353 K), the values of *H* and *S* change only slightly, but *T* changes by 55°C. Therefore, the absolute value of ΔG decreases. For a good estimation, assuming no change in the values of *H* and *S*.

$$\Delta G = -285{,}800 \text{ J/mol } - (353 \text{ K})(-163.2 \text{ J/mol.K})$$
$$= -228{,}200 \text{ J/mol}$$

$$\therefore \quad \Delta E = -(-228{,}200 \text{ J}/2 \times 96{,}487 \text{ J/V})$$
$$= 1.18 \text{ V}$$

Therefore, the maximum cell voltage decreases as well (for the standard case of 1 atm), from 1.23 V at 25°C to 1.18 V at 80°C

A further reduction in the cell voltage will take place if oxygen is not pure and humidified air is used as oxidizer and also hydrogen is not a dry gas.

14.7.2.4 Efficiency

The thermal efficiency of fuel cell is given by the ratio of useful electrical energy produced to the enthalpy formation of fuel and is given by Eq. (14.15) and rewritten as follows.

$$\eta_{FC} = \Delta G/\Delta H \tag{14.18}$$

where ΔG is change in Gibbs free energy and ΔH is the enthalpy of formation of the reaction.

In practice, the actual efficiency of fuel cell is given as follows.

$$\eta = 0.83 \times (V_{CELL}/V_{IDEAL}) \tag{14.19}$$

where V_{CELL} = The actual voltage of fuel cell; V_{IDEAL} = The voltage obtained from Gibbs free energy for the ideal case and 0.83 is the theoretical maximum efficiency of fuel cell.

Example 14.1

A 100 cm^2 fuel cell is operating, under typical conditions of 1 atmosphere pressure and 80°C, at 0.7 V and generating 0.6 A/cm^2 of current, for a total current of 60 A. Calculate the excess heat generated.

Solution The excess heat generated by this cell can be estimated as follows:

Power due to heat = Total power generated – Electrical power

$$P_{heat} = P_{total} - P_{electrical} \tag{14.20}$$
$$= (V_{IDEAL} \times I_{CELL}) - (V_{CELL} \times I_{CELL})$$
$$= (V_{IDEAL} - I_{CELL}) \times I_{CELL}$$
$$= (1.16 \text{ V} - 0.7 \text{ V}) \times 60 \text{ A}$$
$$= 0.46 \text{ V} \times 60 \text{ coulombs/sec} \times 60 \text{ sec/min}$$
$$= 1{,}650 \text{ J/min}$$

This cell is generating about 1.7 kJ of excess heat every minute it operates, while generating about 2.5 kJ of electric energy per minute.

14.7.2.5 Degradation of Fuel Cell Performance

Fuel cell performance degradation is caused by primary irreversible losses known as polarization. The effects of irreversible losses on a typical low-temperature hydrogen–oxygen fuel cell is shown in Figure 14.5.

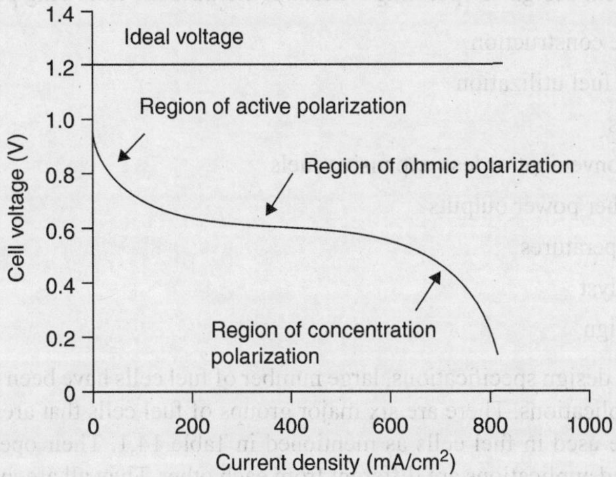

Figure 14.5 Ideal/actual voltage and current characteristics of typical fuel cell

14.7.2.6 Polarization

When electric current passes through the fuel cell, it deviates the electrochemical process from equilibrium. This phenomenon is known as polarization. It occurs either at cathode (a common polarization) or at the anode. The three types of polarization are as follows:

1. Activation polarization
2. Ohmic polarization
3. Concentration polarization

They are explained below.

1. *Activation polarization*: It is the potential difference beyond the value of equilibrium needed to generate currents depending on the energy activation. It is also referred as the activation energy that is required to have electrons transferred from electrodes into analyte. Reaction rate is limited at electrodes and it is dominant at low current density.
2. *Ohmic polarization*: It is the potential drop due to either the high resistivity of the electrolyte surrounding the electrode or an insulation effect of the film on the electrode surface formed by the reaction products. This loss is directly proportional to the current density.
3. *Concentration polarization*: Concentration polarization of an electrode is a result of formation of a diffusion layer adjacent to the electrode surface where there is a gradient of the ion concentration. It is caused by a loss of concentration of the fuel or oxidant at the surface of the electrodes. These losses are present over the entire current density range but become prevalent at high limiting currents where it becomes difficult to provide enough reactant flow to the cell reaction sites

14.8 TYPES OF FUEL CELLS: DESIGN

As already discussed in Chapter 1 (NCER Overview), fuel cells come in many varieties; however, they all work in the same general manner.

A range of fuel cell designs using the necessary electrochemical conversion has been developed to meet different design or operating criteria. It includes the following performance criteria:

1. Less expensive construction
2. More efficient fuel utilization
3. Faster start-ups
4. Use of more convenient or less expensive fuels
5. Achieving higher power outputs
6. Operating temperatures
7. Choice of catalyst
8. Electrodes design

Based on the above design specifications, large number of fuel cells have been developed and used for commercial applications. There are six major groups of fuel cells that are classified based on fuel and electrolyte used in fuel cells as mentioned in Table 14.1. Their operating temperature, costs, efficiency, and applications are different from each other. They all are, however, made up of three common segments that are sandwiched together and placed in a tank. They are as follows:

1. Anode
2. Electrolyte
3. Cathode

14.9 SCHEMATIC AND WORKING OF DIFFERENT TYPES OF FUEL CELL

14.9.1 Alkali Fuel Cells

Alkali fuel cell is shown in Figure 14.6. It operates on compressed hydrogen and oxygen. They use aqueous electrolytes of potassium hydroxide. They were some of the earliest practical cells developed in 1960 and were used in the Apollo space programme for producing drinking water and generating electrical power. Although they are inexpensive compared with PEM cells, operating efficiencies of 60% are possible. Unfortunately, they have a low power output, and the catalyst is prone to poisoning from carbon dioxide in the atmosphere. They require pure hydrogen as a fuels and expensive platinum electrode catalyst. Power output of these cell stacks ranges from about 300 W to 5 kW.

The working and performance evaluation of these fuel cells are presented in Section 14.7. A few salient features of alkaline cells are as follows:

1. There is an excess of OH^+ ions and these play a key role in the reaction in cell.
2. At the anode, hydrogen gas reacts with OH^+ ions, producing water and releasing electrons.
3. The electrons go round the external electrical circuit and reach the cathode.
4. At the cathode, they react with the oxygen and water, thereby producing more OH^+ ions.

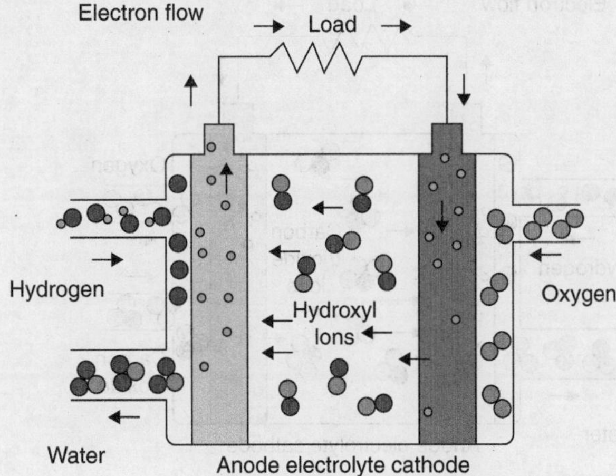

Figure 14.6 Alkali fuel cell

5. The OH^+ ions move through electrolyte and electrons move round the electric circuit.
6. Water is produced at the anode twice as fast as it is used at the anode.

Fuel cell reaction is given as follows:

Anode reaction : $2H_2 + 4OH^+ \rightarrow 4H_2O + 4e^-$
Cathode reaction : $O_2 + 2H_2O + 4e^- \rightarrow 4OH^+$
Cell reaction : $2H_2 + O_2 \rightarrow 2H_2O$

14.9.2 Molten Carbonate Fuel Cells

Molten carbonate fuel cell (MCFC) is represented in Figure 14.7. It uses high-temperature compounds of salt (such as sodium or magnesium) carbonates (chemically, CO_3) as the electrolyte. Efficiency ranges from 60% to 80%, and operating temperature is about 650°C.

1. When heated to about above temperature, these salts melt and become conductive to carbonate ions (CO_3^-).
2. These ions flow from the cathode to the anode.
3. At anode, they combine with hydrogen to give electrons, carbon dioxide, and water.
4. Electrons flow through external circuit and reaches to cathode.
5. Therfore, electricity is generated and heat is produced.

The chemical reaction taking place in the fuel cell is given as follows:

Anode reaction : $CO_3^- + H_2 \rightarrow H_2O + CO_2 + 2e^-$
Cathode reaction : $CO_2 + ½O_2 + 2e^- \rightarrow CO_3^{--}$
Cell reaction : H_2 (g) + ½ O_2 (g) + CO_2 (Cathode)
$\rightarrow H_2O$ (l) + CO_2 (Anode)

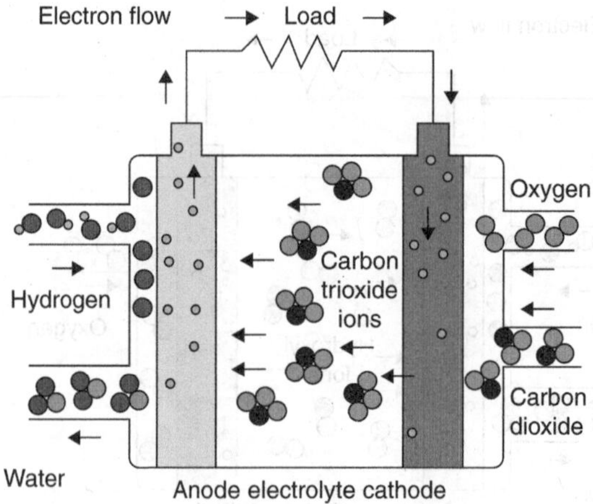

Figure 14.7 Molten carbonate fuel cell

Molten alkaline carbonate such as sodium bicarbonate is used as the electrolyte. The nickel electrode catalysts are inexpensive compared to the platinum used in other cells. Their unique chemistry needs carbon dioxide from the air as a part of the process. They can produce high powers up to 100 MW and can also be operated at high temperatures range of 650°C to 1,000°C. Because of their high working temperature, they can operate directly with hydrocarbon gases which are reformed that less expensive within the cell and do not need a separate hydrogen supply. They are not so expensive in production and hence can be used for commercial uses. They have an efficiency of about 45%–55%.

14.9.3 Phosphoric Acid Fuel Cells

Phosphoric acid fuel cells (PAFC) shown in Figure 14.8 use phosphoric acid (H_3PO_4) as the electrolyte. The ionic conductivity of phosphoric acid is poor at low temperature. Other salient features are as follows:

1. The charge carrier is hydrogen ions.
2. At anode, hydrogen molecule is split into hydrogen ions (protons) and electrons.
3. The electrons flow through external circuit and produce electric power.
4. On the other hand, protons travel through electrolyte and combine with oxygen (usually from air) at the cathode.
5. At cathode, water is released.
6. Reactions in this fuel cell gives electricity and generate heat.

In this cell, following chemical reaction takes place.
At anode

$$2H_2 \rightarrow 4H + 4e^-$$

At cathode

$$O_2 + 4H^+ + 4e^- \rightarrow 2H_2O$$

Cell reaction

$$2H_2 + O_2 \rightarrow 2H_2O$$

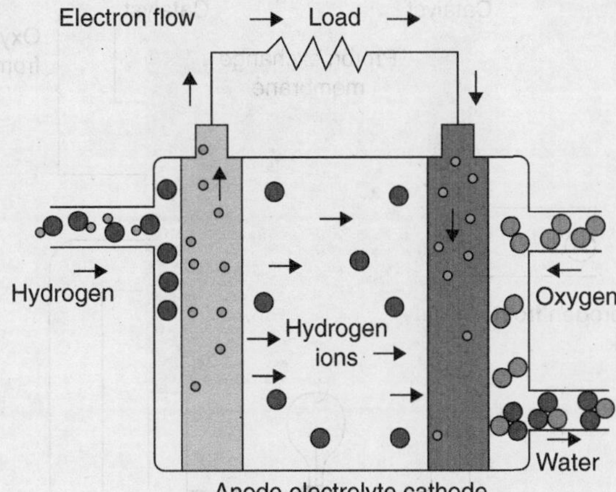

Figure 14.8 Phosphoric acid and PEM fuel cell

Their name is derived from the high concentration phosphoric acid, which is used as the electrolyte material, and they became the first fuel cells to be sold commercially in the 1970s. The efficiency is in the range of 40%–80%, and they operate at temperatures up to 220°C. Unlike other fuel cells that operate at this temperature, they are not susceptible to small amounts of carbon dioxide impurities in the hydrogen fuel. This makes them better suited for large-scale usage. The primary application of these fuel cells is power supply for individual buildings, such as small businesses, hospitals, nursing homes, hotels, schools, and utility. Their advantages include high efficiency, low susceptibility to carbon dioxide poisoning, and stability over extended periods of time. The main disadvantage is that start-up is difficult because phosphoric acid is solid at room temperature but liquid at operating temperature.

14.9.4 Proton Exchange Membrane Fuel Cells

They have been in development since the 1960s and are available in the stack of output range of 50–250 kW and operating efficiency of 40%–50%. As they operate at about 100°C, they widely preferred fuel cell for the automobile industry and hand-held equipments. Low operating temperature allows for much faster start-up time. Advantages include fast start-up, compact design, and easy sealing. The main disadvantage is that cogeneration is not beneficial at such a low operating temperature. In addition, platinum catalyst used on both sides of the membrane, however, increases the cost of fuel cells.

Schematic arrangement of hydrogen–oxygen proton-exchange membrane fuel cell (PEMFC) is given in Figure 14.9. Its working principle can be understood as follows:

1. Hydrogen is pumped to anode electrode where it diffuses to the anode catalyst and dissociates into protons and electrons.
2. Oxygen is pumped into cathode from air for reaction with hydrogen.
3. Multi-facilitated proton membranes are created when protons react with oxidants. The protons are conducted through the membrane to the cathode.

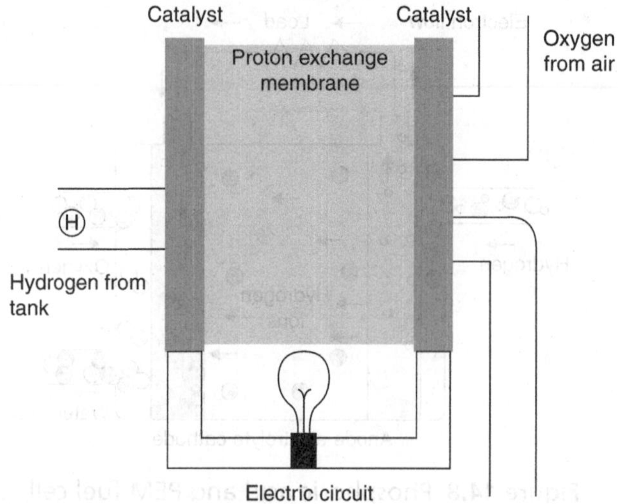

Figure 14.9 Proton-exchange membrane fuel cell

4. Electrically insulating property of membrane forced the electrons to flow through external load circuit and go to cathode.
5. Oxygen molecules with electrons and protons to form water at cathode electrode.

The chemical reaction of cell is as follows:

At anode

$$2H_2 \rightarrow 4H^+ + 4e^-$$

At cathode

$$O_2 + 4H^+ + 4e^- \rightarrow 2H_2O$$

Cell reaction

$$2H_2 + O_2 \rightarrow 2H_2O$$

14.9.5 Solid Oxide Fuel Cells

Schematic arrangement of solid oxide fuel cells (SOFC) is shown in Figure 14.10.

It uses calcium or zirconium oxide (hard ceramic compound of metal) as electrolyte and oxygen as charge carrier. The oxygen from air joins with electrons in the cathode to create oxygen anions. The oxygen ions are transported through the solid electrolyte to the anode. At anode, the oxygen ions combine with hydrogen to create water and release electrons. The electrons go through the circuit to power the load and then go back to the cathode to restart the cycle. Advantages include high efficiency when combined with a turbine and a high tolerance to impurities in the fuel. Disadvantages include expensive materials and failure of seal materials at the high operating temperatures. They deliver power output up to about 100 kW at the operating efficiency of about 60%. They are used in big, high-power industrial applications and in large scale, central electricity-generating stations.

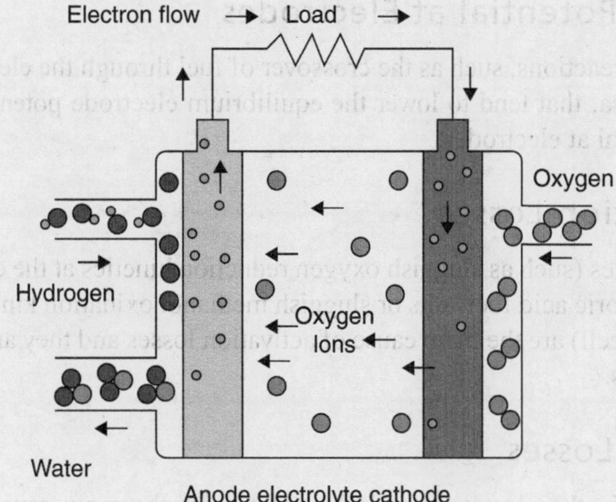

Figure 14.10 Solid oxide fuel cell

Its working principle is explained as follows:

1. Oxygen molecules from air are broken into oxygen ions with the addition of electrons at the cathode electrode.
2. The oxygen ions are conducted through electrolytes and reach the anode.
3. At anode, oxygen ion combines with hydrogen and electrons are released.
4. Free electrons travel through external load circuit and produce electric power and heat.

Chemical reaction taking place in fuel cell is given as follows:

At anode

$$2H_2 + 2O^- \rightarrow 2H_2O + 4e^-$$

At cathode

$$O_2 + 4e^- \rightarrow 2O^-$$

Cell reaction

$$2H_2 + O_2 \rightarrow H_2O$$

14.10 SOURCES OF OVERVOLTAGE IN A FUEL CELL

Prominent sources of overvoltage in a fuel cell are as follows:

1. Mixed potential electrodes
2. Activation losses
3. Ohmic losses
4. Mass transport losses

14.10.1 Mixed Potential at Electrodes

Unavoidable parasitic reactions, such as the crossover of fuel through the electrolyte from anode to cathode or vice versa, that tend to lower the equilibrium electrode potential are the primary cause of mixed potential at electrodes.

14.10.2 Activation Losses

Kinetics at the electrodes (such as sluggish oxygen reduction kinetics at the cathodes of polymer electrolyte and phosphoric acid fuel cells or sluggish methanol oxidation kinetics at the anode of a direct methanol fuel cell) are the main cause of activation losses and they are more pronounced at low current densities.

14.10.3 Ohmic Losses

The resistance losses in the electrodes and electrolytes are the main causes of ohmic losses. Intermediate current densities largely increase losses.

14.10.4 Mass Transport Losses

Mass transportation losses are highly pronounced at high current densities. They occur either by reacting diffusion in the electrode layers or by non-reacting diffusion in the gas diffusion layer.

14.11 FUELS FOR FUEL CELLS

The primary fuels, given in Table 14.2, that can be directly utilized within fuel cell stacks depending upon the fuel cell types are as follows:

1. Hydrogen
2. Carbon monoxide
3. Methanol
4. Dilute light hydrocarbons such as methane

Table 14.2 Gaseous Reactants and Their Effects on Fuel Cells

Gas	PAFC	MCFC	SOFC	PEFC
H_2	Fuel	Fuel	Fuel	Fuel
CO	Poison (>0.5%)	Fuel[a]	Fuel	Poison (>10ppm)
CH_4	Diluent	Diluent[b]	Fuel	Diluent
CO_2 & H_2O	Diluent	Diluent	Diluent	Diluent
S as (H_2S & COS)	Poison (>50 ppm)	Poison (>0.5 ppm)	Poison (>1.0 ppm)	Nothing available

[a]In reality, CO, with H_2O, shifts to H_2 and CO_2, and CH_4, with H_2O, reforms to H_2 and CO faster than reacting as a fuel at the electrode.

[b]A fuel in the internal reforming in MCFC.

It is worthwhile or not that

1. Hydrogen is the main fuel for majority of fuel cells. Methanol is also used in a few cells (direct methanol fuel cell) as a fuel.
2. Carbon monoxide is also used as a fuel in high-temperature fuel cells (such as MCFC and SOFC). However, CO is a poison for low-temperature fuel cells.
3. Carbon monoxide is consumed in the gas phase through the water gas shift reaction, and hydrogen, thus, formed is used in electrochemical process. The water gas shift reaction is given as follows:

$$CO + H_2O \rightarrow CO_2 + H_2$$

4. Dilute concentration of methane can be oxidized directly using a solid oxide fuel cell; however, high concentrations of CH_4 lead to severe coking problems. Methane is reformed within the cell by steam reforming.

14.12 SERIES PARALLEL CONNECTION OF FUEL CELL

A single-stack fuel cell can produce an emf of about 1.25 V at 1 atm and at 25°C temperature. However, it is possible to obtain 100 V to 1,000 V and power level of 1 kW to 100 MW by connecting a number of cells in series (for voltage increase) and in parallel (for current increase).

The current depends upon the physical size of the cell. The output of the fuel cell varies directly with pressure. Therefore, for increasing cell output, the gas pressure is increased. The optimum size of fuel cell currently is 0.027 m^2/kW.

14.13 ADVANTAGES OF FUEL CELL

The main advantages of fuel cells are as follows:

1. It is highly reliable, compact in size, light weight, long operational operating life and has no moving parts.
2. This has the lowest pollution rate (almost pollution-free) when compared to batteries and gasoline-powered devices.
3. Efficiency of fuel cells are higher (60%–85%) as compared to battery, combustion engines, and other power-converting equipments.
4. Modular installations are very flexible and can be used to match the load and increase reliability of the system.

14.14 DISADVANTAGES OF FUEL CELLS

The main disadvantages are as follows:

1. High cost
2. Less durability and bad infrastructure
3. Available sources of fuel for fuel cells

SUMMARY

- The principle of fuel cell was discovered in 1839 by Sir William Grove who is acknowledged as the father of the fuel cell.
- Fuel cells cleanly and efficiently convert chemical energy from hydrogen-rich fuels into electrical power and usable high quality heat in an electrochemical process that is virtually in the absence of pollutants.
- Since there is no combusting of fuel, virtually no harmful emissions are generated by the fuel cells. This results in power production that is almost entirely in the absence of nitrogen oxide (NO_2), sulphur dioxide (SO_2) or particulate matter (PM).
- Fuel cells operate much such as a battery, except they do not require electrical recharging.
- Fuel cells can generate power almost indefinitely, as long as they have fuel to use.
- Fuel cells are versatile and scalable and provide everything from milliwatts to megawatts of power in a variety of uses ranging from cell phones to cars and to entire neighbourhood areas.
- Fuel cells are used presently into three broad areas of portable power generation, stationary power generation, and power for transportation.
- Each fuel cell by itself does not produce very much power, but the voltage provided by each fuel cell individually adds up, yielding a voltage (and a power) that is large enough for practical applications.
- Fuel cells are reliable, efficient, and have no moving parts.
- An electrolyte is any substance containing free ions that make the substance electrically conductive. The most typical electrolyte is an ionic solution, but molten electrolytes and solid electrolytes are also possible.

REVIEW QUESTIONS

1. What is a fuel cell? Mention the basic principle of fuel cells with the reference to H_2–O_2 fuel cells.
2. Explain the essential features of a hydrogen–oxygen cell. Draw a suitable diagram of this cell and give the reactions took place at the electrodes.
3. What are the advantages of fuel cell power sources?
4. Classify the several types of fuel cells.
5. How will you illustrate the performance of a fuel all by the cell voltage V. Electrode current density? Draw a V_{ideal}, $V_{actual,}$ and I curve (polarization curve).
6. Discuss the fuels for fuel cells.
7. What is cell efficiency of fuel cell?
8. Describe
   (a) PAFC power plant
   (b) PAE system
   (c) MCFC
   (d) SOFC

9. Explain the principle and functioning of a typical fuel cell. What are the various reactions tried in experimental cells?
10. Mention the characteristics of an ideal fuel cell.
11. Justify 'fuel cells' may be the future power source of 2002–2010. Explain the development in this area to justify the same.
12. Describe hydrogen–oxygen fuel cell and reaction taking place at anode and cathode and the voltage generated and also, write the material aspects of cell.

Chapter 15

Magnetohydrodynamic (MHD) Power Generation

Magnetohydrodynamic (MHD) power generation is a direct energy conversion directly from a fast moving ionized gases or plasma, without the need of any turbine and electrical generator. It works on the basis of Faraday's laws of electromagnetic induction. Despite the facts that several countries throughout the world initiated the development of MHD systems in the mid of twentieth century, but associated techno-economical constraints hampered the development of practical MHD systems. Therefore, this system is considered as non-conventional energy conversion system. Owing to growing needs of energy supply to sustain social development and industrial and agricultural production, and depletion of fossil and nuclear fuels at rapid rate, efforts have been continuously made for commercial production of electrical energy from economical MHD generation systems. Significant progress has been made in the design and developments of several components and technology of MHD systems.

KEY CONCEPTS

- Electromechanical energy conversion versus MHD
- MHD generator and its Working
- Open-circuit voltage and power output
- The MHD system
- The Plasma
- Schematic diagram of liquid metal MHD
- Types of MHD system
- Advantages of MHD power

15.1 GENERAL

The concept of MHD power generation was introduced for the very first time by Michael Faraday in 1832 in his Bakerian lecture to the Royal Society. A magnetohydrodynamic (MHD) generator produces electrical power following the basic principles of Faraday's laws of electromagnetic induction.

Thermal ionization is the process that produces the plasma in the MHD systems. In this process, gas temperature is increased to such a point that the electrons are no longer remain bound to the gas atom. Plasma gas then becomes electrically conductive with these free electrons. If plasma is required to be created based on temperature only by thermal ionization, then significantly high temperatures are required. Since the alkali metals (like potassium nitrate) ionize easily at low temperatures, seeding of gas by alkali metal lowers the gas temperature. Salt water or liquid metal or an ionized gas (plasma) is used as electrically conducting fluid. The electrodes are made up of high temperature ceramic materials such as carbides (SiC, ZrC, MbC), bromides (ZrB_2, TiB_2, LaB_2), and silicides (WS and $MOSi_2$).

Magnetohydrodynamic generator produces electrical energy directly from a very fast moving stream of ionized gas through a strong magnetic field. It has no mechanical moving parts such as turbine and electrical generator that are necessary in conventional generators. Electrically conductive ionized gas or plasma acts as conductors and are equivalent to electrical conductors in conventional electrical generators. It has no moving parts. Heat sources for MHD generators are combustion of fossil and nuclear fuels, solar, geothermal, or waste heat in the form of hot gas. Superconducting magnets are used to provide a strong magnetic field of 2–7 T for electrical power generation in MHD. The MHD power development was started by several countries way back 1960, but high costs and complexity could not permit the development of practical and readily available commercial systems. Efforts are directed to minimize the cost and increase the fluid density and pressure for getting maximum efficiency of the system.

The power output per length (P) of MHD generator is approximately given as

$$P = \sigma u B^2 / \delta \tag{15.1}$$

where σ = the electrical conductivity of conducting fluid; u = the fluid velocity; B = the magnetic field density; and δ = fluid density.

15.2 ELECTROMECHANICAL ENERGY CONVERSION VERSUS MHD

A comparison of electromechanical energy conversion and MHD generators is given in Table 15.1.

Table 15.1 Comparison of Electromechanical Energy Converters and MHD Generators

Electromechanical Energy Converters	MHD Generator
They work on Faraday's laws of electromagnetic induction.	They work on Faraday's laws of electromagnetic induction.
They are three-stage indirect methods of energy conversion.	They are single-stage direct methods of energy conversion.
They use electrical conductor of copper or aluminium as windings.	Electrically conductive ionized gas or plasma is used as conductor.
Input energy sources are either heat or mechanical energy.	Heat energy from the fossil and nuclear fuels. Solar, geothermal or waste heat.
Normal magnetic field electromagnetic field of about 1 T.	Electromagnetic super conductors of strong magnetic field of 2–7T.

(*Continued*)

Table 15.1 Continued

Electromechanical Energy Converters	MHD Generator
Thermo-mechanical–electrical conversion.	Thermo-electrical process.
Turbines and generators are moving systems.	No moving parts.

15.3 MHD GENERATOR AND ITS WORKING

The essential elements of a simplified MHD generator are shown in Figure 15.1 and a simplified arrangement in Figure 15.2. This type of generator is referred to as a continuous electrode Faraday generator. The construction and principle of MHD generator working can be understood as follows:

The two main parts of MHD generators are the combustion chamber and the generator chamber. Both combustion chamber and generator chamber are surrounded by a heat resistance material and water cooler. The temperature in the combustion chamber is maintained around 2,000°K–2,400°K by burning the fossil fuels and nuclear fuels or other heat sources. The gaseous (fluid) conductor is passed into the combustion chamber through inlet. Seeding materials such as potassium and caesium are used to reduce the ionization temperature. The heat of combustion chamber removes the outermost electrons in the fluid conductor, and thus, the gas

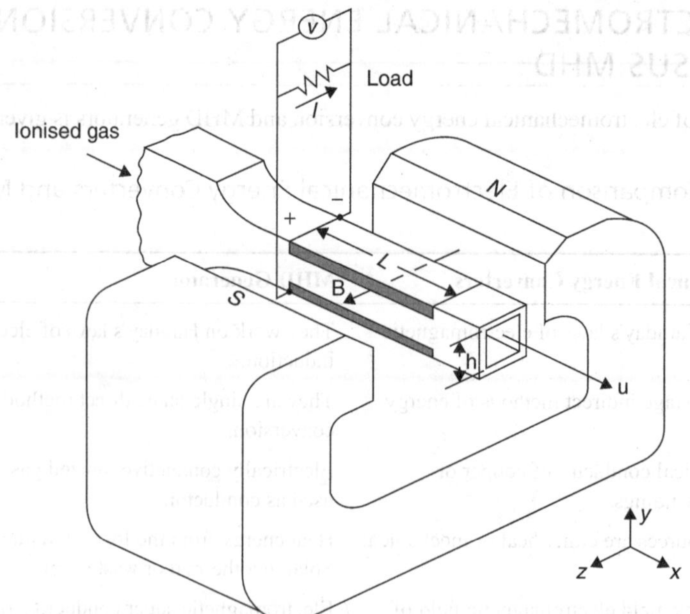

Figure 15.1 A simple MHD generator

particles acquire the charge and get ionized. The charged gas particles with high velocity (about 1,000–2,000 m/s) enter into the generator chamber via nozzle. The generator chamber has a powerful magnet (super conducting magnet) to create a strong magnetic fields of 3–7 T. Further, a number of oppositely located electrode pair is placed in the channel for electrical current flow in load circuit. Based on the principles of Faraday's laws of electromagnetic induction, the high velocity ionized conducting gas particles experience the magnetic field at right angles to their motion of direction, and hence, the potential (current) is produced. The positive and negative charge move to corresponding electrodes (anode and cathode) and constitute the DC current flow in the external load circuit.

15.4 PRINCIPLE AND PERFORMANCES

The Lorentz force law has the basis of development of the basic mathematical equation governing the working of a MHD generator. It states that when a particle is projected with certain velocity (u) in an area of magnetic field intensity (B), then the force acted on the charged particle is given in vector form as

$$F = Q\,[E + (u \times B)] \tag{15.2}$$

where F = force acting on the particle; Q = charge on the particle; v = velocity of the particle; B = magnetic field; and E = electric field

The important thing to remember is that the direction of force vector is perpendicular to the plane of velocity and magnetic field.

15.4.1 Open-circuit Voltage and Power Output

From the simplified version of MHD generator shown in Figures 15.2 and 15.3, its working is explained as follows:

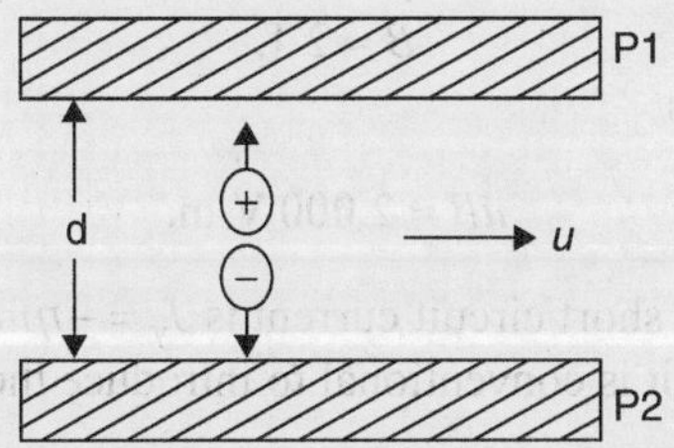

Figure 15.2 MHD output equation

When an electrically conductive (ionized) gas moving with a velocity (u) in an insulated duct is traversed through a strong magnetic field having magnetic field density B, the gas charged particles experience an induced electric field ($u \times B$). The electric current is driven in the direction perpendicular to the magnetic field and speed. The DC current collected by electrodes placed on opposite side of duct flows through externally connected load.

The magnitude of the current density (neglecting the Hall effect) is given by

$$J = \sigma(E + u \times B) \tag{15.3}$$

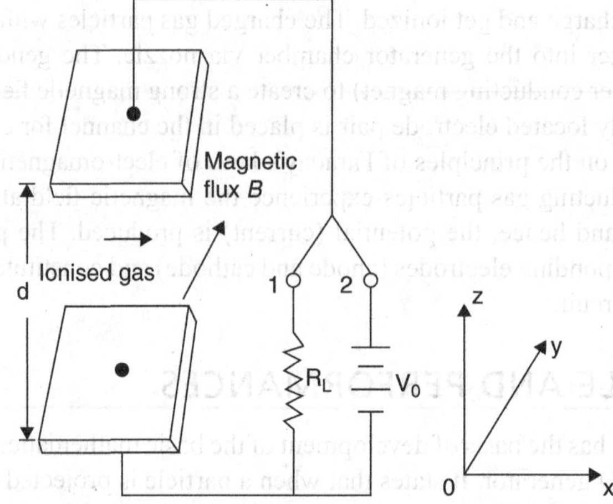

Figure 15.3 Simplifies representation of MHD generators

The electric field E, which is added to the induced field, results from the potential difference between the electrodes. For the purposes of initial discussion in this section, it is assumed that both velocity (u) and electrical conductivity (σ) are uniform.

In terms of the coordinate system shown in Figures 15.1–15.3, it can be written that

$$J_Y = \sigma(E_Y - uB) \tag{15.4}$$

At open circuit, $J_y = 0$, and therefore, the open-circuit electric field is uB.

For the characteristic conditions,

$$u \approx 1{,}000 \text{ m/s and}$$
$$B \approx 2 \text{ T},$$

The open-circuit electric field is

$$uB \approx 2{,}000 \text{ V/m}.$$

At short circuit, $E_Y = 0$, and the short circuit current is $J_y = -\eta uB$.

For general load conditions, it is conventional to introduce the loading parameter

$$K = -E_Y/uB$$

where $0 \sim K \sim 1$, and $J_y = -\sigma\, uB(1 - K)$. (15.5)

The (–) sign indicates that conventional current flows direction is opposite to electron flow. The upper electrode acts as anode and bottom electrode as cathode. The electrical power delivered to the load per unit volume is given by

$$P = -JE. \tag{15.6}$$

For the generator shown in Figure 15.1,

$$P = \sigma\, u^2 B^2 K(1 - K) \tag{15.7}$$

This power density has a maximum value for $K = 1/2$

$$P_{max} = \sigma u^2 B^2/4 \tag{15.8}$$

The rate at which directed energy is extracted from the gas by the electromagnetic field per unit

$$\text{volume is} - u\,(J \times B). \tag{15.9}$$

Hence, the electrical efficiency of an MHO generator is defined as

$$\eta_e = JE/u\,(J \times B) \tag{15.10}$$

For the generator being discussed,

$$\eta_e = K. \tag{15.11}$$

Faraday generator attains higher efficiency when operating near open-circuit conditions. Further, output delivery above 10 MW results in acceptable size of MHD generators.

Using the preceding characteristic values for u and B, this requirement means that the electrical conductivity must be such that

$$\sigma \approx 4P_{max}/u^2B^2 \approx 10 \text{ mhos/m} \tag{15.12}$$

The calculation of open-circuit voltage is straightforward. In a properly functioning MHD generator, open-circuit voltage is given by

$$V_0 = Bdu \tag{15.13}$$

where B = flux density between the electrodes in Wb/m^2; u = gas speed in m/s; and d = electrode separation in m.

15.4.2 Maximum Power Output

Let R_G is the internal resistance of MHD generator and R_L is the load resistance. Then, maximum power output can be obtained when $R_G = R_L$. Therefore,

$$P = V_0 \times I = I^2R_G \tag{15.14}$$

Further, $I = V_0/(R_G + R_L)$

Therefore, $P = [V_0/(R_G + R_L)]^2 \times R_G$

For maximum power transfer, $R_G = R_L$.

Hence, $$P_{max} = V_0^{\,2}/4\,R_G = \sigma u^2 B^2 dA/4 \tag{15.15}$$

Since $R_G = d/\sigma A$,

A = area of the plate.

Therefore, $$P_{max} = u^2B^2d^2/4R_G \tag{15.16}$$

MHD is the reversible process. If instead of resistance, an emf greater than V_0 is applied in opposite to the direction of the flow of current, energy would be supplied to the gas and the particle of the gas would be accelerated. Therefore, the ejected gas is at a higher velocity than the inlet gas.

Example 15.1

An MHD generator has following parameters:
Plate area = $0.2\ m^2$
Distance between plates = 0.4 m
Flux density = $2\ Wb/m^2$
Average gas velocity = 800 m/s
Gaseous conductivity = 10 mho/m
Calculate the open-circuit voltage and maximum power output.

Solution Open-circuit voltage, $V_0 = Bud = 2 \times 800 \times 0.04 = 640$ V.

MHD generator resistance, $R_G = d/\sigma A = 0.4/10 \times 0.2 = 0.2\ \Omega$
Maximum power, $P_{max} = u^2B^2d^2/4\ R_G = V_O^2/4R_G = 640^2/4 \times 0.2$
$= 512$ kW

15.5 MHD SYSTEM COMPONENTS

The typical schematic diagram of an MHD system is shown in Figure 15.4, which shows possible system components and parameters.

1. *Combustion chamber*: it is used to ionize fluid gas at high temperatures to make fluid gas electrically conductive. High temperature heat is obtained by burning fossil and nuclear fuels or solar and geothermal heat or wastes heat from processes.

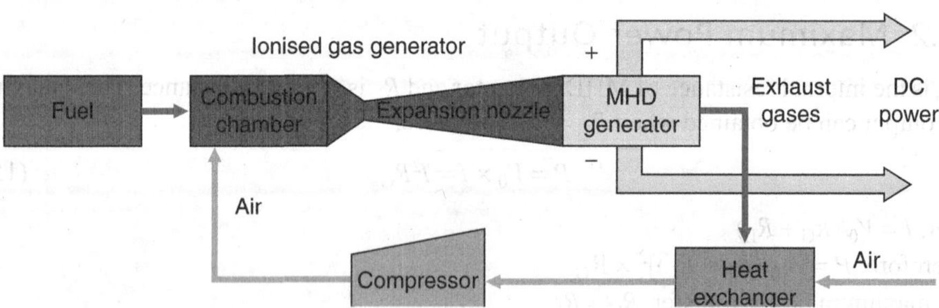

Figure 15.4 Open-cycle MHD system

2. *Expansion nozzle*: a converging–diverging magnetic field is typically employed to create a nozzle contour similar to de Laval nozzles used in chemical rockets. The thrust generation process helps in energy conversion to directed kinetic energy. Transfer of momentum from the plasma to the thruster and detachment of the plasma from the initially confining magnetic field lines. The expansion nozzle reduces the high gas pressure, and consequently, increases the plasma speed following the Bernoulli's Law through the generator duct to increase the power output. It also reduces the requirements of high temperature in combustion chamber.

3. *Seeding*: the hot gases from combustor are then seeded with a small amount of ionized alkali metal (caesium or potassium) to increase. The electrical conductivity of the fluid gas is increased by adding a small amount of ionized alkali metal (such as caesium or potassium). Normally, potassium carbonate is injected in the combustion chamber. The seed material, generally, potassium carbonate is injected into the combustion chamber.
4. *Superconducting magnets*: an electromagnet made from superconducting wires is referred to as superconducting magnet. It is used to provide a strong magnetic field of the order of 2–7 T in the generator chamber. As the ionized gas or plasma passes through the channel, an electromotive force is developed by the strong magnetic field.
5. *Electrodes*: the electrodes are made up of high ceramic material, which is the channel duct walls. The anode and cathode electrodes are connected by an electrical load.
6. *Working fluid*: air is the working fluid for open cycle and argon or helium is used as working fluid for closed-cycle MHD.
7. *Compressor*: compressor is used to pump back the fluid gas exhaust heat in combustion chamber to enhance the combustion rate.
8. *Powerful electromagnet*: when plasma flows through the powerful magnet, electrical output voltage is generated across the two electrodes placed on the opposite side of the plasma. The current flowing through plasma in between the two electrodes is called Faraday current, which is responsible for the main electrical output of MHD generator.
9. *Hall effect current*: the Faraday output current flowing through the load circuits interacts with the applied magnetic field and give rises to Hall effect current perpendicular to the Faraday current. This constitutes an energy loss and the total current generated is obtained as the vector sum of the Faraday current (transverse current) and the Hall current (axial current). The overall conversion efficiency of MHD is enhanced by capturing both Faraday and Hall current. For the purpose, electrodes are split into insulated segments placed side by side and are connected in series. They produce high voltage and low current. Further, they are skewed slightly and placed in the direction of resultant current to permit the extraction of maximum energy.

15.6 PLASMA

A hot ionized gas consisting of approximately equal numbers of positively charged ions and negatively charged electrons is called plasma. It is made up electrically conductive charged particles and its characteristics are very much different from those of ordinary neutral gases as they are strongly influenced by electric and magnetic fields. Gases derived from combustion, noble gases, and alkali metal vapours are the suitable working fluids.

15.6.1 Gas Plasma

Plasma formation is required to heat the working fluid at extremely high temperatures so that electrons are detached from the atoms and only positively charged ions remain in the working fluid (or gas). The plasma flows through the strong magnetic field at significantly high speed and acts as fluid electrical conductors.

15.6.2 Methods of Ionizing the Gas

Different available methods of gas ionization are designed to provide significant and sufficient energy to working fluid or gas. High temperature gas heating or irradiating the gas by X-rays or gamma rays are common methods of gas ionization. The addition of small amounts of seeding materials increases the ionization and improves the conductivity of ionized gas. The variation of electrical conductivity with ionization is shown in Figure 15.5.

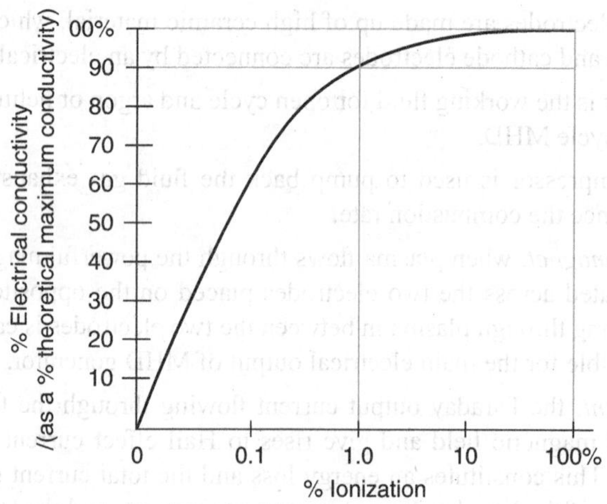

Figure 15.5 Gas electrical conductivity as a function of ionization

15.6.3 Containment

The ionized working fluids of MHD generators operate at significantly high temperatures, and therefore, channel ducts are constructed with high temperature ceramic materials that are electrically non-conducting. Electrode, however, must be heat resistant and electrically conducting.

15.6.4 Power Output

The MHD generator output is directly proportional to the square of the flux density, directly proportional to speed of plasma, plasma electrical conductivity and cross-sectional area and inversely proportional to fluid density. The output power is a DC power.

15.6.5 Efficiency

Heat lost through the high temperature exhaust is the main cause of low operating efficiency of MHD generators. Efficiency can be largely increased by pumping back the exhaust hot gas in combustion chamber and an energy source for steam generation in thermal power plants.

15.6.6 Operating Experience

The principle of MHD generation was very well understood long time back but they have not yet been commercially developed because of their high cost and lack of knowledge on certain associated technology. They have been, therefore, built as demonstration plant (up to about 50 MW) in many countries and are used for peak demand saving applications. Seeding the working fluid by alkali metals to enhance the ionization is one of the problem areas of MHD systems. Cost, corrosion, and recovery (specifically in open-cycle systems) of seeding material are essential. Design and development of cheap and highly strong magnetic field properties can reduce running costs. The design and selection of electrodes and channel duct insulation materials, to withstand high temperatures of about 2,500°K, are another areas of developments.

15.7 SCHEMATIC DIAGRAM AND WORKING OF LIQUID METAL MHD

Schematic diagram of a typical liquid metal MHD is shown in Figure 15.6. Liquid metal MHD generator operates at moderately low temperatures and has high heat transfer and high electrical conduction properties. Any low temperature heat source (such as solar, geothermal, process waste heat, and hot exhaust gas) can be used. Electrical conductivity of liquid metal is about 10^6 times that of an ionized gas or plasma. Seeding of liquid metal (helium) with alkali metal (small caesium) further increases electrical conductivity.

Perhaps, the first liquid metal MHD generator plant (named Elgar-3) with 20 MW capacities was built in Israel.

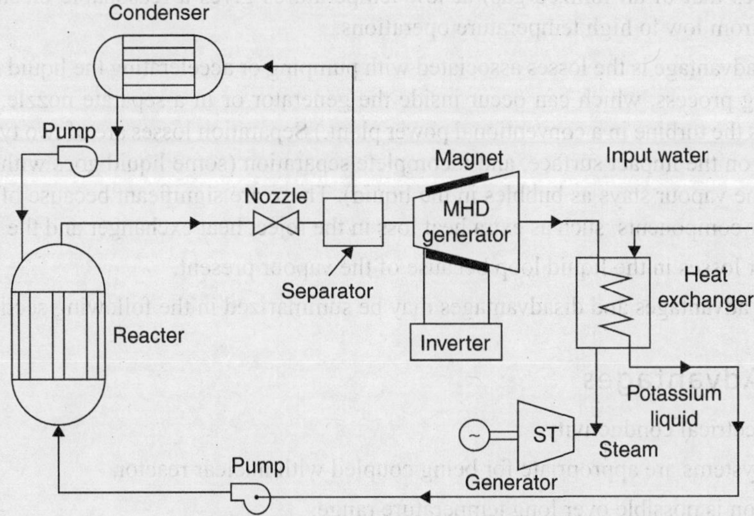

Figure 15.6 Liquid metal MHD

15.7.1 Working

The liquid metal as a working fluid in the MHD generator is heated directly or indirectly to make it flow through the MHD channel. Liquid potassium is pumped into breeder reactor where it is heated and ionized vapour passes through the nozzle to increase its velocity. The flow of high velocity liquid metal through strong magnetic fields produces DC power. Mixed vapours are separated in separator and steam is condensed and pumped back to breeder reactor. The liquid potassium outlet from MHD generator is routed through heat exchanger of conventional steam plants to produce steam for running turbine.

15.7.2 Features and Liabilities

1. Single-phase flow and two-phase flow liquid metal MHD energy conversion systems are very common systems and consist of a liquid metal as an electro-dynamic fluid and some suitable vapour or gas as a thermodynamic fluid. They have better performances, require less maintenance, and economically viable.
2. Best use of low grade thermal energy sources (such as solar, geothermal and waste flue gas exhaust) is feasible. The advantages of an LM-cooled solar collector are expected to be more evident in the case of concentrating collectors with a heat flux of high density.
3. The liquid metal plays an important role of a liquid electrical conductor. It also serves as a heat store for the expanding vapour. Such a system is very attractive in space applications where a long life is required.
4. Simplicity of design and control, isothermal expansion, direct contact heat transfer, high cycle efficiency, simple and robust components are the added features of liquid metal MHD systems.
5. The high electrical conductivity of the liquid metal (as compared to the plasmas used in other MHD systems) is the important advantage. Having high electrical conductivity (about 10^6 times that of an ionized gas) at low temperatures gives a reasonable electrical power output from low to high temperature operations,
6. The disadvantage is the losses associated with pumping or accelerating the liquid metal. (The pumping process, which can occur inside the generator or in a separate nozzle, effectively replaces the turbine in a conventional power plant.) Separation losses are of two types namely friction on the impact surface, and incomplete separation (some liquid goes with the vapour and some vapour stays as bubbles in the liquid). These are significant because of the impact on other components, such as extra heat loss in the reject heat exchanger and the next item.
7. Diffuser losses in the liquid loop because of the vapour present.

In brief, the advantages and disadvantages may be summarized in the following sections.

15.7.3 Advantages

1. High electrical conductivity.
2. These systems are appropriate for being coupled with nuclear reactor.
3. Operation is possible over long temperature range.
4. Low thermal energy source can be used.

15.7.4 Disadvantages

1. Due to low operating temperatures, thermal efficiency is low.
2. Liquids are practically incompressible, and high velocity cannot be produced by expansion.
3. To achieve high velocity, efficiency jet pump or two-phase system with vapour bubble is used.
4. Liquid metals are excellent conductors, but their vapours are poor conductors.
5. Material confinement and erosion–corrosion problems are associated.
6. They generate high current and low voltage DC.

15.8 TYPES OF MHD SYSTEM

The MHD systems are broadly classified into two types:

1. *Open-cycle system*: in this system, the working fluid after generating electrical energy is discharged to atmosphere through the stack. The working fluid used is air.
2. *Closed-cycle system*: in this system, the working fluid is recycled to the heat source. and thus, it is used again and again. Helium or argon is used as the working fluid. This system is further subdivided into
   (a) Seeded inert gas system
   (b) Liquid metal system

15.8.1 Open-cycle System

Working of typical open-cycle MHD system, as shown in Figure 15.7, is the same as explained in Section 15.4. The fuel is burnt into combustor (or combustion chamber). The gasses in the

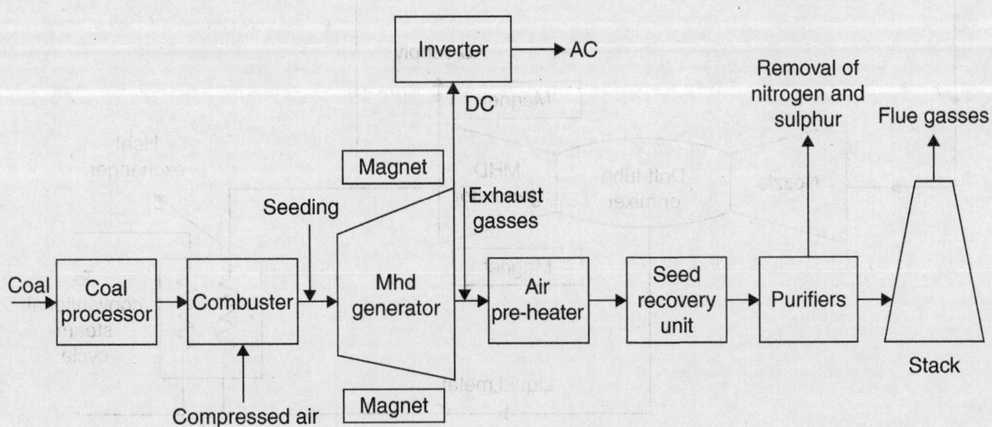

Figure 15.7 Open-cycle MHD system

combustion chamber is seeded to increase the electrical conductivity of ionized gas. The seed material (potassium carbonate) is injected into the combustion chamber, which is then ionized by hot combustion gases at temperatures of roughly 2,300°C to 2,700°C. The compressed air is used to burn the fuel to attain high temperatures. An alternative is used to compress oxygen alone for combustion of fuel, little or no preheating is then required. The additional cost of oxygen might be balanced by saving on the preheater. The hot pressurized and ionized gas coming out of combustion chamber flows through an expansion nozzle and gains high velocity. The ionized gas enters the generator chamber in channel ducts surrounded by powerful magnet. During the motion of gas, ions and electrons are separated and move towards the electrodes and electrical power output is provided to externally connected load.

15.8.2 Closed-cycle Liquid Metal System

Figure 15.6 is redrawn in Figure 15.8.

An inert gas is a good carrier that is heated and pressurized by passing through a heat exchanger within the combustion chamber itself. The hot gas is then mixed into hot sodium (liquid metal) to form working fluid. The working fluid passing through a nozzle acquires high velocity and introduced in the MHD generator. Interaction between fast moving ionized fluid and strong magnetic field of superconducting magnet, electrical power is produced.

Finally, the carrier gas is cooled, compressed, and returned to the combustion chamber for reheating and mixing with the recovered liquid metal. The working fluid temperature is usually around 800°C as the boiling point of sodium even under moderate pressure is below 900°C.

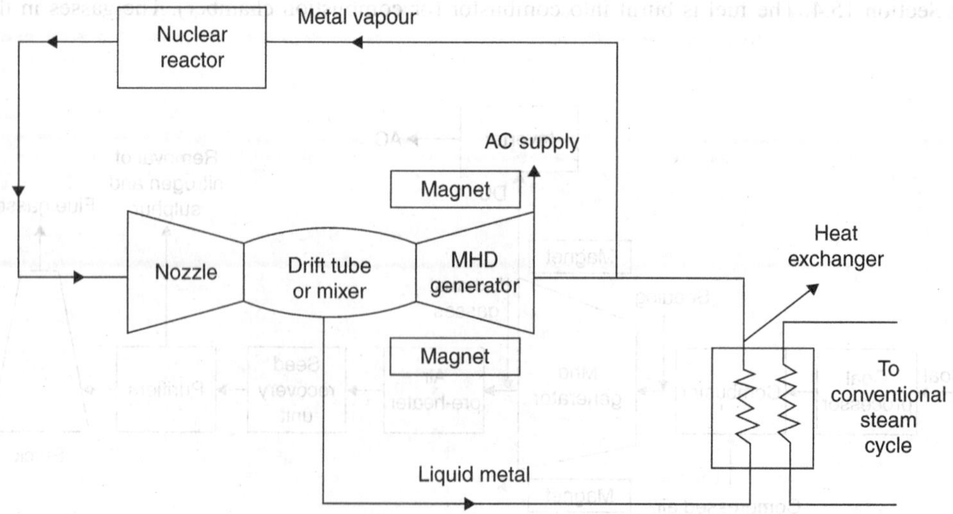

Figure 15.8 Closed-cycle MHD system

15.8.3 Closed-cycle Plasma MHD System

The three distinct but interlocked loops, as shown in Figure 15.9, define the complete arrangement of the MHD system. On the left is the external heating loop, where coal is gasified and the gas is burnt in the combustor to provide heat. A liquid metal provides the conductivity.

The heat is then transferred to carrier gas (such as helium or argon) in the primary heat exchanger. The combustion products after passing through the air preheater and purifier. They are discharged to atmosphere.

The centre loop is the generator loop; carrier gas seeded with alkali metal is passed through the generator chamber at a very high speed. The DC power is then generated for end user. On the extreme right, the third loop is used to remove the outgoing generator heat from the gas by a cooler, then compressed and pumped back to reheating.

Since the combustion system and working fluid system are entirely different, the flue gases are used to preheat the combustion air and then treated for fly ash and sulphur dioxide and finally their discharge in atmosphere.

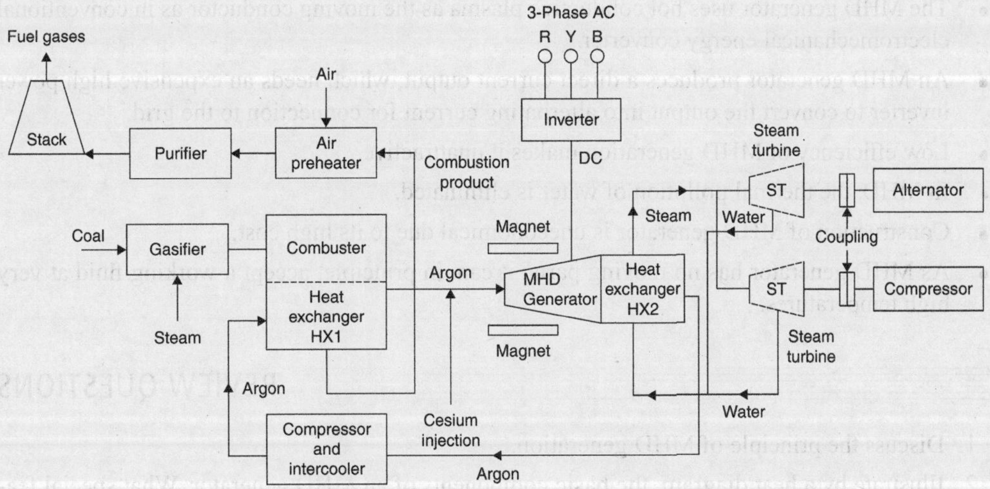

Figure 15.9 Closed-cycle plasma MHD system

15.9 ADVANTAGES OF MHD SYSTEM

1. It has no moving parts.
2. Conversion efficiency is better.
3. Plant size is smaller than conventional power plants.
4. It has ability to full power level generation immediately when started or small start-up time.
5. It is a direct method of heat to electrical energy conversion.

6. Less pollution.
7. Good candidate for peak power savings and for emergency.

15.10 DISADVANTAGES OF MHD SYSTEM

1. Material confinement and erosion–corrosion problems are associated.
2. They generate high current and low voltage DC.
3. Expensive.

SUMMARY

- MHD generation is the production of electrical power by utilizing conducting plasma moving through a strong magnetic field.
- A magnetohydrodynamic generator (MHD generator) converts thermal energy and kinetic energy into electrical energy.
- The MHD generator uses hot conductive plasma as the moving conductor as in conventional electromechanical energy converter.
- An MHD generator produces a direct current output, which needs an expensive high power inverter to convert the output into alternating current for connection to the grid.
- Low efficiency of MHD generation makes it unattractive.
- In MHD, the thermal pollution of water is eliminated.
- Construction of MHD generator is uneconomical due to its high cost.
- As MHD generator has no moving parts, it can, in principle, accept a working fluid at very high temperatures.

REVIEW QUESTIONS

1. Discuss the principle of MHD generation.
2. Illustrate by a heat diagram, the basic components of an MHD generator. What special features must such a system have for efficient application?
3. Derive equations for the voltage and power output of an MHD generator.
4. Why is it necessary to add seed material to the gas?
5. Mention types of MHD systems and compare them.

Chapter **16**

Thermoelectric Converters

Large amount of waste heat generated from industrial processes, equipments, and machinery are the biggest contributor to the global warming. Attempts have been continuously made to minimize waste heat by recovery and reuse by converting them to useful energy forms as a measure against global warming. Conversion to electrical energy is considered as one of the important methods of waste recovery and reuse. Thermoelectric modules (or thermoelectric converters) have found wide applications in this regard.

From the date of discovery of the Seebeck effect (thermoelectric effects) in 1821, thermoelectric modules have undergone design improvement modification resulting in present commercially available prototype. The low thermal efficiency, development of good thermoelectric materials, and high costs are some major parameters still remain to be investigated.

Thermoelectric effect is a reversible phenomenon by which either a temperature difference between the junctions of two dissimilar metals creates an electrical potential or an electrical potential connected across the two junctions creates temperature difference. The phenomena of conversion of heat energy into electrical energy are known as the Seebeck effects. The conversion of electrical energy to heat energy is referred to as Peltier effects. A third effect related with conductor heating and cooling is known as Thompson effects.

Thermoelectric generators are widely used as electrical power source in medical, military, and deep space applications, where relatively high cost and low generating efficiency are insignificant as compared to the combinations of their desirable properties. Thermoelectric coolers are very

KEY CONCEPTS

- Basic configuration of thermoelectric generators
- Major heat source-based thermoelectric generators
- Seebeck, peltier, and thomson Effects
- Principles of operation
- Basic metallic and semiconductor thermoelectric Generators
- Basic theory of operation and performance
- Figure of merit, efficiency, and power output
- Thermoelectric materials
- Thermoelectric converter modules and applications

popular when used in optical communications and infrared detectors. Their applications also include data collection, remote sensing, and offshore engineering.

Known heat sources (such as solar, geothermal, waste heat, and natural gas) are continuously supplied to hot junction for reliable electric power generation.

16.1 BASIC CONFIGURATION OF THERMOELECTRIC CONVERTERS

Thermoelectric power generator converts heat directly into electricity based on thermoelectric effects involving interactions between the flow of heat and the flow of electricity through solid bodies. They have the same basic configuration, as shown in Figure 16.1.

A heat source provides the high temperature (T_H) and the heat flow through a thermoelectric converter to a heat sink (T_C), which is maintained at a temperature below that of the source. The temperature coefficient across the converter ($\Delta T = T_H - T_C$) produces direct current (DC) to a load (R_L) having a terminal voltage (V) and a terminal current (I). As there is no intermediate energy conversion process, thermoelectric power generation is known as direct energy–power conversion. The amount of electrical power generated is given by

$$P = I^2 R_L \tag{16.1}$$

A uniqueness of thermoelectric energy conversion lies in the fact that the direction of energy flow is reversible. If the load resistance is removed and replaced by a DC power supply, the thermoelectric device can be used to draw heat from the heat source element and lower its temperature. Electrical power is then used to pump heat and produce refrigeration.

The process of reversibility exhibited by thermoelectric converter is not seen in many thermo-electrical systems like thermionic converters and it is one of the distinguishing features of thermoelectric converters when compared to other similar devices. They are, however, designed and optimized for a specific defined purpose. For example, either for conversion of thermal energy input to device into electrical energy for lighting, the operation of electrical motors and many other equipments and appliances, etc., or conversion of electrical energy input to the device in thermal energy for heating and refrigeration.

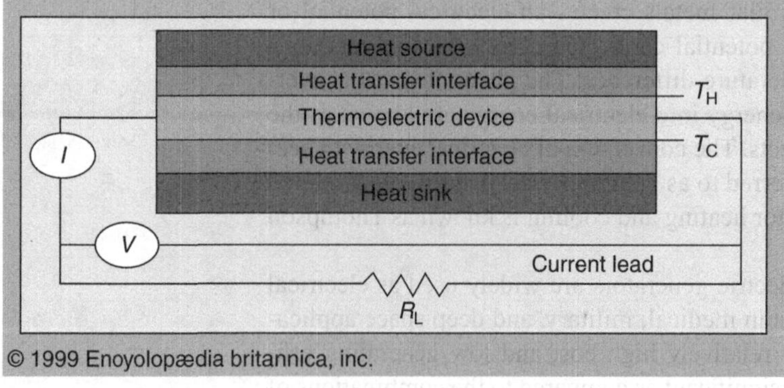

Figure 16.1 Basic configuration of thermoelectric converters

16.1.1 Historical Developments

The field of thermoelectricity development began in the early 1800s with the discovery of the thermoelectric effect by Thomas Johann Seebeck. He discovered that when a temperature difference is maintained between the junctions formed by two dissimilar electrically conducting metals separated with each other along their length but joined together to form junctions at their ends, a magnetic field, and hence, an electrical potential is developed around the conductor's loop. Systematic development of thermoelectric converters were carried out between 1885 and 1910 when a German scientist, Edmund Altenkirch, identified and specified the parameters of the materials needed to build practical devices. He also calculated the efficiency of thermoelectric generators satisfactorily. Metallic conductors were the only materials available at the time to build thermoelectric generators with an efficiency of hardly 1%. A semiconductor-based generator with a conversion efficiency of 4% had been developed only by about 1940. Thermoelectric generators were used to provide electrical power to portable communication transmitters during World War II. Significantly high improvements were made in semiconductor materials and in electrical contacts between 1950 and 1965 that expanded the practical range of application. Conversion efficiency of about 10% could only be achieved by 1980. The development of much better thermoelectric materials were found to be the major parameters identified for improving the performance level of thermoelectric converters.

Thermoelectric generators of low power varieties have their practical importance. They are the most versatile, reliable, radioactive isotopes fuelled electrical power source for isolated, remote sites, and space applications.

16.2 MAJOR HEAT SOURCES-BASED THERMOELECTRIC GENERATORS

Heat source and heat sink in addition to the type of application and electrical power requirement determine the size and geometry of thermoelectric converters. The heat source is required to provide thermal energy to thermoelectric converters for converting it to electrical power at usable voltage. Thermoelectric converters are also named on the basis of heat sources, providing heat energy input to the generators. They are discussed in the following sections.

16.2.1 Fossil Fuel Generators

The thermoelectric converters are designed and constructed to use natural gas, propane, butane, kerosene, jet fuels, and wood, etc., as a source of heat. They are commercially available in low power range (10–100W) for remote area applications (such as data collection and communications from space, navigational aids, and cathode protection).

16.2.2 Solar Source Generators

Thermoelectric converters using solar heat and ocean thermal heat as a source of heat have also been developed with partial success. Power conditioning of small irrigation pumps in remote areas and underdevelopment region has been found attractive for solar source-based thermoelectric converters. They have also been designed to provide electric power to orbiting spacecraft, but they have no competitiveness yet with solar cell systems. In ocean thermal heat-based thermoelectric converters, warm sea surface water is used as heat source, and cold water of deep sea is used as heat sink.

16.2.3 Nuclear-fuelled Generators

High-temperature heat source for thermoelectric generators are the decay products of radioactive isotopes. These sources lost for a long period of time and generators are relatively immune to nuclear radiation. Such nuclear-fuelled thermoelectric generators are, therefore, considered as a useful source of power for many unattended and remote applications (such as radioisotope thermoelectric generators provide electric power for isolated weather monitoring stations, deep-ocean data collection, various warning, communications systems, and spacecraft). They have also been used as a power source in pacemakers. Radioisotope thermoelectric generators are available in the power range of microwatts to 199 watts.

16.3 PRINCIPLES OF OPERATION

Operating principles of thermoelectric devices is based on the introduction to the phenomena of thermoelectricity (thermoelectric effect) discovered by Seebeck in 1821 and named in his honour as Seebeck effect.

16.3.1 Seebeck Effect

Figure 16.2 represents the schematic arrangements of his experimental setup. One metallic conductor of antimony and another metallic conductor of bismuth are joined at their ends to form junctions at both ends. The two junction ends are connected through a sensitive galvanometer. One junction was heated and the other was maintained cool. The temperature difference between the two junctions gave a deflection in galvanometer. The emf generated in the circuit is called thermoelectric emf or thermo-emf and the resulting current is known as thermoelectric current.

The two-junction circuit shown in Figure 16.2 is called a thermocouple.

The Seebeck effect, therefore, explain the creation of an electromotive force (emf), and consequently, an electric current in a loop of material consisting of at least two dissimilar conductors when two junctions are maintained at different temperatures. The conductors are common metals, though they need not even be solids. The Seebeck effect is used to measure temperature with great sensitivity and accuracy and also to generate electric power for special applications.

Seebeck determined that when one junction of two dissimilar conductors are heated to a high temperature (T_H) and the other junction has a low temperature (T_C), a voltage (V) is generated

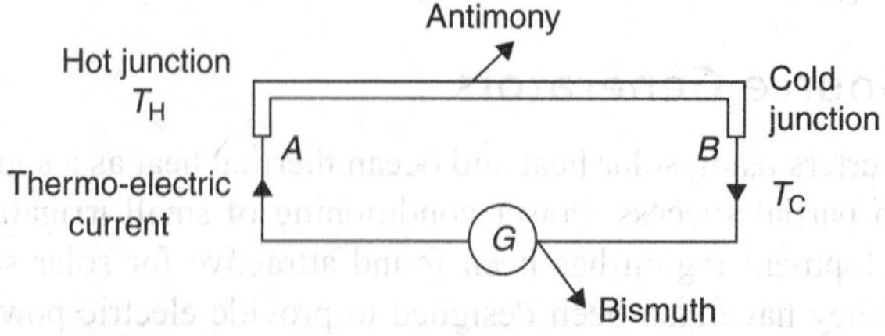

Figure 16.2 Thermocouple developed by Seebeck in 1821

proportional to $\Delta T = (T_H - T_C)$ and is given by

$$V \alpha \Delta T \tag{16.2}$$

The proportionality constant (S) is the Seebeck coefficient or thermopower and is given as

$$S = -\Delta V/\Delta T. \tag{16.3}$$

When the two junctions are connected through an electrical load, electrical current flows through the load circuit. It represents the direct conversion of thermal energy of source into electrical energy.

16.3.1.1 Basic Metallic Thermoelectric Converter

Metallic thermoelectric converter based on Seebeck effect have been developed for many years. When a temperature difference is established between the hot and cold junctions of two dissimilar materials (metals or semiconductors), a voltage is generated, that is, Seebeck voltage. In fact, this phenomenon is applied to thermocouples that are now used for electrical power generations. A schematic diagram of a typical simple thermoelectric power generator developed based on Seebeck effect is shown in Figure 16.3. Thermoelectric power generator operation is as follows:

1. A high temperature heat source supplies heat energy at the rate of (Q_H) to the hot junction, which is maintained constant at a temperature (T_H).
2. It is then rejected at a rate of $Q_C = Q_L$ to a low-temperature sink maintained at $T_C (=T_L)$ from the cold junction.

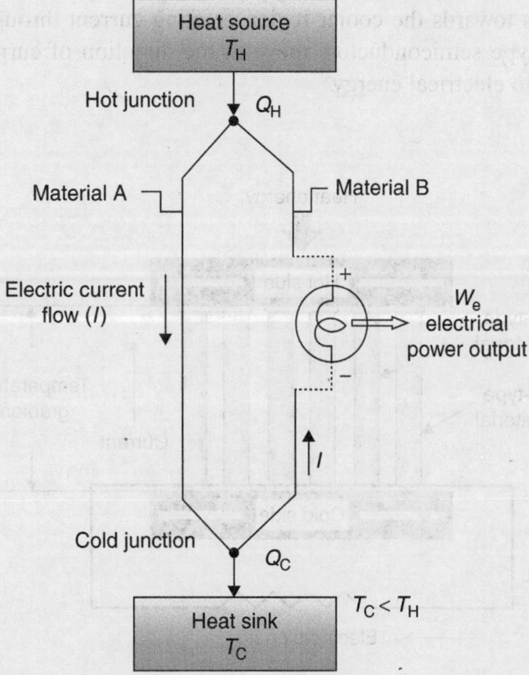

Figure 16.3 Basic metallic thermoelectric converters

3. The heat supplied at the hot junction causes an electric current to flow in the circuit and electrical power is produced.
4. From the principle of energy conservation (first law of thermodynamics), the difference between heat energy Q_H and Q_C will be the electrical power output (W_e).
5. Since the power cycle of thermoelectric converter resembles the power cycle of a heat engine, they may also be thought of as a heat engine.

16.3.1.2 Basic Semiconductor Thermoelectric Converter

Thermoelectric devices were all metallic in early days, but presently they are constructed from p-type and n-type semiconductors connected by metallic conductors. Power generating thermoelectric converter largely uses semiconductor junction and metallic junctions are commonly used in temperature measurement devices.

It consists of a number of alternate n- and p-type semiconductor thermoelements, which are connected electrically in series by metal interconnects as shown in Figure 16.4, sandwiched between two electrically insulating but thermally conducting ceramic plates to form a module.

When temperature difference is maintained between the hot side and the cold side, an electrical current and power will be delivered to an external load and converter is said to be operated as generator.

Electrons move in the opposition of current in the n-type semiconductors and holes move in the direction of current in the p-type semiconductors and removes heat from one side of converters. Charge flowing through n-type semiconductor crosses metallic connection and moves into p-type semiconductors. When a heat source is provided, heat source drives the electrons in the n-type semiconductors towards the cooler region creating current through the external circuit. Then, holes in the p-type semiconductors move in the direction of current. Thus, the thermal energy is converted into electrical energy.

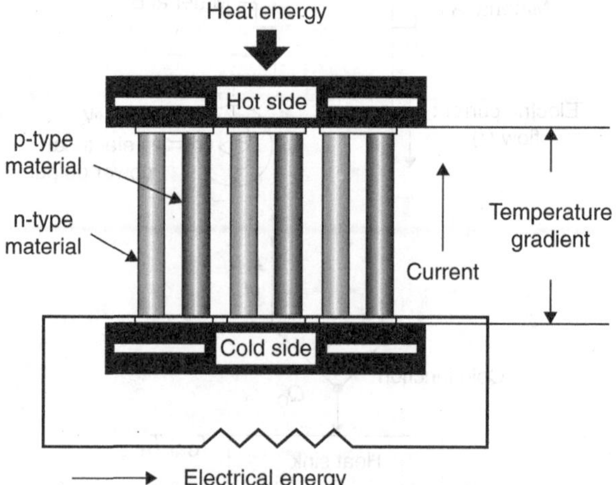

Figure 16.4 Semiconductor thermoelectric converter

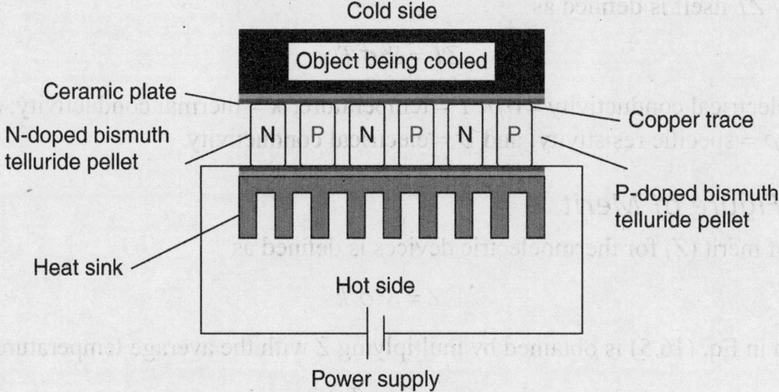

Figure 16.5 Thermoelectric refrigerator

Conversely, when an electric current is passed through the module, heat is absorbed at one face of the module and rejected at the other face; thus, the device operates as a refrigerator as shown in Figure 16.5.

16.3.2 Basic Theory and Operation

A schematic diagram of a simple thermoelectric power generator based on Seebeck effect is shown in Figure 16.6.

In this arrangement, p-type and n-type semiconductor legs are sandwiched between two ceramic hot or cold plates and connected thermally in parallel and electrically in series as depicted in Figure 16.4. A voltage difference induces across the load because of the temperature difference across the two legs.

The thermoelectric converter performance is characterized by its figure of merit (ZT).

Heat is carried by both electrons (κ_e) and photons (κ_{ph}), and

$$\kappa = \kappa_e + \kappa_{ph} \tag{16.4}$$

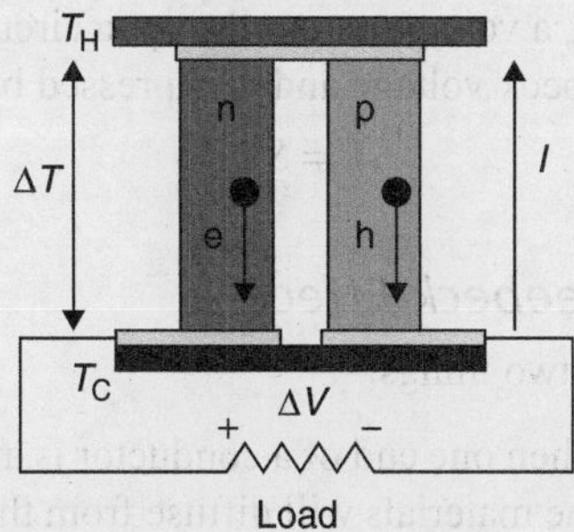

Figure 16.6 Schematic diagram of semiconductor-based thermoelectric generators

The quantity ZT itself is defined as

$$ZT = S^2 \sigma T / \kappa \tag{16.5}$$

where σ = electrical conductivity $= 1/\rho$; T = temperature; κ = thermal conductivity; S = Seebeck coefficient; ρ = specific resistivity; and σ = electrical conductivity.

16.3.2.1 Figure of Merit

The figure of merit (Z) for thermoelectric devices is defined as

$$Z = S^2 \sigma / k \tag{16.6}$$

ZT as shown in Eq. (16.5) is obtained by multiplying Z with the average temperature (T) as

$$T = (T_H + T_C)/2 \tag{16.7}$$

The dimensionless figure of merit defines the device potential efficiency and is used to compare devices made of different materials. The devices with the value of ZT equals to 1 are considered as good from efficiency viewpoint. For comparing thermoelectric devices with electromechanical converters, the values of ZT should be in the range of 3 and 4. Commercially available thermoelectric devices have ZT values in the range of 2 and 3. Worldwide attention has been presently given to increase the value of Seebeck constant (S) and decreasing the value of thermal conductivity (k) to achieve the desired values of ZT.

16.3.2.2 Device Efficiency

The efficiency of a thermoelectric device for electricity generation is given by η, and is defined as

$$\eta_{TE} = \text{energy provided to the load/heat energy absorbed at hot junction}$$

$$= \eta_C \left[\{(1 + ZT)^{1/2} - 1\}/\{(1 + ZT)^{1/2} + T_C/T_H\}\right] \tag{16.8}$$

where η_C = Carnot efficiency;

$$\eta_C = (T_H - T_C)/T_H = \Delta T / T_H \tag{16.9}$$

A significant difference in temperature (and hence large ΔT) is required to generate sufficient electrical energy. The infrared (IR) region of the solar spectrum (which is wasted in solar cells) can be used to supply the necessary hot temperature, T_H.

If electrical external circuit is somehow remained open and temperature difference between two junctions remain maintained, a voltage across the open circuit will be developed. This generated voltage (V) is called as Seebeck voltage and is expressed by

$$V = S \Delta T \tag{16.10}$$

16.3.2.3 Causes of the Seebeck Effect

The Seebeck effect is caused by two things:

1. *Charge-carrier diffusion*: When one end of a conductor is at a different temperature than the other, the charge carrier in the materials will diffuse from the hot end to the cold end; the hot carriers at cold junction will have lower density and vice versa. If the rate of diffusion of hot and cold carriers in opposite directions is equal, there will be no net change in charge. On

the other hand, if the scattering is energy dependent, the hot and cold carriers will diffuse at different rates, creating a high density of carriers at one end of the material and an electrostatic voltage. The diffusing charge are scattered by imperfections, impurities, lattice vibration, etc. The movement of heat in the form of hot charge carrier is a heat current or electric current. Electric field produced opposes the uneven scattering of carriers and equilibrium is reached when net number of carriers diffusing in one direction is cancelled by the same in the opposite direction.

2. *Phonon drag*: Phonons move against the thermal gradient and are not always in local thermal equilibrium. When they interact with electrons (or other carriers) and if imperfections is present in the crystal, they lose momentum. For the predominant phonon–electron interaction, the phonons will tend to push the electrons to one end of the material, thereby losing momentum and contributing to the thermoelectric field. When phonon–electron scattering is predominant in the temperature region, the abovementioned contribution is the most important.

   This happens for

$$T = 1/5(\theta_D) \tag{16.11}$$

   where θ_D = Debye temperature. Under a magnetic field, this region of the thermopower versus temperature function is highly variable.

16.3.3 Peltier Effect

Conversion of electrical energy to heat energy in a thermoelectric device is referred to as Peltier effects. It is the presence of heat at an electrified junction of two different metals and is named after the inventor. A French physicist Jean Charles Peltier discovered it in 1834 that when a current is made to flow through a junction composed of two dissimilar materials A and B, heat is generated at the upper junction at T_H and absorbed at the lower junction at T_C. The Peltier heat Q_P absorbed by the lower junction per unit time is equal to

$$Q_P = \pi_{AB} I = (\pi_B - \pi_A) \tag{16.12}$$

where π_{AB} = Peltier coefficient for the thermoelectric device composed of materials A and B.

π_A and π_B the Peltier coefficient of materials A and B. The π varies with the material's temperature and its specific composition. Peltier effects π should not be confused with mathematical symbol $\pi = 22/7$.

The Peltier coefficients define the amount of heat current carried per unit charge through a given material. Its value is positive below −550 K for p-type semiconductors and negative for n-type semiconductors. Peltier effect is very widely utilized in the design and development of thermoelectric cooling and thermoelectric heat pumps.

16.3.4 Thomson Effect

Thomson effect was discovered in 1851 and it describes the heating and cooling of current carrying conductor with different temperatures at any two points. When a temperature difference is maintained between two points on a current carrying conductors, then for the material used for conductors, the conductors either absorbs or emits heat.

Let us assume that
J = current density of current passing through homogeneous conductors;
Aρ = resistivity of conductor material;
Aτ = Thomson coefficient;
dT/dx = temperature gradient along the length of conductor wire;
The equation for heat produced (q) per unit volume (q) is:

$$Q = \rho J^2 - \tau J dT/dx \qquad (16.13)$$

The first term of Eq. (16.13) is the Joule heating, which does not change in sign and the second term of Eq. (16.13) is the Thomson heating, which follows J changing sign. For the conductors made of metal such as zinc and copper, the temperature is directly proportional to their potential. The heat is generated when current flows from hotter end to the colder end and the positive Thomson effect occurs.

Conversely, for the conductors made of metals such as cobalt, nickel, and iron, the temperature is inversely proportional to their potential. The absorption of heat takes place when current moves from the hotter end to the colder end, the negative Thomson effect occurs.

Though the Thomson coefficient of lead is stated to have a value of zero; in fact, it is non-zero having a very small value. In contrast, the thermoelectric coefficients of all known superconductors are zero. The Thomson coefficient is the only one directly measurable coefficient for individual materials.

The Peltier and Seebeck coefficients can only be determined for pairs of materials and there is no direct methods available for determining absolute Seebeck or Peltier coefficients for an individual material. Lord Kelvin found relationships between the three coefficients, implying that only one could be considered unique in 1854.

The first Thomson relation is

$$\tau = T dS/dT \qquad (16.14)$$

where T = absolute temperature; τ = Thomson coefficient; and S = Seebeck coefficient.

The second Thomson relation is

$$\pi = ST \qquad (16.15)$$

Considering the thermoelectric capacity of both thermoelectric materials being used in the device, the modified dimensionless figure of merit (ZT_{av}) is expressed as

$$ZT_{av} = (S_P - S_N)^2 Tav / [(\rho_N k_N)^{½} + (\rho_P k_P)^{½}]^2 \qquad (16.16)$$

where ρ = electrical resistivity; and T_{av} = average temperature between the hot and cold surfaces; further, the subscripts N and P denote properties related to the n- and p-type semiconducting thermoelectric materials, respectively.

The coefficient of performance (COP) of modern commercial thermoelectric refrigerators ranges from 0.3 to 0.6, which is hardly one-sixth the value of traditional vapour-compression refrigerators

Lord Kelvin (earlier name William Thomson) established the connection between the Seebeck and Peltier Effects in 1855, which were the first significant contribution to the understanding of thermoelectric phenomena. He showed that the Peltier heat or power (Q_P) at a junction was proportional to the junction current (I) through the relationship

$$Q_P = \pi I \qquad (16.17)$$

where π = Peltier coefficient.

However, the π should not be confused with 22/7. Through thermodynamic analysis, Thomson also showed the direct relation between the Seebeck and Peltier effects, namely that $\pi = ST$ (see Eq. 16.15), where T is the temperature of the junction.

Thomson effect predicts that heat power ($Q\tau$) is absorbed or evolved along the length of a material rod whose ends are at different temperatures on the basis of thermodynamic considerations, which is proportional to the flow of current and to the temperature gradient along the material rod. The proportionality factor τ is known as the Thomson coefficient.

16.4 THERMOELECTRIC MATERIALS

Low-cost materials having a sufficiently strong thermoelectric effect (and other required properties) are suitable for design construction of thermoelectric converters for applications in power generation and refrigeration.

A commonly used thermoelectric material in such applications is bismuth telluride (Bi_2Te_3).

The primary criterion for thermoelectric device viability is the figure of merit (see Eq. 16.6) and is given by

$$Z = \sigma S^2/k \tag{16.6}$$

where S = Seebeck coefficient (Volt/Kelvin); k = thermal conductivity (W/meter Kelvin); and σ = electrical conductivity (A/Volt-meter)

The product (ZT) of Z and the used temperature, T, serve as a dimensionless parameter to evaluate the performance of a thermoelectric material. Only a relatively few are identified as thermoelectric materials from the vast number of materials known till date. They are categorized as conventional (established) and novel (new) materials.

Example 16.1

Following properties are given for two different thermoelectric materials.

S. No.	Properties	Material A	Material B
1	Seebeck coefficient (S)	-196×10^{-16}	-224×10^{-6}
2	Specific resistivity (ρ)	1.07×10^{-3}	1.05×10^{-3}
3	Thermal conductivity (k)	14.3×10^{-3}	16×10^{-3}

Calculate the figure of merit (Z) for both materials A and B.

Solution Figure of merit (Z) for thermoelectric material is given by Eq. (16.14) as

$Z = S^2/\rho k$

Hence, for both materials A and B, the figure of merit (Z) is calculated as:

$Z_A = S_A{}^2/\rho_A k_A = (-196 \times 10^{-16})^2/(1.07\times10^{-3} \times 14.3 \times 10^{-3}) = 0.00251/K$

$Z_B = S_B{}^2/\rho_B k_B = (-224 \times 10^{-6})^2/(1.05\times10^{-3} \times 16 \times 10^{-3}) = 0.002986/K$

16.4.1 Conventional Materials

Today's most thermoelectric materials, such as

1. Bismuth telluride (Bi2Te3)-based alloy has a dimensionless figure of merit (*ZT*) value of around unity at room temperature.
2. PbTe-based alloy has a dimensionless figure of merit (*ZT*) value of around unity at 500°K–700°K.
3. Development of thermoelectric generators with thermoelectric materials having *ZT* in the range of 2–3 will have competitiveness with other power generation systems.

As the charge carriers are the interactive correlation of the thermal and electrical conductivity, minimizing the contradiction between high electrical conductivity and low thermal conductivity is highly desirable by adopting new methods. Based on the temperature range of operation, the thermoelectric materials for commercial applications can be conveniently divided into three groups:

1. *Low temperature materials up to around 450°K*: Alloys based on bismuth (Bi) in combinations with antimony (An), tellurium (Te), or selenium (Se). The highest figure of merit for a thermocouple fabricated from n- and p-bismuth telluride is around 2.0×10^{-3} K^{-1}. They are universally employed in thermoelectric refrigeration.
2. *The intermediate temperature range up to around 850°K*: It is the regime of materials based on alloys of lead (Pb).
3. *The highest temperature range up to around 1,300°K*: Thermoelements employed at the highest temperatures are fabricated from silicon germanium alloys (SiGe).

The abovementioned materials are still dominating as widely used materials for commercial or practical applications in thermoelectric generation. All the efforts for developments of new thermoelectric materials are concentrated on improving the figure of merit and conversion efficiency by reducing thermal conductivity.

16.4.2 New Materials

Typical of these are the filled skutterudites and the clathrates.

1. *Semiconductor compound (β-Zn4Sb)*: A material that is a promising candidate to fill the temperature range in the *ZT* spectrum between those based on bismuth telluride and lead telluride is the semiconductor compound β-Zn4Sb. It is relatively inexpensive, stable in operating temperature range, low thermal conductivity, and the maximum figure of merit of 1.3 at a temperature of 670°K.
2. *Silicides*: It is being given considerable attention to improve the thermoelectric properties.

This is a semiconductor at temperatures below 1,259°K and a metal above this temperature. The energy gap is around 0.88 at room temperature.

16.5 ANALYSIS OF A CONCENTRATED SOLAR THERMOELECTRIC CONVERSION

Seebeck effects in the junctions of metallic thermoelectric converters are small. However, in semiconductor thermoelectric converters, a much large Seebeck effect is achieved at the semiconductor junction.

16.5.1 Arrangement of Semiconductor Thermoelectric Generator

A thermoelectric converter (TEC) comprises p-type and n-type semiconductor legs sandwiched between two ceramic hot or cold plates and connected thermally in parallel and electrically in series, as depicted in Figure 16.6 and it is again represented in Figure 16.7. The temperature difference across the two legs induces a voltage difference due to the Seebeck effect.

The TEC performance is characterized by its figure of merit, $ZT = S^2T/(\rho k)$, where S = Seebeck coefficient; ρ = electrical resistivity; and κ = thermal conductivity.

Figure 16.7 shows p-type and n-type semiconductor legs sandwiched between a heat source and a heat sink with an electrical power load of resistance R_L connected across the low-temperature ends. A practical thermoelectric device can be made up of many p-type and n-type semiconductor legs connected electrically in series and thermally in parallel (as shown in Figures 16.4 and 16.5) between the common heat source and the heat sink. Its behaviour can be discussed considering only one couple.

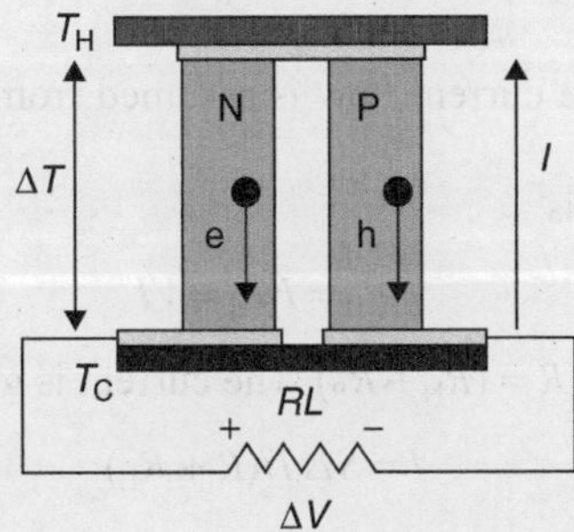

Figure 16.7 Semiconductor thermoelectric generator

The Seebeck coefficient (S) is used to describe the thermoelectric property of a thermoelectric device. From the Thomson second relation (16.15), the Peltier coefficient at a junction is equal to the Seebeck coefficient multiplied by the operating junction temperature. The Thomson effect is comparatively small, and it is generally neglected. The thermal and electric power flow in a thermoelectric device involving the following two factors in addition to the Seebeck effect.

1. The heat conduction in the two semiconductor legs between the source and the sink. The thermal flow down these two legs is given by

$$2k\,(A/L)\,\Delta T \tag{16.18}$$

where κ = their average thermal conductivity in watts per meter-Kelvin; A = area in square meters of the base of each leg; L = length of each leg in meters; and ΔT = temperature differential between the source and the sink in Kelvin.

2. Electrical resistance ohmic loss (heating) that occurs in both of the legs. The heat power produced in each leg is given by

$$\rho I^2\,(L/A) \tag{16.19}$$

where ρ = average electrical resistivity of the semiconductor materials in ohm-meters; and I = electric current in amperes.

Ohmic loss heat in each of the two legs flows towards the source and the sink approximately in the ratio of half.

16.5.2 Efficiency and Power Output

The important design parameters for a power generator device are the efficiency and the power output. The efficiency is expressed as the ratio of the electrical power output (P_0) to the thermal power input (Q_H) to the hot junction as

$$\eta = P_0/Q_H \tag{16.20}$$

The power output is the power dissipated in the load. The thermal power input to the hot junction is given by

$$Q_H = ST_H I + \tfrac{1}{2} I^2 R + K\Delta T \tag{16.21}$$

where S = Seebeck coefficient; T_H = the hot side temperature of the thermoelectric module; I = current; R = electric resistance of semiconductor legs; K = total thermal conductance of the thermoelectric cooling module; and ΔT = temperature difference between hot and cold sides = $(T_H - T_C)$.

The positive direction of the current flow is assumed from the p-type semiconductor to the n-type at the cold junction.

The electrical power output is

$$P_0 = I^2 R_L = VI \tag{16.22}$$

where R_L = load resistance and $R = (R_N + R_P)$. The current is given by

$$I = S\Delta T/(R + R_L) \tag{16.23}$$

Example 16.2

A thermal electric generator has the following properties.

S. No.	Properties	Material A	Material B
1	Seebeck coefficient (S)	-190×10^{-6}	190×10^{-6}
2	Specific resistivity ρ (ohm-cm)	1.45×10^{-3}	1.8×10^{-3}
3	Figure of merit (Z)	2×10^{-3}	1.7×10^{-3}
4	Cross-sectional area (cm²)	1	1.14
5	Length of leg (cm)	1	1

Calculate thermal conductivity of each semiconductor and also the total resistance of both the semiconductor.

Solution From Eq. (16.18), figure of merit $Z = S^2/(\rho k)$.

For N-type semiconductor,
$Z_N = S_N^2/\rho_N k_N$ or $k_N = S_N^2/\rho_N Z_N$
Therefore, $k_N = (-190 \times 10^{-6})^2/(1.45 \times 10^{-3} \times 2 \times 10^{-3}) = 0.01245$ W/cmK

Similarly,

$$k_P = S_P^2/(\rho_P Z_P) = (190 \times 10^{-6})^2/(1.8 \times 10^{-3} \times 1.7 \times 10^{-3}) = 0.011797 \text{ W/cmK}$$

Resistance of N-type semiconductor

$$R_N = \rho_N L_N/A_N = 1.45 \times 10^{-3} \times 1.0/1.0 = 0.00145\ \Omega$$

Similarly,

$$R_P = \rho_P L_P/A_P = 1.8 \times 10^{-3} \times 1/1.14 = 0.001579\ \Omega$$

Therefore, total resistance of thermoelectric converter

$$R = R_P = R_N = 0.00145 + 0.0018 = 0.00325\ \Omega$$

Since the open-circuit voltage is $S\Delta T$, the efficiency is

$$\eta = I^2R_L/[ST_H I + ½ I^2R + K\Delta T] \tag{16.24}$$

The operating design, which maximizes the efficiency, will now be calculated. Let us take

$$Y = R_L/R = R_L/(R_N + R_P) \tag{16.25}$$

$$\eta = \{Y(\Delta T/T_H)\}/[(Y + 1) - (\Delta T/2T_H) + \{(Y + 1)^2RK/S^2T_H\}] \tag{16.26}$$

The efficiency will be a maximum for RK minimized. The shape ratio, which maximizes the efficiency, is given by

$$\gamma_N/\gamma_P = [\rho_N k_N/\rho_P/k_P]^{½} \tag{16.27}$$

With this shape ratio, the efficiency is

$$\eta = \{Y(\Delta T/T_H)\}/ [(1 + Y) - (\Delta T/2T_H) + \{(1 + Y)^2/ZT_H\}] \tag{16.28}$$

The optimum load is calculated by setting the derivative of the efficiency with respect to Y equal to zero. The efficiency with both the geometric and load resistance optimized is

$$\eta = (m + 1)(\Delta T/T_H)/[m + (T_C/T_H)] \tag{16.29}$$

where

$$m = (1 + ZT_{av})^{½} \text{ and } T_{av} = (T_H + T_C)/2 \tag{16.30}$$

Under optimum load, the output current is

$$I = S\Delta T/R(m + 1) \tag{16.31}$$

The output voltage is

$$V = S\Delta T/(m + 1) = S(\Delta T) - IR \tag{16.32}$$

The output power is

$$P_0 = (m/R)[S\Delta T/(m + 1)] \tag{16.33}$$

The internal resistance R is the same as for a refrigerator and given by

$$R = [S/Z^{½}][1/\gamma_P][\rho_P/k_P] + [S/Z^{½}][1/\gamma_N][\rho_N/k_N] \tag{16.34}$$

or approximately by

$$R = (2L/A)\,[\rho_N + \rho_P] \tag{16.35}$$

The load resistance, which maximizes the power output, is obtained by setting to zero. The derivative with respect to the load resistance of the power output given by Eqs (16.22) and (16.23). The well-known result $R_L = R$ is obtained. With this load resistance, the output voltage is

$$V = ½\, S\Delta T \tag{16.36}$$

The current is

$$I = V/R = S\Delta T/2R \tag{16.37}$$

Further, the power output is

$$P_0 = (S\Delta T)^2/4R \tag{16.38}$$

Example 16.3

A thermoelectric generator with dimensions and properties given in Example 16.2 operates between the temperature limits of 100°C and 300°C. Using the optimum value for the product of internal resistance and overall thermal conductance, calculate the maximum generator efficiency and the efficiency for maximum power. Further, calculate the power output for both cases.

Solution

From Example 16.2,

$k_N = 0.01245$ W/cmK; $k_P = 0.011797$ W/cmK

$R_N = 0.00145\ \Omega$; $R_P = 0.001579\ \Omega$; $R = 0.00303\ \Omega$

From Eq. (16.27), $m = (1 + ZT_{av})^{½}$ and $T_{av} = (T_H + T_C)/2$

$T_H = 300 + 273 = 573°K$ and $T_C = 100 + 273 = 373°K$; therefore, $T_{av} = (570 + 373)/2 = 471.5°K$

From Eq. (16.5),

$$Z_{max} = (S_N - S_P)^2/([\rho_N K_N)^{1/2} + [\rho_P K_P]^{1/2}$$
$$= (-190\times10^{-6} - 190\times10^{-6})^2/\,[(1.45\times10^{-3}\times0.0124)^{1/2} + (1.8\times10^{-3}\times0.01179)^{1/2}]$$
$$= 0.00184$$
$$m = [1 + 0.0184\times471.5]^{1/2} = 1.37$$

For maximum efficiency, $\eta = [\eta_C(m-1)/(m + T_C/T_H)]\,100$

$$\eta_C = (573 - 373)/573 = 0.349\text{: } T_C/T_H = 373/573 = 0.650$$

Therefore, $\eta = [0.349\times(1.37-1)/(1.37+0.650)]100 = 6.4\%$

$$R_L = mR = 1.37\times0.00303 = 0.00415\ \Omega$$

$$P_L = 1.37/(1.37+1)^2\times[(-190\times10^{-6} - 190\times10^{-6})^2\,(573-373)^2/0.00303$$
$$= 0.4649 \text{ W}$$
$$P_{Lm} = [S^2_{P,N}\,(T_H - T_C)^2]/4R = (-190\times10^{-6} - 190\times10^{-6})^2(573-373)^2/4\times0.00303$$
$$= 0.477 \text{ W.}$$

16.6 THERMOELECTRIC CONVERTER MODULES AND APPLICATIONS

Owing to the relatively low heat-to-electricity conversion, efficiencies approaching 5% for $Z = 1$; thermoelectric converters have been mainly used in space applications. These are given in Table 16.1.

Table 16.1 Application of Thermoelectric Converters

Use Sector	Application Areas
Military	Missiles, radar, submarines, infra-red detection, guidance systems, etc.
Dedicated devices	Low-temperature oil products tester, biochemical products cryogenic test instrument, bacterial incubators. thermostatically developing tank and computers.
Medicine	Cold, cold with cataract extraction device like blood analyser
Laboratory appliances	Well cold, cold boxes, cold trough, low-temperature testing electronic devices, high temperature control equipment
Daily use appliances	Air conditioning, dual-use hot and cold boxes, drinking fountains, electronic mailbox

The development of new functional ceramic materials has opened new high-temperature application areas like solar electricity generation. It is now possible for thermoelectric converter modules or arrays with concentrated solar radiation to obtain large temperature differences ΔT across the plates, and consequently, increases the energy conversion efficiency.

These thermoelectric devices are then connected electrically in series (see Figure 16.8) and/or parallel (see Figure 16.9) forming an array of multiple thermocouples (thermopile). When heat and cold are applied, this device then generates electricity. Almost any heat source can be used to generate electricity, such as solar heat, ocean heat, geothermal heat, even body heat, or any other heat source. Recovery of wastes heat from the thermoelectric generators and its utilization for other applications can further enhance the conversion efficiency,

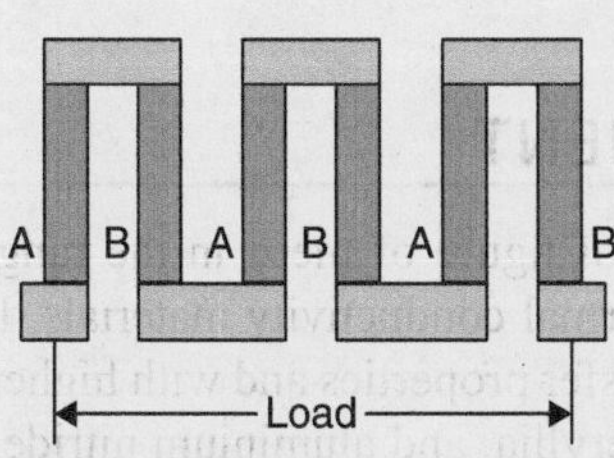

Figure 16.8 Series connection of couples of thermoelectric converters

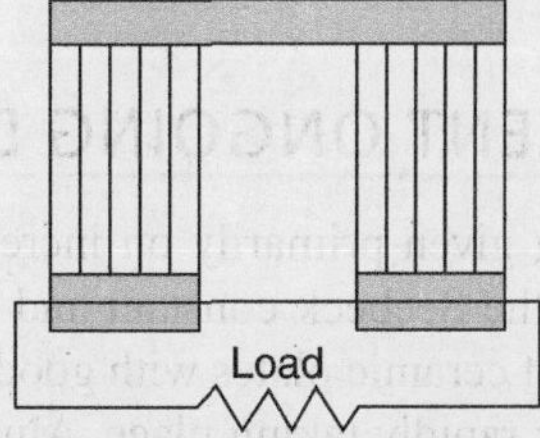

Figure 16.9 Parallel connection of couple of thermoelectric converters

The voltage and electric power output are increased by increasing the temperature difference between the hot and cold ends. In practical to increase the power output, parallel and series connections are used as shown in the Figure 16.8 and 16.9. The direct current generated can be changed into alternating current by inverter. The alternating voltage and current is then increased to the desired value with a transformer.

16.7 ADVANTAGES AND DISADVANTAGES

16.7.1 Advantages

Thermoelectric power generators offer several distinct advantages over other technologies:

1. They are extremely reliable (typically exceed 100,000 h of steady-state operation).
2. They are silent in operation since they have no mechanical moving parts.
3. They are require considerably less maintenance.
4. They are simple, compact, and safe.
5. They have very small size and virtually weightless.
6. They are capable of operating at elevated temperatures.
7. They are suited for small-scale and remote applications typical of rural power supply, where there is limited or no electricity;
8. They are environmentally friendly.
9. They are not position-dependent.
10. They are flexible power sources.
11. Unlimited application areas.
12. Free scalable from mW to kW.
13. Neither toxic nor dangerous.

16.7.2 Disadvantages

1. The major drawback of thermoelectric power generator is their relatively low conversion efficiency (typically ~5%).
2. Still rather expensive fabrication and materials.

16.8 RECENT ONGOING DEVELOPMENT

Focus is being given primarily on increasing the value of figure of merit in the range of 2–3 by increasing the Seebeck constant and reducing the thermal conductivity materials. Improved development of ceramic plates with good lateral heat transfer properties and with higher thermal conductivity is rapidly taking place. Alumina (Al_2O_3), beryllia, and aluminium nitride are tried as materials for ceramic plates. Material research is directed towards the use of other materials to improve the efficiency and power output of thermoelectric modules in large operating temperature range.

SUMMARY

- Thomas Johann Seebeck first identified in 1821 that if two different conductors are joined and the two junctions are maintained at different temperatures, an electromotive force is developed in the circuit.
- The Seebeck effect is a phenomenon in which a temperature difference between two dissimilar electrical conductors or semiconductors produces a voltage difference between the two substances. If the two conductors or semiconductors are connected together through an electrical circuit, and direct current (DC) flows through that circuit
- In 1834, Jean Charles Athanase Peltier first identified that if a current flows in a circuit consisting of two different conductors, then one of the junctions is heated and the other is cooled.
- When electric current is maintained in a circuit of material consisting of two dissimilar materials, the cooling of one junction and the heating of the other is called Peltier effect. The effect is even stronger in circuits containing dissimilar semiconductors.
- Peltier effect is the reverse of Seebeck effect. Seebeck effect is used in thermocouples and Peltier effect is used in refrigeration.
- On a later date around 1854–1855, William Thomson first deduced and demonstrated that when a temperature difference exists between two points in a single electrical conductor, an electrical potential is established between the points. Simultaneously, he also established that if a current passes through a conductor in which a temperature gradient exists, this current causes a flow of heat from one part to the other.
- The evolution or absorption of heat when electric current passes through a circuit composed of a single material that has a temperature difference along its length is known as Thomson effect.
- The voltage of a single thermoelectric converter by Seebeck effect is small of the order of few mV/K of temperature difference at the junction.
- Large number of thermoelectric devices can be connected in series to increase the output voltage or in parallel to increase the maximum deliverable current.
- Large arrays of Seebeck-effect devices can provide useful, small-scale electrical power, if a large temperature difference is maintained across the junctions.
- A thermoelectric module is an array of thermocouples connected electrically in series but thermally in parallel.
- Good thermoelectric materials should have the following characteristics:
  (a) High electrical conductivity (σ) to minimize Joule heating rise in temperature from resistance to electric current flowing through it.
  (b) Large Seebeck coefficient (S) for maximum conversion of heat to electrical power or electrical power to cooling performance.
  (c) Low thermal conductivity (λ) to prevent thermal conduction through the material.
- Figure of merit of thermoelectric generator, $Z = S^2\sigma/\lambda$

  where α is the Seebeck coefficient of the material (V/K).
  σ is the electrical conductivity of the material (A/V/m), and
  λ is the thermal conductivity of the material (W/m/K)

REVIEW QUESTIONS

1. Briefly describe (i) Seebeck effect, (ii) Peltier effect, and (iii) Thomson effect
2. What is the basic difference between thermoelectric and thermionic conversion systems?
3. Explain the working of thermoelectric generators.
4. Derive the expression for efficiency of a thermoelectric generator.
5. Write short notes on:
   (a) Multistage thermoelectric generator.
   (b) Selection of thermoelectric material.
   (c) Figure of merit of a thermoelectric generator.

Chapter 17

Thermionic Converters

Thermionic generator converts heat directly into electricity using thermionic emission rather than first changing it to some other form of energy. A thermionic power converter has two electrodes. One of these is raised to a sufficiently high temperature to become a thermionic electron emitter. The other electrode called a collector because it receives the emitted electrons and it is operated at a significantly low temperature. The space between the electrodes is sometimes a vacuum, but it is normally filled with a vapour or gas at low pressure. The thermal energy may be supplied by chemical, solar, or nuclear sources. The emission of electrons from emitter is analogous to the liberation of steam particles when water is heated. These emitted electrons flow toward the collector and the circuit can be completed by interconnecting the two electrodes by an external load. Part of the thermal energy that is supplied to liberate the electrons is converted directly into electrical energy, while some of the thermal energy heats the collector and must be removed. Thermionic converters are solid-state devices with no moving parts. They can be designed for high reliability and long service life. Thus, thermionic converters have been used in many spacecrafts.

KEY CONCEPTS

- Thermionic energy converter
- Working of thermionic generator
- Performance analysis of thermionic converter
- Efficiency of a thermionic converter
- Types of thermionic converters
- Material development and research
- Applications

17.1 GENERAL

Thermionic converter is a device that directly converts heat energy into electrical energy by the thermionic emission phenomenon and it does not involve any intermediary energy form. They are highly reliable and has long service life and used on spacecraft and similar applications.

It consists of two electrodes and has no moving parts. One electrode is kept at a high temperature for thermionic electron emission and is called emitter. The other electrode is called collector, which is maintained at significantly low temperature, and it receives electron emitted by emitter. The inter-electrode gap is either vacuum or vapour or low pressure gas. Chemical, nuclear, or solar energy source are used to supply thermal energy to emitter.

Thermionic emission is a process of electron emissions from a heated electrode. The valence electrons in electrode material (like tungsten) having large kinetic energy to escape from the electrode but attractive force of nuclei hold them in the electrode itself. Some valence electrons pick up enough thermal energy when the electrode is heated to sufficiently high temperature. Electrons then break away from heated electrode and go to inter-electrode space. These liberated electrons in the space flow towards the collector. Electric current flows when an electrical load is connected externally between the two electrodes.

The amount of thermionic emission depends on Fermi energy level of metals and temperature. The additional energy that a Fermi-level electron is required to escape from heated electrode is called work function and is of the order of volts. Only a part of thermal energy supplied to emitter is converted to electrical energy by thermionic emission and remaining part of the energy heats the collector. The heat coming from the collector needs to be removed for efficient operation.

The basic phenomena characteristics for thermionic energy conversion can be divided into two classes, which are as follows:

1. *Emission phenomena*: They are related to process by which particles are emitted from or collected at the surface of the thermionic energy converters.
2. *Transport phenomena*: They are related to the process of migration of and interaction between the particles in the inter-electrode space.

17.2 THERMIONIC ENERGY CONVERTER

The schematic diagram of thermionic converter is given in Figure 17.1. The five components of a basic thermionic energy converter are as follows:

1. Thermionic emitter
2. Collector
3. Working fluid, which may be an electron gas or partially ionized plasma or vacuum
4. A small gap between two electrodes
5. An electrical load connected externally to electrodes to complete the closed circuit

One electrode is the emitter (cathode) having low work functions, which is supplied with heat (Q_H) at high temperatures (1,500°K–2,000°K).

The other electrode is the collector (anode) that is kept at low temperatures (800°K –1,000°K), and the electrons condense on it. Part of the heat (Q_C), which is removed from the emitter by evaporating the electrons, is transported to the collector by the electrons. The remaining part ($Q_H - Q_C$) is converted into electrical energy. The collector is cooled to remove the output heat. The newly emitted electrons resist the movement of electrons from the emitter to the collector and this effect is called space-charge effect.

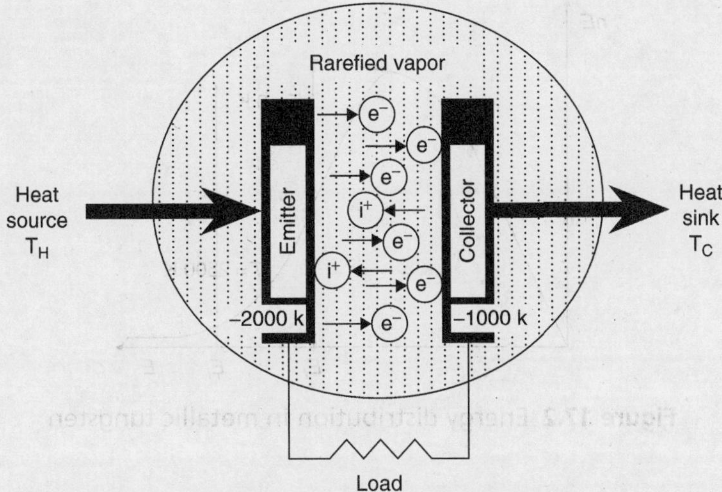

Figure 17.1 Schematic diagram of thermoelectric converter

17.3 WORKING OF THERMIONIC GENERATOR

It can be understood from the following steps:

1. When the emitter of thermionic converter is heated at sufficiently high temperatures, electrons escape from it.
2. They move in vacuum or gas in inter-electrode space and go towards collector maintained at low temperatures.
3. Collector collects emitted electrons.
4. When the two electrodes are externally connected to electrical load, an output voltage is produced and current flows through the load.
5. Large temperature difference and surface work function between the two electrodes maintain the flow of electrons through the load.

17.3.1 Fermi Level

The highest energy level attainable by an electron at absolute zero temperature is called Fermi level. It lies in between the valence band and the conduction band. It occupies a place inside the conduction band for metals, in between the valence band and the conduction band for semiconductors, and inside the valence band for insulators.

Fermi energy (μ) corresponds to the highest energy of all free electrons at absolute zero. An expression for Fermi energy may be obtained on the basis of completely degenerate function.

A plot of the energy distribution for metallic tungsten at 0°K and at 2,500°K is shown in Figure 17.2. The area under the curve represents the total number of free electron nE (always given as number of electron per eV per cubic meter of metal). Thus,

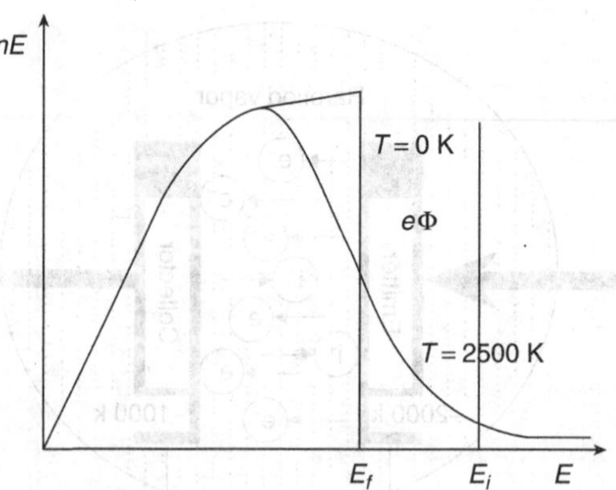

Figure 17.2 Energy distribution in metallic tungsten

$$n = \int_0^{E_f} \gamma E^{1/2}.\mathrm{d}E \tag{17.1}$$

$$= 2/3.\gamma E_f^{3/2} \tag{17.2}$$

Or $$E_f = [3/2.(n/\gamma)]^{3/2} \tag{17.3}$$

where γ = Constant = $4\pi[2me]^{3/2}/h^3 = 6.82 \times 10^{27}$; m = mass of electron in kg; e = electron charge = 1.6×10^{-19} Coulombs. (17.3a)

Therefore, $E_f = 3.64 \times 10^{-19}.n^{2/3}$ (17.4)

Since the number of free electrons (n) Varies from metal to metal, Fermi-level electron (μ) will also vary from metal to metal. For most of metals, $\mu < 10$ eV.

Example 17.1

The specific gravity of tungsten is 18.8 g/cm and its atomic weight is 184. Assume that there are two free electrons per atom. Calculate the numerical value of n and μ.

Solution A mole is a quantity of any substance and its molecular weight is given in g.

One mole of any substance contains the same number of molecules as one of any other substance.

This number is called Avogadro's number = 6.02×10^{23} molecules/mole.

$n = 6.02 \times 10^{23}$ (molecules/mole) $\times$ 1/184 (mole/g) $\times$ 18.8(g/cm^2) $\times$ 2 (electron/atom).
= 12.3×10^{22} (electron/cm^2) = 1.23×10^{29}(electron/m^3)

For tungsten, atomic and molecular weights are same. Therefore,

$$E_f = 3.64 \times 10^{-19} \times (1.23 \times 10^{29})2/3 = 8.59 \text{ eV}.$$

17.3.2 Thermionic Work Function

Thermionic work function is defined as the amount of energy required to separate electron from a metal surface and to move it from Fermi level to surrounding medium. In a simple way, it can be said that it is the minimum energy needed to remove electron from conduction band of metals. The work function plays very important role in thermionic emission and the main source of energy is the heat energy.

17.3.2.1 Richardson's Law

The Richardson's law (or Richardson–Dushman equation) defines the current density of thermionic emission in terms of the emitter material and temperature as

$$J = A_G T^2 \text{EXP}\left[-(\Phi/kT)\right] \tag{17.5}$$

where Φ = work function of the metal; k = Boltzmann constant;

$$A_G = \lambda_R A_0 \tag{17.6}$$

where

λ_R = material-specific correction factor that is typically of order 0.5.

A_0 is the proportionality constant known as Richardson's constant, and it is also given by

$$A_0 = 4\pi m_e k^2 e/h^3 = 1.20173 \times 10^6 \text{ (A/m}^2\text{°K}^2\text{)} \tag{17.7}$$

where m_e = mass of the electron = 9.108×10^{-28} (g); h = Planck's constant = 4.140×10^{-15} (eVs); e = electronic charge = 1.6×10^{-7} (coulombs); k = Boltzmann constant = 1.38066×10^{-23} J/°K or 8.6173324E^{-5} (eV/K).

In 1911–1930, as physical understanding of the behaviour of electrons in metals increased, various different theoretical expressions (based on different physical assumptions) were put forwards for A_G, by Richardson, Saul Dushman, Ralph. H. Fowler, Arnold Sommer feld, and Lotger Wolfgang Nordheim.

Thermionic emission (meaning electrons escaping from the heated negatively charged filament, hot cathode) is important in the operation of vacuum tubes. Let the potential energy difference across the surface due to effective surface dipole be ψ_T. Further, let E_f be the Fermi energy calculated for the finite solid without considering surface distortion effect (see Fig. 17.3), when taking the convention that the potential at $r \to \infty$ is zero. Then, the correct formula for work function is as follows:

The work function Φ of the surface is defined as

$$\Phi = -E_f + \psi_T \tag{17.8}$$

Thus, the electron work function (Φ) would be defined as

$$\Phi = \psi_T - E_f \tag{17.8a}$$

where E_f is negative, which means that electrons are bound in the solid and ψ_T is the inter-electrode motive, which is defined as a scalar, whose negative gradients equals the force exerted on electrons.

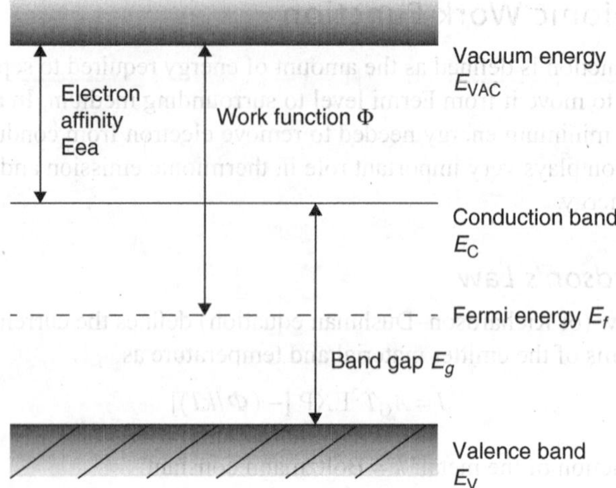

Figure 17.3 Simple energy band diagram

17.3.2.2 Measurement of Work Function

It is based on the thermionic emission of electrons from an emitter. The current density J of the electrons collected by the sample depends on the work function Φ of the sample and is given by the Richardson–Dushman equation as

$$J = A_G T^2 \text{EXP}\,[-(\Phi/kT)]$$

The work function of a sample can be determined by the temperature variation of the sample.

Rearranging Eq. (17.5) yields

$$\ln(J/T^2) = \ln(A_G) - \Phi/kT \tag{17.9}$$

The line produced by plotting $\ln(J/T^2)$ against $(1/T)$ will have a slope of Φ/k, allowing to determine the work function of the sample.

Electron excitation is achieved by absorption of photons in the photoelectric effect. The photoelectric emission takes place when photon energy exceeds the work function of substance and electron is liberated from the surface of substance. The photoelectric work function is given by

$$\Phi = hf_0 \tag{17.10}$$

where h = Planck's constant and f_0 = minimum (threshold) frequency of the photon required to produce photoelectric emission.

17.3.2.3 Work Functions and Richardson's Constants for Various Materials

In Table 17.1, Richardson's constants = Ab (A cm^{-2} K^{-2}(b is the material correction factor)).

Table 17.1 Work Functions and Richardson's Constants for Various Materials

Material	Work Function	Richardson Constant	Material	Work Function	Richardson Constant	Material	Work Function	Richardson Constant
Molybdenum	4.15	55	Platinum	5.32	32	Thoriam	2.54	3.0
Nickel	4.61	30	Rhenium	4.85	100	BaO+ SrO	0.95	~10^{-2}
Tantalum	4.12	60	Thorium	3.38	70	Cs-oxide	0.75	~10^{-2}
Tungsten	4.54	60	Thorium	3.38	70	TaC	3.14	0.3
Barium	2.11	60	Ba on W	1.56	1.5	LaB_6	2.70	9
Cesium	1.81	160	Cs on W	1.36	3.2	———	———	
Iriduim	5.40	170	Th on W	2.63	3.0	———	———	

17.3.3 Inter-electrode Charge Distribution

Figure 17.4 gives a schematic representation of the electron charge distribution in the inter-electrode space of a thermionic converter.

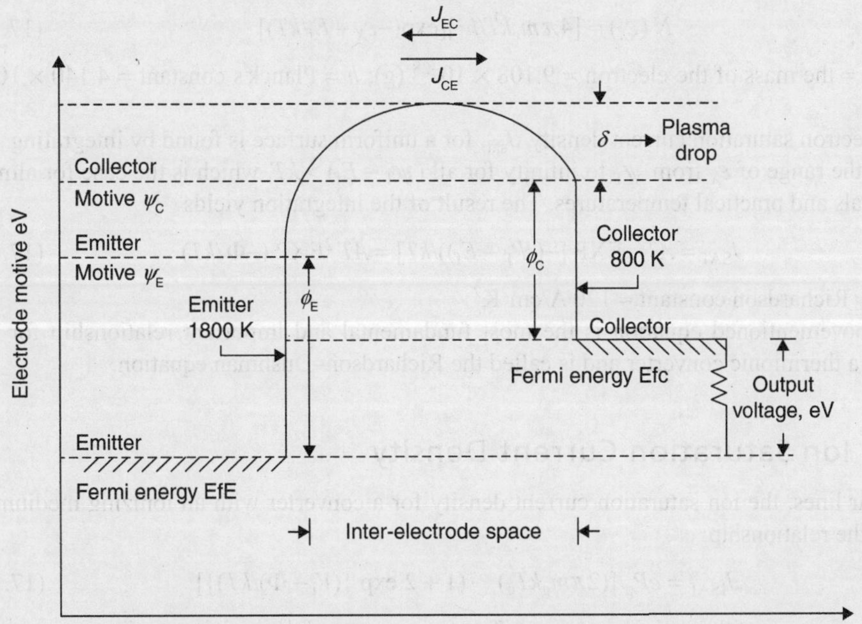

Figure 17.4 Electron charge distribution in inter-electrode gap of thermionic generator

Under ideal conditions of particle transport, the motive force varies linearly from the motive just outside the emitter (ψ_{EM}) to the motive outside the collector surface (ψ_{CO}). The net current exchanged between the two electrodes is

$$= J_{EC} - J_{CE} - J_{IEC} \tag{17.10a}$$

where J_{EC} = emitter–collector current; J_{CE} = collector–emitter current; and J_{IEC} = ion current.

The internal voltage drop (ΔV) is given by

$$\Delta V = (\psi_{EM} - \psi_{CO})/e \tag{17.11}$$

Neglecting lead losses and partial interaction losses, the output voltage is given by

$$V = (V_{CO} - V_{EM})/e \tag{17.12}$$

17.3.4 Electron Saturation Current

It is defined as the electron ejected from heated electrode into the vacuum ionized medium in between inter-electrode space without any external energy source.

As this quantity depends on the number of free electrons $N(\varepsilon_X)$, Fermi–Dirac statistics provide the means to compute the number of free electrons, $N(\varepsilon_X)\,d\varepsilon_X$, incident on a unit area within the metal in unit time with energies corresponding to the motion normal to the area between ε_X and $\varepsilon_X + d\varepsilon_X$.

For energies greater than the Fermi energy, the functional dependence of $N(\varepsilon_X)$ on ε_X is given by

$$N(\varepsilon_X) \approx [4\pi m_e kT/h^3][\exp(-\varepsilon_X - E_f/kT)] \tag{17.13}$$

where m_e = the mass of the electron = 9.108×10^{-28} (g); h = Planck's constant = 4.140×10^{-15} (eVs)

The electron saturation current density, J_{sat}, for a uniform surface is found by integrating $N(\varepsilon_X)$ in the range of ε_X from ψ_T to infinity for all $(\psi_T - E_f) > kT$, which is the case for almost all materials and practical temperatures. The result of the integration yields

$$J_{SAT} = AT^2.\ \mathrm{EXP}\ [-(\Psi_T - E_f))/kT] = AT^2\mathrm{EXP}\ (-\Phi/kT) \tag{17.14}$$

where A = Richardson constant ≈ 120 A/cm^{2}K^2.

The abovementioned equation is the most fundamental and important relationship for the design of a thermionic converter and is called the Richardson–Dushman equation.

17.3.5 Ion Saturation Current Density

On similar lines, the ion saturation current density for a converter with an ionizing medium is given by the relationship:

$$J_{ISAT} = eP_g/[(2\pi m_g kT_g)^{1/2} \cdot (1 + 2\exp\{(V_i - \Phi)/kT)\}] \tag{17.15}$$

where P_g= pressure of the ionizing medium; T_g = temperature of the ionizing medium; m_g = mass of the ionizing medium; V_i = first ionization energy; J_{ISAT} = ion saturation current density; and e = electronic charge.

17.3.6 Performance Analysis of Thermionic Converter

Figure 17.4 is redrawn in Figure 17.5 for thermionic generator. The emitter (cathode) is heated by external source. When the temperature increases, then thermionic emission takes place.

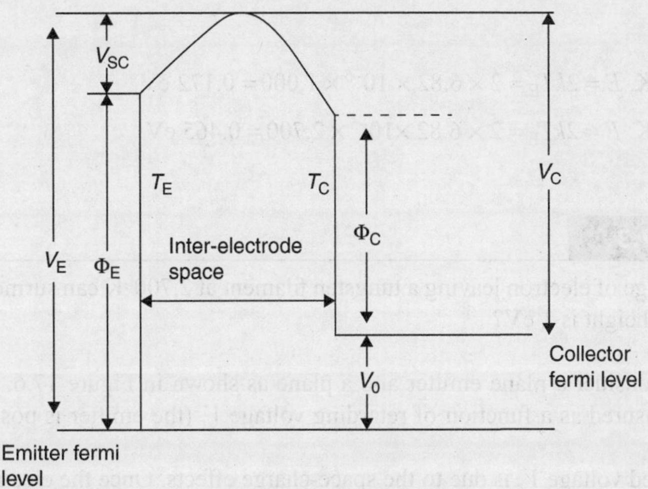

Figure 17.5 Energy band diagram of thermionic converter

As shown in the figure, let V_E = electron motive cathode (emitter) in V; Φ_E = cathode (emitter) work function; V_{SC} = space charge barrier of cathode, V; T_E = cathode temperature, °K; V_C = electron motive of collector, V; Φ_C = collector work function; and T_C = collector temperature, °K; Then, current densities of cathode and anode are, respectively,

$$J_E = AT_E^2 \exp(-V_E/kT_E)\ \text{A/cm}^2 \tag{17.16}$$

$$J_C = AT_C^2 \exp(-V_C/kT_C)\ \text{A/cm}^2 \tag{17.17}$$

Thus, for electron to leave the cathode (emitter), the energy required will be greater than Φ_E.

Space-charge effect is considered, which retard the flow of electrons from cathode (emitter) by an additional potential V_{SC}.

The output voltage (V_0) across the load is

$$V_0 = V_E - V_C = \Phi_E - \Phi_C \tag{17.18}$$

Further, since $V_{SC} = V_E - \Phi_E$;

$$V_E = \Phi_E + V_{SC} \tag{17.19}$$

Therefore, energy required for electron to escape from the cathode is

$$Q_1 = (Q_E + V_{SC})J_E\ \text{W/cm}^2 \tag{17.20}$$

Each electron also carries away its kinetic energy. A statistical analysis provides an empirical formula for average kinetic energy of the escaping electron as $E = 2kT_E$ for the motion of electrons.

Example 17.2

Calculate the average energy of the escaping electrons from an emitter surface at the temperature (i) 1,000 °K and (ii) 2,700 °K. Boltzmann constant is 6.82×10^{-5} eV/°K.

Solution

(i) At 1,000 °K, $E = 2kT_E = 2 \times 6.82 \times 10^{-5} \times 1{,}000 = 0.172$ eV

(ii) At 2,700 °K, $E = 2kT_E = 2 \times 6.82 \times 10^{-5} \times 2{,}700 = 0.465$ eV

Example 17.3

What percentage of electron leaving a tungsten filament at 2,700°K can surmount space charge barrier whose height is 1 eV?

Solution Consider a plane emitter and a plane as shown in Figure 17.6. Parallel collector current is measured as a function of retarding voltage V_r (the emitter is positive with respect to collector).

This retarded voltage V_r is due to the space-charge effects. Once the electron cloud is build up between the electrodes, the flow of electron from emitter is retarded.

Thus,
$$I = I_{th} \cdot \exp(-V_r/V_T) \tag{17.21}$$

where V_T = kT//e = $T/11, 600$ and k = Boltzmann constant, that is, 6.82×10^{-5} eV/°K

From Eq. (17.21),
$$I/I_{th} = \exp(-V_r/V_T)$$

From the given data, $V_r = 1$ eV; $V_T = T/11{,}600 = 2{,}700/11, 600 = 0.2327$

Therefore, $I/I_{th} = \exp(-V_r - V_T) = \exp(-1/0.2327) = \exp(-4.29) = 0.0136 = 1.36\%$

Hence, only 1.36% of electrons have a surface-directed energy in excess of 1 eV.

$$Q_2 = (2kT_E/e) J_E \text{ W/cm}^2 \tag{17.22}$$

$$Q_1 = J_E V_E \tag{17.23}$$

$$\text{Therefore, } Q = Q_1 + Q_2 = J_E V_E + (2kT_E/e) J_E = J_E [V_E + 2kT_E/e] \tag{17.24}$$

The back emission from the anode (collector) will similarly carry energy to the cathode.

The net energy exchanged between the two electrodes is

= cathode – anode energy + average kinetic energy of escaping electrons – anode to cathode back emission energy

$$= J_E [V_E + 2kT_E/e] - J_C [V_C + 2kT_C/e] \tag{17.25}$$

$$\text{The output power } (P_L) = V_0 (J_E - J_C) \tag{17.26}$$

Hence, thermal efficiency of thermionic generator (η) is obtained as

$$\eta = P_L/Q \tag{17.27}$$

Substituting Eqs (17.26) and (17.25) in Eq. (17.27), we get

$$\eta = P_L/Q = [V_0 (J_E - J_C)]/\{J_E [V_E + 2kT_E/e] - J_C [V_C + 2kT_C/e]\} \qquad (17.28)$$

Let us define

$$D_E = V_E/kT_E$$
$$D_C = V_C/kT_C \text{ and}$$
$$T = T_C/T_E$$

Substituting the abovementioned expressions in Eq. (17.19) result in

$$\eta = (D_E - TD_C).\ [1 - T^2 \exp(D_E - D_C)]/\{(D_E + 2) - T^2 (D_C + 2T) \exp(D_E - D_C)\}] \qquad (17.29)$$

For maximum efficiency, defined constant D_E and D_C must be equal. Let $D = D_E = D_C$. Then,

$$\eta_{max} = (1 - T)D/(2 + D) \qquad (17.30)$$

17.3.7 Other Efficiency Equations of a Thermionic Converter

The thermionic converter is basically a heat engine in which the thermodynamic efficiency is limited by the efficiency of a Carnot cycle. Thus, the higher the temperature of the heat input stage of the thermodynamic engine, the greater the efficiency that may be theoretically achieved, since this efficiency is, of course, a function of the difference in temperature between the input and the output stage. However, in a working engine, other loss factors are also introduced.

The efficiency of a thermionic converter is defined as the ratio of the output power to the input power. Many authors have formulated expressions for this efficiency and for the optimum operating conditions in a converter, taking into account the losses that are associated with both electrical and thermal factors.

As a very general statement of the efficiency (η) of such a converter, the expression can be written.

$$H = \Phi_E - \Phi_C - V_{ext}/(\Phi_E + P_L/J) \qquad (17.31)$$

where Φ_E and Φ_C = work functions of the emitter and collector, respectively; and J = net current.

P_L is all the thermal losses inherent in the operation of a thermionic converter lumped together. It includes (i) the radiative energy transfer from the emitter to the collector, (ii) the kinetic energy of the electrons that results in a rise of temperature of the collector, (iii) the thermal conduction through lead wires, and (iv) the gaseous heat conduction.

V_{ext} is the voltage drop in the external circuit determines the electrical loss resulting from the resistance of the electrical circuit. A compromise has to be reached in choosing the proper diameter of the leads to the external circuit since leads of small diameter would reduce the heat loss, but at the same time, the effective voltage in Eq. (17.31) would be lowered because of the presence of the high electrical resistance.

$$P_L = \sigma(T_E^4 - T_C^4)\,[1/\varepsilon_E + 1/\varepsilon_C - 1]^{-1} \qquad (17.32)$$

where σ = Stefan-Boltzmann's constant; ε_E = thermal emissivity of the emitter; ε_C = thermal emissivity of the collector; and T_E and T_C = operating temperatures of the emitter and collector, respectively.

The saturation current, *Js*, at a temperature, T °K, generated from an electron emitter is given by the well-known Richardson–Dushman equation.

$$Js = AT^2 \exp(-\Phi_E/kT_E) \quad (17.33)$$

where k = Boltzmann constant; and A = Richardson A that has a theoretical value of 120 A/cm² °K².

For the semi ideal case, scientist Mr Hernquist substituted Eq. (17.23) into Eq. (17.21) and set the net current, J, equal to the saturated emission current of the emitter as given by Eq. (17.23). The maximum efficiency then becomes

$$\eta = (1 - \Phi_C/\Phi_E)\,[1 + (\sigma/A_1)\,\{1/\varepsilon_E + 1/\varepsilon_C - 1\}^{-1}\,[(T_E^4 - T_C^4)/(\Phi_E T_E^4)]\,\exp(-\Phi_E/kT_E) \quad (17.34)$$

Thus, it can be seen that a low collector work function and a low collector thermal emissivity favour high conversion efficiencies. The efficiency is strongly dependent upon the work function of the emitter because Φ_E occurs in the exponential.

In general, if the other quantities in Eq. (17.24) remain constant, the efficiency at first increases with increasing Φ_E. It reaches a maximum for some value of Φ_E and then decreases for higher Φ_E.

Other more complex expressions for the maximum efficiency of a thermionic converter, in which most of the heat loss factors are taken into account, have been determined by Rasor, Houston, Hatsopoulos , Kaye, and Schock among others. Most of these expressions were derived by assuming arbitrary values, but reasonable values for certain of the parameters and optimizing the efficiency equations with respect to various physical properties of the materials concerned, such as dimensions and the thermal and electrical characteristics. For example, Rasor derived a maximum efficiency equation in which the collector work function was arbitrary but the emitter work function and collector temperature were optimized. This is shown as

$$\eta_{max} = \beta/(1 + 4E\Phi_E)^3 \quad (17.35)$$

where Φ_E = emitter work function; β= constant ≈ 0.8; and $E = D\sigma\alpha^4\,(A_r\varepsilon/A_E J_s)$ (17.36)
where

$$D = 1/(1 + 2k\alpha/\varepsilon) \approx 0.9 \quad (17.37)$$

k, σ and *Js* have been previously defined, α is a function of Φ_E and Φ_C , and the temperatures that are chosen for the electrodes and is approximately 640 °K/eV.

For the system considered by Rasor, A_r is the effective area for heat radiation from the emitter, A_E is the effective emitting area of the emitter, and ε is the effective emissivity of the emitter–collector system.

17.4 TYPES OF THERMIONIC CONVERTERS

Depending on the presence of an ionizing medium in the inter-electrode gap, thermionic converters can be broadly classified as in the following sections.

17.4.1 Vacuum Thermionic Converter

The vacuum thermionic converter has a very small inter-electrode gap for reducing the effects of space charge. The inter-electrode space is evacuated so that the space is free of particles other than electrons, thereby neutralizing the negative space charge build-up on the electrode surface and reducing the total number of electrons in transit. Although this type of converters are developed and tested long time back, they had only limited commercial applications owing to manufacturing difficulty of very small spacing between electrodes.

Since the thermionic emission involves hot electrons, these electrons carry excess energy that can be converted into electrical power. The operation may be described in an electromotive diagram, as shown in Figure 17.6.

In this diagram, the operation is in a regime where space-charge effects can be neglected.

Electromotive diagrams for thermionic energy conversion devices shown in Figure 17.6 represents a system that employs a flat surface emitter and collector, and Figure 17.7 represents a system with an emitter that has field enhancing structures on the surface.

The symbol Φ_E represents work function of the emitter electrode material, Φ_C represents work function of the collector electrode material, and V represents electric voltage developed.

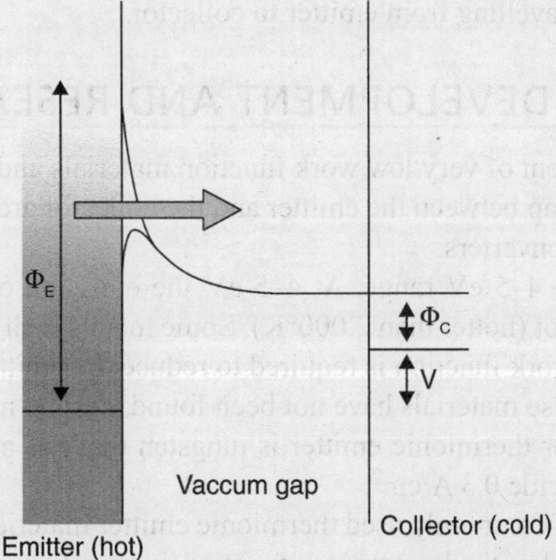

Figure 17.6 Electromotive diagram of thermionic energy conversion device with flat surface emitter and collector

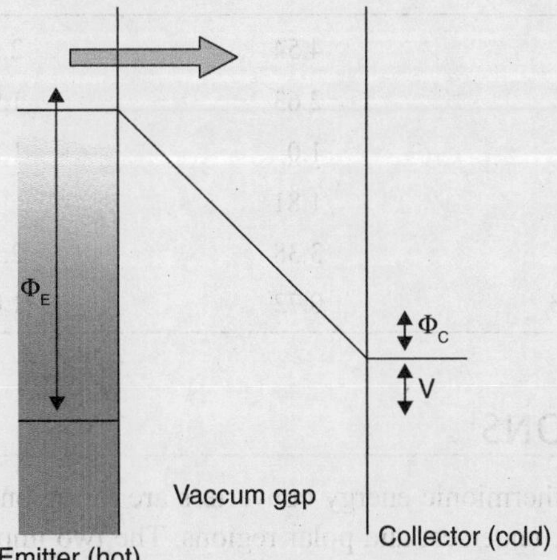

Figure 17.7 Electromotive diagram for thermionic energy conversion device with an emitter that has field enhancing structures on the surface

17.4.2 Vapour Thermionic Converters

Owing to manufacturing limitations, vacuum thermionic converters are getting replaced by gas filled or plasma converters. In vapour-filled thermionic converters, the inter-electrode space is filled with a rarefied ionizing medium at a vapour pressure, generally, on the order of 1–10 Torr (1 mm to 10 mm of Hg at 0°C). Since Caesium has a low ionization potential, it is very widely used in efficient thermionic converters. Other alkali metals (such as potassium and rubidium) are also used.

They are designed in such a way that the positively charged ions are continuously generated. When positively charged ions mix with negatively charged electrons, it results in plasma with a relatively neutral space charge. Thus, liberated emitter electrons experience very little electrostatic resistance in their travelling from emitter to collector.

17.5 MATERIAL DEVELOPMENT AND RESEARCH

The design and development of very low work function materials and manufacturing techniques for obtaining very small gap between the emitter and the collector are very essential for efficient operation of thermionic converters.

Most metals are in the 4–5 eV range. At 4–5 eV, the emission of electrons does not occur until the cathode is very hot (hotter than 2,000°K). Some metals melt before they emit electrons. Materials with very low work function is required to reduce the emitter operating temperature.

The problem is that these materials have not been found, despite much searching.

The best pure metal for thermionic emitter is tungsten that has a work function of 4.54 eV which at 2,500°K can provide 0.3 A/cm^2.

Basic characteristics of commonly used thermionic emitter materials are shown in Table 17.2. Caesium (an alkali metal) has the lowest work function approaching unity.

Table 17.2 Important Characteristics of Some Thermionic Emitter Materials

Material	A (A/cm^2.K)	Φ (work function) (eV)	Temperature (°K)	J (A/cm^2)
Tungsten	60	4.54	2,500	0.3
Thoriated W	3	2.63	1,900	1.16
Mixed oxides	0.01	1.0	1,200	1.0
Caesium	162	1.81		
Tantalum	60	3.38	2,500	2.38
Cs/O/W	0.003	0.72	1,000	0.35

17.6 APPLICATIONS

Principal applications of thermionic energy converters are in regions that are not easily accessible, such as outer space, undersea, and polar regions. The two important heat sources are the sun and nuclear reactors. They are widely used as a source of electric power for spacecraft, submarines and boat, and for domestic, agricultural, and industrial purposes.

SUMMARY

- Thermionic generator is a device for the conversion of thermal energy into electrical energy on the basis of the phenomenon of thermionic emission.
- Thermionic emission is the emission of electrons from a heated filament or substance.
- The emission of particles from a hot filament was first discovered by Thomas Edison in 1883 but the effect was left unexplained until the discovery of the electron by J.J. Thomson in 1897.
- The minimum amount of energy required to emit electrons from a metal surface is called the work function or threshold energy of that metal.
- In this type of emission, the electron emission is achieved by heating the electrode. Due to heating, the electrons get enough energy that they emit from the surface of that material. An electron emitted from a hot cathode comes out with a velocity that presents difference between the kinetic energy possessed by electron just before emission usually used in cathode of diode, triode, pentode, CRT, etc.
- Emission of electrons is fundamental to thermionic power conversion.
- The advantages of thermionic generators include absence of moving parts, high reliability, compactness, and possibility operation for longer time without regular servicing.

REVIEW QUESTIONS

1. What is thermionic emission effect?
2. Describe the principle of working and constructional details of basic thermionic generator.
3. Derive the expression for power and efficiency for a thermionic generator.
4. Explain in brief
   (a) Potential barrier
   (b) Thermionic emission
   (c) Work function
5. What is the basic principle of thermoelectric power generation?
6. Draw schematic of thermoelectric converter. Analyse the performance.
7. Describe the operation of thermionic converter.

Chapter 18

Concept of Energy Conservation and Energy Management

Till date, energy is recognized as a key measure of techno-socio-economic development of a nation. Terms energy conservation, energy consumption, and energy management have found wide publicity globally, and in many cases, people get confused with these terms and consider them as one. The energy is always conserved and consumed. The two terms energy conservation and energy consumption cannot be recommended as they are technically inaccurate terminology although they are very frequently used.

The term energy management is accepted globally more accurate as it includes performing a task with minimum quantity of energy at an appropriate quality without affecting the personnel working in the vicinity with capital expenditure avoidance. It is presently finding wide publicity and is being continuously recommended term as it is technically accurate.

Energy conservation or energy management is a term that has a number of meanings, but it is mainly concerned with the one that relates to saving energy in all the energy activity sectors.

KEY CONCEPTS

- Definition and necessity of energy conservation or management
- Concept of energy conservation or management
- Opportunities of energy management
- Fundamental principles of energy management
- Strategy for energy management
- Energy savings tips
- Constraints and considerations for implementing energy conservation and management program

18.1 INTRODUCTION

Because of the energy scarcity and fuel shortages that began to manifest themselves in the early 1970s as a result of Arab–Israel conflict, the cost of energy began its asymptotic rise. Government regulatory agencies throughout the world were formed and a new discipline in energy conservation and management were focused to tackle the energy problems.

Rising population and rapid pace of economic growth make its energy needs particularly challenging and most of the country is facing a series of challenges to expand its energy supply while increasing efficiency. In the light of energy crunch being faced all around the world, India too is facing varying associated difficulties for its techno-socio-economic development because of energy shortages.

Therefore, each government has pressure for intervention through the introduction of energy management opportunities to enhance efficiency in the industrial and other energy activity sector.

To cope up with the rapid depletion of fossil and nuclear fuel reserve because of exponentially rising large energy consumption in industrial and other sectors, implementation of energy management program to reduce energy usage, increase energy efficiency, and cost reduction is given high priority.

The energy availability problems have further been complicated by its energy dependence on foreign country. Each country has recognized energy as a key input for techno-socio-economic growth and the prosperity of a nation and its people is indexed by the per capita energy consumption. Nevertheless, consequent upon the fast increasing population, financial scarcity, improper coordination, unsystematic approach, etc., for reducing the energy demand-supply gap are a necessity of time for accepting the challenges of energy scarcity.

The majority of country worldwide, no doubt, is endowed with multifarious natural and other energy resources; further, increase in energy generating capacity is an expensive proposition associated with environmental and social problems.

Planning and operating energy-related production and consumption activity, resource conservation, environmental protection, and cost savings are the key elements included in the energy management.

Till date, energy is recognized as a key measure of techno-socio-economic development of a nation, the following three terms have found wide publicity globally. Further, in many cases, people get confused with them and consider them as one term. These three terms are as follows:

1. *Energy conservations*: From the principle of conservation of energy, energy can neither be created nor destroyed, but it can be transformed from one form to another. The objective of this principle is simply means in present day context is to obtain maximum useful energy and least wasteful energy.
2. *Energy consumptions*: Quality of life, standard of living, and prosperity of a country are indexed by per capita of energy consumption of its people. It is, therefore, a parameter used to identify between developed, developing, and underdeveloped nations. Minimizing wasteful energy was also included in this context.
3. *Energy management*: This term is accepted globally more accurate as it includes performing a task with minimum quantity of energy at an appropriate quality without affecting the personnel working in the vicinity with capital expenditure avoidance.

The two terms energy conservations and energy consumptions cannot be recommended as they are technically inaccurate terminology although they are very frequently used. Energy management term is presently finding wide publicity and is being continuously recommended term as it is technically accurate and universally accepted methodology of reduction of energy consumption.

Energy management measures are carried out to perform a task or job with minimum energy use and with minimum expenditure to achieve increased efficiency with minimum energy wastage, minimizing energy consumption, and development of new substitute energy resources.

18.1.1 Necessity of Energy Management

Energy management is finding wide publicity as a subject of current interests because of its multiplied effects on energy savings as well as on improving the national economy. Effective energy management appears to be the most significant strategy for alleviating the inevitable energy crunch represented by a widening energy supply-demand gap. The principal goal of energy management is the reduction in the energy wastage for a particular task at an appropriate quality together with the capital expenditure avoidance. Therefore, in the present chapter, an attempt has been made to highlight several aspects related with energy management strategy. Concept of energy management, introduction of possible energy management opportunities, and elements of energy strategy are salient features covered in the book. It is also emphasized that end-use energy management program can be implemented either as housekeeping, maintenance, and operational improvement exercise or as retrofit (replacement) exercise or as new design improvement and installation exercise.

18.2 CONCEPT OF ENERGY MANAGEMENT

The term energy management is widely used to signify the utilization of minimum quantity of energy required for the task at an appropriate quality together with the capital expenditure avoidance. To be more specific, energy management involves the following steps:

1. Utilization of minimum quantity of energy required for the task
2. Appropriate quality, neither better nor worse than needed
3. No damage to environment and morale of personnel
4. Minimum additional investment

In most practical applications, energy management signifies preserving the quality of the energy form. The quality of an energy source is measured in terms of its available work, which is consumed in the process of utilizing energy, and thus, conservation of available work is a meaningful goal of energy management policies.

The energy system includes

1. Infrastructures like buildings and technologies such as fittings, appliances, and transportation deliver energy services.
2. Infrastructures and technologies such as oilfields, mines, pipelines, solar collectors, and power stations that produce, process, and deliver the fuels and electricity required for energy services.
3. The energy system also includes the markets, policies, guidelines, etc., and regulations that co-ordinate and govern how the system operates.

Energy flow diagram in any energy activity system is shown in Figure 18.1, where energy is made available to energy users in a usable form. Raw resource harnessed and tapped from energy

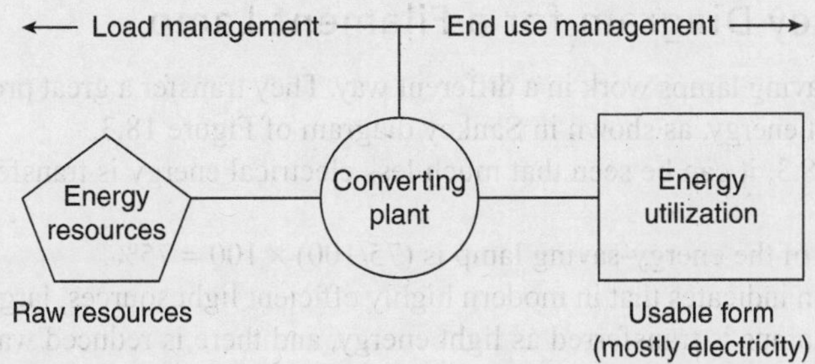

Figure 18.1 Energy flow diagram in an energy activity system

resource is transported to converting plant that converts energy to usable form. Ultimately, this usable energy is supplied to energy users.

Electricity, being the main usable form of energy, is used in every walk of life ranging from household sector to industrial sector for the purpose of lighting, motive power, and to produce heat. Nevertheless, energy in other forms is also used.

Energy wastage and dissipation are always associated with energy activities starting from energy resources up to energy end use. Energy that is wasted, like the heat energy from an electric lamp, does not disappear but it is spread up in surroundings in a scattered way.

18.2.1 Wasted Energy in Electric Lamps

Ordinary electric lamps contain a thin metal filament that glows when electricity passes through it. However, from the Sankey diagram for a typical filament lamp, as shown in Figure 18.2, most of the electrical energy is transferred as heat energy instead of light energy

The efficiency of electric lamp can be calculated using the equation:

$$\text{Efficiency} = (\text{useful energy transferred/energy supplied}) \times 100 = (10/100) \times 100 = 10\%.$$

The abovementioned calculation shows that only 10% of total electricity is input to the bulb and it is utilized for illumination purposes and rest of the energy is wasted as heat.

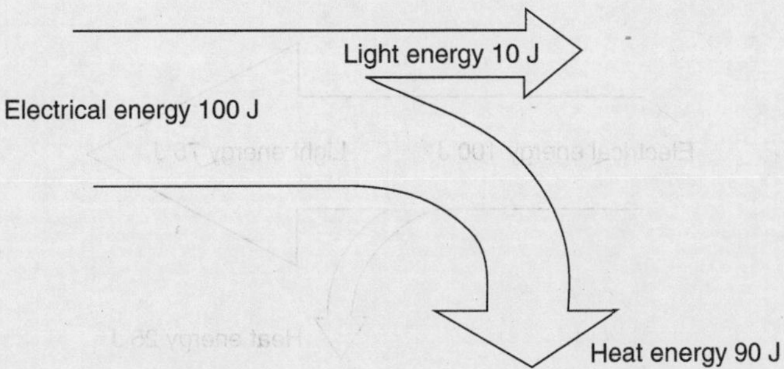

Figure 18.2 Sankey diagram for a filament lamp

18.2.2 Sankey Diagram for a Filament Lamp

Modern energy-saving lamps work in a different way. They transfer a great proportion of electrical energy as light energy, as shown in Sankey diagram of Figure 18.3.

From Figure 18.3, it can be seen that much less electrical energy is transferred, or wasted, as heat energy.

The efficiency of the energy-saving lamp is (75/100) × 100 = 75%.

This calculation indicates that in modern highly efficient light sources, large share of electricity input to light lamps is transferred as light energy, and there is reduced wastage of energy as heat.

A light bulb is said to have the theoretical maximum luminous efficacy of 683.002 lumens when all the electrical energy input is transferred into light and no energy is wasted as heat. It may be noted that the efficiency of a device will always be less than 100%.

The luminous efficiency is defined as the ratio of luminous efficacy to the maximum luminous efficacy.

The results for incandescent lamp and CFL are

Incandescent lamp = 14.16/683.002 = 2% and

CFL = 69.23/683.002 = 10%

Thus, in the case of CFLs, 10% of the power consumed is being converted into light. Ideally, we want 100% of the input energy to be converted into light, but 10% is 5 times better than the incandescent bulbs at 2%.

It is now an established fact that 1 unit of energy saved at user's end conserves 2.5 to 3 units of energy available from energy resources. Reduction and minimization of energy wastage at each point of energy flow diagram is a necessity for bridging the gap between the supply and the demand of energy in view of techno-socio-economic development and population of a nation.

Therefore, energy management includes the following aspects

1. load management
2. efficient end use
3. fuel conservation
4. more efficient equipment
5. more efficient process

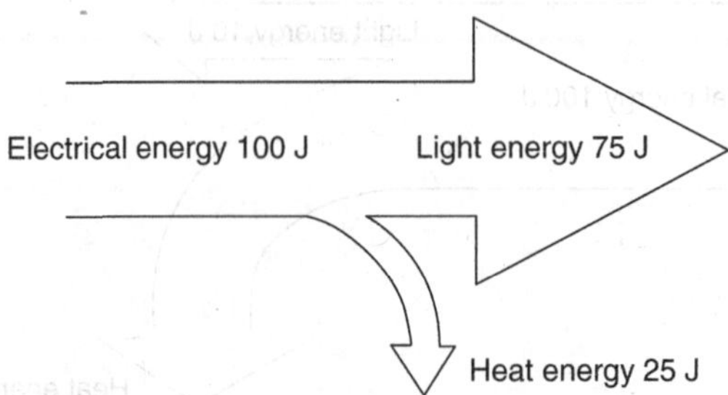

Figure 18.3 Sankey diagram for a lamp

Load management refers to the supply side of the energy system. It includes those activities taken by energy utility that is used to manipulate the load seen by their generating system in such a way to meet energy demands so that most favourable and economic operating conditions can be achieved.

End-use management relates to reduction in and control of energy at the user's own convenience. Such an attempt can help in reducing the needed capacity and can help in obtaining more favourable energy pricing.

A new terminology, the demand-side management (DSM) has been now universally accepted as a combination of load and end-use management. It is a program implemented for reducing peak load demand, and thereby postponement of further plant capacity addition when investment cost is high and sites are scarce. It gives significant financial, environmental, and operational benefits.

When DSM is applied to the consumption of energy, in general, and not just electricity but fuels of all types, it can also bring significant cost benefits to energy users and less damage to environment. Opportunities for reducing energy demand are numerous in all sectors and many are low cost, or even zero, or no cost, and items that most enterprises or individuals could adopt in the short term, if good energy management is practiced.

All kinds of energy users are motivated to lower their energy demand to the extent possible following the DSM and to act as virtual power plants. Such activities at energy end-user improve the stability of electricity grid and energy cost savings.

From the abovementioned discussion, it may be said that energy management, although not a physical resource but is only a program, may be considered as the cheapest and environment friendly energy.

18.2.3 Benefits

The motivation behind the implementation of energy management, which is presently addressed as DSM, is obviously different for the various segments involved in energy activity system.

1. *Energy suppliers or utilities*: For utility companies, the reduction or shift of a customer's energy demand could mean avoiding or delaying building additional generating capacity when sites are scarce and huge investments are required. In some situations, this would avoid or defer energy price increases that would otherwise be imposed on customers to help finance new investments in system capacity.
2. *Energy users*: Demand-side management offers several benefits to energy users in different sectors. Industrial sector users reduce their production costs and maintain their competitiveness in national and global market. Money saved by residential and household sector energy user can be utilized on other commodities with added benefits of improved comfort and living style.
3. *Government organization, manufacturers, enterprise, and owners of buildings*: They are largely interested in DSM to lower their own energy consumption and costs, and thus extending their support to electric supply company to maintain a good power quality and reliable energy supply.

Benefits or cost analysis and pay back calculation of any proposed program is essential to determine benefits of program and selection of alternative solution provided a reasonable return on investment can be assured. The enterprise management should take prompt action, and in some cases, with the technical advice of the utility company experts.

From the abovementioned discussion, benefits of energy management can be concluded as follows:

1. *Saving money*: Saving money is the principal benefit that users derive from energy management program.
2. *Improving the utility load factor*: Another important beneficial aspect of energy management is the improvement of utility load factor. Because of the properly conducted energy management programs, there will be least consumption of energy for the same task, and therefore, utility can postpone the addition of new energy generating capacity at a time when the sites are scarce and capital costs are high.

18.2.4 Key Issues

These issues are very simple to complex that can be tackled with the coordinated efforts of energy utilities, researchers, engineers, scientists, technologist government bodies, manufactures, suppliers, and lastly energy users. Principal costs requirements, highly involved time consuming R&D activities, technology transfer from research laboratories to ground implementation, environmental acceptability, and off-the shelves availability of developed products constitute the main issues of energy management. These issues for planning and carrying out energy management program and can be divided into two categories:

1. *Techno-economic issues*: These issues may include:
   (a) The improvement in existing technology (retrofitting exercises).
   (b) Development and application of new components, sub-system, and systems (installation and new design and systems).
   (c) Identification and application of preferred system, etc.
2. *Programmatic and institutional issues*: Several aspects in this category are as follows:
   (a) identification of energy users' need
   (b) achieving balance between end-use energy management and load management
   (c) identification of priorities between long-term, medium-term and short-term programs
   (d) establishing the relative emphasis between R&D, demonstration, testing, and commercial applicability programs
   (e) establishing the role of the government agencies, the energy suppliers (utilities and corporations) with equipment manufacturers and suppliers
   (f) evaluation of energy management program on the basis of benefits and costs with due regards to the risk and probability of success

To be more specific, any energy management implementation criteria should be based on the following questions

1. Does it satisfy the need?
2. Does it have a favourable impact on utilities or end users?

3. Does it involve R&D or demonstration or testing?
4. Who should carry it out?
5. Will it lead to identification of preferred systems in the areas of energy use which is involved?

18.2.5 Opportunities of Energy Management

In general, end-use energy management opportunities can be looked from different directions based on relative costs, relative time, and complexity and relative energy savings. They are grouped as follows.

18.2.5.1 Procedural Energy Management Opportunities

Procedural energy management opportunities (PEMOS) involve housekeeping and maintenance type actions with little or no cost involved. Although housekeeping and maintenance measures will reduce energy consumptions and energy costs, continuous efforts must be sustained; otherwise, these savings are lost by going back to the traditional ways.

Energy management programs with housekeeping and maintenance measures can be implemented by reduction of demand

1. either through self-awareness
2. self-denial
3. through enforcement and regulatory actions
4. through economic measures

The first two ways of implementations are perhaps the better one and last two may cause chaos, political, and social problems.

As it is well-established fact that electric passenger trains consume about 175 kW to about 190 kW power, while diesel-driven passenger trains consume 25 to 35 gallons of high speed diesel for their start up, unauthorized stoppage of trains started creating large energy wastage. Railway in cooperation with district administration tried to curb it through enforcement and regulatory actions, but it introduced some other types of social problems. Ultimately, self-awareness minimized the unauthorized stoppage to large extent.

Following are some examples in this category that can be approached at no or zero cost simply to implement almost in no time and energy can be saved up to 10%–15%:

1. Unauthorized stoppage of trains must be avoided.
2. Timers or switches to turn light on only when and where needed.
3. Cleaned light fixtures transmit more light. Hence, fixtures and light sources have to be cleaned frequently. Layers of dust can absorb up to 30% of the light from the lamps.
4. Use low wattage bulbs in areas that do not need bright lights, for example, in storage rooms and bathrooms.
5. When damaged fluorescent lamp is removed, which is not intended to be replaced immediately, the starter, or else the choke should also be removed. Otherwise, they will consume electricity at the rate of 12 to 15 W.

6. Garment in bulk should be ironed. After the ironing work is completed, iron should be turned off and plug from the socket should also be removed without fail.
7. Room temperature of 24°C–26°C is comfortable enough for normal work or relaxation. Air conditioner should be set for abovementioned temperature range.
8. Mobile phone chargers could also consume up to 10 watts if left on, even though the phone may not be connected.
9. Working off at rated and properly balanced supply voltages reduce electric losses.
10. Well-lubricated equipment has reduced frictional losses.
11. Use of properly sized motors to run at full load minimizes electric losses and run only when needed.
12. Optimize use of thermostat settings when space heating and cooling are required.
13. Maintenance of heat transfer surfaces and air conditioning.
14. Inspection of unit followed by appropriate repairs.
15. Keep refrigerators away from direct sunlight.
16. Defrost the freezer compartment regularly since ice build-up causes excessive electricity wastage.
17. Install improved bearings and lubricate regularly.
18. Maintain all equipment regularly.

18.2.5.2 Equipment Modifications, Additions, or Replacement EMOS (Retrofit EMOS)

This can be implemented using available off-the-shelf hardware and technology to achieve the desired end results with great efficiency of course, capital expenditure involved may vary from moderate to high level. The following are the examples:

1. Use of more efficient lamps (replacement of incandescent lamps by fluorescent tubes: replacement of fluorescent tubes by metal halide or sodium lamps)
2. Use high efficiency motors
3. Use task lighting
4. Use of solid state devices (such as solid state ballasts and solid state fan regulators), which consume less energy
5. Use of low energy consuming household electric appliances without sacrificing their functional requirements
6. Use of solid state controlled variable speed electric drives if loads vary like in flow control, thereby avoiding the throttling losses
7. Employ method of regeneration to recover and return energy to the source in industrial and transportation sector, etc.
8. Replace aging appliances with new energy efficient ENERGY STAR models
9. Use cogged belts or improved gears; smooth belts often slip and are not efficient

Table 18.1 Comparison of Different EMOS

EMOS	Relative Cost	Relative Complexity	Relative Time to Implement	Relative Benefits
Housekeeping and maintenance	Low	Low	1 Year	10%–15%
Retrofitting	Moderate	Low–moderate	1–2 Years	15%–30%
R&D	High	High	2–10 Years	20%–50%
Resource substitution	Mod to High	High	3–5 Years	20%–40%

18.2.5.3 Research and Development and New Installation (R&D EMOS)

This not only involves high capital investment but also involves R&D activities. Examples are as follows:

1. Redesign of production processes
2. Use of efficient processes like continuous steel rolling mills avoiding energy loss involved in cooling and reheating in batch production, etc.
3. Improvement in design and manufacture of less energy consuming equipment and appliances
4. Design and implementation of new system of transportation.

18.2.5.4 Resource or Fuel Substitution (Substitution EMOS)

Substitution of one fuel or energy source for another may or may not lead to net energy savings but can shift the energy requirement from one fuel to another, thus saving scarce resources. Following are the examples:

1. Interconnection of solar energy system of heating and cooling with conventional system. Solar energy will reduce the burden on coal or electricity used for the abovementioned purpose and can be utilized for other use.

18.2.5.5 Comparison of Energy Management Opportunities (EMOS)

A brief comparison of different EMOS is given in Table 18.1.

Implementation of the energy management program based on the abovementioned approaches depends largely on the availability of technology and economic justification. It will be good practice to analyse each EMOS in the form that may be easily understood by the individual with the ultimate authority for deciding whether or not to implement the EMOS.

18.3 FUNDAMENTAL PRINCIPLES OF ENERGY MANAGEMENT

Energy management signifies as task- or job-based energy use of proper quality of energy form where, when, and how much it is needed as per its basic definition. Judicious selection of quality of energy for a process or product controls the misuse of potential value of energy or fuel.

In order to obtain hot water or low temperature process stream, the best use would be to employ a topping cycle of gas turbine rather than high quality energy form (oil or natural gas or electricity). Gas turbine is used to perform primary task and the low quality waste heat for process steam production and water heating.

Irrespective of wide variety and diversity of energy end-use technology, fundamental principles of energy management can be identified and grouped together that applies to wide variety of applications.

It is difficult to be exhaustive in reviewing several energy management principles because of the enormous diversity of processes in industrial, transportation, residential, and agricultural sectors, and in technologies, equipment, etc., having unique features and thus calling for unique energy management principles. Effective energy management principle can, therefore, be applied by taking variety of steps depending upon their suitability for a particular energy activity. The cost effectiveness of these principles depends upon the energy cost together with the detrimental effects on production and/or employee, general principles of energy management can be grouped as given in Table 18.2.

Table 18.2 Fundamental Principles of Energy Conservation and Management

S. No.	Fundamental Principles	Relative Cost	Relative Complexity	Relative Time to Implement (Year)	Expected Energy Savings (%)
1	Review of past energy use	Low	Low	1	5–10
2	Review of current energy use (energy audit)	Low	Low	1	5–10
3	Improvement in housekeeping and maintenance	Low	Low	1	10–15
4	Energy use analysis	Low to Moderate	Moderate to high	1–2	10–15
5	Economic calculation	Moderate to high	Moderate to high	Several years	10 –20
6	Material substitution	Moderate	Low	1–2	10 –20
7	Material economic	Low	Low to high	1–3	15–50
8	Good quality materials	Low to Moderate	Low to Moderate	1–2	15–20
9	Use of efficient equipment and appliances	Moderate to high	Moderate to high	More than 3 years	10 –30
10	Use of efficient process	Moderate to high	Moderate to high	Several years	10 –30

(Continued)

Table 18.2 Continued

S. No.	Fundamental Principles	Relative Cost	Relative Complexity	Relative Time to Implement (Year)	Expected Energy Savings (%)
11	Energy containment	Moderate to high	Moderate to high	Several years	15–50
12	Aggregation of energy uses	Moderate to high	Moderate to high	Several years	15–50
13	Fuel and energy Substitution	Moderate to high	Moderate to high	Several years	15–30
14	Cascading of energy uses	Moderate to high	Moderate to high	Several years	15–50
15	Energy conversions	Moderate to high	Moderate to high	Several years	15–50
16	Energy storage	Moderate to high	Moderate to high	Several years	15–50

18.3.1 Review of Past Pattern of Energy Use

Review of past pattern of energy use brings out unrecognized scheduling discontinuities and seasonal variation in energy use. Although the picture provided is incomplete, they suggest the point of effective savings.

18.3.2 Review of Current Pattern of Energy Use

The need to reduce energy costs is a crucial business practice for any successful establishment and review of current practice of energy use (energy audits) has begun to play a more significant role in managing energy expenses in this regards. Energy audit of an equipment, appliance, and machines and establishment is a method of obtaining current practice of energy use with instrumentation that aims to evaluate the existing energy consumption to identify the potential energy savings.

Energy audit of a premise provides points of inefficiency and energy wastage, which can be immediately solved by implementing energy conservation measures. It also provides a unique pathway for energy users to save money. Energy conservation and cutting energy supplier costs are extremely important as energy prices rise.

18.3.3 Improvement in Housekeeping and Maintenance

Improving the housekeeping and maintenance program in establishment equipment, appliance, and machines lead to saving energy with minimum cost complexity. This principle is included in the following section.

18.3.3.1 Proper or Optimal Control of Equipment and Appliances

Proper or optimal control of energy equipment and appliance is the first principle of effective energy management program. Examples are as follows:

1. Improved lighting controls: a system of ON–OFF switches on individual or specific group of light fixtures (called selective switching) to turn light on only when and where needed. Automatic switching using either photo control or clock timers are well suited for lighting controls such as for street lighting, outdoor, security lighting, and vehicle parking places.
2. Idle operation of electric motors should be avoided; turn off motors when not in use.
3. Low cost microcomputer or microprocessor permit a much broad range of control function that could be used to optimize the equipment and process functioning, thereby reducing energy consumption.
4. Solid state controlled variable speed drive can provide better control of working machine in the industrial and other sectors, thereby minimizing energy consumption.

18.3.3.2 Optimization of Excessive Capacity of Equipment and Appliances

Since excess capacity of equipment and appliance for particular application leads to inefficiencies, it is better to optimize capacity of equipment and appliances. Consider the following examples:

1. Excess capacity in electric motors leads to inefficiencies. Electric motors are inefficient at partial loads and since power factor also decreases at partial loads, electrical distribution system incurs high line losses.
2. Excessive illumination should be avoided. Required illumination for visual task in the working area only and appropriate lower levels in the general areas (such as corridors, storage, and circulation areas) should be provided. However, loss of productivity, employee morale, and human discomfort should not be ignored while implementing the reduction of illumination levels.
3. Minimization of heat losses. For reducing heat losses in electric heating, proper insulation of furnace walls, ducts, pipes and proper covering of tanks or vats help in optimizing the furnace capacity.

18.3.3.3 Reduction in Non-essential Loads

Non-essential loads should be avoided to the possible extent.

1. A good example of this principle is related with lighting system. Since office building almost invariably have excessive illumination levels, wastage of energy is recognized in two ways:
   (a) first, excessive electricity is consumed to provide excessive illumination
   (b) second, extra energy consumption required in air conditioning system to remove extra heat in the summer
2. Reduce peak demand by rescheduling electric motor operation.
3. Since air-conditioners do not work effectively when the outside air temperatures is close to the temperature of the conditioned space, air-conditioner should be shut down under these conditions.

18.3.4 Analysis of Energy Use

Energy data obtained by energy audit is analysed to determine effectiveness of equipment and appliances in establishment and to determine energy consumption or savings if output parameter changes (flow control in fluid flow control equipment) or to simulate operations.

18.3.5 Economic Calculation

Economic evaluation is an essential tool for comparing alternative option. Simple economic calculation, benefits or cost analysis, and payback time calculation based on present and future costs are used for identification of preferred system leading to energy savings.

18.3.6 Material Substitution

Substitution of proper material for a particular application has an advantage of energy savings.

1. Low melting point alloys can be substituted for high temperature materials used for low temperature applications.
2. Substitute materials for improving hazardous environment because of home appliances, electronics, and information equipment.
3. Substituting a material that is easy to machine or that involves less energy for manufacturing.
4. Water-based paint without baking.
5. Saw dust is a good material to combine with clay or cement mixtures and used for walls as compared to dry clay walls. These walls turn out surprisingly sturdy and effectively recycle any trees that may need to be excavated from the building area. Hardwood dust in the walls absorbs moisture content and avoids cracking of walls during freezing cycle.
6. Amorphous silicon as substitute of the glass frame in the solar panels in which the solar cells are enclosed in an array. Amorphous silicon, although less abundantly found in nature, serves all the requirements of solar panel manufacturing just like silicon in its crystal form. The solar panels made out of amorphous silicon are sturdy and also powerful.

18.3.7 Material Economy

Material economy means recovery of scrap, reduction of waste, and salvage value. The powder metallurgy is one example. Product design such as electric equipment and mechanical transmission gears permits salvage or recovery of reusable parts. Building structure designed for easy dismantling and shifting to new location and reuse.

18.3.8 Material Quality

Unnecessary quality almost always means high cost and large energy use. Therefore, material quality selection is extremely important. Purity of chemicals and process stream has an important impact on energy use. Whether distilled water is needed for a task or deionized water is sufficient?.

18.3.9 Use More Efficient Equipment and Appliances

Energy consumption for a particular work can be reduced by using more efficient equipment and appliances. Consider the following examples

1. Use of more efficient lamps: replacement of incandescent lamps by fluorescent tubes and replacement of fluorescent tubes by sodium vapour and metal halide lamps should be promoted. However, this may be noted that all light sources are not appropriate to any particular application and lumens per watt (efficacy) should not be the only factor to be considered.
2. Use of highly efficient, high power factor, and/or high voltage motors.
3. Use of highly efficient, electric motor-driven household electrical appliances.
4. Use of solid state components and devices that consume less energy; electronic regulators with ceiling and table fans and solid state ballasts with lamps can perform same functions with reduced electrical energy consumption.
5. Use of solid state controlled electric drives if load varies such as flow control in large rating pump and fan drives avoiding the throttling losses.

18.3.10 Employing Special Techniques to Reduce Losses

It includes the following losses

1. Line loss correction by the use of power factor correcting capacitors and resizing of the distribution system, etc., would lead to further reductions in energy consumption.
2. Better insulation of high temperature devices or buildings can reduce losses.

18.3.11 Use More Efficient Processes

Technological changes have made possible the use of more efficient processes that are not only more effective, but also reduce energy consumption. The following are the examples:

1. Use of continuous rolling mills avoiding energy loss involved in cooling and reheating in batch production.
2. Use of solid state controlled variable speed drives in transportation and industrial sectors with the provision of energy recovery; use of three-phase technology rather than single-phase or direct current (dc) technology may be preferable.
3. Use of dielectric heating rather than resistance heating because of the fact that in dielectric heating, heat is deposited directly in the heated materials as opposed to other heating processes.
4. Use of powder metallurgy rather than machining will help in reducing process energy consumption.

18.3.12 Energy Containment

Energy containment means to confine energy in a restricted energy flow path avoiding energy and heat losses. The following are the examples.

1. Repair of steam leaks and better insulation on boilers and piping to recover heat.
2. Repair of compressed air system leaks to reduce energy losses.

3. Energy in the form of heat is available at a variety of sources in industrial operation, many of which are not normally derived from primary heat sources. Such sources include electric motors, transformers, crushing and grinding operations, air compressors and air thickening, and drying processes.
4. Recovery of waste heat from process operations can contribute to increase in overall energy efficiency.
5. The heat pump in refrigerator system effectively contains energy and reduces losses.

18.3.13 Aggregation of Energy Uses

Electric load wave shaping to reduce the peak load demand is perhaps the main advantage of aggregation of energy use. Aggregation may be thought as combining loads of different consumers with different load factors in such a way that peak load demand is greatly reduced. Care should be taken that benefits should almost equally shared by all energy user in aggregation pool.

A variety of methods have been proposed based on energy price, energy (defined as the measure of the useful work obtainable from an energy resource or material), Community Choice Aggregation, etc., but none is accepted universally. Aggregation of energy uses helps in achieving high efficiency in many cases. For example, proper location of process steps in a manufacturing plant and energy used in transportation of product material from one process step to other is minimized. Another example includes the time sequencing of energy use like using heat generated by one step of the process to provide preheating needed by the other step.

18.3.14 Fuels and Energy Substitutions

1. Cogeneration plants should be promoted where a large amount of heat and electricity are used. Energy savings through cogeneration is best explained by its alternate name of combined heat and power (CHP).
2. As the name suggests (whether gas turbine or steam turbine CHP), electric power and heat are generated simultaneously. Electricity generated is used to improve stability and reliability of electric supply system. The process heat is recovered in the form of steam or hot water and utilized for process heat applications. CHP has the advantages of energy efficiency and energy savings. Cheap renewable fuels are available as fuel substitution for CHP and provides great flexibility.
   (a) Interconnection of solar energy system of heating and cooling with the conventional system may not lead to net energy savings but can shift the requirements from one fuel to another, thus saving scarce energy.
3. Interconnection of electricity generated from renewable energy resources to the national grid.
4. Captive power generation near the workplace.

18.3.15 Cascade Energy Use

The use of the residual heat in liquids or steam from one process to provide heating, cooling, or pressure for another process is referred to as cascading of energy use. Energy use efficiency can be further improved by cascading energy use. For example

1. The use of steam from an electric power plant in a district heating system.
2. Processing of biomass in bio-based final product and utilization of resulting product for material or energy is called cascading of biomass. As an example, biorefining of grass yields biofuels, fibres, sugar, and proteins.
3. Wood biomass is used in construction and furniture materials; it can be reused and recycled and final product is used to produce energy as wood pellets.
4. Geothermal energy can be cascaded for electricity generation and direct thermal heating applications.

18.3.16 Development of Simple and Highly Efficient Energy Conversion Principle

This is also a practical method of effective energy management. For example,

1. Solid state controlled variable speed electric drive with regenerative facility can eliminate mechanical drive train, thereby reducing losses.
2. Air-driven equipment might be replaced with electric drives.

18.3.17 Development of Energy Storage System

The development of bulk energy storage system is the most important components of energy management program. If energy storage systems are effectively developed, they will improve the reliability of energy delivery to energy consumers, improving the efficiency and stability of energy distribution and development of equipment and appliances.

The way energy is stored depends primarily on the source of energy. Presently, for limited capacity, direct or indirect energy storage are battery and hydro pump storage, chemical and electro chemical storage, sensible and latent heat storage, mechanical and flywheel storage and hydrogen energy storage, etc.

18.4 STRATEGY FOR ENERGY MANAGEMENT

Energy management system should be designed in such a way that it strengthens self-reliance at sector, local, state, or nation level. It should also be in harmony with environment to insure long term sustainability of the development process and it has to be economically efficient so that it results in economic growth. The broad spectrum of ideas, factors, and problems involved in developing and implementing any energy management program requires some type of methodological framework to relate and integrate several complex parts into a unified whole by following problem solving approach. The problem solving approach is a combination of engineering and management methodology and it deals specifically with how to attack the problem of formulating, developing, and conducting an energy management strategy that need careful consideration. Strategic planning for a comprehensive energy management program include following essential components.

18.4.1 Decision to Undertake Program

The initiative to start an effective energy management program must come from somewhere. This may be as follows:

1. In response to the information from the energy utility that the establishment is likely to face energy curtailment.
2. Energy cost escalation may create financial incentive to initiate a program.
3. Since line personnel may be aware of opportunities for saving energy, this may form possible origin to start a program.

Whatever may be possible reason to initiate a program (whether incentive was passed down from top management or was passed up the line from the operations end of the establishment or organization), it is established that time is appropriate to begin the electrical energy management program.

18.4.2 Commitment by Management

Identification of energy management opportunities begins with management commitment to formulate and allocate resources for the energy conservation program. In order to ascertain the advisability of understanding further development on the program and to propose terms of reference for it, logical steps are: appointment of an energy management coordinator and creation of an energy management committee.

Depending on the size and complexity of the establishment or organization, an energy management committee is formulated by assembling representatives from each division or department using energy in addition to energy coordinator appointed by the top management. The management committee identifies the objectives. This committee is given responsibility of carrying out the program objectives by organizing, staffing, directing, and controlling the numerous required tasks.

18.4.3 Statement of Objectives

For a formal assessment of a program idea, it is mandatory to define and clarify the objectives of the program and to identity all alternative approaches that also meet these objectives. If there are multiple objectives that a program must meet, the relative importance of each must be decided as it will aid to subsequent evaluation. System analysts offer a principle of objectives which requires that they should be clearly defined (preferably quantified and measurable), realistic they are attainable with some reach or difficulty; specific and understandable (known by all members of establishment affected by them). It is also essential to distinguish between absolute and relative objectives.

18.4.4 Database and Information Collection

Quantification of the problem statement by defining criteria that provide measurable requirements to determine when the goal has been satisfied and also identifying resources and constraints. Resource availability such as cost and performance information and sources of energy, human factors such as physiological and psychological, technical, and engineering tools and techniques that control the problem solution to achieve objectives constitute the database of concerned establishment or organization. Information relating to energy consumption pattern, technical and non-technical barriers to the design, and development and implementation of energy management alternatives are also prerequisite for energy management strategy.

Collection of data and technical information is hardly a new activity. Methodology of data collection can be evolved through a simple walk through surveys and review of historical energy use. Such preliminary surveys and review of historical energy use thus identify the area(s) where energy can be saved by simple housekeeping and maintenance measures or simple changes in operating practices, etc., with nominal investment. Making correct energy management decision requires the availability of a realistic energy information database.

18.4.5 Energy Audit

Energy audit is a technique for investing current practice of energy use and provides insight into the point of inefficiency and energy wastage. Energy audit is conducted with the help of instrumentation.

Simple walk through survey and historical energy use is performed with essentially no measurements, whereas energy audit is performed with a wide range of measurement depending upon the size of establishment, number of process and equipment involved, and the desired degree of details. The detailed energy audit is carried out to provide realistic answers to the following questions:

1. How effectively and what functions does the equipment serve?
2. Possible elimination of replacement by a more energy efficient equipment; if the equipment has to be retained, how can its energy performance be improved?

In other words, energy audit involves analysis of establishment to determine the forms, quantities, costs, and purpose for which energy is used. Further, the identification of energy conservation opportunities is established.

18.4.6 Computer Analysis and Simulation

Computer plays a major role in the management of energy. Its application can be divided in two major categories.

18.4.6.1 Actively Interact and Control Operations

The first principle in this category is that the computer provides a versatile means to regulate energy use raging from microprocessors controlling a single fan to mainframe computer controlling fuel usage in a steel mill.

Process optimization is the second principle in the controlled operation category. Computer receives signals and data from a number of sensors and then evaluates these data under the prescribed boundary conditions until some predetermined best operation is achieved.

18.4.6.2 Analysis and Simulation

Computers can afford to accommodate a wide range of input information (database) that can be sorted and used when data processing and data analysis are needed to carry the task of engineering or economic investigations, analysis, simulation, forecasting, implementation, monitoring, and maintenance of any energy management strategy. This category of computer application provides an important source of information to the energy management coordinator.

18.4.7 Energy Efficiency Analysis

Efficiency is an important element of any energy management strategy and concept of useful work provides useful measure of efficiency. A clear understanding of the concept of useful work suggests equipment parts and process components or areas where improvements in efficiency are possible. Factors that contribute to inefficiency are possible. Factors that contribute to inefficiency are losses that result in loss of energy or loss of availability. These types of losses are present in any energy using process or equipment. Measurements or data provided by manufacturers will provide an estimate of overall efficiency of equipment and appliances. The total energy lost in a particular industrial process is, however, the sum of the energies dissipated by several equipment, appliance, and mechanism. In order to determine the relative importance of each loss term, detailed analysis of energy efficiency is essential that will then suggest the potential and priorities for establishing corrective action either as a simple measure or as a R&D measure. Goal or mission oriented terms include effectiveness (for window air-conditioners) and efficacy (light power sources).

18.4.8 Energy Economics

The calculation of energy savings and hence return on investment for energy management evaluation is a broad subject area. Money has time value associated with its use (or yield return), if saved or invested. This is often neglected. The assessment of economic potential of energy management approaches for different energy activities requires databases that include:

1. Design cost and performance information on the system development.
2. Identification or technical and non-technical barriers to the implementation of alternatives.
3. Identification, quantification, and characterization of process, establishment, and equipment energy requirements.
4. Present and projected cost and cost escalation.

18.4.9 People's Involvement

It is ultimately human being who is entrusted with these marvels of engineering that are supposed to save all this energy and money. More efficient process or program and equipment are only half the battle. The human element is vital and is all too often ignored. Obviously, it makes life different; how efficient are the equipment, process, and programs, if the operating personnel do not

1. understand the need of efficiency
2. believe in the necessity of energy management
3. like to take pain in implementing the program

Therefore, to promote the continuing awareness of energy management and involvement of personnel is another important element of energy management strategy. The only way to create energy management awareness and involvement of people is through multimedia publicity campaign. This may be achieved by the following steps:

1. Distribution of printed folders or booklets emphasizing management pros and cons for several energy activity sectors.
2. Conduct of seminars, workshops, exhibitions, fairs, etc.
3. Comprehensive campaign through electronic media.

18.4.10 Implementation

Once the best energy management alternative is selected, it is necessary to develop a new problem statement to decide how effectively this alternative can be implemented. This is the most important step to be incorporated in the strategy for energy management. Implementation includes aspects of making developed techniques or programs operational, construct facilities for plans and specifications, and test all work implemented as per the objectives and problem defined.

18.4.11 Verification

It includes monitoring of program while in operation, synthesis of monitored data, optimization of program details, and evaluation whether problem was solved and objective was attained.

In order that the program must finally succeed, it must be reviewed periodically to determine its strengths and weakness. It should be flexible and capable of responding to changing economic conditions and to changing program needs.

18.5 ENERGY SAVINGS TIPS

Energy is mostly used and consumed in domestic sector, agricultural sector, transportation sector, and industrial sector. There is ample scope to conserve and save energy from 10% to 50% by implementing energy management principles. Following are some simple tips that can save energy and reduce the energy bill cost:

1. Turn off the lights when not in use.
2. Use day light as long as it is available.
3. Clean lighting fixtures to maintain illumination.
4. Use task lighting instead of brightly lighting an entire room and focus the light where it is needed.
5. Use electronic chokes in place of conventional copper chokes.
6. Replace conventional regulators with electronic regulators for ceiling fans.
7. Install exhausts fans at a higher elevation than ceiling fans.
8. Select iron boxes with automatic temperature cut off.
9. Use appropriate regulator position for ironing.
10. Do not put more water on clothes while ironing and avoid ironing wet clothes.
11. Switch off the power when TV, computers, audio systems, etc., are not in use. Idle operation leads to an energy loss.

12. Setting computers, monitors, and copiers to use sleep mode when not in use helps cut energy costs.
13. Remove the plug from socket of electrical and electronic appliances and save energy.
14. Sufficient distance between building walls and refrigerators should be maintained to allow fresh air circulation around refrigerator.
15. Refrigerators regular defrost reduces electricity consumption.
16. Ensure that refrigerator door seals are airtight and avoid frequent door openings.
17. Always wash only with full capacity loads in washing machines.
18. Use optimal quantity of water in washing machines.
19. Use timer facility to save energy in washing machines.
20. Use the correct amount of detergent in washing machines.
21. Use hot water only for very dirty clothes in washing machines.
22. Always use cold water in the rinse cycle in washing machines.
23. Prefer natural drying over electric dryers in washing machines.
24. Use air-conditioners having automatic temperature cut off.
25. Keep regulators at 'low cool' position.
26. Seal the doors and windows of air-conditioners properly.
27. Use windows with sun films or curtains.
28. Avoid dry grinding in food processors (mixers and grinders) as it takes longer time than liquid grinding.
29. Use microwaves ovens as it consumes 50% less energy than conventional electric or gas stoves.
30. Turn off electric stoves several minutes before the specified cooking time.
31. Use flat-bottomed pans on electric stove so that they make full contact with the cooking.
32. When cooking on a gas burner, use moderate flame settings to conserve LPG.
33. Regularly maintain and clean burners to obtain blue flame for efficient operation of gas stove.
34. Use pressure cookers as much as possible.
35. Use solar water heater – a good replacement for an electric water heater.
36. Use highly efficient electric motors and mechanical transmission gear
    (a) When purchasing a new motor choose the most energy efficient one with affordable cost. High efficiency motors cost about 20% more, but can have a relatively short pay-back to off-set these costs.
37. Properly size the motor with load.
38. Use variable speed drives in place of throttling operation for fluid flow control.
39. Avoid use of compressed air for cleaning purposes.
40. Carry out preventive maintenance and condition monitoring schedule regularly.

41. Replace motor–generator (M–G) sets by power semiconductor drives.
42. Fuel consumption in diesel generator set is optimal when all the parts are properly cleaned and maintained.
43. Optimize the boiler combustion process by monitoring the flue gases. Only careful monitoring allows operators to strike a balance between supplying too much air, which carries heat away up the flue, and insufficient air, resulting in incomplete combustion.
44. Make sure that boiler duty is at optimum efficiency.
45. Clean burners, nozzles, strainers, etc., of boilers.
46. Fix steam leaks and condensate leaks.
47. Recycle steam condensate. Reduce hot water wastage to drain.
48. Use variable speed drives on large boiler combustion air fans with variable flows.
49. Undertake regular energy audits of furnace.
50. Use of low air pressure 'film burners' helps save oil up to 15% in furnaces.
51. Recover heat from incinerator off-gas.
52. Use waste heat for fuel oil heating, boiler feed water heating, outside air heating, etc.
53. Properly seal water leakage of pipes, repair and maintain taps, and minimize water wastage that leads to energy savings in pumping water.
54. Using standard fuel efficient pump sets and proper installation of pump system in agricultural sector.
55. Through public awareness in farmers and workers.
56. Strictly following the norms and standards for each agricultural equipment.
57. Use of highly efficient equipment and machinery and replacement of existing transportation system by a new highly efficient one.
58. Car pooling.
59. Fuel-mix, environment friendly vehicle drive.
60. Ethanol blending.
61. Strict following of environment standards with regular check-up of the vehicle.
62. Avoid unauthorized stoppage of trains.

18.6 CONSTRAINTS AND CONSIDERATIONS FOR IMPLEMENTING ENERGY CONSERVATION AND MANAGEMENT PROGRAM

From the discussion presented in the chapter, constraints and considerations for implementing energy management program can be summarized as follows:

1. Decision and commitment
2. People involvement

3. Problem formulation and statement of objective
4. Data and information collection
5. Energy audit
6. Analysis, computer simulation, and investigations
7. Economic analysis to decide alternative option
8. Implementation
9. Periodic review and monitoring
10. Evaluation and maintenance
11. Risk and benefits and probability of success
12. Computer applications and management information system

SUMMARY

- The term energy management is widely used to signify the utilization of minimum quantity of energy required for the task at an appropriate quality together with the capital expenditure avoidance.
- Load management refers to the supply side of the energy system. It includes those activities taken by energy utility that is used to manipulate the load seen by their generating system in such a way to meet energy demands so that most favourable and economic operating conditions can be achieved.
- End-use management relates to reduction in and control of energy at the use's own convenience. Such an attempt can help in reducing the needed capacity and can help in obtaining more favourable energy pricing.
- Demand-side management (DSM) has been introduced as a combination of supply side and end-user side management. It is considered as a means of reducing peak electricity demand so that utilities can delay further plant capacity addition.
- DSM has various beneficial effects, including mitigating electrical system emergencies, reducing the number of blackouts and increasing system reliability. Possible benefits can also include reducing dependency on expensive imports of fuel, reducing energy prices, and reducing harmful emissions to the environment.
- Saving money and improving the utility load factor are the important beneficial aspects of energy conservation or management.
- Energy audit means for investigating energy use by specific processes and machines and provides insight into inefficient operations.
- Energy audit is performed with a wide range of measurement depending upon the size of establishment, number of process and equipment involved, and the desired degree of details.
- To promote continuing awareness of energy management and involvement of personnel is an important element of energy management strategy.

REVIEW QUESTIONS

1. 'Energy management can be considered as a cheapest and environment friendly energy'. Explain.
2. Define and explain energy management and explain its significance.
3. Describe the benefits and key issues of energy conservation and management.
4. State and explain different approaches of energy conservation and management. Mention at least three examples in each case.
5. Describe strategic planning for energy management.
6. State and explain energy saving tips in domestic sector or agricultural sector or transportation sector or industrial sector.
7. Mention and explain at least five fundamental principles of energy conservation and management and compare them.
8. Explain Sankey diagram of lamps and mention its importance.
9. Briefly explain zero leak philosophy in plants with examples and mention its advantages.
10. Mention energy saving tips with furnace.

Chapter 19

Energy Conservation and Management in Different Energy Activity Sector

Energy is heavily consumed and wasted knowingly or unknowingly in various sectors of human activity. However, an industrial sector uses more energy than any other end-use sectors, and currently, this sector is consuming about 40%–45% of the world's total delivered energy. Energy is consumed in the industrial sector by a diverse group of industries including manufacturing, agriculture, mining, and construction and for a wide range of activities, such as processing and assembly, space conditioning, and lighting. Energy saving by management including energy audit, training programs, and housekeeping beside some energy management practices in the world have been reviewed. Tips on energy saving technologies include the use of high efficiency motors, variable speed drives, economizers, leak prevention, reducing pressure drop, etc. Based on the results of the energy saving technologies, it has been found that in the industrial sectors, a sizeable amount of electric energy, emissions, and utility bill can be saved using these technologies. Payback periods for different energy saving measures have been identified and found to be economically viable in most cases.

KEY CONCEPTS

- Energy management with electric power supply systems
- Energy management opportunities with lighting systems
- Energy saving tips with light source
- Energy management opportunities with electric motors
- Energy management opportunities with household electric appliance
- Energy saving tips with household appliances
- Energy management opportunity with HVAC systems
- Energy saving tips for other industrial processes

19.1 GENERAL

Industry is the major user of energy in modern society, accounting for roughly 40% of final energy use. However, household sector also consumes large portion of energy.

Fossil fuels are widely used in industrial sectors specifically in manufacturing processes, chemical and refinery, and refining industries. Since gas is a cleaner fuel than coal, it is gradually replacing coal to protect environmental damage. The secondary energy resource or the usable electrical energy form is used for driving electric motor to impart motion to working industrial machines. It constitutes 25%–30% of total industrial energy requirements.

The major sub-sectors within industry sector can be sub divided as follows:

1. *Manufacturing*: This includes the processing of primary raw materials and resources into consumer products. Large energy-use industries are mineral and oil refining and chemical manufacturing and all these industries can be restructured for considerable energy savings.
2. *Power generation*: Fossil fuels (such as coal, oil, and gas) and nuclear fuels (such as uranium and thorium) based power generating plants severely damage the environment and are the major source of greenhouse effects. Many of them are also less efficient. Environmental protection and development of highly efficient power plant have considerable impact on energy savings. They are often a cause of air pollution as well.
3. *Mining*: Exploration, extraction, and transportation of raw resources and materials consume large energy. Energy management activities to minimize wastage of energy in several mining activities lead to energy savings and protection from environmental damage.
4. *Agriculture*: Another major user of primary and secondary energy that takes place in rural areas are agriculture and rural cottage industries.
5. *Construction*: Construction activities use petroleum, diesel, and electricity and numerous raw materials. Energy management and conservation opportunity exist for large energy savings with due importance attached to environment.
6. *Transportation*: Methods of transport are a term used to distinguish substantially different ways to perform mobility of passengers and goods. The three main types of transportation system are as follows:
   (a) *Land transportation*: It includes road and rail transportation. Both of these methods of transportation consume large space to construct road and lay down the railway tracks.
      (i) *Road transportation*: Historically road transportation was developed to support conventional non-motorized forms of transportation such as walking, bullock carts, and cycling at the end of the 19th century. Fuel (oil and gas)-powered vehicles have highly improved the efficiency and performance of road transportation.
      (ii) *Rail transportation*: Railways are composed of a tracked path on which wheeled vehicles move. In light of more recent technological developments, rail transportation also includes monorails and magnetically levitated (maglev).

      Rail transportation is the large consumer of fossil fuels and electricity. Since consumption of fuels and electricity is large in land transportation systems, there do exist the opportunity of energy management and conservation by developing highly efficient vehicles and systems.
   (b) *Water transportation*: Shipping transportation is the most effective transportation of goods, materials, and passengers in different water routes covering a long distance. Main water routes are composed of oceans, coasts, seas, lakes, and rivers. Despite the advantages of limited water friction and tendency to float in water, movement of large quantities of cargo and ships for a longer distance, water transportation system is also a large consumer of fossil fuels. Improvements in ship design and driving and driven mechanism can save considerable energy in addition to improvement in seaport buildings' construction and maintenance.

(c) *Air transportation*: Air routes are practically unlimited. Air transport constraints are multidimensional and include the site (aerodrome), the climate, fog, and aerial currents.

Nevertheless, air transportation is the fastest mode of transport of large quantities of goods and materials and passengers. Improvements in aerodrome infrastructure and design and development of improved airplane can save considerable amount of energy or fuels. The various equipments and process that consume large energy are as follows:

(i) *Compressed air*: Compressed air is air kept under pressure that is more than atmospheric pressure. It serves many domestic and industrial purposes. It consumes 10%–15 % of industrial electrical energy consumption, and thus, it is a candidate for energy savings.

(ii) *Process heat*: Heat energy requirements in industrial sectors constitute a major portion (about 70%) of total energy. Process heat is obtained by the burning of fossil fuels and biomass and electrical heating systems. A large energy savings and environmental protection can be achieved by improved high efficiency design of process heating systems and heat recovery and exchange systems.

(iii) *Pump and pumping systems*: Liquid pumping is very important necessity of any manufacturing. Processing conditioning and all other commercial, industrial, agriculture and life comfort residential applications.

Pumps are mainly of two types:

- *Positive displacement or reciprocating pump*: In these pumps, pressure is generated either by displacing liquid with a mechanical rotary device or by reciprocating action of a moving system.

  Force is created by forcing or drawing the liquid through mechanical motion and it produces the same flow at a given rpm, regardless of the discharge pressure. Reciprocating pumps are used for handling small quantity of liquid requiring a very large delivery pressure.

  For examples, they are commonly used in pneumatic pressure systems, feeding small boilers, condensate return, gasoline and light oil pumping, irrigation, liquid transfer systems for soaps, tars, paints and varnishes. Since the output of this type of pump is constant, reduction in flow is achieved through discharge or suction throttling or recirculation systems. Such mechanical methods of flow control involve considerable wastage of energy (throttling loss).

- *Dynamic (centrifugal) pump*: Working of centrifugal pump is simple and relies on impeller that has a series of curved vans fitted inside the insulated shroud plates. The liquid is given as input to impeller where it is forced into circular motion by impeller vanes. The rotational mechanical energy of impeller is transferred to liquid. The liquid is then discharge out of impeller vanes with increased pressure and velocity. Centrifugal pump, having steep descending pressure flow (*H–Q* curve), is a good candidate for energy savings by replacing throttling method of fluid flow control with variable speed drive control. Centrifugal pumps are the common types of pumps used in fluid flow control in every walk of life. Positive displacement (reciprocating) pumps are used when high outlet pressure is required and dynamic (centrifugal) pumps are used when high discharge is required. Based on this and many other issues, pumps are selected for specific applications.

- *Electricity*: Electricity is a powerful usable form of energy that is essential to the operation of virtually every facility in the application sectors. It is also an expensive form of energy that can represent a significant portion of a manufacturing facility's cost of production.

  Electrical energy management through proper maintenance scheduling, optimal control, the use of highly efficient equipment, appliances, and processes can lead to energy savings. Thus, similar energy-saving strategies can be applied to any size of energy activity sector and establishment.

19.2 ENERGY MEASUREMENT SYSTEMS IN INDUSTRY

Before devising a strategy to minimize electricity use, one must answer the following three questions:

1. How and where are the power outlets that consume energy and how much is consumed at each outlet?
2. How much should each power outlet consume?
3. How can energy consumption be reduced?

Development of an effective energy audit system is the only answer to abovementioned questions. Energy audit with the help of highly effective instrumentation measures the energy and power consumption and helps in implementing energy management program to minimize the wastage of energy. Setting up of an energy measurement system (energy audit program) is a key component of any energy management strategy.

The appointment of an energy measurement coordinator in each industry is the first step. He is a link between top management and different shop or floor officials to interact on points of energy wastage and methods of minimizing the energy wastage. Energy measurement coordinator can fulfil his responsibility of providing accurate energy and power measurement data by implementing the following steps:

1. Formulation and development of a strong energy data measurement system.
2. Procurement, maintenance, calibration, standardization, and safety of measuring and recording instruments and improved instrumentation systems.
3. Evaluation of energy efficiency of existing equipments and machinery and suggestions on retrofitting, equipment modifications, and process modernization.
4. Ensuring the timely availability of accurate energy data.

19.3 ENERGY MANAGEMENT WITH ELECTRIC POWER SUPPLY SYSTEMS

The technical knowhow for generation, transmission, distribution, and utilization of electrical energy is well established. Bulk storage of electrical energy is not likely to be developed in near future. It is, therefore, essential to consume all the generated electrical energy; otherwise, they will be wasted as heat in transmission and distribution systems may cause malfunctioning of electrical energy system.

On the other hand, if the electrical load on the electrical power grid approaches the maximum electrical generating capacity connected to grid, the grid system will be unstable. If either the timely supply of additional energy to meet the electrical load demand is not met or reducing the energy load is not carried out, grid fails and hence complete blackout can occur. This situation can be avoided by implementing load management or supply-side management.

Implementation of load management are discussed in the following sections.

19.3.1 Reduction of Peak Demand or Maximum Demand

A steady supply of electrical power is generally quite straightforward to produce with a typical coal, gas, or nuclear plant where simply thermo generator, provided with steady supply of fuel is maintained. However, the demands are not steady and fluctuate 24 h a day. This is referred to as a load profile as shown in Figures 19.1 and 19.2. It is actually a broad term that can refer to a number of different forms of data namely regression and profile coefficients or demand and consumption data. Figures 19.1 and 19.2 depict a typical daily pattern of demand for the average domestic unrestricted custom.

There may be more demand during morning and evening hours and on hot afternoons when the air conditioners are on as shown in Figure 19.2. In order to maintain continuity of electric supply throughout the entire day, electrical supply company shifts electrical loads from high peak demand periods to low-demand, off-peak periods.

As already mentioned, the management program to carry out this are collectively called load management (or supply-side management). It prevents grid failure and unstable operation of power system in addition to control of energy pricing, improved safety, and protection from associated environmental and health problems.

19.3.1.1 Peak Shaving or Load Shifting

It is considered as the best method for reducing electrical load demand during maximum or peak load demand period. It is a method used to reduce electricity consumption during maximum load demand periods.

The load shifting techniques were used to select the load or processes, which can be shifted or rescheduled during off-peak periods. Understanding of 24 h a day profile is necessary to reduce peak demand. By doing so, cost of peaking charges is reduced significantly.

Load management by reduction of peak demand or maximum demand generally falls into one of six categories of load profile shaping. They are load clipping, load shifting, valley filling, conservation, load building, and flexible utility load shape, as shown in Figure 19.1

1. *Peak clipping*: It is a method (see Figure 19.1a) of the reduction of energy supply company load primarily during peak demand periods. Maximum demand reduction of electric power from an electric power company is often achieved by direct control of consumer loads by signals directed to consumers' appliances.
2. *Valley filling*: It is a technique for the improvement of system load factor by building load in off-peak periods as shown in Figure 19.1(b).
3. *Load shifting*: It is a method of reducing utility loads during periods of peak demand as shown in Figure 19.1(c). Simultaneously, building load in off-peak periods. This, however, does not influence sales of electricity.

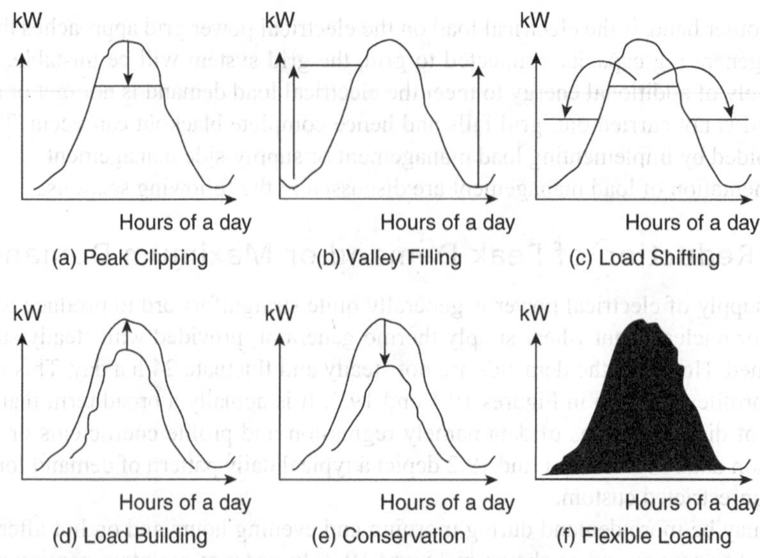

Figure 19.1 Load profile shaping of reducing peak demand

4. *Load building*: It defines the increase of utility loads, more or less equally, during all or most hours of the day as shown in Figure 19.1(d).
5. *Conservation*: Peak demand charges reduction are linked with kWh reduction. Therefore, if energy efficiency measure is to be initiated in areas that may be unrelated to high amounts of power draw like lighting, it is necessary to potentially lower peak demand charges simply because overall less electricity is used (conservation). The reduction of utility loads (see Figure 19.1e), more or less equally, during all or most hours of the day reduce the peak demand charges.
6. *Flexible utility load shape*: The provision of a more flexible utility load shape (see Figure 19.1f) refers to program that set up utility options to alter customer energy consumption on an as-needed basis, as in interruptible or curtailable agreements

   Demand charges are used to cover the cost of electrical conductors and equipment to provide electricity to energy consumers. In order to prevent overloading of the system if working with higher peak demand, it means to have higher electricity cost.

   Let us consider that electricity consumers A and B are using the same overall energy. However, user B is drawing its electricity at slower rate by adopting techniques of reducing peak demand as shown in Figure 19.2. This means that it has a lower peak demand when compared to user A. As a result, user B will have a lower demand charge on his bill.

19.3.1.2 Load Shedding

Load shedding is the term used for disconnecting some electrical loads from the electrical power grid during peak load demand periods, thereby trimming the peak demands. There exists a commitment between the electric supply company and the energy consumer not to exceed a fixed

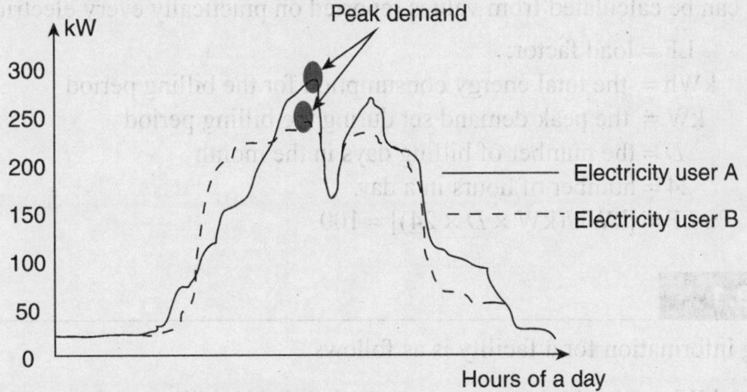

Figure 19.2 Reduction in demand charge by reducing maximum demand

kWh demand. Energy consumers do this normally either by switching non-essential or non-critical loads during maximum demand period or only operating the loads during off-peak periods or shifting all his load to a generator installed for the purpose.

19.3.1.3 Peak Load Sharing

This method is designed in such a way that the energy supply company asks the energy consumers to synchronize their captive power generation with electrical power grid by running their generator and take over a portion of the grid-connected electrical load in addition to their own loads. The following terms are also important to understand the concepts of peak demand.

1. *Average (or annual average) megawatt*: It is defined as 1 MW of energy produced continuously over an entire period of one year and expressed as

   $$1 \text{ aMW} = 1 \text{ MW} \times 365 \times 24 = 8{,}760 \text{ MWh}.$$

2. *Load factor*: It is defined as the ratio of average load demand to the maximum load demand over a finite period and is expressed as

   Load factor = average load demand/maximum load demand

   It dictates the cost of electricity per unit generated. Higher the load factor, the lesser will be the cost of electricity per unit generated.

3. *Conservation load factor*: It is defined as the ratio of average annual load savings to the peak demand savings and expressed as

   CLR = Average annual energy savings/(peak demand savings × 8,760 h)

   Conservation Load factor = energy savings/peak demand savings × 8,760 h

4. *Loss of load probability* (*LOLP*): It is a measure of overall probability of loss of load time period during which energy shortages are experienced.

   Load factor of any energy activity facility is defined as the ratio of average to peak energy demand. The demand management is measured by calculating the load factor of energy activity facility from the available energy data.

Load factor can be calculated from values reported on practically every electric bill.

Let LF = load factor;

kWh = the total energy consumption for the billing period

kW = the peak demand set during the billing period

D = the number of billing days in the month

24 = number of hours in a day.

% LF = [kWh/(kW × D × 24)] = 100.

Example 19.1

Sample billing information for a facility is as follows:

kWh = the total energy consumption for the billing period = 1,132,000

kW = the peak demand set during the billing period = 2,880

D = the number of billing days in the month = 30

Calculate the load factor of the facility.

Solution

% LF = [kWh / (kW × D × 24)] × 100. = [1,132,000/(2,880 × 30 × 24)] × 100 = 54.6%

19.3.1.4 Power Factor Correction

Power factor (PF) is defined as the ratio of real power to apparent power as shown in Figure 19.3. Power factor = (real power)/(apparent power) = cos (φ)

In a circuit with no reactive power, the ratio of real power and apparent power is equal to 1.

When the power factor of the system is 0.7, real power equals the reactive power. Power factor provides the criteria to understand the real power used from the electrical energy system for a given kVA. The operating performance of electrical power system is improved by employing power factor correction. It can reduce electricity costs, increase efficiency, and reduce environmental hazards.

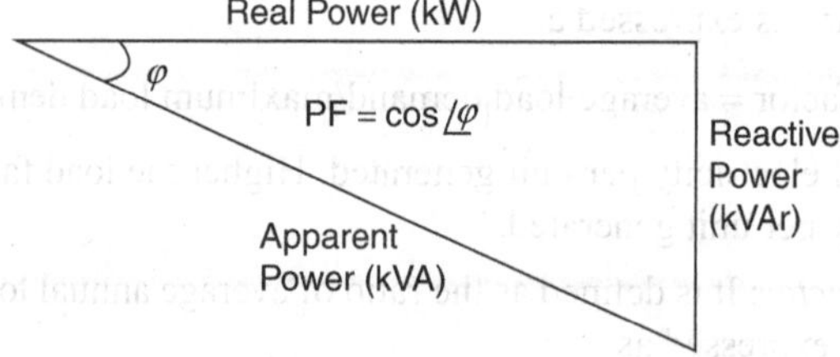

Figure 19.3 Power triangle diagram

Example 19.2

If a transformer supplies power at 480 V and the voltage at motor terminals is 470 V. Original power factor of distribution system is 0.80. The system power factor is improved to 0.95 by the installation of capacitors very near the motor load. Show that the total power loss in the in-plant distribution system upstream of connected load equipment seldom exceeds 2% of the load requirement.

Solution The voltage drop at motor terminal = 10 V or approximately = (10/480) × 100 = 0.02 = 2% of transformer supply voltage.

Let $PF_{initial}$ = original power factor = 80% (given), and
PF_{final} = improved power factor of the system = 95% (given)
The loss fraction saved through the installation of capacitors at the motor terminal is

$$[1 - (PF_{initial}/PF_{final})^2] = 1 - 0.709 = 0.291 = 29.1\%$$

Then, the resistance or I^2R losses in in-plant distribution wiring

$$= 0.291 \times 0.02 \times 100 = 0.58\% \text{ of load requirement.}$$

Example 19.3

A warehouse factory has an average monthly electrical bill of 50,000 kWh and the power factor is 0.92. There are two possible options for a warehouse consumer to get billed.

(1) Option 1: ₹5.85 per kWh.

(2) Option 2: ₹5.26 per kVAh

(a) Which option will provide reduced electricity bill to warehouse owner? and

(b) If the power factor of warehouse factory is improved to 0.99, calculate the annual savings.

Solution

(1) If option 1 is chosen, then the bill = kWH × electricity charges/kWh = 50,000 × 5.85 = ₹292,500.

Since the company is not penalized for low power factor, there is no obvious incentive for improving it.

(2) If option 2 is chosen, then the bill = (kWH/PF) × electricity charges/kVAh = (50,000/0.92) × 5.26 = ₹285,869.

(a) Monthly savings in electricity bill by choosing tariff option 2 and without any investment

= 292,500 – 285,869 = ₹6,631.

Annual savings = 6,631 × 12 = ₹79,572.

(b) If option 2 is chosen and factory power factor is improved to 0.99, then the bill = (kWH/PF) × electricity charges/kVAh = (50,000/0.99) × 5.26 = ₹265,656.

Therefore, monthly savings in bill by choosing option 2 of tariff = ₹(285,869 – 265,656)
= ₹20,213.0

Therefore, annual savings by installing power factor corrector = 12 × 20,213 = ₹242,556. Obviously, this will involve some cost of installation of power factor corrector, but the payback of such investment is usually less. It can, therefore, be seen that even though there is no penalty involved, there is good incentive for maintaining high PF.

Example 19.4

For a company having monthly consumption of 60,000 kWh, the monthly energy charges would be 267,000. The charges and rebates are as follows:

Energy charges: ₹4.45 per kWh

PF rebate for each percent excess of 0. 95 is 2.4%.

PF penalty for each percent drop below 0.90 up to 0.85 is 1%.

PF penalty for each percent drop below 0.85 is 2%.

Calculate the rebate and penalty when working below the power factor of 0.9.

Solution The possible rebates or penalties based on different power factors can be calculated as

Power factor at 0.99

Rebate = 2.5% of 267,000 = ₹6,675 per month or ₹80,000 annually

PF from 0.9 to 0.95: no rebate or penalty.

PF at 0.85: penalty = 5% of 267,000 = ₹13,350 or ₹160,200

Thus, there is a clear incentive to maintain power factor at least 0.9. Above that, the rate of return for investment in capacitor banks may differ from case to case.

Example 19.5

An industrial plant is consuming 400 kW of power with a maximum demand of 470.5 kVA. The demand charge cost per month is ₹100/kVA. Determine the savings possible by improving power factor to 0.95 and simple payback period if investment on capacitor bank is ₹75,000.

Solution Existing power factor of plant:

PF = kW/kVA = 400/470.5 = 0.85.

Existing demand charges = demand charge/kVA × kVA demand = 100 × 470.5 = Rs. 47,050.

When power factor of plant was improved by installing capacitor bank = 0.95 kVA, demand with PF of 0.95 = kW/PF = 400/0.95 = 421 kVA.

Modified demand charge costs = 100 × 421 = ₹42,100.

Electricity cost savings per month = 47,050 − 42,100 = ₹4,950

Electricity cost savings per year 4,950 × 12 = ₹59,400.

Simple payback period = capacitor installation cost/yearly savings = 75,000/59,400. = 1.3 years.

19.3.1.5 Power Quality

Good power quality of electrical power system is constant magnitude and fixed frequency sinusoidal voltage waveform for an ideal supply system. Power systems having good power quality provides successful operation of electrical power system with any kind of load with reduced running costs and pollution. A large variety of loads, abnormal phenomena, and other parameters deviate the electrical power system from the ideal system and power quality of practical system is expressed in terms of harmonic pollution, reactive power, and electrical load unbalance. They give rise to increased electrical losses and if they are not compensated, involve cost for additional power. Distinguishing features of good and bad power quality are summarized in Table 19.1.

Table 19.1 Distinguishing Features of Good and Bad Power Quality

Good Power Quality	Bad Power Quality
1. Continuity of power supply without any interruption.	1. Voltage fluctuations causing flicker.
2. Balanced, constant voltage supply with specified fluctuations.	2. Voltage dips and short interruption
3. Fixed frequency operation with specified fluctuation.	3. Unbalanced voltages on three-phase system.
	4. Transient over voltages.
	5. Harmonics and frequency fluctuations

If the power quality of the electrical power system is bad, it has the followings adverse effects:

1. Circuit breaker tripping, fuse blowing off, etc., may cause unusual power supply failures.
2. Malfunctioning or failure equipment will increase.
3. Overheating of connected motors, transformers, etc., reduces their lifetime.
4. Production line sensitive control equipment and appliances, PLC, PC, etc., may experience frequent damage.
5. There will be increased interference with communication systems.
6. System losses increase.
7. Addition of excess capacity installation is required to cope up with increased system losses.
8. Danger of penalty imposition on energy consumers by electricity supply company.
9. Disconnection of energy supply to the concern industry causing supply network pollution.
10. Flickering of light sources may cause unusual sensation to human beings.
11. Health issues with and reduced efficiency of personnel.

There is, therefore, more attention is being paid on finding energy savings and ensuring appropriate power quality than ever before for good reasons. The quality electrical power has become an important issue for electric supply company and the electrical energy users. It is often considered as synonymous of voltage quality as electrical equipments are designed to operate under defined supply specifications. Optimizing the power quality and reduced energy usage will have large energy cost savings in addition to protect the connected equipment and machinery from damage and for increasing their lifespan.

19.3.1.6 Distributed Energy

Once, when a big power plant completed its expected life, it was replaced by larger power plant when compared to the existing power plant. However, it is not the case anymore. Large investment on adding new generating plants are very crucial for electrical supply company with due care to electricity demand growth and environmental regulations. Instead, electrical energy supplier is emphasizing the need for reducing the demand for more power by implementing electrical energy management opportunity and encouraging the distributed generation. Curtailment of power consumption does not necessarily mean to use less power. However, it only means that less power from the grid.

Distributed energy resource means distributed or dispersed generation (such as small combined heat and power and small renewable energy resources). The distributed generation systems are mass produced, small, and less site-specific and their development took place because of

1. Distributed generation works on the concept of micro-grids, eliminating the necessity of long transmission line for bulk power transmission, and thus, minimizing transmission and distribution losses to a large extent. Further, high costs involved in laying down long transmission systems are also reduced.
2. A distributed generation system localizes problems associated in the case of centralized grid failure during the natural disaster, thereby reducing number of affected localities and people. Use of renewable power (such as solar photovoltaic panels, fuel cells, and wind power) installed by a few individual can provide electricity to nearby localities.
3. Costs involved in construction of distribution generation power plants and smaller ratings converting machinery as compared to the centralized grid systems is greatly reduced.
4. Although costs per unit of electricity generation of distribution generation system may be higher than centralized grid power generation, heavy cost and complexity of maintenance, administrative and financial and trade bill section will reduce considerably.

Because of financial crisis, shortage of land space to set up new power plants and depletion of fossil and nuclear fuels, rapid transition from a big centralized power grid system to distribution generation of micro-grid is taking place at large scale.

Power that is generated in-house does not accrue a monthly service charge. If natural gas is being used for space heating, system can be upgraded to combined heating and power technology. This uses excess heat to generate electricity, which can then be used to power other parts of the operation.

Very large facilities in several energy activity sectors are beginning to add distributed generation to their electrical energy management programs; this is because the drastic curtailment of their loads connected to centralized power grid system will fail or will have a reduced lifetime and the efficiency of the electrical installation will considerably reduce, installation running costs and environmental hazards will be high, and/or operation may not be possible at all.

19.4 ENERGY MANAGEMENT OPPORTUNITIES WITH LIGHTING SYSTEMS

The energy consumption in a residential and office blocks of industries and establishment is about 40% of total energy. A modern lighting installation can, however, halve the energy consumption. Lighting system, therefore, represents an attractive saving opportunity, specifically if lighting systems have not been upgraded or improved in the past few years. This requires a combination of energy efficient light sources and reflectors as well as a systematic management of light system. Since energy efficient lighting system produces less heat, it reduces the requirement of air conditioners and improves the indoor climate in the residential and office buildings.

The most cost-effective approach for energy savings in lighting system is to carefully implement the following housekeeping and maintenance principles:

1. Turn off lights during the times when they are not needed. Lighting controls work better than people. Certainly, it is cheaper to have a person turn off a light, but persons forget and they may not have access to circuit breakers controlling large banks of industrial lighting fixtures; sophisticated lighting control systems have proven to be much more cost-effective.

   Use of remote operated circuit breakers and microprocessors are very effective for centralized lighting systems' control. Further, light sources can be controlled with other control devices such as motion sensors or photocells.
2. Reduce light levels to match the requirements for the tasks being performed in the area.
3. While decorating and beautifying office and residential buildings, damage to environment is restricted by lighting controls.
4. Replacement of less efficient lamps by highly efficient one, electromagnetic ballast by electronic or solid state ballast and use of fixtures with LED sources. The facility's lamp replacement practices usually fall into one of two categories:
   (a) Spot replacement: replace individual lamps as they fail.
   (b) Group replacement: replace all lamps at a predetermined point in time, even though many of those lamps are still glowing perfectly.
5. Light control systems and dimmers allow the required amount of luminance level. Such control reduces electrical consumption and extends the life span of light sources.
6. Light level reductions are also a method of lighting conservations.
7. Procurement and use of eco-friendly lighting system products in different premises lead to considerable energy savings without having any adverse effects on human comfort and style of living.

19.4.1 Efficacy of Light Lamps

Efficacy of a light source defines how well it turns input power (Watts) into the desired light output (lumens). It is expressed as lumens/watt. It will never be a number as efficiency. Efficacy is the rating of light source given in lumens/watt. For a light bulb, if electrical energy input is 100 W and brightness output obtained is 1,500 lumens, the ratings on light bulb packages will be given as 15 lumens/watt.

19.4.1.1 Distinction between Efficiency and Efficacy

The efficiency is unitless. It is defined to be a ratio of one thing (output) to another (input), where both numerator (output) and denominator (input) are represented in same unit. Thus, efficiency is a dimensionless quantity.

For example

1. An electric motor takes input power of 3 kW and delivers an output equivalent of 2.4 kW, and hence, efficiency is 2.4 kW/3 kW = 0.8.
2. A cow is fed 25 kg equivalent of food, water, and other enriched materials that produce 15 kg of milk a day. Efficiency of milk produce per day per cow is 15 kg/25 kg = 0.6.

3. The daily energy input to a distribution transformer is 30 kWh, while it is delivering daily energy output of 29 kWh. The all-day or energy efficiency of distribution transformer is 29 kWh/30 kWh = 0.967.

19.4.1.2 Efficacy of Different Types and Wattage Lamps

Efficacy, wattage ratings, and application areas of commonly used light sources are given in Table 19.2.

Table 19.2 Efficacy and Application of Different Types and Wattage Lamps

Light Sources	/Wattage	Ballast Watts	Efficacy (l/W)	Application
Incandescent	25–200	–	8–30	Indoor
Fluorescent	25,40,65	13–18	30–38	Indoor, general lighting
Low pressure sodium vapour	70–140	30–35	70–140	Outdoor, trunk road and street lighting
High pressure sodium vapour	70–138	30–35	92–129	Indoor/outdoor in buildings, airfield, super market
High pressure mercury vapour	80–1,000	10–30	40–60	Outdoor in streets and parking lots, landscape lighting, factories, and gymnasium
High pressure metal halide	400–2,000	–	85–100	Flood lighting in stadium and sport field
LED	5–23	–	37–98	Indoor/outdoor in office, hotels and outside space.

19.4.2 Factors Affecting Selection of Light Sources (Lamps)

Lighting designers consider several factors when selecting light sources for a given application. Some of the important criteria considered for selecting the best light source for a particular application are as follows:

1. *Efficacy*: It dictates that how much light is delivered per watt of power to the bulb. This is the headline number for news reports on trendy light sources, as if efficacy were the only thing that matters. Sodium vapour lamps are amongst the most efficient light sources. On the other hand, incandescent lamps are less efficient one. It should be noted that even the most efficient LED lamps have an efficacy of about 60 lumens/watt.
2. *Colour performance*: It means that how faithfully colours are reproduced when compared to daylight. Colour performance affects the comforts and living style of the human and animals of the illuminated space. It is defined as follows:
   (a) *Colour rendering index* (*CRI*): It is a useful quantitative measure of the light sources' ability to show any object colour look natural as it is seen in daylight.
   (b) *Correlated colour temperature* (*CCT*): It specifies the luminance colour appearance of a light as warm or cool and is expressed in degrees Kelvin and is an useful single quantity

to decide the installation of electrical lighting system for human comfort equivalent to natural daylight.

3. *Controllability*: It is another important consideration for lighting designers, which has multiple aspects.

   It signifies whether the light can be dimmed, or controlled with sensors and or timers. Switched on and off of light with timers or sensors comes into play with minimal used rooms and rooms that receive a lot of natural daylight.

   Timers and sensors are widely used for residential light sources control without any damage to them except CFL's light sources as they can be damaged by the on/off cycle operations of sensors and timers. The intensity of the light can be controlled through dimming. This is especially important in multi-use spaces such as conference facilities, meeting spaces, houses, etc. Simple phase-controlled dimmers are satisfactorily used to control incandescent and halogen bulbs from full brightness to meagre glow.

4. *Lifetime cost*: Procurement and operating costs of electrical lighting system over the entire lifetime is another important consideration. It includes the combination of light source initial cost and running cost to use over entire operating life.

5. *Start-up time and steady operation*: Other factors include performance issues such as instant start-up of the lamp when it is on, operates steadily without flickering, and whether it is easy to install throughout the home.

The presence of mercury contents in the light sources and their disposal problems play important roles in determining the sustainability of such light sources. Since some of these light sources have certain drawbacks, all of them are not appropriate to all applications and efficacy should only be the factor to be considered. For example, sodium vapour lamp produces more efficacy, they produce intense single-point source as against fluorescent tube. It is, therefore, used at least 3 m above the work place.

However, they suffer from the drawbacks of emitting low decibel humming sound and requires 5 to 15 min of warm-up period before reaching brightness at full intensity.

Selection and design of illumination system for particular premises depend on many varying parameters. For example, colour performance plays important role in museum and gallery, controllability is important criteria in hotels and conference complexes and efficacy, and lifetime costs are considered as most important in large-scale commercial applications. Of course, residents of residential premises dictate the illumination system design in their premises.

Example 19.6

In a building room space, adequate illumination was provided by 175 incandescent bulbs with specifications of 60 W and 14 lumens/watts for each bulb. Purchase cost of each bulb was ₹25 and cost of installation was ₹3 per point.

Following the energy savings awareness, building owner proposes to replace the entire incandescent bulb by compact fluorescent lamp with 13 W having efficacy of 68. The price of each CFL is ₹225 and a charge of installation per point is ₹40. The salvage value of each bulb is ₹8. Operating hours of light sources is 10/day. The cost of electricity is ₹3.85. Neglecting interest, depreciation, and inflation, calculate:

1. Number of CFL required to maintain same illumination
2. Yearly energy cost savings
3. Simple payback time

Solution

With incandescent bulb:

Total luminosity provided = 175 × 60 × 14 = 147,000 lumens
Total wattage of all bulbs = 175 × 60 = 10,500 W
Total electricity consumed per day = 10,500 x 10/1,000 = 105 kWh
Total electricity consumed per year = 105 × 365 = 38,325 kWh
Electricity charge = 38,325 × 3.85 = ₹147,551/per year

With CFL:

1. Total luminosity required = 147,000 lumens
2. Number of CFL required to maintain same illumination
   = total number of light source point = 175
3. Total electricity consumed per year = 175 × 13 × 10 × 365/1,000 = 8,304 kWh
   Electricity charge = 8,304 × 3.85 = ₹31,970 per year
4. Yearly energy cost savings = 147,551 − 31,970 = ₹115,580.
5. Total cost of purchase of CFL = 175 × 225 + 175 × 40 − 175 × 8 = ₹40,600.
   Simple payback time = total cost/yearly energy cost savings
   = 40,600/115,580 = less than 5 months

19.4.3 Energy Management Opportunities in Lighting Systems

Efforts are directed worldwide to attend to the following interacting lighting variables as opportunities to energy management in lighting systems. They include:

1. *Reduction in demand for artificial light*:
   (a) Implement daylighting strategies to make effective use of daylighting wherever possible.
   (b) Improve colour and reflectivity of walls, ceiling, and floor to reduce lighting needs.
   (c) Use light coloured surfaces or window glazing.
2. *Optimizing the use of existing lighting system*
   (a) De lamp means reducing the excessive illumination level. Reduction in illumination level should be provided with regards to no adverse effect on production and employees morale.
   (b) Re lamp means providing sufficient illumination level by using low wattage lamps.
   (c) Maintaining light fixture and fittings. Cleaning of light sources. Fixtures and fittings increase illumination levels.
   (d) Turning light off when not needed.
   (e) Using task lighting means keeping the light sources as close to the work place as possible.
   (f) Implementing and maintaining automated control system.
   (g) Disconnecting ballast where lamps have either been damaged or removed. This avoids energy dissipation in ballast similar to the case of transformer.
   (h) Using selective switching or zone switching by using time clocks or photocell controls.

3. *Upgrading lighting systems*:
   (a) Using more efficient lamps and luminaries. Replacing incandescent lamp by fluorescent lamp and replacing fluorescent lamp by metal halide or sodium vapour lamp.
   (b) Utilizing sensors and lighting controls.
   (c) Using electronic ballast in place of electromagnetic ballast consume 30%–40% less energy.

19.4.4 Energy Saving Tips with Light Sources

1. Turn off the lights when not in use.
2. Taking daylight advantages should be encouraged to maximum extent.
3. Clean lighting fixtures to maintain illumination.
4. Use of task lighting for focusing the light on work place only.
5. Use electronic chokes in place of conventional electromagnetic chokes.
6. Replace inefficient lamp by highly efficient lamps.

Example 19.7

Two 40 W rapid start fluorescent lamps with conventional electromagnetic ballast (consumption 15 W) are used for lighting a space for 12 h a day. By conducting energy audit for illumination requirement of space, it was determined that one lamp can be removed from the circuit without detrimental to illumination level in space. Calculate the energy saved

1. When one lamp is removed from the circuit, but both ballasts remain energized.
2. When one lamp and corresponding both ballasts are removed.
3. When in circuit of part (b), conventional electromagnetic ballast is replaced by electronic ballast (consumption 5 W). Further, mention the drawbacks of electronic ballast if any.

Solution

When both the lamps are in operation:

Total connected load = $2 \times (40 + 15) = 110$ W

Electricity consumed per day = $110 \times 12 \times 10^{-3} = 1.32$ kWh

1. When one lamp is removed from the circuit, but both ballasts remain energized.

   Electricity consumed per day = $[40 + 2 \times 15] \times 12 \times 10^{-3} = 0.84$ kWh

   Energy saved per day = $[(1.32 - 0.84)/1.32] \times 100 = 36.36\%$

2. When one lamp and corresponding ballast both are removed

   Electricity consumed per day = $[40 + 15] \times 12 \times 10^{-3} = 0.66$ kWh

   Energy saved per day = $[(1.32 - 0.66)/1.32] \times 100 = 50\%$

3. When in circuit of part (b) conventional electromagnetic ballast is replaced by electronic ballast.

Electricity consumed per day = $[40 + 5] \times 12 \times 10^{-3} = 0.54$ kWh

Energy saved per day = $[(1.32 - 0.54)/1.32] \times 100 = 59\%$

The following are the drawbacks:

1. Initial replacement cost
2. Low power factor operation
3. Increase in kVA demand
4. Harmonic noise

19.5 ENERGY MANAGEMENT OPPORTUNITIES WITH ELECTRIC MOTORS

Electric motors are used to impart motion as drive motors to industrial machines and represent 60% of electrical energy consumptions in industrial and transportation sectors. They are, therefore, represented as potential areas of energy savings. Certain important principles of energy management are as follows:

1. *Operational improvement of electric motors*: It includes
   (a) *Working with balanced rated voltage*: It gives improved operation and satisfactory motor performance. Electric motor current will also be balanced as voltage applied to motor is balanced rated voltage. Voltage unbalance, if at all any, should be minimum to minimize its adverse effects on motor performance (such as torque pulsation, increased vibration and mechanical stresses, increased motor losses, and overheating.)
   (b) *Scheduling regular maintenance*: Continuous observation and auditing of running electric motors for operating data (such as temperature, vibration, and efficiency) is the key to scheduling regular maintenance. It provides information either overhauling of motor or its replacement prior to electric motor failure in addition to increase of motor life span.
   (c) Regular greasing and oiling of bearings, etc.
2. Reducing peak energy demand by properly scheduling the timing of operation. Switching off motors when not in use by proper control will help in considerable energy savings.
3. *Soft starting of electric motors*: Using a soft starting reduces the wear and tear at every start, and a properly specified installed soft start will reduce the strain on mechanical and electrical systems by as much as 60%–70%. Soft starting can also be achieved by using variable speed drives, although it is efficient but more costly.
4. *Retrofitting or replacement of electric motors*: With the availability of off-the-shelf equipment and accessories, a part of electric motor drive system can be replaced to save electric energy. The following are some of the examples:
   (a) Replacement of conventional regulators of ceiling fans by electronic regulators. It can save 12 W to 14 W.
   (b) Replacement of old inefficient motors by high performance, highly efficient motors.
   (c) Replacement of an oversized motor for a given drive load by a properly sized electric motor. Oversized motors are more expensive. They operate with reduced efficiency and less power factor at partial loads. Working with reduced efficiency causes more energy consumption

when compared with properly sized motor. Reduced power factor operation when motor operates at reduced load increases the reactive power demand, and hence, additional kVA demand charges.

It is good engineering practice to slightly oversize a motor for a particular application. This will extend motor life and provide some extra capacity when it is required, and if a motor is oversized (larger than the required), the motor should be re-evaluated.

5. *Improvement of electric motor power factor*: The power factor improvement can be achieved as follows:
   (a) Electric motor should be operated very close to its rated full load condition to maintain its efficiency and power factor high and closed to design rated value.
   (b) Replacing the existing motor by a high power factor and highly efficient electric motor.
   (c) Installation of capacitors in AC circuit to minimize the reactive kVA.
6. *Replacement of conventional drive installation by a new systems*: It includes high cost of R&D and application of newer systems with emphasis on energy savings and drive performance improvements. Examples are as follows
   (a) Use of variable speed semiconductor drives specifically when load varies. For fan-type load, savings may be 30%–40% of conventional throttling control.
   (b) Variable speed drives (power semiconductor drives) are costly equipments and are a source of harmonic pollution, they should be implemented based on benefit/cost analysis. Example of applications for VSDs: hoisting, positioning in machine tools, closed-loop control, circular pumping or ventilation (without throttle) or booster pumps, complex control electronics, etc.
   (c) Use of high-voltage electric motors in drive applications.
   (d) Use of three-phase motors in place of single-phase motors.
   (e) Employing heat recovery from large motors.

Example 19.8

A 100 hp totally enclosed force cooled electric motor is driving industrial machines at 75% of rated load with an efficiency of 94.5%. Energy awareness and savings' program determine the cost savings by purchasing an energy efficient motor of same ratings and replacing the existing motor. Seventy-five percentage of rated load operation of this energy efficient motor gives an efficiency of 95.7%.

Calculate

(a) The annual energy savings by replacing existing motor by energy efficient motor, if a yearly operating hour of electric motor is 8,000 h.

(b) Payback period, if electricity charges are
Electricity charge = ₹5.85 per unit
Demand charge = ₹700 per kW per month

Solution

(a) Let hp = horse power of motor = 100 hp.
LF = load factor or percentage of full load = 75%
HRS = Annual operating hours = 8,000 hours

η_{St} = %Efficiency of standard motors = 94.5%
η_{eff} = %Efficiency of standard motors = 95.7%
A_{ES} = Annual energy savings

The annual energy savings obtained by replacing standard motor with an efficient motor can be calculated by the equation:

$$= \times 100$$
$$= 100 \times 0.75 \times 0.746 \times 8{,}000 \times [(1/94.5) - (1/95.7)] \times 100$$
$$= 5{,}939.0 \text{ kWh}$$

Example 19.9

Replacement of an oversized electric motor by a properly sized motor. The efficiency and power factor of an electric motor decrease when it operates on partial load shown in Figure 19.4. As can be seen from Figure 19.4, an electric motor when operated at partial load capacity is operating at reduced efficiency and power factor. It can be seen that efficiency curve is approximately flat in the load range close to full load. However, below 55%–60% of full load, efficiency drops very rapidly.

It is evident that input power of an oversized motor to drive a load is higher than a properly sized motor. Actual power required by an electric motor to drive a load is calculated by the task done by working machine.

Understanding the actual load will help in replacement of oversized electric motor by a properly sized motor. One of the methods of load calculation on three-phase induction motor is slip method. The formula for load calculation is carried out as follows:

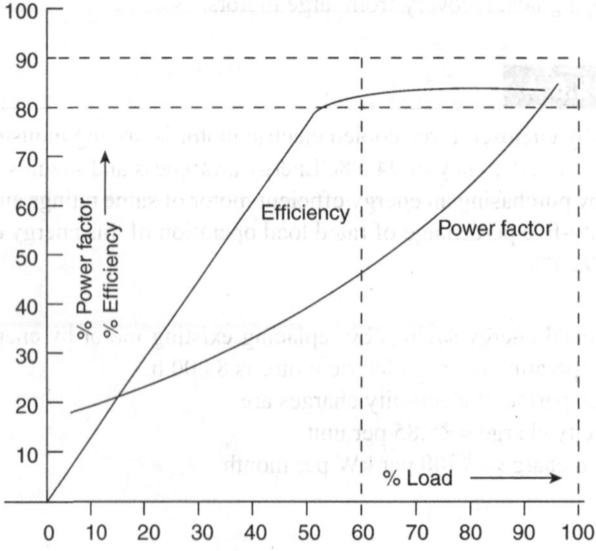

Figure 19.4 Efficiency and power factor curves of electric motor

Let N_S = synchronous speed of motor in rpm = 1,500 rpm

N_R = name plate full load in rpm = 1,440 rpm

N_A = actual measured speed in rpm = 1,464 rpm

hp = name plate rated horse power = 15 hp

Actual load horse power on electric motor = hp $[(N_S - N_A)/(N_S - N_R)]$

$= 15\ [(1{,}500 - 1{,}464)/(1{,}500 - 1{,}440)]$

$= 9$ hp = 60% of full load.

Example 19.10

An irrigation pump is driven by a 15 hp, 400V, 50 Hz, three-phase induction motor has name-plate full load speed of 1,440 rpm. Its efficiency and power factor curve are shown in Figure 19.4. When the speed of motor was actually measured by stroboscope, it is recorded as 1,464 rpm. Energy saving awareness emphasize the need of actual load on the motor and replacement of oversized (if any) motor by properly sizing the motor with load.

Calculate

(a) The actual load of irrigation pump
(b) Selection of a new electric motor and its operating efficiency
(c) Percentage of power saved

Solution

(a) Actual load horse power on electric motor

$= \text{hp}\ [(N_S - N_A)/(N_S - N_R)]$

$= 15\ [(1{,}500 - 1{,}464)/(1{,}500 - 1{,}440)]$

$= 9$ hp = 60% of full load

Hence, motor is oversized by 40% and its operating efficiency is 82%.

(b) For 9-hp pump load, suitable electric motor rating will be 10 hp and its full load efficiency is 92% (say from the motor technical specification).

(c) For existing oversized motor, input power = 1/0.82 = 1.22 pu

For properly sized 10 hp motor, input power = 1/0.92 = 1.08 pu

Power saved = [(1.22 – 1.08) / 1.08] 100 = 12.96%

19.5.1 Energy Management Opportunities with Ceiling Fan Regulators

19.5.1.1 Ceiling Fans

In tropical country like India, common cooling comfort home appliance is ceiling fan and there exists ample scope for considerable energy savings. Annual production of ceiling fans in India is about 6 billion with an average input per fan including electromagnetic regulator is 70–80 W at rated voltage. The power demand for fan product only works out to be approximately 375 MW/year. If line and other system losses are assumed to be 20%, an extra generation of 375 + 75 = 450 MW is required to be added each year for fan load only. The addition of new converting plant of 450 MW each year is impracticable as sites are scarce and large financial investment is required. Two options of energy management need to be implemented to reduce the burden of electricity shortages:

1. Standardization of efficient design of ceiling fans as per established norms. As per my experience, most of the branded fans fail to meet specified energy input and other norms. This topic is beyond the scope of this book.
2. Development and replacement of conventional electromagnetic fan regulator by electronic or other type of energy saving regulator. Electronic regulators are the latest type of regulators available in the market. These are much smaller in size than the electromagnetic regulators. Step-less smooth control and step types of electronic regulators are available in the market. A brief discussion on energy saving by replacing conventional resistance-type regulator by an electronic regulator is included in next section.

19.5.1.2 Energy Savings by Replacing Conventional Resistance-type Regulator by an Electronic Regulator

A typical measured value of electrical quantities with both regulators is given in Table 19.3.

Input supply voltage = 225V

From Table 19.3, following observations can be made:

1. The power saving is maximum at step 1 (lowest speed of fan). Electronic regulator helps in saving of about 44% of electrical power.
2. The volt-amp value has increased. The power factor of load reduces and this may cause penalty for kVA demand charges on the installation where other loads are also lagging power factor load.
3. Waveform distortion and noise emission are an added drawback of electronic regulator.

Table 19.3 Measurement on Ceiling Fan

Speed Step	Resistance Regulator			Electronic Regulator			% Power Savings	Increased VA
	V_{MT}	W_M	PF	V_{MT}	W_M	PF		
1	108	45	1.0	128	25	0.463	44.4	216
2	126	54	1.0	147	35	0.556	35.2	180
3	142	58	0.998	166	44	0.631	24.1	159
4	164	64	0.94	183	55	0.75	14.9	131
5	225	78	0.93	221	75	0,93	–	–

V_{MT} = Voltage at motor terminal; W_M = Input power to motor; PF = Power factor.

Example 19.11

Assuming that a ceiling fan declared input is 78 W and is operated for 10h/day for 340 days/year. The purchase cost of a good quality electronic regulator is ₹225 and electricity charge is ₹3.5 per unit. There will be an average reduction (based on 5-step averaging) in energy consumption by 23.7% with an electronic-type regulator as against a conventional-type regulator. Calculate the energy cost saving for one ceiling fan and simple payback period.

Solution

Total hours of fan operation in a year = 10 × 340 = 3,400 h.
Yearly energy saved = 78 × 0.237 × 3,400/1,000 = 62.85 kWh.
Cost of electricity charged saved per year = 62.85 × 3.5 = ₹220.
Simple payback time = cost of regulator/saving per year = 225/220 = 1 year approx.

Example 19.12

Energy audit conducted in a residential colony gave the following information:
Number of flats = 250
Number of ceiling fans installed = 600 (1,200 mm sweep)
Electricity charge = ₹3.5/kWh (unit)
kVA demand charge = ₹100/kVA

Solution

(A) For resistance regulator

Operational Speed Step	Step 3	Step 4	Step 5
Number of fans	300	200	100
Operating hours/day	18	12	10
Input to fan with resistance regulator	58	64	78
Number of days in a year of operation	300	300	300
Energy consumed per year (kWh)/fan	313.2	230.4	234
Total Energy consumed per year (kWh)	93,960	46,080	23,400
Power factor	0.998	0.94	0.93
kVA demand	17.43	13.6	8.38

Total electricity consumed per year = 163,440 kWh
Total kVA demand = 39.41 kVA

(B) For electronic regulator

Operational Speed Step	Step 3	Step 4	Step 5
Number of fans	300	200	100
Operating hours/day	18	12	10
Input to fan with electronic regulator	44	55	78
Number of days in a year of operation	300	300	300

(*Continued*)

(Continued)

Operational Speed Step	Step 3	Step 4	Step 5
Energy consumed per year (kWh)/fan	237.6	132	78
Total energy consumed per year (kWh)	71,280	39,600	23,400
Power factor	0.631	0.75	0.93
kVA demand	21	17.1	8.38

Total electricity consumed per year = 134,260 kWh
Total kVA demand = 48.48 kVA
Hence, energy cost savings per year = 3.5 [163,440 – 134,260] = ₹102,130.
Extra kVA demand charge per year = 100 × [48.48 – 39.41] = ₹907.
Therefore, net electricity cost savings = 102,130 – 907 = ₹10,100.0 approximately.

19.6 ENERGY MANAGEMENT OPPORTUNITIES WITH HOUSEHOLD ELECTRIC APPLIANCE

About 20%–30% of total energy consumption of most of the country goes to residential or household sector, and there do exists large scope for energy conservation and savings. The cost of electricity consumption per household appliance and equipment depends on several factors (such as family size and their living habits and life and usage hours of equipment and appliances). Consumer can calculate the electricity cost for different appliances, if the following are known:

1. Power rating of the electrical appliance and its efficiency.
2. Number of hours the appliances being used.
3. The domestic tariff rate per kWh.

The energy department also works with the environmental protection agency on providing energy star, which is a program designed to help consumers to lower their energy costs by using energy-efficient products.

In spite of abovementioned information, variation in power ratings of many household appliances differ by 10%–25% from the declared ratings. Electrical energy consumption characteristic should be given significant importance by manufacturers and suppliers in addition to cost, safety, and performance. There are, therefore, several opportunities available to conserve electricity used in the house. In addition to electrical energy conservation, efficient and proper use of electricity protect from the environmental damage and save money.

19.6.1 Energy Saving Tips with Household Appliances

A large number of energy saving tips with household appliances is as follows:

1. **Air conditioners:**
   (a) Use of air conditioner is suggested only when ventilation is inadequate.
   (b) Air conditioner in unoccupied rooms should be switched off.
   (c) Air conditioners having automatic temperature cut off should be preferred for use.

(d) Keep regulators at 'low cool' position.
(e) All doors and windows of conditioned space should always be closed when operating an air conditioner.
(f) Windows with sun films or curtains should be covered.
(g) Set thermostat of air conditioners as high as comfortably possible in the summer.
(h) When the temperature difference between inside and outside temperature of conditioned space is less, air conditioner must be switched off.

2. **Cooking:**
   (a) Less efficient cooking appliances such as chulhas, gas and kerosene stove, conventional electric stove, and microwave ovens to save energy and increasing human comfort.
   (b) Full-contact flat-bottomed pan electric stove minimizes wastage of heat while cooking and saves energy.
   (c) Switching off electric stove earlier than specified cooking time saves electricity.
   (d) Moderate flame settings to obtain blue flame of gas burner indicate efficient operation of gas stove and hence LPG is conserved. Yellowish colour of gas burner flame is an indicator of gas burner cleaning and maintenance.
   (e) Pressure cookers should be used as for as possible.
   (f) Use of pots and pans with bases of the same size as the hotplate should be considered. Induction hob is particularly efficient as the heat is generated in the pan base and not on the hob. Hobs with cooking sensors prevent boiling over. Considerable energy can be saved.
   (g) Use of damaged pots and pans should be avoided. Dented or uneven bases cause a considerable loss of energy.
   (h) Automatic frying program should be purchased when buying a new oven. This ensures perfect results and optimizes the use of energy.
3. **Dish washer:** Economy cycle for everyday washing should be used and dishwasher should only be operated when it is full. This helps save water too.
4. **Electric iron:** Excess water on clothes should not be spread while ironing and avoid ironing wet clothes.
   (a) Iron boxes with automatic temperature cut off should be selected.
   (b) Use of appropriate regulator position for ironing is essential.
5. **Electronic appliances:**
   (a) Idle operation of television and audio system involves wastage of energy, and hence, these entire devices should be turned off when not in use.
   (b) Sleep mode settings of computers, monitors, photo copier, etc., during non-use period save electricity and energy costs.
   (c) As for as possible plug should be removed from the socket at wall, when these appliances are not in use.
   (d) Removing plug of battery charger for laptops, digital camera, cell phone, etc., from the socket when not in use saves energy.
6. **Hot water:**
   (a) Electric kettle for boiling water rather than the stove should be preferred.
   (b) Energy use can be reduced substantially by water conservation in water heating.
   (c) Use of solar water heater is a good replacement for an electric water heater.

7. **Refrigerators:**
   (a) Manual defrost refrigerators and freezers should be regularly defrosted because of the fact that frost build-up increases the amount of energy needed to keep the motor running.
   (b) Sufficient space between the refrigerator and the walls should be kept so that air can easily circulate around the refrigerator.
   (c) Ensure that the refrigerator door seals are airtight. Rubber seals around the door of fridge should be regularly maintained. If they are damaged, they should be replaced to prevent the cold air from escaping.
   (d) Warm or hot food should never be kept inside fridge. Wait until they have cooled to room temperature and cover them and then put them in the fridge. This will minimize the unnecessarily wastage of energy.
   (e) Frequent opening of refrigerator door increases the wastage of energy. Door should not remain open for any length of time.
   (f) If possible refrigerator should be placed in a cool area and protected from direct sunlight, there should be some space between fridge and walls.
8. **Washing machines:**
   (a) Washing machines should be operated to wash only with full capacity loads.
   (b) Optimal quantity of water and detergent should be used in washing machines.
   (c) Timer facility should be used in washing machines to save energy.
   (d) Always cold water should be used in washing machines' rinse cycle. Hot water should be used only for very dirty clothes.
   (e) Natural drying should be encouraged over electric dryers for energy savings.
   (f) Most economical energy use in addition to the best washing quality for specific fabric can be achieved by adopting optimized special program designed for the purpose.
   (g) Use of modern automatic load sensor, whether 1 or 3.5 kg of washing load saves energy. Machine sets the optimum combination of resources for a perfect wash.
   (h) Speed setting for spin drying should be at the highest speed.

19.7 ENERGY MANAGEMENT OPPORTUNITY WITH HVAC SYSTEMS

A heating, ventilation, and air conditioning (HVAC) system provides comfortable temperature in building space, in addition to fresh and filtered air together with comfortable humidity level. They are the key elements in building energy and comfort management. HVAC equipment takes the large share of building energy consumption, estimated at over 40% of consumption for large buildings. Therefore, there exists large scope for energy savings.

19.7.1 HVAC Systems

HVAC has three functions and are often combined into one system. It controls the ambient environment in residential, office, and other premises. The most convenient and economical system to cool entire building is a centralized HVAC system. Warm or cool or fresh dehumidified air is pumped and distributed in all rooms and space through a series of ducts.

19.7.1.1 Heating System

Heating units can be categorized into furnaces that provide heated air through ductwork, which is a popular type of heating system like boilers in many countries that heat water for steam radiators or forced-water systems with baseboard radiators, electric heat, and heat pumps. Generally, furnaces use natural gas or propane for fuel, while boilers can use gas or oil. Heating system mainly consists of three activities and related energy saving tips are as follows:

1. **Heat generation**
   (a) Controlling properly the air to fuel ratio saves energy.
   (b) Preheating combustion air uses less energy.
   (c) Use of oxygen-enriched combustion air saves energy.
2. **Heat transfer**
   (a) Oxygen is used in furnace to improve heat transfer. It has a significant effect on reducing the flue gas heat losses.
   (b) Use of proper burner equipment and their control improve heat transfer, thereby saving fuel and energy.
3. **Heat recovery:** It is the process of collection and reuse of low temperature waste heat of processes and buildings for fluid heating, absorption cooling, electricity production to provide heat for other processes, etc.

19.7.1.2 Cooling System

Cooling systems is an air-conditioning system of window or split phase type.

1. *Window air conditioner*: It is compact equipment having its entire component in a single box and is used for air conditioning of single room.
2. *Split air conditioner*: It is a unit of two different parts, namely, the outdoor and the indoor unit. The compressor, condenser, and expansion valve are the part of outdoor unit and are placed outside the conditioned premises. It is the largest components of HVAC system. The evaporator or the cooling coil and the cooling fan are housed in the indoor unit.

19.7.1.3 Ventilation

It is an inexpensive and energy efficient method of providing comfortable cooling to inmates of residential and office premises. Natural ventilation is achieved by opening the doors and windows of rooms and additionally supplemented by spot ventilation (use of exhaust fans) and use of ceiling and window fans.

Use of fans to circulate air into buildings can provide continuous ventilation, regardless of climate and without openings doors and windows of buildings. This maintains air quality in building and helps in removing moisture generated from daily living activities. HVAC systems are available in different types. There are even hybrid gas furnace or heat pump systems that can be used to obtain better energy savings.

19.7.1.4 Heat Pumps

It is a device capable of transferring heat from a lower temperature fluid to another fluid at higher temperature as against the natural phenomena of heat transfer from high temperature to low temperature. In HVAC system, heat pumps are referred to as reversible vapour compression refrigeration devices and they are bi-directional heat transfer equipment to either heat or cool a conditioned space. It uses outside air for heating a home in winter and cooling it in summer as an integral part of centralized part of HVAC system. Such a system also consists of two parts: air handler as indoor unit and heat pump as an outdoor unit. Further, as it is powered by electricity, it results in saving substantially on fuel consumption.

19.7.1.5 Vent Duct

Another main component of an HVAC system is the ductwork. These are the lines that bring the conditioned air to each room of building.

19.7.2 General Principles of Energy Management in HVAC Systems

Following principles (tips) of energy management can be implemented to save energy without detrimental to comfort and cost to building owners:

1. *Reducing heating and cooling demand of building*
   (a) This can be achieved by creating awareness in inmates of buildings to make habits of consuming less energy.
   (b) Lighting sources and other home appliances also generate heat inside the buildings. Minimizing such internal heat generation and perfect sealing of air leakage reduce heating and cooling loads.
   (c) Loads on HVAC system can also be reduced by minimizing the external environment impacts on internal temperatures and the energy consumption of the building.
2. *Improved maintenance of buildings*: Improved building insulation, high-performance window glazing, natural ventilation, external window shading and proper window coverings and white painting of roofs reduce air-conditioning loads and save energy. Perfectly sealed, high performance glazing, external window shading, and proper window coverings result in significant energy savings. The use of double-glazing and/or low-emissivity windows can drastically reduce heat loss during winter and minimize the amount of heat entering during the summer.
3. *Improve proper operation and control*:
   (a) Unoccupied and unused space should be closed for avoiding unnecessarily heating and cooling it.
   (b) Air and water systems should be balanced that are out of balance.
   (c) Loose fan belts should be tightened and malfunctioning of variable speed drive should be examined and corrected.
   (d) Proper checking and perfect sealing of ductwork minimize the leaks. Further, repairing leakage and maintenance of damper, variable air volume boxes, and control valves reduce energy consumptions.
   (e) Temperature and humidity 'set points' should be optimized to save energy.
4. *Properly sized HVAC unit should be selected*: By selecting properly sized HVAC system, reliability and efficiency will be maintained and cost will be less. Further, there will be considerable energy savings as compared to oversized systems.

5. *Scheduling a regular maintenance programme*: Improvement in energy efficiency, providing better comfort, increasing life span of equipment and machinery, and minimizing the system component failure are some of the benefits obtained by scheduling regular maintenance of equipment, process, and systems.
6. *Replacement of inefficient equipment and technology by highly efficient equipment and improved technology*:
   (a) If HVAC system not performing efficiently and not maintaining building comfortably, replacing of and installing existing system by a new energy star-qualified heating and cooling equipment should be considered.
   (b) High-efficiency fans with variable speed motor drives rather than belt drives save significant amounts of energy.
   (c) Variable speed motor drives enable fan speed to match the required air flow.

   Proper commissioning and scheduling regular maintenance is essential with power semiconductor controlled variable drive to minimize their failure rate.
7. *Selection of most efficient source for providing heating and cooling*:
   (a) Free or low-cost energy sources such as solar and geothermal energy should be used first, and then high cost sources as necessary.
   (b) If electric prices are time scheduled, high demand loads should be used in the cheapest time schedule.

19.8 ENERGY SAVING TIPS FOR OTHER INDUSTRIAL PROCESSES

It includes

1. *Creating energy management awareness in staff and employees of industry*:
   (a) Educating and providing training to all and personnel on water and electricity conservation.
   (b) Offering incentives to employees in the form of prizes, certificates, etc., for helping to conserve and improve efficiency should be considered.
   (c) Due recognition should be given them to participate in suggestion schemes that would help save water and energy.
   (d) Staff and other personnel should be encouraged to use stairs instead of elevators if they are not carrying loads.
   (e) Regular checking and implementation of energy and water conservation methods save energy.
   (f) Conducting inspections of all departments regularly to make sure that energy and water conservation methods are being followed
   (g) By properly closing the car windows while driving and use of multi-grade oil help in reducing drag force on vehicle, fuel consumption can be reduced to a larger extent.
2. *Recycling and disposal of wastes*: even small inefficiencies can result in energy waste that can build up unnoticed. Perfect sealing of water, gas, fuels, and heat leaking regularly minimizes the wastage of energy, reduces the energy consumption, and lengthens the life of equipment and machinery.

(a) Methods of minimizing the production of waste and hazardous emission should be implemented by proper operation and control to the extent possible. Pollution control measure should be taken to dispose hazardous wastes.
(b) Recycling of waste or scrap materials such as glass, plastic, aluminium, cans, and waste materials from production, etc., conserves energy.
(c) Low temperature heat coming out of several processes and equipment can be recycled and utilized for other applications. Thus, energy can be saved.
(d) Recycling of steam condensate reduces hot water wastage.
(e) Recovery of waste heat from industrial equipment and process and its use saves considerable energy.

It can be used for fuel oil heating, boiler feed water heating, outside air heating, etc.

3. *Scheduling continuous monitoring, regular repair and preventive maintenance of all equipment, machinery, and other accessories*:
(a) Electrical wiring in the entire industry should be perfect and tight as per standard norms.
(b) Unused equipment and machinery should be switched off.
(c) Monitor external doors, windows, and outside walls for gaps; cracks and unnecessary openings should be monitored to minimize unnecessary escaping of conditioned space air. Sealing of weather-strip doors and windows that leak air save energy.
(d) Continuous monitoring of operations of arc furnaces, circulating water pumps, and other equipment; further, the quick action taken towards fixing and sealing air and water leakages result in considerable energy savings.
(e) The boiler combustion process should be optimized by monitoring the flue gases.
(f) Regular and proper maintenance of nozzle, filter, and injection pump of diesel generator set avoid excessive fuel combustion.
(g) Fuel consumption per kWh of electricity generated should be monitored regularly in DG set. If this shows a rising trend, corrective action should be taken.
(h) Burners, nozzles, strainers, etc., of boilers should be regularly cleaned and maintained. Steam and condensate leakage should be avoided by perfect sealing.
(i) All oil leakage in furnace should be completely sealed or plugged. Leakage of one drop of oil per second amounts to a heavy loss of oil.
(j) Compressors should be switched off when not in use. Reducing pressure by 10% can lead to 5% savings in energy. It should be checked that operations are not affected by reducing the pressure.
(k) It should be ascertained that compressed air is really required. Compressed air should not be used for cleaning, drying, and ventilation when other methods are available.
(l) Preheating the furnace oil gives proper combustion; incomplete combustion leads to wastage of fuel in furnace. Hazy brown smoke coming out of chimney is an indicator of proper combustion, black smoke is an indicator of fuel combustion because of improper combustion and smoke is an indicator of excess air, and hence, loss of heat.
(m) Properly sealing water leakage of pipes, repair, and maintenance of water taps minimize water wastage that leads to energy savings in pumping water.
(n) Car pooling is a method of energy and fuel savings.
(o) Modal shifting ensures that goods (freight) and passenger transport is carried out in the most energy efficient mode such as considering switching from road to rail and encouraging public transport over individual vehicles.
(p) Improved road or track maintenance in transportation sector saves energy.

4. *Replacing old and inefficient equipment, machinery, process, and systems by efficient one*:
   (a) Retrofitting of old machinery and equipments by new and highly efficient one conserve energy.
   (b) Despite the fact that cogged V-belt is expensive than standard belt used as mechanical transmission gear, they reduce losses in mechanical transmission gear. It is used for speed and torque variation between drive motor and driven equipment.
   (c) Monitoring of external doors, windows, and outside walls for gaps; cracks and unnecessary openings in buildings should be monitored to minimize unnecessary escaping of conditioned space air. Sealing of weather-strip doors and windows that leak air saves energy.
   (d) Installation of solar water heaters and outdoor solar lighting should be considered. Other solar-powered equipment and applications or ones that run on other renewable energy resources should be examined.
   (e) Air-cooled chillers should be replaced by water chillers.
   (f) Light emitting diodes (LED) should be used in areas where the lighting is on continuously for long periods.
   (g) Drip irrigation systems in place of conventional irrigation system for agricultural work for increasing water efficiency.
   (h) Water purifying devices should be installed to treat water for use in other factory areas.
   (i) Replacement of multiple production lines with single lines that are able to produce various vehicles decreases energy consumption.
   (j) Using modern and energy efficient technologies consume less energy.
   (k) Solid state devices consume less energy. Electromagnetic ballast for lamps and resistance fan regulators should be replaced by electronic ballast and electronic fan regulators as they consume less energy.
   (l) Solid state controlled drives in place of conventional drives for fan-type load save energy.
   (m) Variable speed drives on large boiler combustion air fans with variable flows should be used as it consumes less energy.
   (n) Installation of automatic on/off switching of cooling tower fans saves large amount on electricity costs.
   (o) Drift losses is reduced by replacing wooden drift eliminator by PVC drift eliminator in cooling tower, thereby considerable energy savings in fan power.
   (p) Using standard fuel efficient pump sets should be selected and proper installation of pump system in agriculture sector should be done.
   (q) Energy efficient technologies with strict following of norms and standards in transportation system should be implemented.
   (r) Replacing a fuel efficient fleet by choosing more fuel efficient vehicles may have fuel cost savings but most important is the maintenance and proper driving of vehicles as per set standard (such as smooth driving in correct gear and regular maintenance of tyres) to reduce fuel consumptions.

Example 19.13

Explain the preheating of combustion air for improving boiler efficiency.

A 36,000 kg/h capacity boiler plant has a total combined efficiency of boiler and combustion of 77.8%. If the combustion air is preheated by 20°C, total efficiency of boiler system increases to 78.9%. The cost of retrofitting of steel duct for preheating is ₹50,000. Calculate simple payback time if annual fuel cost used is ₹3,000,000.

Solution

Heating combustion air (or combustion air preheating) is a method used to improve boiler thermal efficiency by about 1% when the combustion air temperature is increased by 20°C and thermal combustion efficiency increases by 1%.

The most common method to preheat is with the help of a heat exchanger placed on the flue exhaust. This technology commercially developed in the early days of 1900 century for large coal-fired boilers that are widely used in present days also.

Any device designed to heat air prior to its use in another process (like combustion in boilers) to increase the process thermal efficiency is called air preheater. The potential heat energy available in the top area of boiler room is utilized in preheating air for use in combustion process of boilers. The two common approaches for air preheating are as follows:

1. Hot air from boiler room ceiling is drawn by laying down a pipe duct from the close vicinity of boiler room ceiling and up to the boiler combustion air inlet.
2. In another approach, heat is recovered from flue gases by combustion air preheater.

This represents an effective and inexpensive energy savings' opportunity, if the warm air is ducted directly to the combustion intakes and utilized for combustion.

Let us assume

Existing boiler efficiency = 77.8%

Proposed boiler efficiency = 78.9%

Annual fuel cost = ₹3,000,000

Retrofitting cost = ₹50,000

Annual cost savings = fuel cost/per year × fuel savings

= Fuel cost per year × [increase in efficiency/proposed efficiency]

= 3,000,000 × (78.9 − 77.8)/78.9

= 41,825/year.

Simple payback time = 50,000/416,825 = 1.2 years.

It should be noted that the simple savings analysis used here could be applied to any actions that would influence boiler efficiency and for which existing and proposed boiler efficiencies were known.

5. *Improving energy efficiency of systems*: there are several points of inefficient energy efficiency in the energy use activities in industrial, residential, and other energy activity sectors that can be improved by properly conducting energy management program. For example
   (a) Better management of material handling by properly arranging equipment in sequence of operation reduces losses, and thus, results in considerable energy savings.

(b) By properly scheduling regular maintenance of power station, efficiency is largely enhanced, leading to considerable reduction in energy losses.

(c) Reduction of heat losses will lead to energy savings in hot water system simply by better thermal insulation of its component parts such as piping, storage tank, and improved design of water heater.

(d) Upgrading of electric supply system by using proper size of conductors, and installation of equipment and appliances to achieve good power quality reduce transmission or distribution losses to a large extent.

SUMMARY

- The best electrical supply would be a constant magnitude and frequency sinusoidal voltage waveform.
- Load factor is the ratio of average energy savings to peak energy savings. This is also known as peak coincidence factor. More generally, load factor is the average demand divided by any number of peak demands, such as load factor at the time of system peak and load factor at the time of non-coincident peak.
- Load shifting or peak saving was considered as the best method for reducing consumer demand during the peak period.
- The power quality of a system expresses to which degree a practical supply system resembles the ideal supply system. Power quality is characterized by parameters that express harmonic pollution, reactive power, and load unbalance.
- Electric motors operating at partial loads operate at reduced efficiency and power factor.
- Efficacy of a light source defines how well it turns input power (watts) into the desired light output (lumens). It is expressed as lumens/watt. It will never be a number as efficiency.

REVIEW QUESTIONS

1. State and explain methods of reducing peak demand on electric supply systems?
2. An industrial plant is consuming 400 kW of power with plant power factor of 0.77. The demand charge is ₹150/kVA/month. Determine the savings possible by improving power factor to 0.95 and simple payback period, if investment on capacitor bank is ₹100,000. [saving = ₹14,650/month; simple payback time = 0.56 years].
3. State and explain operational improvement principles of energy management with electric motors.
4. Calculate the annual energy savings and simple payback time by replacing motor A with highly efficient motor B. Motor specification and operational details are as follows:

Specifications	**Motor A**	**Motor B**
kW rating	15 kW	15 kW
% Motor efficiency at 75% loading	88.3	93.5

(Continued)

(*Continued*)

Specifications	Motor A	Motor B
Yearly working hours	8,000	8,000
Purchase price of motors in Rs	–	40,000
Average cost of rewinding in Rs	5,000	–
% loading of motors	75	75
Electric cost in Rs	4.0	4.0

[**Ans.** Energy cost savings = ₹22,675.0; payback time = 1.54 years]

5. State and explain factors affecting the selection of light lamps.
6. Discuss general principles of energy management opportunity with light sources.
7. Differentiate between electronic ballast and electromagnetic ballast according to working principle and application.
8. Discuss the two important options of energy management opportunities with ceiling fan.
   (a) Is it more energy efficient to keep turning fluorescent lighting on and off all day when not in use or to just leave it on? Explain.
   (b) Does it make a difference what type of light it is whether incandescent or compact fluorescent?
9. Does leaving a computer on consume less energy than switching it on and off? Explain.
10. Discuss in brief methods which energy can be saved in an air conditioning system.

Appendix A

Multiple Choice Questions Chapter Wise

CHAPTER 1

NCER—an Overview

1. Energy can be conserved by
   (a) cogeneration in power plants, fuel-efficient transportation, and reengineered appliances
   (b) cogeneration in automobiles, especially in SUV
   (c) replacing compact florescent with tungsten bulbs
   (d) replacing mass transit with more and larger roads for cars and trucks
2. Identify the non-renewable energy resource from the following:
   (a) Coal (b) Solar power
   (c) Wind power (d) Wave power
3. Which of the following is a disadvantage of most of the renewable energy sources?
   (a) Highly pollution (b) High waste disposal cost
   (c) Unreliable supply (d) High cost
4. Photovoltaic energy is the conversion of sunlight into
   (a) chemical energy (b) biogas
   (c) electricity (d) heat energy
5. Solar cell converts
   (a) solar energy to heat energy
   (b) solar energy to mechanical energy
   (c) solar energy to electrical energy
   (d) solar energy to liquid fuels
6. Fuel cells are
   (a) carbon cell (b) hydrogen battery
   (c) nuclear cell (d) solar cell

7. The outermost layer of the earth is
   (a) magma
   (b) mantle
   (c) crust
   (d) solid iron core
8. Common energy source in Indian villages is
   (a) electricity
   (b) coal
   (c) sun
   (d) wood and animal dung
9. The one thing that is common to all fossil fuels is that they
   (a) were originally formed in marine environment
   (b) contain carbon
   (c) have undergone the same set of geological processes during their formation
   (d) represent the remains of one living organisms
10. The process that converts solid coal into liquid hydrocarbon fuel is called
    (a) liquefaction
    (b) carbonation
    (c) catalytic conversion
    (d) cracking
11. Lignite, bituminous, and anthracite are different ranks of
    (a) nuclear fuel (b) coal (c) natural gas (d) biogas
12. Crude oil is
    (a) colourless
    (b) odourless
    (c) smelly yellow to black liquid
    (d) violet liquid
13. BTU is measurement unit of
    (a) pressure (b) head (c) heat content (d) temperature
14. Boiling water reactor and pressurized water reactors are
    (a) nuclear reactor
    (b) solar reactor
    (c) OTEC
    (d) biogas reactor
15. Which of the following is a non-renewable resource?
    (a) Coal
    (b) Forests
    (c) Water
    (d) wind
16. Which among the following is a renewable source of energy?
    (a) Solar energy
    (b) Coal
    (c) Nuclear energy
    (d) Oil
17. Identify the secondary form of energy from the following:
    (a) coal
    (b) water
    (c) electricity
    (d) sea wave energy
18. Horizontal axis and vertical axis are the terms used with
    (a) nuclear reactor
    (b) wind mills
    (c) biogas reactor
    (d) solar cell
19. Which among the following is not an adverse environmental impact of tidal power generation?
    (a) Interference with spawning and migration of fish
    (b) Pollution and health hazard in the estuary due to blockage of flow of polluted water the sea

(c) Navigational hazard
(d) None of the above

20. Steam reforming is currently the least expensive method of producing
(a) coal (b) biogas
(c) hydrogen (d) natural gas

21. In order to produce electricity, a fuel cell uses
(a) helium (b) nitrogen
(c) hydrogen (d) carbon

22. Fuel cells are
(a) carbon cell (b) hydrogen battery
(c) nuclear cell (d) solar cell

23. Cooking gas and manure both are provided by
(a) nuclear plants (b) thermal plants
(c) biogas plants (d) hydroelectric plant

24. The tidal waves are caused by the periodic rise and fall of oceans. It is associated with the position of
(a) moon (b) sun
(c) earth (d) sea

25. The energy available when the temperature of water at the surface of ocean is different from that at deeper level is called
(a) ocean thermal energy (b) sea water energy
(c) ocean water energy (d) none of these

26. Which of the following energy originate from the ocean?
(a) Tidal energy (b) Sea energy
(c) Wind energy (d) Hydropower

27. Ocean and sea waves are indirectly caused due to
(a) pressure gradients (b) solar energy
(c) geothermal energy (d) none of these

28. The ocean waves are caused by the periodic rise and fall of oceans. It is associated with
(a) kinetic energy (b) heat energy
(c) light energy (d) potential energy

29. Hydrogen can be produced from
(a) soil (b) silver
(c) iron (d) water

30. Energy derived from hot spots beneath the earth is called
(a) bioenergy (b) geothermal energy
(c) nuclear energy (d) hydrogen energy

31. In rural areas, the locally generated gas from cow dung used for cooking and lighting is called
(a) biogas (b) oxygen (c) ammonia (d) carbon dioxide

32. Solar energy cannot be stored in following mediums:
    (a) Water (b) Iron (c) Gas (d) Wood
33. Which of the following is NOT a renewable source of energy?
    (a) Geothermal (b) Propane (c) solar (d) wind
34. The disadvantage of renewable sources of energy is
    (a) high cost (b) low energy density
    (c) intermittency (d) all of these

CHAPTER 2

Energy from the Sun

1. Sunlight light reaches the earth through
   (a) direct radiation (b) diffuse radiation
   (c) scattered radiation (d) all of these
2. The angle measured from directly overhead to the geometric centre of the sun's disc is
   (a) declination angle (b) hour angle
   (c) latitude angle (d) zenith angle
3. The angle in minute subtended by sun on the earth's surface is approximately
   (a) 53 minutes (b) 42 minutes (c) 31 minutes (d) 23 minutes
4. Select the correct statement from the following:
   (a) Global radiation = = beam radiation − diffuse rad
   (b) Global radiation = = beam radiation + diffuse radiation
   (c) Global radiation = [beam radiation − diffuse radiation]^2
   (d) Global radiation = [beam radiation + diffuse radiation]^2
5. The solar constant, which is defined as the rate of energy received from the sun on a unit area perpendicular to the rays of the sun at the mean distance of the earth from the sun outside of the atmosphere, has a standard value of
   (a) 1.0 kW/m^2 (b) 1.251 kW/m^2
   (c) 1.367 kW/m^2 (d) 1.398 kW/m^2
6. The solar hour angle is zero at
   (a) sunrise (b) sunset (c) solar noon (d) midnight
7. The declination angle of the sun is the angle between the equator and a line drawn from the centre of the earth to the centre of the sun. It has a value of −23.457 on
   (a) March 21 (b) September 22
   (c) June 21 (d) December 22
8. The hour angle at 0900 hour (local apparent time) is
   (a) +45° (b) –45° (c) +15° (d) = −15°

9. If LST is local solar time, then solar hour angle (HRA) is expressed as
   (a) HRA = 30° × (LST − 12)
   (b) HRA = 15° × (LST − 12)
   (c) HRA = 30° × LST
   (d) HRA = 15° × LST
10. In the zenith angle (θ_z) range between 0° to 70°, air mass (m) at sea level can be calculated as
    (a) $m = \sin(\theta_z)$
    (b) $m = \cos(\theta_z)$
    (c) $m = \sec(\theta_z)$
    (d) $m = \text{cosec}(\theta_z)$
11. For the zenith angle of 60° at sea level, air mass is
    (a) 0
    (b) 1
    (c) 2
    (d) 3

CHAPTER 3

Solar Thermal Energy Collectors

1. Harmful radiation emitted from the sun is
   (a) visible radiation
   (b) infrared radiation
   (c) ultraviolet radiation
   (d) none of the above
2. Which type of dryer can be used to dry fruits and vegetables using renewable energy?
   (a) Oil furnace
   (b) Solar dryer
   (c) Coal furnace
   (d) Wood-based furnace
3. A solar pond is a combination of which of the following combinations?
   (a) Solar energy collection and heat storage
   (b) Solar energy storage and heat collection
   (c) Solar energy collection and energy storage
   (d) None of these
4. Which of the following system is an application of solar thermal energy?
   (a) Internal combustion engine
   (b) Biogas generation
   (c) Solar water heating
   (d) Solar lighting
5. The value of solar constant is approximately
   (a) 6.5 kW/m^2
   (b) 1.36 kW/m^2
   (c) 3.64 kW/m^2
   (d) 10 kW/m^2
6. A typical insulation material used in a solar collector is
   (a) fibre glass
   (b) cotton
   (c) glass wool
   (d) none of the above
7. In a box-type solar cooker, the solar radiations that enter and leave the box are of
   (a) short and long wavelength, respectively
   (b) long and short wavelength, respectively
   (c) similar wavelengths
   (d) none of these

8. Conversion of solar thermal energy into electrical energy using reflecting mirrors is called
   (a) diffuser (b) heliostat
   (c) reflector cookers (d) solar cell array
9. Thermal storage of energy is done in the form of
   (a) chemical reaction for chemically changing a medium
   (b) latent heat
   (c) sensible heat
   (d) all of these
10. In a flat plate collector, the following fluid is not used as a medium of heat exchange,
   (a) air (b) ethylene glycol and water
   (c) fuel oil (d) water
11. Use of thin plastic sheet as glazing material in flat plate collectors introduces serious problems of
   (a) Plastics have limited life span due to the effect of UV radiation that reduces its transmissivity.
   (b) Plastics are transparent to long-wavelength radiation and are, therefore, less effective in reducing radiated heat losses from the absorber plate.
   (c) Plastics cannot withstand high temperature encountered in collector especially when the collector is idle.
   (d) All of the above
12. The ideal glazing materials for solar collector have the following properties:
   (a) Transmit light very well and has high temperature withstand capability.
   (b) Long life when exposed to UV and high temperatures, and also opaque to long wave-length infrared to reduce heat loss.
   (c) Low cost having good impact resistance.
   (d) All of these
13. Which of the following statement is correct for using water as a heat exchange fluid in flat plate collectors?
   (a) Water is non-toxic and inexpensive.
   (b) Water has high specific heat and a very low viscosity.
   (c) Water has a relatively low boiling and high freezing point.
   (d) All of these
14. Which of the following is the drawback of using water as a heat exchange fluid in flat plate collectors?
   (a) Air is available in abundance and inexpensive.
   (b) Air will neither freeze nor boil.
   (c) Air has a very low heat capacity.
   (d) Air is non-corrosive.
15. Small rock pebbles are used for optimum storage of heat as
   (a) They give a large surface area to mass ratio.
   (b) They give a small surface area to mass ratio.
   (c) Cost of the rocks is reduced.
   (d) They improve the specific heat.

16. Selective coating (surface) in flat plate solar collectors
    (a) maximize the absorption of solar energy and emission of radiative loss
    (b) maximize the absorption of solar energy and minimize emission of radiative loss
    (c) minimize the absorption of solar energy and emission of radiative loss
    (d) minimize the absorption of solar energy and maximize emission of radiative loss
17. Following statements are given related to solar energy:
    I. The Equation of Time (EOT) in minutes is an empirical equation that corrects for the eccentricity of the Earth's orbit and the Earth's axial tilt.
    II. Since the Earth rotates 15° per hour, each hour away from solar noon corresponds to an angular motion of the sun in the sky of 15°.
    III. The solar constant is the average extraterrestrial insolation at the edge of atmosphere.
    IV. Local Time equals Local Standard Time irrespective of the eccentricity of the earth's orbit and because of human adjustments such as time zones and daylight saving.

    Select the correct answer using the code given below.
    (a) I, II, and III are correct
    (b) II, III, and IV are correct
    (c) III, IV, and V are correct
    (d) I, II, and IV are correct
18. Solar pond is a saline water pond in which salt concentration
    (a) decreases from top surface to the bottom surface
    (b) increases from top surface to the bottom surface
    (c) remain the same in the entire pond
    (d) none of these
19. Working fluid used in solar Rankine cycle is
    (a) air (b) hydrogen
    (c) helium (d) steam

CHAPTER 4

Solar Cell

1. The word photovoltaic comes from words meaning
    (a) wind energy (b) brightness
    (c) light and electricity (d) picture that moves
2. A PV module is
    (a) dozens of photovoltaic cells connected together
    (b) wired in series
    (c) wired in parallel
    (d) all of these

3. Solar PV systems can be
   (a) connected to the power grid
   (b) used to sell power to the grid
   (c) a standalone source of electricity
   (d) all of these
4. Improving the efficiency of a PV cell can be done by
   (a) placing the PV cell angle facing light all day
   (b) placing coloured acetates on the cell
   (c) heating the cell
   (d) changing its direction to north
5. In a series connection of photovoltaic,
   (a) the positive terminal is connected to the positive terminal
   (b) the negative terminal is connected to the negative terminal
   (c) the positive terminal is connected to the negative terminal
   (d) all of these
6. What is the most commonly used material in making solar cells?
   (a) silver
   (b) iron
   (c) aluminium
   (d) silicon
7. The electrical output of a solar cell depends on the
   (a) intensity of solar radiation
   (b) heat component of solar radiation
   (c) ultraviolet radiation
   (d) infrared radiation
8. Solar constant throughout the year varies in the range
   (a) 1.0–1/153 kW/m^2
   (b) 1.16–1.295 kW/m^2
   (c) 1.3–1.34 kW/m^2
   (d) 1.353–1.395 kW/m^2
9. The solar constant at normal incidence outside the atmosphere at the mean sun earth distance is approximately
   (a) 1 kW/m^2
   (b) 1.4 kW/m^2
   (c) 1.367 kW/m^2
   (d) 1.399 kW/m^2
10. Solar photovoltaic cell converts solar energy directly into
    (a) mechanical energy
    (b) electricity
    (c) heat energy
    (d) transportation
11. What does SPV stand for with respect to solar energy?
    (a) Solar photovoltaic
    (b) Solid plate voltaic
    (c) Solar plate voids
    (d) None of these
12. Which of the following appliances use solar photovoltaic technology?
    (a) Solar lantern
    (b) Biogas plant
    (c) Solar water heater
    (d) Solar air heater
13. Pyranometer is an instrument used for measuring
    (a) temperature
    (b) solar irradiance
    (c) wind speed
    (d) efficiency
14. The presence of CO_2 and H_2O in the atmosphere results in absorption of
    (a) long wave infrared radiations
    (b) short wave ultraviolet radiations
    (c) visible wavelengths of spectrum
    (d) all of these

15. Pyrheliometer instrument measures
    (a) wind speed at an elevation
    (b) total electromagnetic radiation emitted by the sun
    (c) direct beam solar irradiance
    (d) biogas pressure
16. Solar cells are made of
    (a) aluminium (b) germanium (c) silicon (d) silver
17. Open circuit voltage of a single solar cell is approximately
    (a) 2 V (b) 1.5 V (c) 1 V (d) 0.5 V
18. A photovoltaic module is a
    (a) series connection of flat plate collectors
    (b) parallel connection of fuel cells
    (c) series connection of thermionic converters
    (d) series–parallel connection of solar cells
19. Conversion of solar light into electricity has the following major drawbacks:
    (a) Practical efficiency of solar cell is very less.
    (b) Solar energy is available for limited period in a day.
    (c) High cost
    (d) All of these
20. The maximum theoretical efficiencies of solar cells in a controlled chamber is approximately
    (a) 10% (b) 20% (c) 45% (d) 75%
21. The band gap energy of silicon is
    (a) 1.6 eV (b) 1.4 eV (c) 1.2 eV (d) 1.1 eV
22. Fill factor of solar cell is defined as the ratio of the maximum useful power to the ideal power. Its typical values range from
    (a) 0.1–0.2 (b) 0.21–0.42 (c) 0.45–0.55 (d) 0.6–0.8
23. Electron volt (eV) is the unit of
    (a) current (b) energy (c) power (d) voltage

CHAPTER 5

Hydrogen Energy

1. Hydrogen can play an important role as an alternative fuel to conventional fuel as
   (a) an energy carrier (b) an energy device
   (c) a fossil fuel source (d) an energy system
2. What is the main by-product of burning hydrogen?
   (a) H_2O (b) ash (c) O_2 (d) steam

3. Hydrogen is produced by
   (a) electrolysis (b) incineration
   (c) aerobic digestion (d) decomposting
4. Which of the following is not an application of hydrogen energy?
   (a) Fuel cell (b) Generator set
   (c) Running a two wheeler (d) Producing biofuel
5. Hydrogen can be generated commercially by
   (a) aerobic digestion (b) steam methane reforming
   (c) anaerobic digestion (d) incineration
6. Fuel reforming is a process of
   (a) transforming fossil fuel into CO
   (b) transforming fossil fuel into H_2
   (c) removing carbon compounds from fossil fuel
   (d) none of these
7. When hydrogen burns in air, it combines with which of the following gases to form water?
   (a) O_2 (b) N_2 (c) CO_2 (d) O_3
8. Which form of energy source yields only water?
   (a) Hydrogen (b) Biogas (c) Methane (d) Oxygen
9. Hydrogen can be stored as a
   (a) compressed gas (b) liquid
   (c) metal hydride (d) both (a)and (c)

CHAPTER 6

Wind Energy

1. Which of the following statement is not correct for wind power?
   (a) It releases no greenhouse gases or acid forming emissions.
   (b) It has been used from centuries for water pumping.
   (c) It can be used to produce electricity.
   (d) It is a continuous and uninterruptible source of energy.
2. The maximum energy conversion efficiency of a wind turbine for a given swept area is
   (a) 25.1% (b) 50.4% (c) 59.3% (d) 99.9%
3. If the velocity of wind is doubled, then the power output will increase by
   (a) 10 times (b) 8 times (c) 2 times (d) 6 times
4. The terms Darrieus and Savonius rotors are related to
   (a) small hydropower (b) wind energy
   (c) turbine (d) coal extraction mechanism

5. Power output from a wind energy generator is directly proportional to
   (a) wind velocity (b) square of wind velocity
   (c) cube of wind velocity (d) square root of wind velocity
6. A place where many wind turbines are installed together to produce electricity is called a
   (a) wind farm (b) propeller collection
   (c) wind station (d) wind turbine station
7. What type of energy is associated with wind?
   (a) Potential energy (b) Chemical energy
   (c) Kinetic energy (d) Rotational energy
8. Wind turbines using aerodynamic lift produce more energy for a given area than wind turbines using aerodynamic drag as the
   (a) lifting force pushes the blade in the direction of the wind
   (b) lifting force roughly perpendicular to the local flow fields
   (c) lifting force produces more torque
   (d) drag services capture more energy because of greater friction on the blade surfaces
9. The relationship between power available from wind '*P*' and wind velocity '*v*' is
   (a) *P* ά *v* (b) *P* ά v^2 (c) *P* ά v^3 (d) $P = v$
10. An anemometer is an instrument used for measuring
    (a) solar radiation (b) wind speed
    (c) temperature gradient (d) depth in ocean
11. Lower speed wind turbines are mainly driven by
    (a) drag forces (b) lift forces
    (c) push forces (d) none of these
12. The torque causing the rotation of a rotor is due to the
    (a) drag force (b) gravitational force
    (c) force of lift (d) axial thrust
13. With increase in height, wind speed
    (a) increases (b) decreases
    (c) remains the same (d) none of these
14. Wind power plants are required to have a large rotor size for large power output because of
    (a) low power density of air stream
    (b) lift force acting perpendicular to the direction of wind flow
    (c) lift force being more than drag force
    (d) drag force acting perpendicular to lift force
15. Which of the following forces act on the blades of wind turbine rotor?
    (a) Lift force (b) Drag force
    (c) Both (a) & (b) (d) None of these
16. Wind machine with Darrieus type of rotor is a
    (a) vertical axis machine (b) horizontal axis machine
    (c) machine that can spin in one direction only (d) none of these

17. Air density at standard conditions is about
    (a) 1.885 kg/m^3 (b) 2.55 kg/m^3 (c) 1.226 kg/m^3 (d) 3.267 kg/m^3
18. The main disadvantage of wind power is that
    (a) it is unreliable energy source as winds are uncertain and unpredictable
    (b) wind energy systems are noisy when in operation
    (c) large land open area is required
    (d) all of these
19. Wind energy conversion devices based on drag force
    (a) move faster than wind
    (b) move slower than wind
    (c) move with equal velocity as wind
    (d) do not depend on the velocity of wind
20. The power generated by a wind turbine can be approximated as,

    $$\text{Power} = (1/2) \times (\text{air density})(\text{blade area})(\text{wind velocity})^3$$

    As per the above equation, the greatest increas\e in the power generated by a wind turbine will be obtained by
    (a) decreasing the area of the turbine blades half
    (b) increasing the area of the turbine blades 3 times as large
    (c) shifting the turbine to a location having 1/4th wind velocity
    (d) shifting the turbine to a location having double the wind velocity
21. Betz law of theory of momentum finds application in
    (a) geothermal plant (b) MHD plant
    (c) solar plant (d) wind power plant
22. Energy of wind blowing on large sea surface is stored in sea waves as
    (a) chemical energy (b) electrical energy
    (c) mechanical energy (d) thermal energy
23. Following is the vertical axis wind turbine:
    (a) Multi-blade type (b) Propeller type
    (c) Sail type (d) Savonius type
24. The speed of operation of propeller type wind turbine is about,
    (a) 50–80 rpm (b) 100–150 rpm
    (c) 150–250 rpm (d) 300–400 rpm
25. Darrieus wind rotor has the following properties:
    (a) It has two or three blades shaped like aerofoil.
    (b) It requires less surface area as compared to Savonius type windmill.
    (c) It runs independently of the direction of the wind
    (d) All of these

CHAPTER 7

Geothermal Energy

1. The correct statement related with geothermal energy is
   (a) the use of naturally derived heated water or steam to heat building or drive turbo generators
   (b) the heat derived from the crushing of large quantities of rocks
   (c) the heat derived from molten fluid
   (d) it is a heating system that uses hot water pipes embedded in flooring within a home
2. Which of the following are considered to be drawbacks of geothermal energy?
   (a) It is not available everywhere. It is available only in areas where hot rocks are Present near the earth's surface.
   (b) High cost of exploration and drilling
   (c) Land pollution and seismic hazards
   (d) All of these
3. Which type of energy does volcanoes posse?
   (a) Mechanical energy (b) Geothermal energy
   (c) Electrical energy (d) Nuclear energy
4. The molten mass of earth is called
   (a) magnus (b) magma (c) hot cake (d) magmus
5. Geothermal energy reservoirs are
   (a) liquid-dominated reservoirs (b) steam-dominated reservoirs
   (c) hot rocks with no water (d) all of these
6. No moving parts are required in
   (a) geothermal power plant (b) thermionic power plant
   (c) OTEC power plant (d) tidal power plant
7. In geothermal power plants, waste water is
   (a) re-circulated after cooling in cooling towers
   (b) evaporated in ponds
   (c) discharged back inside the earth
   (d) none of these
8. Geysers steam is continuously vented through fissures in the ground. These vents are called
   (a) fumaroles (b) pot holes
   (c) sun spot (d) pores
9. In saline water-dominated geothermal energy, power is developed using
   (a) binary systems (b) flashed steam systems
   (c) total flow systems (d) all of these
10. Geothermal energy is
    (a) a renewable energy resource (b) an inexhaustible energy resource
    (c) alternative energy resource (d) all of these

CHAPTER 8

Solid Wastes and Refuse and Agricultural Refuse

1. Match the list I with list II and select the correct answer using the codes given below.

List I	List II
A. Pyrolysis	1. When coupled with high temperature, waste treatments are recognized as thermal treatments.
B. Incineration	2. Thermochemical decomposition of organic material at elevated temperatures in the absence of oxygen
C. Economizer	3. It consists of a circular honeycomb matrix of heat absorbing material, which is slowly rotated within the supply and exhaust air streams of an air handling system.
D. Thermal wheel	4. Waste heat in the boiler exhaust gas is passed through it that carries the inlet fluid for the boiler and thus decreases thermal energy intake of the inlet fluid.

	A	B	C	D
(a)	1	2	3	4
(b)	2	3	4	1
(c)	2	1	4	3
(d)	4	1	2	3

2. Select the correct statement using the codes given below.
   I. Incineration is a waste treatment technology that includes the combustion of waste for recovering energy.
   II. Many hazardous wastes can be recycled into new products.
   III. Recuperators are different types of heat exchanger that the exhaust gases are passed through, consisting of metal tubes that carry the inlet gas and thus preheating the gas before entering the process.
   IV. eat pipes are low thermal conductors.

   Codes:
   (a) I, II, and III are correct
   (b) I, II, and IV are correct
   (c) II, III, and IV are correct
   (d) I, II, III, and IV are correct
3. Following are the advantages of waste heat recovery systems:
   (a) Thermal and air pollution decrease since less flue gases of high temperature are emitted from the plant as most of the energy is recycled.
   (b) The recovery process will add to the efficiency of the process and thus decrease the costs of fuel and energy consumption needed for that process.

(c) As Fuel consumption reduces, so the control and security equipment for handling the fuel decreases as fuel combustion reduces in the process.
(d) All (a), (b), and (c) are the advantages.

4. Waste material is reduced to ashes in the following process:
(a) Biodegradation
(b) Composting
(c) Incineration
(d) Recycling

5. Environment can be kept clean by the following waste management methods (three R's):
(a) Recreate, Recycle, and Remember
(b) Recreate, Recycle, and Reuse
(c) Recycle, Reuse, and Reduce
(d) Reuse, Reduce, and remember

6. Method of final disposal of waste is
(a) landfills (b) recycle (c) reuse (d) sanitation

7. By the fermentation of sugar cane, following fuel can be produced:
(a) Biodiesel (b) Biogas (c) Ethanol (d) Producer gas

8. Urban wastes are used for the following:
(a) Heat generation
(b) Electricity generation
(c) Composting
(d) All of these

9. Exposure to hazardous toxic waste is dangerous to health. The safest disposal method of hazardous waste is the following:
(a) Briquetting
(b) Fermentation
(c) Landfills
(d) Incineration

10. Following statements are given related with waste management:

I. A landfill is a site for the disposal of waste materials by burial of waste in ground.

II. Recyclate is a raw material processed in a waste recycling plant or materials recovery facility, which will be used to form new products.

III. Recycling is a process to change waste materials into new products and thus preventing wastage of potentially useful materials and minimizing the consumption of fresh raw materials.

IV. Solid waste management is the methodology of generation, prevention, characterization, monitoring, treatment, handling, reuse, and residual disposition of solid wastes.

Using the codes given below, select the correct answer:
(a) I is correct.
(b) I and II are correct.
(c) I, II, and III are correct.
(d) I, II, III, and IV are correct.

11. Match List I (Plastic Resin ID Code) with List II (Plastic Type) and select the correct answer using the code given below.

List I (Plastic Resin ID Code)	List II (Plastic Type)
A.1	P. Nylon and polycarbonate (PC)—these products are the hardest for recycling and they go directly to landfills.
B.2	Q. Polystyrene (Styrofoam) items—it is widely useful and accepted for recycling.
C.3	R. Polyethylene terephthalate (PETE)—the easiest and the most common plastics to recycle.
D.4	S. High-density polyethylene plastic—it is widely accepted at recycling.
E.5	T. Vinyl or polyvinylchloride (PVC)—it has a low rate of recyclability.
F.6	U. Low density polyethylene (LDPE)—it is a less commonly recycled Plastic.
G.7	V. Polypropylene (PP)—it is a less commonly recycled plastic

Codes:

	A	B	C	D	E	F	G
(a)	R	S	T	U	V	Q	P
(b)	P	Q	R	S	T	U	V
(c)	P	R	Q	U	V	S	T
(d)	R	S	T	P	Q	U	V

12. Match List I (Technology) with List II (Major Products) and select the correct answer using the codes given below.

List I (Technology)	List II (Major Products)
A. Anaerobic digestion	1. Crude oil and metal
B. High temperature gasification	2. Synthetic gas, and electricity
C. Pyrolysis	3. Biogas and compost
D. Thermal de-polymerization	4. Synthetic gas, electricity, and mixed metals

Codes:

	A	B	C	D
(a)	1	2	3	4
(b)	3	4	2	1
(c)	1	2	4	3
(d)	3	1	2	4

13. Following statements are related with availability of plastic wastes:

    I. Polyethylene terephthalate (PETE) plastics are used to make soda and water bottles, medicine containers, and many other common consumer product containers. Once it has been processed by a recycling facility, PETE can become fibrefill for winter coats, sleeping bags, and life jackets. It can also be used to make bean bags, rope, car bumpers, tennis ball felt, combs, cassette tapes, sails for boats, furniture and, of course, other plastic bottles.

II. High-density polyethylene plastics are used in the production of heavier containers of laundry detergents and bleaches as well as milk, shampoo, and motor oil jugs and bags. It is often recycled into toys, piping, plastic lumber, and rope. It can also be processed to crude oil.

III. Polyvinylchloride is commonly used in plastic pipes, shower curtains, medical tubing, vinyl dashboards, and even some baby bottle nipples.

IV. Polystyrene (Styrofoam) plastic is used to manufacture items such as coffee cups, disposable cutlery, meat trays, packing "peanuts", and insulation. It is widely accepted because it can be reprocessed into many items including cassette tapes and rigid foam insulation.

Using the codes given below, select the correct answ.er:

(a) I is only correct.
(b) I and II are correct.
(c) I, II, and III are correct.
(d) I, II, III, and IV are correct.

14. The following statement is not correct:

(a) Plastic recycling is the process of recovering scrap or waste plastics and reprocessing the material into useful products, sometimes completely different in form from their original state.
(b) Thermal de-polymerization is a process that involves the conversion of assorted polymers cracked into crude oil.
(c) The easiest and the most common plastics to recycle is nylon and polycarbonate (PC).
(d) Plastic wastes can also be processed into wax-free hydrocarbon such as naphtha and diesel oil.

CHAPTER 9

Biomass Energy

1. Using biomass as a source of energy for heating includes heating with

(a) coal (b) natural gas (c) petroleum (d) wood stove

2. Biomass can be converted into

(a) liquid fuel (b) hydrogen
(c) producer gas (d) all of these

3. Gasification of biomass is a

(a) biochemical conversion process (b) chemical conversion process
(c) thermochemical conversion process (d) biological conversion process

4. Biomass can be used as fuel through
   (a) combustion (b) fermentation
   (c) digestion (d) all of these
5. Biomass can be converted into
   (a) solid fuel (b) liquid fuel
   (c) gaseous fuel (d) all of these
6. Which of the following fuel does not give ash as residue when burnt?
   (a) Wood (b) Charcoal (c) Biogas (d) Coal
7. The gas produced by burning wood in an insufficient supply of oxygen is called
   a) Producer gas (b) Biogas
   (c) Natural Gas (d) Nitrogen gas
8. Bagasse is
   (a) a type of coal (b) wood product
   (c) sugarcane product (d) rice straw

CHAPTER 10

Biogas Energy

1. Biogas is produced when anaerobic bacteria digest organic matters in absence of oxygen. This process is called
   (a) anaerobic reduction (b) anaerobic digestion
   (c) anaerobic oxidation (d) anaerobic drying
2. Which gas has a major share in biogas?
   (a) N_2 (b) CH_4 (c) CO_2 (d) H_2
3. What is the biggest source of biogas in rural India?
   (a) Kitchen waste (b) Leaves
   (c) Cow dung (d) Industrial waste
4. Under normal conditions, the maximum yield of biogas can be obtained at
   (a) 7°C–10°C (b) 11°C–15°C
   (c) 16°C–25°C (d) 26°C–35°C

CHAPTER 11

Tidal Energy

1. The interaction between sun, earth, and moon causes the strongest phenomena for the availability of the following energy:
   (a) ocean thermal (b) sea waves
   (c) tidal (d) wind

2. The tidal energy of a sea wave is the sum of its
   (a) mechanical energy and thermal energy
   (b) potential energy and hydropower energy
   (c) potential energy and kinetic energy
   (d) kinetic energy and thermal energy
3. The tidal range is the difference between
   (a) water movement speed and high tide
   (b) water movement direction and high tide
   (c) Water elevation at high tide and low tide
   (d) mean sea level and low fields elevation
4. Bulb turbines used in tidal power plants are
   (a) high head turbines
   (b) high pressure turbine
   (c) high speed
   (d) low head turbine
5. Tides are created primarily by
   (a) absorption of solar radiation in the atmosphere
   (b) gravitational attraction between the earth and moon
   (c) interaction of winds with the surface of oceans
   (d) rotation of the earth about its axis and its motion around the sun
6. Following advantages are given for tidal power:
   I. Tidal power belongs to renewable energy sources meaning it cannot be depleted like fossil fuels.
   II. Tidal power belongs to the most efficient energy sources by having efficiency of approximately 80%.
   III. Tides are predictable, and this predictability is also one of the advantages that tidal power has over other energy sources because rise and fall of tides are much more cyclic than random weather patterns.
   IV. Harnessing tidal power has positive impact on climate change because it produces no greenhouse gas emissions.
   V. Tidal power plants maintain cost-competitiveness on global energy market.

   Using the codes given below, select the correct answer.
   (a) I, II, III, and IV are correct.
   (b) I, II, III, and V are correct.
   (c) II, III, IV, and V are correct.
   (d) III, IV, and V are correct.
7. Following disadvantages are given for tidal power:
   I. Tidal power plants are connected with high upfront costs needed for construction and therefore lack cost-competitiveness on global energy market.
   II. Tidal power is intermittent source of energy that can only produce electricity during tidal surges.
   III. Harnessing tidal power has positive impact on climate change because it produces no greenhouse gas emissions.

IV. Tidal power needs big funds and plenty of scientific research before reaching the commercial character.

Using the codes given below, select the correct answer.

(a) I, II, and III are correct.
(b) I, III, and IV are correct.
(c) II, III, and IV are correct.
(d) I, II, and IV are correct.

8. For a tidal power plant, h is the vertical tidal range, A is the horizontal area of the barrage basin, ρ is the density of water, and g is the acceleration due to the gravity. The energy (E) available from barrage is
(a) $E = (1/2)\,A\rho gh$
(b) $E = (1/2)\,A\rho gh^2$
(c) $E = (1/2)\,[A\rho gh]^2$
(d) $E = A\rho gh^2$

CHAPTER 12

Sea Wave Energy

1. Because of the following property, a sea wave is likely to bend toward a headland rather than travel straight toward land.
(a) Diffraction
(b) Reflection
(c) Refraction
(d) Wave interference

2. The two swells in phase interfere with each other
(a) destructively
(b) constructively
(c) orthogonally
(d) refractively

3. Shape of movement of sea waves in deep water is
(a) circular
(b) elliptical
(c) horizontal
(d) spherical

4. Water waves are also called as
(a) celestial waves
(b) hydro spherical waves
(c) orbital waves
(d) spherical waves

5. Ocean waves moving with steady speed is called
(a) Tsunami
(b) Progressive wave
(c) spilling breaker
(d) Tidal wave

6. Following factor(s) generally determine(s) the maximum length and height of ocean waves
(a) celerity
(b) duration
(c) fetch
(d) all of these

7. The speed of a sea wave is equal to
(a) the height divided by the frequency
(b) the wavelength divided by the amplitude
(c) the period divided by the wavelength
(d) the wavelength divided by the period

8. A wave with a period of 10 s and a wavelength of 130 m will travel at a velocity of
   (a) 1.3 m/s (b) 13 m/s (c) 130 m/s (d) 1,300 m/s
9. The horizontal distance between two successive crests or troughs of a sea wave is the
   (a) Amplitude (b) Height
   (c) Period (d) Wavelength
10. In a wave train,
   (a) the individual wave crests move toward the back of the group.
   (b) the individual wave crests move toward the front of the group.
   (c) the train carries the energy of the waves.
   (d) both (b) and (c)
11. Following factors work determine(s) the size of wind waves:
   (a) Wave length (b) wave height (c) wave period (d) all of these
12. Sea waves which can travel long distances without losing much energy is called
   (a) capillary wave (b) gravity waves
   (c) rogue waves (d) swell waves
13. On a moderately sloped beach, following occurs:
   (a) Open ocean breaker (b) Plunging breaker
   (c) Spilling breaker (d) Surging breaker
14. Following device is not used for sea wave energy conversion to electrical energy:
   (a) Aquabuoy (b) Bulb turbine
   (c) Oscillating water column (d) Tapchan
15. Given that the significant height of sea wave is H (m), period of wave is T (s), g is the acceleration due to gravity in m/s, and ρ is the specific density of water in kg/m^3. The approximate expression for wave power is given by
   (a) $P = 0.5\, H^2 \times \mathrm{T}$ (kW/m) (b) $P = 0.5\, H^2 \times T^2$ (kW/m)
   (c) $P = 0.5\, H \times T^2$ (kW/m) (d) $P = \rho \times \mathrm{g} \times H \times T$
16. Assertion (A): Capillary waves transfer energy from air to water to drive currents.
   Reason (R): Capillary waves are the first to form when the wind blows.
   Using the codes given below, select the correct answer:
   (a) Assertion (A) and Reason (R) both are correct and R is the correct explanation of A.
   (b) Assertion (A) and Reason (R) both are correct but R is not the correct explanation of A.
   (c) A is correct but R is incorrect.
   (d) A is incorrect but R is correct.
17. Sea waves usually break when wave ratio equals to
   (a) 4/7 (b) 3/7 (c) 2/7 (d) 1/7
18. Steepness of sea wave is given as
   (a) the product of wave height and wavelength
   (b) the ratio of wave height to wavelength
   (c) the ratio of wavelength to wave height
   (d) the square of product of wave height and wavelength

19. In a shallow water wave, depth of water is 4,000 m and wavelength is 50 km. Wave speed is
    (a) 39,240 m/s (b) 198.1 m/s (c) 99 m/s (d) 49.52 m/s

CHAPTER 13

Ocean Thermal Energy Conversion

1. The Ocean thermal energy conversion system that is meant to generate power is most suitable in
   (a) sub-tropical region (b) tropical region
   (c) cold region (d) moderate climate region
2. The temperature gradient of ocean thermal energy conversion system is utilized in
   (a) internal combustion engines (b) heat engine
   (c) water turbines (d) none of these
3. The overall efficiency of an OTEC power plant is
   (a) 2%–3% (b) 10%–15%
   (c) 15%–20% (d) 20%–25%
4. Large amounts of solar energy are stored in oceans and seas. The process of harnessing this energy is called
   (a) ocean thermal energy conversion (OTEC)
   (b) ocean thermal conversion (OTC)
   (c) ocean and sea thermal energy conversion (OSTEC)
   (d) sea thermal energy conversion (STEC)
5. The energy available when the temperature of water at the surface of ocean is different from that at deeper level is called
   (a) ocean thermal energy
   (b) sea water energy
   (c) ocean water energy
   (d) none of these
6. Ocean and sea waves are indirectly caused due to
   (a) pressure gradients
   (b) solar energy
   (c) geothermal energy
   (d) none of these

CHAPTER 14

Fuel Cell

1. The following statement is not correct:
   (a) German scientists Schonbein discovered the principle of the fuel cell in the year 1838.
   (b) Fuel cell works as long as fuel is supplied to it.

(c) A battery cell is a closed system that stores reactant in it.
(d) Fuel cell is a closed system that stores the reactant in it.
(e) A battery cell requires regular replacement.

2. Following statements are given in relation to fuel cell:

   I. Fuel cells operate using hydrogen.

   II. Hydrocarbon fuels yield hydrogen in a process called reforming.

   III. Use of reformed hydrogen does not lead to carbon dioxide emission.

   IV. Fuel cells are open system in which reactants are continuously supplied.

   V. A fuel cell is different from battery.

   Using the code given below, select the correct answer.

   (a) I, II, and III are correct.
   (b) II, III, and IV are correct.
   (c) III, IV, and V are correct.
   (d) I, II, and IV are correct.

3. Intrinsic Gibb's free energy of the reaction in hydrogen–oxygen fuel cell is

   (a) 237 kJ/mole of hydrogen (b) −237 kJ/mole of hydrogen
   (c) 273 kJ/mole of hydrogen (d) −273 kJ/mole of hydrogen

4. Assertion (A): A fuel cell is an electrochemical cell that converts chemical energy into electrical energy without combustion.

   Reason (R): The chemical energy of reactants in fuel cell is converted as low voltage dc as an isothermal process.

   Using the codes given below, select the correct answer.

   (a) Assertion (A) and Reason (R) both are correct and R is the correct explanation of A.
   (b) Assertion (A) and Reason (R) both are correct but R is not the correct explanation of A.
   (c) A is correct but R is incorrect.
   (d) A is incorrect but R is correct.

5. Fuel cells have the following advantage:

   (a) Fuel cells are more efficient than combustion engines.
   (b) Fuel cells emit low emission.
   (c) Fuel cells give quite (silent) operation.
   (d) All of these

6. Following is not the correct statement related with fuel cells:

   (a) Fuel cells require hydrogen and oxygen as fuel to produce electricity and water.
   (b) Fuel cells are simple in construction with a few or no moving parts.
   (c) All fuel cells are very cheap.
   (d) Modular installations of fuel cells can be used to match the load and increase the reliability of system.

7. The ideal voltage of hydrogen–oxygen fuel cell at 1 atmospheric pressure and 25° C temperature is

   (a) 1.18 V (b) 1.23 V (c) 1.32 V (d) 1.81 V

8. Given the following choices, which one has the greatest ability to perform useful work as the energy source quality is high.
   (a) Coal (b) Electricity (c) Oil (d) Solar

CHAPTER 15

Magnetohydrodynamic (MHD) Power Generation

1. In a MHD generator, material used as conductor is
   (a) copper and aluminium
   (b) gas
   (c) liquid metal
   (d) liquid metal and gas
2. Power output of an MHD generator per unit volume is
   (a) proportional to the specific electrical conductivity of gas
   (b) square of the fluid velocity
   (c) square of magnetic field strength
   (d) all of these
3. The electrical power generated by MHD generator is
   (a) AC only
   (b) AC and DC both
   (c) DC only
   (d) triangular waveform
4. Commonly used seeding material in MHD generator is
   (a) aluminium oxide
   (b) cesium
   (c) potassium carbonate
   (d) sodium chloride
5. Following statement is given for MHD generators:
   I. The nature of electric current developed in MHD generator is pure direct current (DC).
   II. Generally, potassium carbonate is used as seed material in MHD generator.
   III. In MHD generator, the seeding material is used to reduce the electric conductivity of gas.
   IV. A magnet producing high magnetic flux density is necessary for efficient operation of MHD generator.

   Using the following codes, select the correct answer:
   (a) I, II, and III are correct.
   (b) I, II, and IV are correct.
   (c) II, III, and IV are correct.
   (d) I, II, III, and IV are correct.
6. Seeding material is injected into the working fluid of MHD generator in order to
   (a) increase the electrical conductivity of gas
   (b) reduce the electrical conductivity of gas
   (c) reduce the temperature of the gas
   (d) increase the magnetic field density

7. MHD generation mainly depends on
   (a) electric conductivity (b) permeability
   (c) permittivity (d) all of these
8. Following statements are given for the application advantage of MHD generators as compared to a conventional fossil fuel generator.
   I. It has no moving parts.
   II. It is a most promising direct energy converter.
   III. It is smaller in size.
   IV. It is very cheap.

   Using the code below, select the correct answer:
   (a) I, II, and III are correct. (b) I, II, and IV are correct.
   (c) II, III, and IV are correct. (d) I, II, III, and IV are correct.

CHAPTER 16

Thermoelectric Converters

1. Electrode materials in thermo electric generators are
   (a) carbon (b) ferrite
   (c) metal conductor (d) semiconductor
2. Following statements are given related to thermo electric generators:
   I. The voltage output of a thermoelectric generator depends on Seebeck constant and temperature difference.
   II. The temperature difference between hot source and heat sink causes flow of AC current through the load in thermo electric generator.
   III. The most commonly used material for thermo electric generator is lead telluride.
   IV. The efficiency of thermo electric generator is defined as the ratio of power developed across the load resistance to the heat flow from the source.

   From the code given below, select the correct answer:
   (a) I, II, and III are correct. (b) I, II, and IV are correct.
   (c) I, III, and IV are correct. (d) I, II, III, and IV are correct.
3. Assertion (A): A thermo electric power converter having hot source at one end and heat sink at the other end produces potential difference.

   Reason (R): The voltage generated by this converter is given by the product of Seebeck constant and temperature difference between hot and cold junctions.

   Using the codes given below, select the correct answer.
   (a) Assertion (A) and Reason (R) both are correct and R is the correct explanation of A.
   (b) Assertion (A) and Reason (R) both are correct but R is not the correct explanation of A.

(c) A is correct but R is incorrect.
(d) A is incorrect but R is correct.

4. Following is not a thermo electric material:

(a) Bismuth Telluride (b) Copper
(c) Lead Telluride (d) Silicon Germanium

5. When an electric current is passed through two dissimilar conductors connected to form a thermocouple, heat is evolved at one junction and absorbed at the other end. The absorption and evolution of heat depends on the direction of flow of current. This effect is called

(a) Joule effect (b) Peltier effect
(c) Seebeck effect (d) Thompson effect

6. Following statements is not correct in describing the advantage of thermoelectric generator:

(a) Thermoelectric generator has quieter operation as there are no moving parts.
(b) Thermoelectric generation is environment friendly.
(c) Thermoelectric generators are highly efficient.
(d) Thermoelectric generators are very useful for waste heat recovery.

7. Seebeck effect was discovered in the year

(a) 1810 (b) 1821
(c) 1838 (d) 1883

8. Select the correct statement from the following:

(a) The direction of energy flow in thermoelectric generator is reversible.
(b) The thermoelectric generators are highly efficient converter.
(c) The thermoelectric generators are suitable for high power applications.
(d) Moving parts in thermoelectric converters require regular maintenance.

9. A difference of temperature between the two junctions of a thermocouple will produce

(a) cooling (b) heating
(c) voltage between junction (d) nothing

CHAPTER 17

Thermionic Converters

1. A thermionic converter converts heat energy directly into electrical energy by using

(a) Peltier effect (b) Seebeck effect
(c) Thermionic emission effect (d) Thompson effect

2. Gas filled in between electrodes of a thermionic converter is

(a) carbon dioxide (b) caesium
(c) hydrogen (d) methane

3. Direct conversion of heat into electrical power is possible through

(a) batteries (b) fuel cell
(c) thermionic converters (d) all of these

4. The common material for electrodes of thermionic converter is Tungsten. It has work function of approximately
   (a) 5.4 (b) 4.54 (c) 3.38 (d) 1.5
5. If k is the Boltzmann constant in eV/°K and T is the temperature at the emitter surface, the average energy of escaping electron escaping from emitter surface is given by the expression
   (a) kT (b) $\sqrt{kT}$ (c) $2kT$ (d) $(2kT)^2$
6. If the Boltzmann constant is 8.62 eV/°K and the temperature at the emitter surface is 1,000 Kelvin , the average energy of escaping electron escaping from emitter surface is given by
   (a) 0.415 eV (b) 0.172 eV (c) 0.086 eV (d) 0.03 eV
7. Individual thermionic converter is a
   (a) high voltage and high current device. (b) high voltage and low current device
   (c) low voltage and high current device (d) low voltage and low current device
8. Assertion (A): A large number of thermionic converters are sequentially arranged to obtain useful voltage.

   Reason (R): Individual thermionic converter is a low voltage and high current device.

   Using the code given below, select the correct answer.
   (a) Assertion (A) and Reason (R) both are correct and R is the correct explanation of A.
   (b) Assertion (A) and Reason (R) both are correct but R is not the correct explanation of A.
   (c) A is correct but R is incorrect.
   (d) A is incorrect but R is correct.
9. Assertion (A): Thermionic converters are used in space power application for spacecraft.

   Reason (R): Separator of fluids is required in thermionic converters.

   Using the code given below, select the correct answer.
   (a) Assertion (A) and Reason (R) both are correct and R is the correct explanation of A.
   (b) Assertion (A) and Reason (R) both are correct but R is not the correct explanation of A.
   (c) A is true but R is false.
   (d) A is false but R is true

CHAPTER 18

Concept of Energy Conservation and Energy Management

1. Following statements are given related to energy conservation and management. Select the correct answer by choosing the code given below:
   I. The objective of conservation of energy is simply, in present-day context, to obtain maximum useful energy and least wasteful energy. It is called energy conservation.
   II. Quality of life, standard of living, and prosperity of a country are indexed by per capita of energy consumption of its people. Energy consumption is, therefore, a parameter that is used to identify the social and techno-economical disparity between different countries.

III. The *Energy Management* term is accepted globally as more technically accurate than energy conservation and energy consumption. It includes performing a task with minimum quantity of energy at an appropriate quality without affecting the morale of personnel working in the vicinity with capital expenditure avoidance.

IV. Energy management is defined as doing a task with minimum quantity of energy at appropriate quality without attention to cost and morale of personnel.

V. The two terms *Energy Conservations* and *Energy Consumptions* cannot be recommended as they are technically inaccurate terminology although they are very frequently used. *Energy Management* term is presently finding wide publicity and is being continuously recommended as it is technically accurate.

Codes:

(a) I, II, III, and IV are correct.
(b) I, II, III, and V are correct.
(c) II, III, IV, and V are correct.
(d) I, II, III, IV, and V are correct.

2. Energy conservation and management do not relate to the following statements:
   (a) Preserving the quality of the energy form.
   (b) No damage to environment and adverse effect on morale of personnel.
   (c) Energy cost saving is the direct savings to end user but it has no effect on utility load factor.
   (d) Load management refers to the supply side of the energy system.
3. Benefits of energy conservation or management is
   (a) setting of new energy generating plants can be delayed
   (b) reduction in energy consumption cost on billing to energy user
   (c) reduced damage to environment
   (d) all of these
4. Any energy management implementation criteria is based on
   (a) necessity satisfaction
   (b) favourable impact on utilities or energy user
   (c) identification of preferred systems
   (d) all of these
5. Energy management programs with housekeeping and maintenance measures can be implemented by reduction of demand in a simple and better way without chaos and politics is
   (a) through self-awareness or self-denial
   (b) through enforcement and regulatory actions
   (c) through economic measures.
   (d) all of these
6. Energy containment is one of the general principles of energy conservation and management. It includes
   (a) repair and sealing of air, steam and water leaks, and recovery of energy from wastes
   (b) proper location of process steps in a manufacturing plant

(c) captive power generation near the workplace
(d) retrofitting exercise

7. When cooking on a gas burner, colour of flame settings to conserve LPG is
(a) blue flame (b) yellowish flame (c) red flame (d) brown flame

8. In the refrigerator system, the following contains energy effectively and reduces losses:
(a) compressor (b) heat pump (c) thermostat (d) none

CHAPTER 19

Energy Conservation and Management in Different Energy Activity Sector

1. The Power Management System (PMS) prevents blackouts and disturbances of power system operations in addition to the following benefits:
(a) Control energy costs (b) Enhance safety
(c) Mitigates both environmental and health impacts (d) All of these

2. The best method for reducing consumer demand during peak period is
(a) peak saving or load shifting (b) valley filling
(c) conservation (d) none of these

3. Power factor correction is an energy saving technology that is used
(a) to improve the operating efficiency of electrical power systems
(b) to reduce electricity costs
(c) to reduce environmental hazards
(d) all of these

4. If LF is load factor, kWh is the total energy consumption for the billing period, kW is the peak demand set during the billing period and D is the number of billing days in the month, then percentage load factor is given by
(a) $\%LF = [kW/(kWh \times D \times 24)] \times 100$
(b) $\%LF = [kWh/(kW \times D \times 24)] \times 100$
(c) $\%LF = [kWh \times D/(kW \times 24)] \times 100$
(d) $\%LF = [kW \times D \times 24)/kWh] \times 100$

5. Actual load calculation on three-phase induction motor by slip method is carried out. Let N_S be motor synchronous speed in rpm, N_R be name plate full-load speed in rpm, N_A be actual measured speed in rpm, and HP be the name plate-rated horse power. Actual load horse power on electric motor (ALHP) is given by
(a) $ALHP = HP \times [(N_S - N_A)/(N_S - N_R)]$
(b) $ALHP = HP \times [(N_S - N_R)/(N_S - N_A)]$
(c) $ALHP = HP \times [(N_R - N_A)/(N_S - N_R)]$
(d) $ALHP = HP \times [(N_S + N_A)/(N_S + N_R)]$

6. The annual energy savings are obtained by replacing standard motor with an efficiency of (η_{st}) by an efficient motor with an efficiency of (η_{ef}). Annual energy saving will be
   (a) proportional to $(1/\eta_{ef})$
   (b) proportional to (η_{st}/η_{ef})
   (c) proportional to $[(1/\eta_{st}) - (1/\eta_{ef})]$
   (d) proportional to $[(1/\eta_{ef}) - (1/\eta_{st})]$
7. Two 40-W rapid start fluorescent lamps with conventional electromagnetic ballast (consumption 15 W) are used for lighting a space for 12 hours a day. When one lamp is removed from the circuit, but both ballasts remain energized, energy saved per day is
   (a) 26.26% (b) 36.36% (c) 50% (d) 59.59%

ANSWERS

Chapter 1

1. (a) 2. (a) 3. (d) 4. (c) 5. (c) 6. (b) 7. (c) 8. (d) 9. (b) 10. (a)
11. (b) 12. (c) 13. (c) 14. (a) 15. (a) 16. (a) 17. (c) 18. (b) 19. (d) 20. (c)
21. (c) 22. (b) 23. (c) 24. (a) 25. (a) 26. (a) 27. (b) 28. (a) 29. (d) 30. (b)
31. (a) 32. (d) 33. (b) 34. (d)

Chapter 2

1. (d) 2. (d) 3. (c) 4. (b) 5. (c) 6. (c) 7. (d) 8. (b) 9. (b) 10. (c)
11. (c)

Chapter 3

1. (c) 2. (b) 3. (a) 4. (c) 5. (b) 6. (c) 7. (a) 8. (b) 9. (d) 10. (c)
11. (d) 12. (d) 13. (d) 14. (c) 15. (a) 16. (b) 17. (a) 18. (b) 19. (d)

Chapter 4

1. (c) 2. (d) 3. (d) 4. (a) 5. (c) 6. (d) 7. (a) 8. (d) 9. (c) 10.(b)
11. (a) 12. (a) 13. (b) 14. (a) 15. (c) 16. (c) 17. (d) 18. (d) 19. (d) 20. (c)
21. (c) 22. (d) 23. (b)

Chapter 5

1. (a) 2. (a) 3. (a) 4. (d) 5. (b) 6. (b) 7. (a) 8. (a) 9. (d)

Chapter 6

1. (d) 2. (c) 3. (b) 4. (b) 5. (c) 6. (a) 7. (c) 8. (c) 9. (c) 10. (b)
11. (a) 12. (c) 13. (a) 14. (a) 15. (c) 16. (a) 17. (c) 18. (d) 19. (b) 20. (d)
21. (d) 22. (c) 23. (d) 24. (d) 25. (d)

Chapter 7

1. (c) 2. (d) 3. (b) 4. (b) 5. (d) 6. (b) 7. (c) 8. (a) 9. (d) 10. (d)

Chapter 8

1. (c) 2. (a) 3. (d) 4. (c) 5. (c) 6. (a) 7. (c) 8. (d) 9. (d) 10. (d)
11. (a) 12. (b) 13. (d) 14. (c)

Chapter 9

1. (d) 2. (d) 3. (c) 4. (d) 5. (d) 6. (c) 7. (a) 8. (c)

Chapter 10

1. (b) 2. (b) 3. (c) 4. (d)

Chapter 11

1. (c) 2. (a) 3. (c) 4. (d) 5. (b) 6. (a) 7. (a) 8. (b)

Chapter 12

1. (c) 2. (b) 3. (a) 4. (c) 5. (b) 6. (d) 7. (d) 8. (b) 9. (d) 10. (d)
11. (d) 12. (d) 13. (b) 14. (b) 15. (a) 16. (a) 17. (d) 18. (b) 19. (b)

Chapter 13

1. (a) 2. (b) 3. (a) 4. (a) 5. (a) 6. (b)

Chapter 14

1. (d) 2. (d) 3. (b) 4. (a) 5. (d) 6. (c) 7. (b) 8. (b)

Chapter 15

1. (b) 2. (b) 3. (c) 4. (c) 5. (b) 6. (a) 7. (a) 8. (a)

Chapter 16

1. (d) 2. (c) 3. (b) 4. (b) 5. (b) 6. (c) 7. (b) 8. (a)

Chapter 17

1. (c) 2. (b) 3. (c) 4. (b) 5. (c) 6. (b) 7. (c) 8. (a) 9. (c)

Chapter 18

1. (b) 2. (c) 3. (d) 4. (d) 5. (a) 6. (a) 7. (a) 8. (b)

Chapter 19

1. (d) 2. (a) 3. (d) 4. (b) 5. (b) 6. (c) 7. (b)

Appendix B

Multiple Choice Objective Questions on Energy Systems

1. For controlling chain reaction in a nuclear reactor, material used is

   (a) beryllium (b) boron
   (c) heavy water (d) thorium

2. The total power of a wind stream is directly proportional to

   (a) velocity (b) square root of velocity
   (c) square of velocity (d) cube of velocity

3. The bio-gasification is a process also referred as

   (a) aerobic combustion (b) anaerobic digestion
   (c) fermentation (d) thermochemical gasification

4. The most widely used bond in the solar cell work is

   (a) covalent (b) hydrogen like
   (c) ionic (d) metallic

5. Fill factor is also known as

   (a) short circuit factor (b) refractive index
   (c) reflection factor (d) efficiency

6. The Betz criterion of maximum performance of wind turbine has the value of maximum power extraction coefficient of

   (a) 0.35 (b) 0.50
   (c) 0.593 (d) 0.666

7. The fluid that cannot be used in OTEC system as working fluid is

   (a) ammonia (b) liquid sodium
   (c) refrigeration fluid (d) steam

8. In anaerobic digestion, the useful end product is

   (a) carbon monoxide (b) hydrogen
   (c) methane (d) Freon

9. Match the quantities indicated in List I with List II, and select the correct answer using the following codes

List I	List II
(A) Aerobic conversion	1. Conversion to glucose
(B) Pyrolysis	2. Out of contact with air
(C) Thermo chemical conversion	3. Composting
(D) Anaerobic conversion	4. Thermal process
(E) Enzymatic fermentation	5. gasification

Codes:

	A	B	C	D	E
(a)	3	4	5	2	1
(b)	3	4	5	1	2
(c)	1	4	5	3	2
(d)	1	2	3	4	5

10. When heat is applied to the junction of two dissimilar metals, an emf is generated across the cold ends, the effect is known as

    (a) Hall effect (b) Peltier effect
    (c) Seebeck effect (d) Thomson effect

11. For hydrogen–oxygen fuel cell, the Gibbs free energy change is $\Delta G = 56.7\,\text{Kcal/mole}$. Two electrons are released at the anode. The emf of the cell is

    (a) 0.21682 V (b) 0.43365 V
    (c) 1.23 V (d) 2.46 V

12. The plasma in a nuclear fusion generator can be contained by

    (a) magnetic field (b) high powered Lasers
    (c) toroidal field (d) all the above

13. The quantity [(Seebeck coefficient)^2 × thermal resistivity × electrical conductivity] is known as

    (a) coefficient of merit (b) coefficient of performance
    (c) figure of merit (d) thermal efficiency

14. The fluid not used as working fluid in an OTEC system is

    (a) ammonia (b) liquid sodium
    (c) refrigeration fluid (d) steam

15. The minimum wind velocity required for wind turbine is called

    (a) cut-in speed (b) cut-out speed
    (c) rated speed (d) tip speed

16. A wind is blowing at a velocity v m/s through an area A m^2 at right angle to it. If the wind density is 'ρ', the kinetic energy of wind is

    (a) $\rho A v^3$ (b) $\frac{1}{2}\rho A v^2$
    (c) $\rho A v^2$ (d) $\frac{1}{2}\rho A v^3$

17. The temperature required to overcome coulomb barrier is fusion power generation is in the range of

    (a) 10^3–10^4 K (b) 10^6–10^7 K
    (c) 10^9–10^{10} K (d) 10^{11}–10^{12} K

18. Conversions principle of solar light into electricity is known as
    (a) thermionic conversion
    (b) thermo-mechanical–electrical conversion
    (c) electro-chemical conversion
    (d) photovoltaic conversion
19. For an efficient solar cell, high values of fill factor (FF), short circuit current (I_{sc}), and open circuit voltage (V_{oc}) are required. However, high value of energy band gap changes above in the following ways:

	FF	I_{sc}	V_{oc}
(a)	decreases	increases	increases
(b)	increases	increases	increases
(c)	decreases	decreases	increases
(d)	increases	decreases	increases

20. Solar cells are made from the following material
    (a) crystalline (b) ferrite
    (c) iron (d) silver
21. For a typical solar cell, open circuit voltage is (V_{oc}) and short circuit is I_{sc}. At the point of maximum power output of cell, the values of voltage and current are V_{mp} and I_{mp}, respectively. The fill factor of solar cell is given by
    (a) $\frac{V_{mp}}{V_{oc}}$ (b) $\frac{I_{mp}}{I_{sc}}$
    (c) $\frac{V_{mp}I_{mp}}{V_{oc}I_{sc}}$ (d) $\frac{V_{oc}I_{sc}}{V_{mp}I_{mp}}$
22. A typical solar cell has the v–I characteristic, as shown in the figure. The maximum power occurs at 0.5 V. The fill factor is

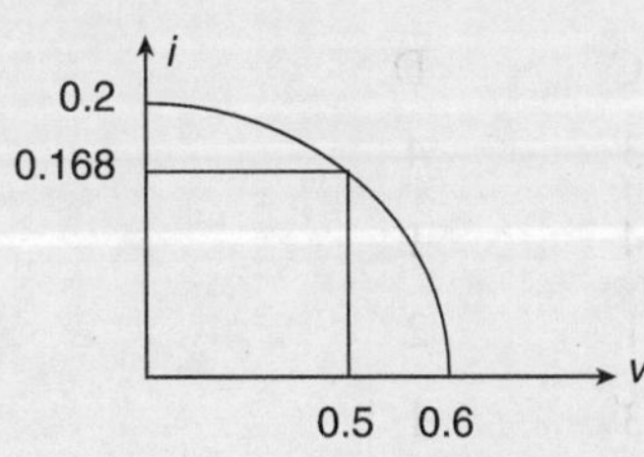

    (a) 0.246 (b) 0.7
    (c) 0.84 (d) 1.44
23. The density of seawater is taken in kg/m^3 approximately as
    (a) 1 (b) 80
    (c) 1,000 (d) 1,025
24. The output of a solar panel depends on
    (a) the insulation level only
    (b) the temperature only

(c) the insulation level and the temperature both
(d) All the above

25. The depletion region or space charge region or transition region in a semiconductor p–n junction diode has
(a) electrons and holes
(b) positive ions and negative ions
(c) no ions, electrons or holes
(d) negative ions and holes

26. p-type semiconductors are obtained by adding a small amount of the following to a pure semiconductor
(a) pentavalent impurity
(b) gallium
(c) arsenic
(d) antimony

27. Global solar radiation flux is measured by
(a) pyrheliometer
(b) anemometer
(c) pyranometer
(d) none of these

28. The surface and deep water temperatures of ocean are 27°C and 7°C, respectively. The maximum possible efficiency of a heat engine operating these temperatures is
(a) 6.67%
(b) 8.37%
(c) 10%
(d) 17%

29. Match the quantities in List I and List II and select the correct answer using the following codes:

List I
A. Wave energy
B. Tidal energy
C. Geothermal energy
D. Storage of energy in the form of kinetic energy

List II
1. Flywheel
2. Kaplan turbine
3. Puga Valley
4. Salter duck

Codes:

	A	B	C	D
(a)	1	2	3	4
(b)	2	3	4	1
(c)	3	4	1	2
(d)	4	2	3	1

30. Match the quantities in List I and List II and select the correct answer using the following codes:

List I
A. Solar collector absorber
B. Seebeck effect
C. Steam turbine
D. Liquid fuels from biomass

List II
1. Ethanol
2. Thermoelectric generation
3. Black chrome
4. Rankine cycle

Codes:

	A	B	C	D
(a)	3	2	4	1
(b)	2	3	4	1
(c)	3	4	1	2
(d)	4	2	1	3

31. Match the quantities in List I and List II and select the correct answer using the following codes:

List I	List II
A. Flat plate solar collector glazing	1. Liquid helium
B. Cooling of superconducting material	2. Biogas plant
C. Anaerobic digestion	3. Plastic film of Tedlar
D. Aerobic digestion	4. Composting

Codes:

	A	B	C	D
(a)	1	3	4	2
(b)	1	3	2	4
(c)	3	1	2	4
(d)	3	2	1	4

32. Match the quantities in List I and List II and select the correct answer using the following codes:

List I	List II
A. Pyrheliometer	1. Net flux of long wave radiation through the horizontal surface during night
B. Pyranometer	2. Intensity of direct solar radiation at normal incidence
C. Pyrgeometer	3. The global solar radiation received from the entire hemisphere.
D. Pyrradiometer	4. Net flux of downward and upward total radiation through a horizontal surface.

Codes:

	A	B	C	D
(a)	2	3	1	4
(b)	2	3	4	1
(c)	3	2	1	4
(d)	3	1	4	2

33. Assertion (A): Solar pond is a simple device for collecting eight solar energy.
Reason (R): Unlike ordinary water ponds, convective and evaporative heat losses are reduced and thermal energy is stored.

In the context of abovementioned two statements, select the correct answer by using the following codes:

Code:

(a) Both A and R are true and R is the correct explanation of A.
(b) Both A and R are true but R is not the correct explanation of A.
(c) A is true but R is false.
(d) A is false but R is true.

34. The device used for converting brackish water into potable water using solar energy is called
(a) solar stills
(b) solar dryers
(c) solar ponds
(d) solar green houses

35. The most commonly used material for photovoltaic cell is silicon. Its energy gap in electron volt is
(a) 0.67
(b) 0.9
(c) 1.12
(d) 1.27

36. Following statement is given for solar thermal energy systems:
I. Thermal energy can be stored as a sensible heat in a porous solid.
II. Thermal energy can be stored in the form of latent heat in a substance when melts and extracted when it freezes.
III. Thermal energy can be stored in the form of latent heat in an insulated container containing water.
IV. A novel device that combines the functions of both collection and storage of solar thermal energy is solar ponds.
V. Porous solid like pebbles or rock for thermal storage is preferred with liquid collector.

Select the correct answer using following codes:
(a) I, II, and V are correct
(b) I, II, and IV are correct
(c) II, IV, and V are correct
(d) III, IV, and V are correct

37. Assertion (A): The declination angle varies from a maximum value +23.45° on June 21 to a minimum value of −23.45° on December 21.
Reason (R): The declination angle variation depends upon the geometry of sun-earth relationship.

In the context of abovementioned two statements, select the correct answer by using the following codes:

Code:

(a) Both A and R are true and R is the correct explanation of A.
(b) Both A and R are true but R is not the correct explanation of A.
(c) A is true but R is false.
(d) A is false but R is true.

38. Following statement is given for solar thermal energy systems:

    I. Thermal energy can be stored as a sensible heat in a porous solid.
    II. Thermal energy can be stored in the form of latent heat in a substance when melts and extracted when it freezes.
    III. Thermal energy can be stored in the form of latent heat in an insulated container containing water.
    IV. A novel device that combines the functions of both collection and storage of solar thermal energy is solar ponds.
    V. Porous solid such as pebbles or rock for thermal storage is preferred with liquid collector.

    Select the correct answer using following codes:

    (a) I, II, and V are correct (b) I, II, and IV are correct
    (c) II, IV, and V are correct (d) III, IV, and V are correct

39. The value of solar constant is taken in kW/m^2 within the following ranges:

    (a) 0.5–0.9 (b) 1.0–1.25
    (c) 1.353–1.367 (d) 1.4–2

40. If L_b is the observer's longitude, L_s is the longitude on which the local standard time is based, and n is the day of the year, the following statements are given for solar time:

    I. $\text{Solar time} = \text{Standard time} \pm 4(L_b - L_s) + E$
    II. $\text{Solar time} = \text{Standard time} \pm 2(L_b - L_s) + 2E$
    III. $E = 9.87 \sin 2B - 7.53 \cos B - 1.5 \sin B$
    IV. $E = 9.87 \sin 2B - 1.5 \sin B$
    V. $B = \dfrac{360(n-81)}{364}$

    Select the correct answers using the following codes:

    (a) I, IV, and V are correct (b) II, III, and V are correct
    (c) I, III, and V are correct (d) II, IV, and V are correct

41. Based on the Cooper relationship for declination angle (δ) in degrees for the day of year is shown in the figure.

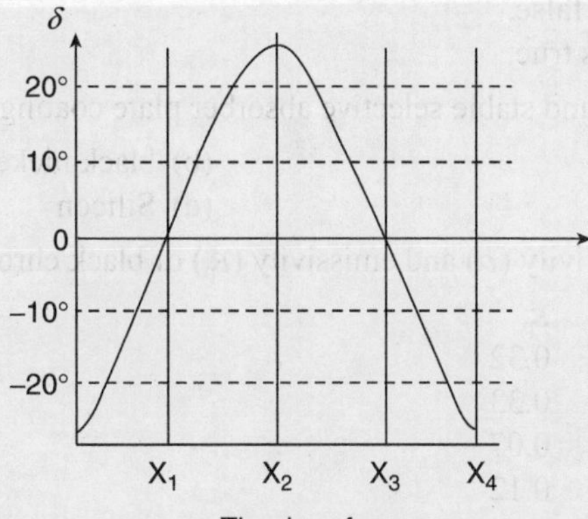

The day of year will be the following

	X_1	X_2	X_3	X_4
(a)	June 21	September 21	December 21	March 21
(b)	March 21	June 21	September 21	December 21
(c)	September 21	December 21	March 21	June 21
(d)	December 21	March 21	June 21	September

42. The angle made by beam radiation with the normal to a flat plate collector on December 1 at 09:00 hours (local apparent time). The collector is located in New Delhi (28°35′N and 77°12′E) and is tilted at an angle of 36° with the horizontal and is pointing to south. The angle of inclination will be the following:

    (a) −23.45° (b) −22.11°
    (c) +22.11° (d) +23.45°

43. If T_r is the transmittivity obtained by considering only reflection and refraction and T_a is the transmittivity obtained by considering only absorption, the transmittivity of the cover system of a collector obtained with adequate accuracy is given by

    (a) $\frac{T_r}{T_a}$ (b) $\frac{T_a}{T_r}$
    (c) $T_a + T_r$ (d) $T_a T_r$

44. Assertion (A): Selective surfaces yield higher collector efficiencies than those obtained when absorptivity and emissivity are equal.

    Reason (R): Selective surfaces maximize the absorption of solar energy and minimize the emission of the radiative loss.

    In the context of abovementioned two statements, select the correct answer by using the following codes :

    Code:

    (a) Both A and R are true and R is the correct explanation of A.
    (b) Both A and R are true but R is not the correct explanation of A.
    (c) A is true but R is false.
    (d) A is false but R is true.

45. The most successful and stable selective absorber plate coating for flat plate collector is

    (a) black chrome (b) black nickel
    (c) natural oxide (d) Silicon

46. The values of absorptivity (α) and emissivity (ε_p) of black chrome is

	α	ε_p
(a)	0.89	0.32
(b)	0.75	0.33
(c)	0.96	0.07
(d)	0.95	0.12

47. Match the quantities in List I and List II and select the correct answer using the following codes:

List I (Semiconductor materials)	List II (Band gap energy in eV)
A. Silicon	1. 2.42
B. Gallium arsenide	2. 1.55
C. Cadmium telluride	3. 1.43
D. Cadmium sulphide	4. 1.12

Codes:

	A	B	C	D
(a)	1	2	3	4
(b)	4	2	3	1
(c)	4	3	2	1
(d)	3	4	1	2

48. Following is not a phase change material desirable for latent heat storage.
   (a) Paraffin wax (b) Calcium chloride hexahydrate
   (c) Sodium nitrate (d) Magnesium oxide
49. Graphite is used in nuclear power plant as a
   (a) coolant (b) electrode
   (c) fuel (d) moderator
50. Sunshine recorder measures the duration of the sunshine
   (a) in a day (b) in a week
   (c) in a month (d) in a year
51. Solar air heaters are used for
   (a) water heating (b) heating buildings
   (c) power generation (d) cooking
52. For wind energy conversion systems, generally, the generator used for power generation is
   (a) induction generator (b) tacho generator
   (c) DC generator (d) servo generator
53. When the sun is at zenith, then at the sea level, air mass is equal to
   (a) 0 (b) 1
   (c) 2 (d) sec θ
54. The combination of wind turbine and generator is
   (a) electric generator (b) aero-generator
   (c) induction generator (d) turbo-generator
55. The primary fuel for fusion is
   (a) deuterium (b) hydrogen
   (c) oxygen (d) nitrogen

56. Lawson criterion depends on the nature of
    (a) fusion reaction (b) plasma temperature
    (c) fusion reaction and plasma temperature (d) fission reaction
57. Solar photovoltaic systems converts solar energy into
    (a) thermal energy (b) electrical energy
    (c) mechanical energy (d) chemical energy
58. Pollution prospects in the process of bio-gasification
    (a) increases (b) decreases
    (c) will not affect (d) maintains
59. The output power delivered by an MHD generator is proportional to
    (a) (magnetic flux density)^2 (b) magnetic flux density
    (c) (magnetic flux density)$^{-1}$ (d) (magnetic flux density)$^{-2}$
60. Hydrogen is a
    (a) primary fuel (b) secondary fuel
    (c) not a fuel (d) fossil fuel
61. Wind mill converts
    (a) kinetic energy of the wind to mechanical energy
    (b) kinetic energy of the wind to potential energy
    (c) potential energy into mechanical energy
    (d) mechanical energy into electrical energy
62. Thermoelectric generator converts
    (a) heat energy into electrical energy
    (b) heat energy into mechanical energy
    (c) mechanical energy into electrical energy
    (d) heat absorbed into chemical energy
63. Practical hydrogen–oxygen fuel cells, generally, have conversion efficiencies in the range of
    (a) 70%–80% (b) 50%–60%
    (c) 40%–50% (d) 30%–35%
64. The following is not considered as the likely fuels used in fusion reactor
    (a) deuterium (b) hydrogen
    (c) plutonium (d) tritium
65. The suitability of a particular material for thermoelectric generation is given by a factor Z, high value of which indicates good thermoelectric material. If S is the Seebeck coefficient (V/°K), p is the electrical resistively (Ω/m), and k is the thermal conductivity of the material (W/m/°K), then factor Z is given by
    (a) $\frac{S^2}{pk}$ (b) $\frac{S}{p^2k}$
    (c) $\frac{S}{pk^2}$ (d) $\frac{S}{(pk)^2}$

66. The force $F = q(E + u \times B)$ on a charge q moving at velocity v in combined magnetic field B and electric field E is called
(a) Einstein's force (b) Lorentz force
(c) Maxwell's force (d) Saha's force

67. A magnetic mirror is used in the following
(a) Ion Engines (b) Lasers
(c) MHD (d) nuclear fusion related work

68. Deuterium and tritium are two isotopes of hydrogen containing, respectively, one and two neutrons and are capable of combining in different ways to produce helium. The reaction that holds the most commercial promise is
(a) ${}^{2}_{1}\text{D} + {}^{2}_{1}\text{D} \rightarrow {}^{3}_{2}\text{He} + \text{n} + 3.2\ \text{MeV}$ (b) ${}^{2}_{1}\text{D} + {}^{2}_{1}\text{D} \rightarrow {}^{3}_{1}\text{T} + {}^{1}_{2}\text{H} + 4.0\ \text{MeV}$
(c) ${}^{2}_{1}\text{D} + {}^{3}_{1}\text{T} \rightarrow {}^{4}_{2}\text{He} + \text{n} + 17.6\ \text{MeV}$ (d) ${}^{2}_{1}\text{D} + {}^{3}_{2}\text{He} \rightarrow {}^{4}_{2}\text{He} + {}_{1}^{1}\text{H} + 18.3\ \text{MeV}$

69. The spectrum density for square of wave height is given approximately by a triangle with maximum amplitude of 7.4 m^2/Hz within the frequency range of 0.05–0.2 Hz. It is zero elsewhere. The significant wave height will be the following
(a) 2.98 m (b) 2.22 m
(c) 0.745 m (d) 0.555 m

70. The spectrum density for square of wave height is given approximately by a triangle with maximum amplitude of 7.4 m^2/Hz within the frequency range of 0.05–0.2 Hz. It is zero elsewhere. If the energy period is 10.75 s, power potential per meter length is
(a) 11.71 kW/m (b) 15.612 kW/m
(c) 23.416 kW/m (d) 46.835 kW/m

71. If wind speed is V, the potential power varies in the following ways causing the wave
(a) proportional to V^2 (b) proportional to V^3
(c) proportional to V^4 (d) proportional to V^5

72. Following is not the working fluid of an OTEC system
(a) ammonia (b) liquid sodium
(c) refrigeration fluid (d) sea system

73. The efficiency of an OTEC system is about
(a) 5% (b) 20%
(c) 30% (d) 40%

74. For all practical purposes in an OTEC system, it is estimated that temperature difference between the ocean surface water and the ocean depth should be at least
(a) 5°C (b) 10°C
(c) 15°C (d) 19°C

75. In open-cycle OPEC system, fluid used is
(a) ammonia (b) Freon
(c) methyl alcohol (d) seawater

76. Steam turbine operates on

(a) Brayton cycle and Rankine cycle
(b) Carnot cycle and Brayton cycle
(c) Joule cycle and Rankine cycle
(d) Brayton cycle and Joule cycle

77. If ρ is the density of water, g is acceleration due to gravity, h is hydraulic head, Q is quality of water flow, and η is the efficiency of turbine–generator unit, then power output (P_0) of a hydraulic generation scheme is given by the formula

(a) $P_0 = \eta\rho Qgh$
(b) $P_0 = \dfrac{\rho Qgh}{\eta}$
(c) $P_0 = \dfrac{\eta\rho Qh}{g}$
(d) $P_0 = \dfrac{\rho Qh}{g\eta}$

78. Following equipment is used for tidal power station

(a) Horizontal Kaplan turbine
(b) Bulb turbine with inclined shaft
(c) Bulb turbine with horizontal shaft
(d) Straight flow turbine with conventional AC generator

79. Isotopes have

(a) same atomic masses
(b) same atomic numbers
(c) same number of protons
(d) none of a, b, and c

80. Following is not valid statement

(a) Hall coefficient = reciprocal of charge density
(b) Hall gradient = velocity of charge flux density
(c) Flux density = Hall voltage × material thickness/Hall coefficient × control current
(d) Lorenz force = electron charge × velocity of charge particles/flux density

81. Following is not a correct statement for a charged particle movement in a uniform magnetic field

(a) Its motion is along the axis of magnetic field
(b) Its motion is perpendicular to the direction of magnetic field
(c) It follows the Larmor orbit
(d) It moves in a circle in a plane perpendicular to the direction of magnetic field

82. When a charged particle of mass m with charge q moves at a velocity u (m/s) in a magnetic field B (T). The cyclotron frequency in radian/s will be

(a) $\dfrac{u}{mqB}$
(b) $\dfrac{qB}{m}$
(c) $\dfrac{mq}{B}$
(d) $\dfrac{B}{mq}$

83. Which of the following is not the correct statement:

(a) Harmful nuclear radiations include α, β and γ particles.
(b) NO_2 emission is removed by fluidized bed combustion.
(c) Reactors using water (light and heavy) are more dangerous than using graphite.
(d) Acid rains, nuclear reactor accidents, and water pollution have become the international issues.

84. Aerosols are removed by the
(a) centrifuging and removal of iron pyrites
(b) electrostatic precipitators
(c) fluidized bed combustion
(d) removal of iron pyrites

85. The following battery works at a temperature range of 300°C–400°C.
(a) Lead acid
(b) Zinc halide
(c) Redox
(d) Lithium-iron sulphide

ANSWERS TO MULTIPLE CHOICE QUESTIONS

1. (b)	2. (d)	3. (b)	4. (a)	5. (d)
6. (c)	7. (b)	8. (c)	9. (a)	10. (c)
11. (c)	12. (c)	13. (c)	14. (b)	15. (b)
16. (b)	17. (c)	18. (d)	19. (d)	20. (a)
21. (c)	22. (b)	23. (d)	24. (c)	25. (c)
26. (b)	27. (c)	28. (a)	29. (d)	30. (a)
31. (c)	32. (a)	33. (a)	34. (a)	35. (c)
36. (b)	37. (a)	38. (b)	39. (c)	40. (c)
41. (b)	42. (b)	43. (d)	44. (a)	45. (a)
46. (d)	47. (c)	48. (d)	49. (d)	50. (a)
51. (b)	52. (a)	53. (d)	54. (b)	55. (a)
56. (c)	57. (b)	58. (b)	59. (a)	60. (b)
61. (a)	62. (a)	63. (b)	64. (b)	65. (a)
66. (b)	67. (d)	68. (c)	69. (a)	70. (d)
71. (d)	72. (b)	73. (b)	74. (d)	75. (d)
76. (d)	77. (a)	78. (c)	79. (c)	80. (d)
81. (a)	82. (b)	83. (c)	84. (b)	85. (d)

Appendix C

Terms and Definition

Absorber: The absorber is the active surface in a solar collector. The absorber can be made of various materials and often the surface is applied with a selective coating that allows solar short wave (solar heat and light) to transmit, while long wave heat is reflected. Upon absorption, the solar energy transforms from short wave into long wave energy.

Acceptor: It is a dopant atom (from group III of periodic table) that when added to a semiconductor can form a p-type region.

Adhesion: It is an extremely thin layer of molecules (as in gases, solutions, or liquids) to the surface of solid bodies or liquids with which they are in contact.

Adsorption: It is a process where a solid is used for removing a soluble substance from the water.

Aerobic fermentation: Fermentation processes that require the presence of oxygen.

Aerobic: Organism able to live, grow, or take place only where free oxygen is present.

Aerodynamic: It is the study of forces and the resulting motion of objects through the air.

Agricultural residue: Agricultural crop residues are the plant parts, primarily stalks and leaves, not removed from the fields with the primary food or fibre product. Examples include corn straws (stalks, leaves, husks, and cobs), wheat straw, and rice straw.

Airfoil: The cross-sectional profile of a wing.

Air mass (sometimes called air mass ratio): A large body of air with only small horizontal variations of temperature, pressure, and moisture and approximated as equal to the cosine of the zenith angle.

Algae: Fast growing photosynthetic plants containing chlorophyll in freshwater, seawater, or damp oils.

Alternating current (AC): The periodically varying flow of electric current.

Alternative fuels: Fuels derived from non-conventional energy resources for replacing conventional fossil fuels.

Ambient temperature: The temperature of the surrounding area.

Amorphous semiconductor: A non-crystalline semiconductor material that has no long range order.

Amorphous silicon: A thin-film, silicon photovoltaic cell having no crystalline structure.

Amplitude: The maximum displacement or distance moved by a point on a vibrating body or wave measured from its equilibrium position.

Anaerobic digestion: A biological process in which degradation of organic matter by microbes takes place in the absence of oxygen to produce methane and carbon dioxide.

Anaerobic: Life or biological processes that occur in the absence of oxygen.

Anemometer: An instrument that measures wind speed.

Angle of attack: It is used to define the angle that forms chord of the profile with a fluid stream.

Angle of incidence: The angle that an incident straight line or ray of light makes with a perpendicular drawn on the surface at the point of incidence line meeting.

Angle of inclination: The angle that a solar collector is positioned above horizontal with reference to solar energy systems.

Angular frequency: It is the frequency of a periodic process (as electric oscillation or sound vibration) expressed in radians per second and given by frequency in cycles per second multiplied by 2π.

Anode: It is the positively charged electrode (or element) that attracts the negative (−) charged particles in a circuit or chemical reaction.

Antireflection coating: A thin coating of material applied to a solar cell surface that reduces the light reflection.

Array current: It is the electrical current produced by a photovoltaic array when it is exposed to sunlight.

Array operating voltage: The voltage produced by a photovoltaic array when exposed to sunlight and connected to a load.

Aspect ratio: For a geometric shape, it is the ratio between its sizes in different dimensions such as numerical ratio of the length (height) to cross-sectional area of a thermoelectric element.

Azimuth angle: The azimuth angle is often referred to when a solar collector surface is installed or engineered. The azimuth angle is a figure that tells you how many degrees out of south, the direction of the solar collectors are.

Band gap energy (E_g): The amount of energy (in electron volts) required to free an outer shell electron from its orbit about the nucleus and allows it crossing from the valence to the conduction level.

Band gap: It is the energy difference between the highest occupied state in the valence band and the lowest unoccupied state in the conduction band. The material is either a semiconductor if the band gap is relatively small or an insulator if the band gap is relatively large.

Barrier energy: The energy given up by an electron in penetrating the cell barrier; a measure of the electrostatic potential of the barrier.

Battery energy storage: Electrical energy stored in electrochemical batteries.

Beam radiation: Another name for direct normal irradiance, the amount of solar radiation from the direction of the sun.

Begasse: Residue remaining after extracting a sugar juice from plants like sugarcane.

Betz coefficient (16/27 = 59.3%): This is the theoretical maximum efficiency limit of a wind generator.

Binary cycle: It is a combination of two cycles adjacent to each other utilizing two different working fluids for power production. The low thermal heat from the first cycle is transferred in a heat exchanger to the working fluid in the second closed cycle that drives a turbine coupled to an electrical generator to produce electricity.

Bioconversion: A general term describing the raw biomass conversion to energy or fuels.

Biodiesel: A biodegradable transportation fuel for use in diesel engines obtained from biomass by biological conversion.

Bioenergy: Energy derived from biomass by means of bio-conversion technologies and is a substitutes of fossil fuels.

Biofuels: Biomass converted to liquid or gaseous fuels such as ethanol, methanol, methane, and hydrogen.

Biogas: A combustible gaseous mixture of methane (50%–60% of total gas) and carbon dioxide produced by the anaerobic digestion of organic matters.

Biomass: Any plant-derived organic matter such as agricultural crops and residue, wood and wood waste, animal waste, aquatic plants and organic components of municipal and industrial wastes.

Biomass fuel: Liquid, solid, or gaseous fuel produced by conversion of biomass.

Biosphere: The portion of the earth and its atmosphere that can support life.

Bismuth telluride: It is a grey powder that is a compound of bismuth and tellurium and exhibits good thermoelectric properties.

Bismuth antimony: It is one of the most important material systems for low temperature thermoelectric and other applications.

Blocking diode: It is a semiconductor diode for limiting the direction of flow of energy flow in one direction only.

Boron (B): The chemical element commonly used as the dopant in photovoltaic device or cell material.

Bottoming cycle: A cogeneration system in which steam is used first for processing heat and then for electric power production.

British thermal unit (BTU): It is a standard measure of thermal energy in UK. A single BTU is the amount of energy required to increase the temperature of one pound of water by one degree Fahrenheit (F); further, 1,000 BTU per hour is equal to 0.293 kW.

Bulb turbines: It is a variation of the propeller-type turbine (similar to the Kaplan turbine) in which the generator is located in a watertight housing called the bulb.

Cadmium (Cd): A chemical element used in making certain types of solar cells and batteries.

CdTe solar cells: A cadmium telluride solar cell uses a cadmium telluride (CdTe) thin film, a semiconductor layer to absorb and convert sunlight into electricity.

Cadmium telluride (CdTe): A polycrystalline thin-film photovoltaic material.

Capacity factor: It indicates how much electricity a generator actually produces relative to the maximum. It could produce at continuous full power operation during the same period.

Carbohydrate: Organic compounds made up of carbon, hydrogen, and oxygen and having approximately the formula $(CH_2O)_n$ includes cellulose, starch, and sugars.

Carbon dioxide (CO_2): A colourless, odourless gas produced by respiration and combustion of carbon-containing fuels and used by plants as a food in the photosynthesis process. It is also called as greenhouse gas.

Carbon monoxide (CO): A colourless, odourless, and poisonous gas produced by incomplete combustion.

Cathode: The cathode is the negatively charged electrode. It attracts the positive charge and is the source of electrons or an electron donor.

Celsius scale: The metric temperature scale for which 0°C is the temperature at which water freezes and 100°C is the temperature at which water boils at standard atmospheric pressure. The conversion from Fahrenheit to Celsius is

$$C = (F - 32)/1.8$$

Central receiver system: A single central receiver such as a boiler, engine, or photovoltaic array (for solar electric power generation) uses a series of tracking mirrors (heliostat) or a paraboloid (three-dimensional parabola or dish) of mirrors to focus solar energy onto it.

Chord line: The imaginary straight line from leading edge to trailing edge of an airfoil, and this distance being the chord.

Coefficient of performance (COP): It is the ratio of performance achieved or useful output or work to the energy or work input. It is a measure of the efficiency of a thermoelectric module, device, or system. Mathematically, COP is the total heat transferred through the thermoelectric device divided by the electric input power.

Co-generation: The technology of producing electric energy and another form of useful energy (usually thermal) for industrial, commercial, or domestic heating or cooling purposes through the sequential use of either rejected heat from industrial processes (bottoming cycle) or surplus heat from an electric generating plant (topping cycle).

Collector fluid: The working fluid in a collector that transfers the heat of solar energy for appropriate applications.

Compressed natural gas: It is natural gas under pressure that remains clear, odourless, and non-corrosive.

Concentrating parabolic trough: It is a large, U-shaped (parabolic) reflectors (focusing mirrors) that have oil-filled pipes running along their centre or focal point. The mirrored reflectors are tilted toward the sun, and focus sunlight on the pipes to heat the oil inside to as much as 400°C. The hot oil is then used to boil water, which makes steam to run conventional steam turbines and generators.

Concentrator: Optical components such as lenses (Fresnel lens) to direct and concentrate sunlight onto a solar cell of smaller area. They can increase the power flux of sunlight hundreds of times.

Condenser: An equipment that condenses turbine exhaust steam into condensate.

Crust: Earth's outer layer of rock, which is also called the lithosphere.

Conduction band (or conduction level): It is an energy band in a semiconductor that accepts electrons from valence band where they can move freely in a solid, thereby producing a net transport of charge.

Copper indium diselenide ($CuInSe_2$ or CIS): It is a polycrystalline thin-film photovoltaic material (sometimes incorporating gallium (CIGS) and/or sulphur).

Cooling tower: A structure in which heat is removed from hot condensate.

Couple: A pair of thermoelectric elements consisting two dissimilar materials (such as one N-type and one P-type) forming junctions and maintained at different temperatures.

Crystalline silicon: It is a slice of single-crystal silicon or polycrystalline silicon crystals having very good thermal, electrical, mechanical, and photon absorbing properties for application in solar photovoltaic, thermoelectric devices.

Cut-in-speed: The lowest wind speed at which a wind turbine begins producing usable power.

Cut-out-speed: The highest wind speed at which a wind turbine stops producing power.

Declination: It is the angular position of the sun at solar noon with respect to the plane of the equator.

Degrees Kelvin: It is a temperature scale designed so that 0°K is defined as absolute zero. In a later date, the name Kelvin (symbol K) instead of degree Kelvin (symbol °K) was used to define the unit of thermodynamic temperature.

It is defined as the fraction (1/273.16) of the thermodynamic temperature of the triple point of water. The Celsius scale 0°C equals 273.15K.

Dehydration: The removal of a substantial portion of the water from any substance.

Delta T: The temperature difference between the cold and hot sides of a thermoelectric module. Delta T may also be expressed as ΔT.

Density: The mass of a material per unit volume. It is often expressed as pounds per cubic foot or grams per cubic centimetre or kilogram per cubic metre.

Diffuse insolation: Sunlight received indirectly as a result of scattering due to clouds, fog, haze, dust, or other obstructions in the atmosphere. The rate at which this energy falls on a unit horizontal surface per second is called the diffuse solar insolation (or irradiance).

Diffuse radiation: Diffuse solar radiation is the result of the atmosphere attenuating, or reducing the magnitude of the Sun's beam. Some of the energy removed from the beam is redirected or scattered towards the ground. The radiation received from the sun after reflection and scattering by the atmosphere and ground.

Digester: It is a huge tank constructed underground in which biodegradable organic matters are decomposed by anaerobic bacteria into methane and carbon dioxide.

Direct beam radiation: The radiation received by direct solar rays. It is measured by a pyrheliometer with a solar aperture of 5.7° to transcribe the solar disc.

Direct current (DC): It is the unidirectional flow of electric current.

Direct Insolation: It is also called as beam radiation that describes solar radiation traveling on a straight line from the sun down to the surface of the earth.

Distillation: It is a process of separating the component substances from a liquid mixture by boiling (evaporating) and re-condensing the resultant vapours.

Distributed energy resources (DER): A variety of small, modular power-generating technologies that can be combined with energy management and storage systems used.

Distributed generation: A popular term for localized or on-site power generation.

Distributed power: Generic term for any power supply located near the point where the power is used.

Distributed systems: Small-scale generating technologies (e.g., solar, wind, CHP, hydro or newer technologies) that are connected to the local electric power grid are identified as distributed systems.

Distributed generation (DG): It allows customers to be powered by their own electric generation systems as opposed to central systems of electricity supply through national grids.

Diurnal, daily, or the daily cycle: It refers to patterns of happening or recurring at regular intervals within a 24-h period.

Downdraft gasifier: A gasifier in which the product gases pass through a combustion zone at the bottom of the gasifier.

Downwind turbine: The downwind turbine refers to a horizontal axis wind turbine that has its rotor on the back side of the turbine. The nacelle typically is designed to seek the wind, thus negating the need for a separate yaw mechanism.

Drying: Moisture removal from biomass to improve serviceability and utility.

Ebb generation: The generation of electricity in a tidal power plant from a head of water created by the incoming tide trapped behind tidal barrage and by allowing the trapped water through turbine during low tides.

Efficiency: It is a ratio of performance achieved to the input to the system, where performance achieved and input both are expressed in same unit of measurement.

Electro deposition: It is an electrolytic process to deposit a dissolved or suspended substance on an electrode by electrolysis.

Electrochemical cell: It is a device used for either producing electrical energy from chemical reactions or causing chemical reactions through the application of electrical energy.

Electrode: It is a conductor that passes an electrical current from one medium to another.

Electrolyte: A substance that dissociates into ions in solution and acquires the capacity to conduct electricity.

Electromagnetic radiation: The energy produced by an oscillating electrical (and magnetic) field and transmitted by photons.

Emissivity: It is defined as the ratio of the energy radiated from the surface of a material to that radiated from a blackbody (a perfect emitter) at the same temperature and wavelength under the same viewing conditions.

Energy: It is the ability to do work. The relationship between power and energy is

$$\text{power} = \text{energy/time}.$$

Equation of time (EOT): It is a formula used in the process of converting between solar time and clock time to compensate for the earth's elliptical orbit around the sun and its axial tilt and is given as

$$(\text{EOT}) = 229.18 \times [0.000075 + 0.001868 \cos(D) - 0.032077 \sin(D)$$

$D = n(360°/365)$, and n is the number of the day counted from January of year (e.g., Feb. 1 makes $n = 32$).

Equinox: It means equal day and night. When the number of hours of daylight equals the number of hours of night in one day (24 h period). The vernal equinox, usually March 21, signals the onset of spring, while the autumnal equinox, usually September 21, signals the onset of autumn.

Ethanol (CH_3CH_2OH): A colourless, flammable liquid produced by fermentation of sugars. It is used as a fuel oxygenate. Further, it is the alcohol found in alcoholic beverages, but is denatured for fuel use. Examples of these systems include central receiver systems, parabolic dish, and solar trough.

Extraterrestrial radiation: It is the amount of global horizontal radiation that a location on earth would receive if there was no atmosphere or clouds (i.e., in outer space).

Feedstock: Any raw material used in energy conversion process for conversion to another form of fuel or energy product.

Fermentation: A metabolic biochemical reaction in which bacteria or yeasts that breaks down (ferment) complex organic molecules (like carbohydrates) into simpler materials (such as ethanol, carbon dioxide, and water).

Figure of merit: It is a measure of the overall performance of a thermoelectric device or material.

Fill factor: The ratio of a photovoltaic cell's actual power to its power if both current and voltage are at their maxima.

Fixed tilt array: A photovoltaic array set in at a fixed angle with respect to horizontal.

Flat plate heat exchanger: A flat plate heat exchanger is an installation where heated water on one side transfers the heat onto the other side of flowing water. Both sides must have a circulating fluid.

Flat plate array: A photovoltaic (PV) array that consists of non-concentrating PV modules.

Flat plate module: An arrangement of photovoltaic cells on a substrate of metal or glass or plastic to provide structural rigid flat surface support on the back, solar cell protecting encapsulate material and a transparent cover of plastic or glass so that cells exposed freely to incoming sunlight.

Flat plate photovoltaic (PV): A PV array of flat plate module that consists of non-concentrating elements.

Fluidized bed: It is a gasifier or combustor in which fuel particles are suspended in a hot, bubbling fluidized bed of ash and other particulate materials (sand, limestone, etc.).

Fly ash: Small ash particles carried in suspension in combustion products.

Focusing collector: A collector that enhances solar energy by focusing it onto a smaller area through mirrored surfaces or lenses.

Forestry residues: It includes the remains of forest trees and plants after the harvest of crop and forestry products such as tops of trunks, stumps, branches, and leaves.

Free electrons: Any unattached electron from an ion, atom, or molecule that is free to move under the influence of an applied electric or magnetic field.

Fresnel lenses: An optical device usually made of glass or plastic that focuses light like a magnifying glass. Concentric rings are faced at slightly different angles so that light falling on any ring is focused to the same point.

Fuel cell: A device that converts the chemical energy of a fuel directly to electricity and heat without combustion.

Furling: The act of a wind generator by which turning rotor out of the wind as wind speed reaches cut out speed to protect itself from high wind speeds.

Fumaroles: A vent or hole in the earth's surface from which volcanic gas, steam, etc., escape into the atmosphere.

Gallium arsenide (GaAs): It is a crystalline, high-efficiency compound semiconductor. It is mixture of two elements gallium (Ga) and arsenic (As) and is used to make certain types of solar cells and semiconductor material.

Gas turbine: It is a type of internal combustion engine that converts the energy of hot compressed gases (produced by burning fuel in compressed air) into mechanical power.

Gasification: A chemical or heat process to convert a solid fuel to a gaseous form.

Gasifier: A device for converting solid fuel into gaseous fuel. In biomass systems, the process is also referred to as pyrolytic distillation.

Geothermal energy: The earth's interior heat made available to man by extracting it from hot water or rocks.

Geothermal gradient: It is the rate of temperature increase in the earth as a function of depth.

Geyser: A hot spring that erupt jets of hot water and steam into the atmosphere.

Global warming: A term used to describe the increase in average global temperatures due to the greenhouse effect.

Greenhouse effect: It is a natural process that warms the earth's surface in which the solar radiant energy is trapped by greenhouse gases produced from both natural and human sources.

Greenhouse gas: Gaseous components in the atmosphere (such as water vapour, carbon dioxide, troposphere ozone, methane, and low level ozone) that contribute to a gradual warming of the planet.

Heat engine: It is an equipment that converts thermal energy into mechanical energy output. They often pick up alternate names, such as gasoline or petrol, turbine, or steam engines.

Heat exchanger: A device for transferring thermal energy from one fluid to another.

Heat pump: It is a device that is able to transfer heat from one fluid at a lower temperature to another at a higher temperature.

Heliostat: They are computer-controlled large flat mirrors that keep the sun reflected on a target as the sun moves across the sky so that it can continuously reflect the sun's rays onto a central receiver.

Horizontal axis wind turbine: A wind turbine design in which the shaft is parallel and the blades are perpendicular to the ground.

Hub: The central part of the wind turbine, which supports the turbine blades on the outside and connects to the low-speed rotor shaft inside the nacelle.

Hydrocarbon: An organic compound that entirely contains only hydrogen and carbon.

Hydrothermal resource: An underground system of hot water and/or steam.

Insolation: Solar radiation on the surface of the earth and is usually expressed as watts per square metre. This term has been generally replaced by solar irradiance because of the confusion of the word with insulation.

Inverter: An equipment that is used to convert a direct current into an alternating current.

Irradiance: It is a measure of total solar power radiation that strikes a surface on the earth. It is also expressed in kilowatts per square metre.

***I–V* curve:** They are curves that show the relationship between the current flowing through an electronic device and the applied voltage across its terminals.

Kelvin: The temperature unit based on the theoretical minimum temperature used in many physical calculations. The relationship between Kelvin and Celsius is $K = C + 273.16$.

Latent heat: Thermal energy required to cause a change of state of a substance such as changing water into ice or steam.

Local apparent time: The time of day based strictly on the longitude of the locality and not on 'blocky' time zones.

Local standard time: The time of day based on the longitude of the zone meridian associated with a locality.

Longitude: The East–West angular distance of a locality from the Prime Meridian. The Prime Meridian is the location of the Greenwich Observatory in England and all points North and South of it.

Magma: A molten rock within the earth, from which igneous rock is formed by cooling.

Mantle: It is one of the three main layers of the earth that lies between the innermost layer (the core) and the thin outermost layer (the crust).

Mass flow rate: It is the mass of a substance that passes per unit of time through a given cross-sectional area.

Methane (CH_4): It is a colourless, odourless, non-toxic, and flammable gas and is the most simple of the hydrocarbons.

Methanol (wood alcohol) (CH_3OH): An alcohol formed by catalytically combining carbon monoxide with hydrogen in a 1:2 ratio under high temperature and pressure.

Methanol, denatured ethanol, and other alcohols, separately or in blends of at least 10% by volume with gasoline or other fuels.

Microorganism: Any microscopic organism such as yeast, bacteria, and fungi.

Municipal solid waste: It includes commercial and residential, garbage, refuge, rubbish wastes generated in municipal or notified areas in either solid or semi-solid form.

Nacelle: A housing structure for rotor shaft, gearbox, and generator at the top of the wind turbine tower just behind (or in some cases, in front of) the wind turbine blades.

Non-renewable resource: Energy resource that is once exhausted cannot be refilled or restored by nature.

Ocean thermal energy conversion (OTEC): The energy conversion technologies used for electricity generation by using the temperature differences (thermal gradients) between warm ocean surface waters and cool ocean depth waters.

Oil shale: It is any sedimentary rock that contains solid bituminous materials (called kerogen) that are released as petroleum-like liquids hydrocarbons called shale oil.

Organic compound: An organic compound contains carbon chemically bound to hydrogen that may also contains other elements particularly O, N, halogens, or S.

Ozone layer: It is the ozone shield of high concentrations of ozone (O_3) refers to a region of earth's stratosphere that absorbs most of the sun's UV radiation.

Ozone: It is most stable form of elemental oxygen in the upper atmosphere and protects the earth from the sun's ultraviolet rays.

Particulates: A fine liquid or solid particle such as dust, smoke, mist, fumes, or smog, found in air or emissions.

Passive solar technologies: Technologies that capture sunlight for heat energy without use of active mechanical systems and convert sunlight into usable heat (water, air) or store heat for future use.

Peltier effect: When an electrical current is passed through a junction consisting of two dissimilar metals, it results in a cooling effect.

Phase change: The change of a substance from a liquid to solid, liquid to gas, etc.

Photovoltaic: It is a process in which the light is converted into electricity through the displacement of electrons and the current created as a result. Photovoltaic are the basis for the operation of solar panels.

Photoelectric: It is the conversion of light (radiant energy) to electricity.

Photometer: An instrument that measures luminance.

Photons: A photon is an elementary particle and the basic unit of light.

Photosynthesis: A process used by plants and bacteria to produce carbohydrates from carbon dioxide and water using solar light energy. Photosynthesis is the key initial step in the growth of biomass and is given by the equation:

$$CO_2 + H_2O + \text{light} + \text{chlorophyll} = (CH_2O) + O_2.$$

Photovoltaic (PV) array: An interconnected system of large number of PV modules to provide high voltage and power unit with common support or mounting.

Photovoltaic (PV) cell: The smallest semiconductor single element (solar cell) within a PV module to perform the immediate conversion of light into electrical energy (direct current voltage and current).

Photo conversion: A process of conversion of light into other forms of energy by chemical, biological, or physical processes.

Photovoltaic (PV) conversion efficiency: It is the ratio of the electrical output of a solar cell to the incident sunlight energy on it.

Photovoltaic (PV) device: A solid-state electrical device (solar cell device) that converts light energy directly into direct current electricity.

Photovoltaic (PV) effect: The phenomenon that occurs when photons strike the atoms to free electrons.

Photovoltaic (PV) generator: It is the total of all photovoltaic panel connected to solar electric grid.

Photovoltaic (PV) system: A complete set of components for converting sunlight into electricity.

Photovoltaic array: A photovoltaic module or set of modules used for converting solar radiation to energy.

Photovoltaic module: An unit comprising several photovoltaic cells that are the principal unit of photovoltaic array. A photovoltaic module's size is on the order of 1 m^2, although its size is governed by convenience and application.

Photovoltaic(s) (PV): It is the direct conversion of light into DC electricity.

Point-focusing concentrator: A solar power generator that uses a series of tracking mirrors (heliostat), Fresnel lenses, or paraboloidal (three-dimensional parabola or dish) mirrors to focus solar energy onto a single central receiver.

Polycrystalline silicon: A material used to make photovoltaic cells.

Polymer: A large molecule made by linking smaller molecules (monomers) together.

Propeller: It is a type of device that transmits power by converting rotational motion into thrust.

P-type material: Semiconductor materials that are doped to have a deficiency of electrons.

Pyranometer: An instrument used for measuring global solar irradiance.

Pyrheliometer: An instrument used for measuring direct beam solar irradiance.

Pyrolysis: It is a thermo-chemical decomposition of organic material at elevated temperatures in the absence of oxygen.

Rankine cycle: It is an idealized thermodynamic cycle of a heat engine converting heat into mechanical energy.

Recycling: The process of converting materials that are no longer useful as designed or intended into a new product.

Renewable energy resource: Any energy resource that is continuously renewed by nature.

Residues, biomass: A by-product (residues or wastes) from processing all forms of biomass have significant energy potential for energy or fuel recovery.

Electrical resistivity: It is the measure of how strongly a material opposes the flow of an electrical current.

Salt gradient solar ponds: They consist of three main layers having different salt concentrations. The top layer is near ambient and has low salt content. The bottom layer is hot (typically 71°C–100°C) and is very salty.

Scrubber: It is an air-stream pollution control device used to remove solid and liquid particulate matter (by washing them out) and gaseous pollutants (either by absorbing or by chemically neutralizing) with the help of liquid spray.

Seebeck effect: When the two junctions formed by two dissimilar metals are maintained at temperature gradients (one junction is hot and the other is cold), an electrical current will flow in a load circuit connected to both junctions.

Selective absorber: A solar absorber surface having high absorbance at solar spectrum wavelengths and low emittance in the infrared range.

Selective surface coating: A material with high absorbance and low emittance properties applied to or on solar absorber surfaces.

Selective coating: Visible light represents different wavelengths that have various energies. In brief, the wavelengths below 380 nm is UV light between 380 nm and 780 nm is visible light, 780 nm and 2,500 nm is near infrared light, and above 2,500 nm is far infrared light. A selective coating transmits light in the range from 360 nm to 2,500 nm, while it will block light or energy above 2,500 nm. In practice, this means that the coating will allow short wave solar light to enter, while long wave energy from the solar collector will be reflected back on the flowing liquid.

Shale oil: It is an unconventional oil produced from oil shale rock fragments by pyrolysis, hydrogenation, or thermal dissolution.

Silicon-germanium: A high temperature (range of 500°C–1,000°C) thermoelectric semiconductor material that exhibits its optimum performance.

Silicon (Si): A non-metallic chemical element in the carbon family that is present in sand and glass and which the best known semiconductor is.

Sluice gate: An artificial passage for water fitted with a valve or gate in barrage or dam for stopping or regulating flow.

Solar altitude angle: It is the angular height of the sun measured from the horizon.

Solar array: A group of solar panels collectively makes up a solar array.

Solar cell or photovoltaic cell: A device that converts solar light energy directly into direct current electricity by the photovoltaic effect.

Solar constant: It is defined as the average amount of solar radiation that reaches the earth's upper atmosphere on a surface perpendicular to the sun's rays, in the range of 1,353–1,367 w/m^2. The currently accepted value is 1,366 W/m^2.

Solar cooling: The use of solar thermal energy or solar electricity to power a cooling equipment.

Solar energy: It is the radiant energy from the sun that influences earth's climate and weather and sustains life.

Solar hours: These are the number of hours with a clear blue sky and sunshine during a timeframe.

Solar irradiance: It is solar power per unit area (kW/m^2) on the earth's surface in the form of electromagnetic radiation.

Solar noon: It is defined when the sun is at the zenith (directly above the head of the observer).

Solar panel: A series–parallel connection of solar cells for desired power output.

Solar pond: A pond full of saltwater that acts as a large-scale solar thermal energy collector with integral heat storage for supplying thermal energy.

Solar power: The power (electricity) generated from the solar energy.

Solar spectrum: It is the intensity of sunlight (irradiance) variation over wavelength for given atmospheric condition.

Solar thermal electric systems: Technologies that convert solar thermal energy into electricity.

Specific gravity: The ratio of the mass of any material to the mass of an equal volume of water at a temperature of 4°C.

Specific heat: It is the amount of thermal energy required to raise the temperature of a given substance by one degree when compared to the energy required to raise the temperature of an equal mass of water by one degree. The specific heat of water is 1.000.

Stand-alone system: A system that provides continuous power without any support from other systems.

Stirling engine: It is an external combustion engine that operates by cyclic compression and expansion of the working fluid at different temperatures,

Subsidence: Cavity formation in the earth's crust owing to fluid and energy raw resource extraction.

Substrate: A supporting material on which a circuit is formed or fabricated or enzyme acts or organism grows.

Surface work function: A measure of the electron-emitting capacity of the surface.

Tar: It is a black mixture of hydrocarbons.

Thermal conductivity: It is the property of a material to conduct heat and its units is W/(mK)

Thermionic energy conversion: An energy conversion process in which heat energy is directly converted into electrical energy by the phenomena of thermionic emission.

Thermoelectric generator: A device that directly converts heat energy into electrical energy when one junction of two dissimilar metals are kept hot and the other junction is kept cool (Seebeck effect).

Tidal barrage: A concrete dam constructed in ocean to store water during tides with sluices.

Tidal or wave farming: A method of harvesting tidal power that relies on underwater turbines.

Thomson effect: It is the absorption or evolution of heat along a conductor having temperature difference at the two ends when current passes through it.

Tidal energy: Tidal energy, tidal power, wave energy, and wave power are terms used interchangeably. These terms describe the methods of creating energy from the movement of water due to the ocean tides, ocean waves, and currents in rivers.

Toxics: Poisonous and environmentally unfriendly.

Turbines: The mechanical component of electricity generating systems that creates electricity. Turbines rotate as water, air, or steam push the blades of a rotor.

Turbulence: Air, water, or other fluid unsteady or violent movement.

Ultraviolet radiation: The range of radiation just beyond the violet in the visible spectrum (at about 4,000 Angstroms or 400 nm) to the X-ray region (at about 40 Angstroms or 4 nm).

Vapour dominated: A hydrothermal system having dominance of vapour or steam.

Vertical-axis wind turbine (VAWT): A type of wind turbine in which the rotor of the turbines rotate vertically around its axis.

Viscosity: A fluid property related to the interaction between fluid molecules that determines the fluids' resistance to sheering forces and flow.

Watt: A watt is a unit of measure for electrical energy. It is equal to one joule of energy per second.

Wind energy: The energy caused by the earth's weather patterns creating high and low pressure centres forcing wind to rush from high to low pressure.

Wood: The hard fibrous material substance obtained from the trunk, branches, and root of a tree or shrub.

Yeast: Any of various single-cell fungi capable of fermenting carbohydrates.

Zenith angle: It is the angle between the geometric centre of the sun and directly overhead of the observer.

Bibliography

[1] Begamudre R. D. *Energy Conversions Systems*, New Age International Publishers, New Delhi, India, 2000.

[2] Chauhan D. S. and Shrivastava S. K. *Non-Conventional Energy Resources (as per GBTU /UPTU syllabus)*, New Age International (P) Limited, New Delhi, India, 2004.

[3] Desai A. K. and Munasinghe M. *Nonconventional Energy*, New Age International, New Delhi, India, 1990.

[4] Garg H. P. and Prakash. *Solar Energy—Fundamentals and Applications*, Tata McGraw-Hill Education, 2000.

[5] Gupta B. R. *Generation of Electrical Energy*, Eurasia Publishing House, Paharganj, Delhi, 1998.

[6] Khan B. H. *Non-Conventional Energy Resources*, Tata McGraw-Hill Education, 2006.

[7] Rai G. D. *Non-Conventional Energy Sources*, Khanna Pub., New Delhi, 3rd Ed.1999.

[8] Rajput R. K. *Non-Conventional Energy Sources and Utilisation (Energy Engineering)*, S. Chand, First Edition 2012; Reprints 2013, Second Revised Edition 2014, New Delhi.

[9] Rao S. and Parulekar B. B. *Energy Technology—Nonconventional, Renewable & Conventional*, Khanna Pub., New Delhi, 2009.

[10] Saeed S. H. and Sharma D. K. *Non-Conventional Energy Resources* (2nd Edition), S. K. Kataria & Sons, New Delhi, 2009.

[11] Sawhney G. S. *Non-Conventional Energy Resources*, PHI Learning Private Limited, New Delhi, 2012.

[12] Singhal R. K. *Non-Conventional Energy Resources*, S. K. Kataria and Sons, New Delhi, 2013.

[13] Smith C. B. *Energy Management Principles*, Pergamon Press, 1981.

[14] Sukhatme P. S. *Solar Energy–Principles of Thermal Collection and Storage*, Tata McGraw-Hill Education, 1996.

[15] Twiddel J. W. and Anthony D. W. *Energy Resources*, ELBS/E. and F. N. Spon UK, 1986.

Index

C

D